Ethics and the politics of food

Ethics and the politics of food

Preprints of the 6th Congress of the European Society for Agricultural and Food Ethics

EurSAFE 2006
Oslo, Norway
June 22 – 24, 2006

edited by:
Matthias Kaiser
Marianne Elisabeth Lien

ISBN-10: 90-8686-008-7
ISBN-13: 978-90-8686-008-1

First published, 2006

Wageningen Academic Publishers
The Netherlands, 2006

EurSafe 2006 Committees

Organizing Comittee
Matthias Kaiser (Chair)
Marianne Elisabeth Lien
Reidar Almås
Sissel Rogne
Stine Wohl Sem
Arild Vatn
Tore Tennøe
Thorleif Paasche
Vonne Lund
Ellen- Marie Forsberg
Eivind Jakobsen
Henrik Stenwig

Scientific committee
Marianne Elisabeth Lien (Chair, Norway)
Helena Röcklingsberg (Sweden)
Peter Sandøe (Denmark)
Andoni Ibarra (Spain)
Kate Millar (UK)
Johan de Tavernier (Belgium)
Ole Doering (Germany)
Maria Eduarda Goncalves (Portugal)
Eugenius Gefenas (Lithuania)
Tom Eide (Norway)
Kjellrun Hiis Hauge (Norway)
Unni Kjærnes (Norway)
Florence Bergaud-Blackler (France/Italy)

The Eursafe 2006 conference is organized by:

- Forskningsetiske komiteer
  (The National Committees for Research Ethics)

In co-operation with:

- The University of Oslo

The organizers gratefully acknowledge financial support by:
- The Research Council of Norway
- The Fishery and Aquaculture Industry Research Fund
- The Foundation for Research Levy on Agricultural Products
- The City of Oslo

Table of contents

Part 2 – Foundational issues in philosophy and ethics

Part 3 – Trust in food

Part 4 – Ethics and safety in food discourse

Part 5 – Sustainability in food production

Part 6 – The tool-box for assessing ethical issues

Part 8 – Emerging technologies in food production

Part 9 – Nutrition, equality and global food security

Part 10 – Who is responsible? Corporate responsibility and political governance

Part 11 – Critical issues in aquatic production systems

Preface: Ethics and the politics of food

It is a paradox of the Western World that amidst of technological innovation, economic welfare and social change many people have developed a serious concern about our primary goods: food and food production. This topic mobilizes people from different backgrounds and cultures. It is the focal point of both high and low politics. On the one hand, there are lively disputes and conflicts about globalization of trade as witnessed for instance in the conflicts about the WTO-rounds or debates about food biotechnology. On the other hand, there is a very concrete level of consumer choice at the shelf in the supermarket where value conflicts cannot easily be resolved. Yet, choice itself is often limited, through monopolising tendencies of corporate power.
At the same time we witness the traditional sectors of food production in a rapid phase of change in nearly every society. In times of globalization questions of micro- and macro-politics become intertwined, and only collective action seems to stand a chance of influencing the development. However, the disputes and political conflicts resist containment and assumed boundaries of the systems are moved only to be replaced by even wider considerations. What starts out as concerns about food risk and safety soon moves to ethics and cultural values. Ethics and cultural values are indeed a common denominator for peoples' concern about food and food production. Slogans like "from farm to fork", "vote with your fork", or "slow food" express a mingling of political, ethical and cultural values.
The challenge to ethics is to relate to these new societal contexts and provide useful concepts for analysis and understanding. The programme of the 6th Congress of the European Society for Agricultural and Food Ethics is designed to capture these complexities in their variety through different thematic sub-sections and multi-disciplinary approaches.
This volume contains the presentations for the conference. Each contribution was peer-reviewed before being accepted to the conference. Not all contributions could be included in this book, since not all authors managed to send their final manuscripts in time for the publication deadline. However, we believe that this collection of papers provides for interesting reading, both for all those who attend the conference, and for those who cannot attend but have an interest in the written records. The collection of viewpoints, approaches, and arguments is in itself witness of the trans-disciplinary character of this emerging field of research. We hope this book will stimulate an interesting dialogue among specialists and open new possibilities for a wider societal debate on ethics and food.

Matthias Kaiser and Marianne Elisabeth Lien

Keynote papers

Political consumerism: Why the market is an arena for politics

Michele Micheletti
Political Science, Karlstad University, Sweden, michele.micheletti@kau.se

Abstract

We are witnessing the blooming of the phenomenon of political consumerism. Citizens are bringing politics to the market and confronting corporate actors in ways and levels unimaginable in the past. They even challenge governments to add political, ethical, and environmental demands to their procurement policies. Citizens and activist groups spice up their discourse with spectacular vocabulary. Less-than-standard working conditions are described as sweatshops; farm laborers providing us with food are called chocolate and fruit slaves, and the term slave labor asks consumers to think twice about their purchasing preferences and choices. Properly defined political consumerism is use of the market for political purposes, to raise political issues, create responsibility-taking, control uncertainty, and solve common problems. Political consumers are people who use their buying power to attempt to change objectionable institutional or market practices regarding issues of sustainability, justice, fairness, and non-economic issues that concern citizen well-being (Micheletti *et al.* 2003). Their shopping choices reflect an understanding of material goods as embedded in the complex social and normative context of "the politics behind products." This paper discusses how political consumers use consumer choice to promote human rights and sustainable development, why consumer choice has evolved into political engagement, and what prompts citizens to find new arenas for setting their political values and beliefs in motion.

Keywords: boycotts, buycotts, labeling schemes, political consumerism, political responsibility-taking, governance, globalization, consumers

Political consumerism's three basic forms

Negative, positive, and discursive are the three basic forms of political consumerism in existence today. They represent different ways for "citizen-consumers" to use the market as an arena for politics. *Negative political consumerism* (boycotts) is the oldest form and has been used extensively around the world to protest perceived corporate and government wrong-doings. Good examples are the decades-long Nestlé boycott and the Shell Oil boycott in the 1990s (Friedman 1999). Boycotts are contentious in nature because they encourage people to protest against wrong-doings by not buying targeted goods or patronizing targeted corporations and stores. *Positive political consumerism* (buycotts) is a more recent form and tends to be more cooperative in nature. Examples are seals of approval labeling schemes (ecological, organic, and fair trade labels), socially responsible investing (SRI), and even alternative products like Black Spot Sneaker and No Sweat Clothing. They encourage people to purchase products, producers, and services that promote sustainable development. The third form does not involve the purchasing or non-purchasing of goods and services. Rather, *discursive political consumerism* is a variety of communicative efforts directed at business and the public at large about corporate policy and practice. It can be as contentious and confrontational as boycotts or represent attempts to engage in dialog with corporate actors and the general public about the politics behind the products offered in the consumer market today.

Many boycotts are well known internationally. Some like the Montgomery Bus Boycott have books written about them, their own web sites, and have even become folklore. For centuries and across the globe, citizens have boycotted products and producers to express their dissatisfaction with business and politics. They have even targeted corporate actors and products to protest government policy, as witnessed in the boycott of South African goods and the more recent one of French export-sensitive products. Boycotts have been used in revolutionary and constitutional struggles, in protests for social justice, for civil and human rights, against environmental pollution and sweatshop working conditions, and by minorities, women, ethnic groups as well as labor unions and other civic associations. Some have a global profile, as illustrated by on-going ones against Nestlé for its violation of the International Code of Marketing of Breastmilk Substitutes, Microsoft for its anticompetitive practices, and Procter & Gamble for its animal testing policies (Micheletti 2003:37-72, 82-9).

Survey research shows that boycotting as a way for people to participate in politics is on the rise. The average level of citizen participation in them for representative postindustrial societies was 5% in the mid-1970s and 15% in the mid-1990 (Norris 2002:198). Boycott increases have also been found in developing countries like Mexico, South Korea, and South Africa (Inglehart 1997:313). Boycotting is a frequent form of participation in the United States (about 20% in 2005) and in Europe. The general European average in 2003 was 16 percent. Northern European countries were over the average and Southern and Eastern ones under it (Stolle and Micheletti forthcoming, Ferrer-Fons 2004). For citizens in some countries (e.g., Nigeria) boycotting is one of the few safe ways to express political dissatisfaction. Survey results also highlight the importance of boycotts for young people. A recent Swedish study reports that almost 100% of all young people between 16 and 29 years of age have actually participated or can consider participating in a boycott (Ungdomsstyrelsen 2003:171). People who boycott for political, environmental, and ethical reasons tend more to be women, more highly educated, politically interested, and more leftist political orientation (Stolle and Micheletti forthcoming).

Positive political consumerism politicizes products by calling on consumers to purchase brands and support corporations that promote good environmental, working, and social conditions. Political consumerist labeling schemes are "beyond compliance" regulatory mechanisms (soft laws) that encourage corporations to do more than what is required in governmental regulation ("command and control" hard laws) (Kollman and Prakash 2001). Several positive political consumerist schemes are in operation today. They include national and international eco-, organic, and fair trade labels. On average slightly less than 25 percent of Europeans and Americans "buycott" for political, ethical, and environmental reasons (Ferrer-Fons 2004, Stolle and Micheletti forthcoming). Another kind of positive political consumerism is socially responsible investing (SRI), which offers people opportunities to place their money in green, social, and ethical funds. These funds are growing in number in Europe; in 2004 here were 354 such funds which are increasing their market share annually (Avatzi, 2004).

In certain quarters the positive political consumerist message is offensive because it maintains that money and morality mix well. This criticism applies to almost all forms of political consumerism, which rely on the capitalist system and market mechanisms to develop a broadened causal horizon of solidarity and political responsibility or what scholars of anti-slavery have called the humanitarian sensibility (Haskell, 1992).

Discursive political consumerism is the least-researched form. It appears to be the most recent and emerging one, and some of its methods (culture jamming, adbusting, billboard liberation) make it quite attractive to young people. Discursive political consumer action can involve contentious or cooperative relationships with business. This form is particularly important for the global anti-sweatshop movement. The reason is the lack of an anti-sweatshop garment labeling scheme and problems with using boycotts to promote the cause. Here discursive political consumerism is showing itself to be an effective method. Garment market restructuring towards buyer-driven commodity chains and lean retailing puts the global corporate giants in an increasing vulnerable setting (Geriffi 2001). To compete, corporations must invest huge resources

in corporate identity (logotype, image, and culture) and rely on the sewing hands of individual garment workers rather than garment-making machinery to provide "fashion and quality at the best price" (the H & M 2005 slogan). They are caught in a bind: "The catch is that the more successful corporations have become at branding our culture and creating a certain reputation for themselves, the more vulnerable they are to disruptions of that image through exposés linking their products to sweatshop conditions" (deWinter 2003, 108). Political consumers use this vulnerably to promote workers' rights in global production in a variety of fashions.
Examples of discursive political consumerism are the now-famous Nike Email Exchange, which re-contextualized the meaning of Nike's marketing slogans when a college student used the innovative prize-winning Nike iD web site to order a pair of customized shoes with the word sweatshop on them. When Nike repeatedly refused his request, he wrote back: "Your web site advertises that the NIKE iD program is 'about freedom to choose and freedom to express who you are.' I share Nike's love of freedom and personal expression. ....My personal iD was offered as a small token of appreciation for the sweatshop workers poised to help me realize my vision. I hope that you will value my freedom of expression and reconsider your decision to reject my order" (Peretti and Micheletti 2003). Other discursive political consumerist actions demand serious replies from corporations to conscientious questions about the politics of brand name products, independent monitoring of production plants, formulation and better implementation of codes of conduct, and engage consumers in discussions on the desirability and necessity of sustainable production. Still others create spectacular events in demonstrations and colorful parades and marches, as illustrated by the singing of "sweatshop" Christmas carols in shopping malls and the staging of alternative fashion shows to make the work behind brand name clothes rather than their stylish fit and status attractiveness the important part of consumer choice.

Why more political consumerism today?

Globalization is the general development that explains the sense of urgency felt by many people for political consumerism. Both the globalization of problems (climate warming and "social dumping" through outsourced manufacturing) and the regulatory vacuums created by fast-moving economic globalization and slow-moving political globalization trigger political consumerism. Globalization has weakened the ability of states, institutions, and other actors traditionally viewed as being in charge of policy-making to solve problems and carry out their responsibilities for securing general citizen well-being. This development has created, and continues to create policy vacuums and responsibility-floating (i.e., difficulty in pinpointing the institutions responsible for specific actions and problems) that lead to more serious problems regarding sustainable development. Economic globalization has opened up new opportunities for corporate business that cannot be effectively regulated by national policy. At the same time, political globalization, or the creation of supranational governmental policy-making institutions, has been slow in developing.
If globalization is the problem, governance, new theoretical models of political responsibility, and post-modernization are the theories used by social scientists to explain the growing importance of political consumerism. Governance theory questions the ability of nation-states (governments) to solve the complex problems facing our more globalized world, whose solution frequently requires coordinated efforts among states and other institutions and actors. The governance perspective on policy-making acknowledges the joint role of governmental, semi-governmental, non-governmental and private institutions as well as individual citizens in providing for the well-being of global citizens. This means new steering capacities in the form of "beyond compliance" and stakeholder institutions and new conceptions of political responsibility-taking. Problem-solving and policy takes on new meaning, and a variety of actors are called upon to take part.

Governance implies also active participation and responsibility-taking by individual citizens and consumers.

Post-modernization, the other general theoretical explanation for the rise of political consumerism, is defined as basic change in the fundamental values of the citizens of industrialized and industrializing societies signaling "...a move away from the emphasis on economic efficiency, bureaucratic authority, and scientific rationality...toward a more human society with more room for individual autonomy, diversity, and self-expression" (Inglehart 1997:12). It highlights the increased individualization of society, political conflicts over values promoting sustainable development (post-materialism), and focus on consumption over production as a potentially powerful steering mechanism for innovative regulatory politics. It erases the distinction between the private and public sphere and relocates "politics" in all societal spheres, including consumer choices traditionally thought of as private in orientation. This implies a blurring of the distinction between the public responsibility of citizens and private desires of consumers as well as the private nature of business and the public responsibility of governments. The claim that consumers and business have public responsibilities is one of the basic—and controversial—ideas behind political consumerism. This claim prompts a growing number of citizens to use the market as an arena for setting their political values and beliefs in motion.

References

Avatzi. 2004. *Green, social and ethical funds in Europe 2004*. Available on http://www.siricompany.com/pdf/SRI-Funds-Report2004.pdf.

DeWinter, R. (2001). "The anti-sweatshop movement. Constructing corporate moral agency in the global apparel industry. *Ethics & International Affairs* 15 (2): 99-115.

Ferrer-Fons, M. (2004). *Cross-national variation on political consumerism in Europe: exploring the impact of micro-level determinants and its political dimension*. Paper for the ECPR Joint Sessions, Uppsala, Sweden. Workshop 24 "Emerging repertoires of political action: toward a systematic study of postconventional forms of participation".

Friedman, M. (1999). *Consumer boycotts. Effecting change through the marketplace and the Media*. New York: Routledge.

Gereffi, G. (2001). Beyond the producer-driven/buyer-driven dichotomy. The evolution of global value chains in the internet era. *IDS Bulletin* 32 (3): 30-40.

Haskell, T.. (1992). Capitalism and the origins of the humanitarian sensibility, Part 1 and 2. In *The antislavery debate. Capitalism and abolitionism as a problem in historical interpretation*, Bender, T., ed. Berkeley: University of California Press. 107-160.

Inglehart, R. (1997). *Modernization and postmodernization. Cultural, economic, and political change in 43 societies*. Princeton: Princeton University Press.

Kollman, K. and Prakash, A. (2001). Green by Choice? Cross-National Variation in Firms' Responses to EMS-based Environmental Regimes. *World Politics* 53 (April):399-430.

Micheletti, M. (2003). *Political virtue and shopping. Individuals, consumerism, and collective action*. New York: Palgrave.

Micheletti, M., Follesdal, A. and Stolle, D. eds. (2003). *Politics, products, and markets. Exploring political consumerism past and present*. Rutgers, NJ: Transaction Publishers.

Micheletti, M. and Stolle, D. (forthcoming). Book manuscript on the political responsibility of political consumers.

Norris, P. (2002). *Democratic Phoenix: Reinventing Political Activism*. Cambridge: Cambridge University Press.

Peretti, J. and Micheletti, M. (2003). The Nike sweatshop email: Political consumerism, internet, and culture jamming.' In *Politics, products, and markets. Exploring political consumerism past and present*, Micheletti, M., Follesdal, A., and Stolle, D. eds. New Brunswick: Transaction Publishers, 127-144.

Stolle, D. and Micheletti, M. 2005. *Political consumerism as political responsibility*. Talk for Chaire de responsibilité sociale et de développement durable. University of Quebec at Montreal. November 17.

Ungdomsstyrelsen. 2003. *De kallar oss unga. Ungdomstyrelsens Attityd- och värdringstudie 2003*. Stockholm: Ungdomsstyrelsen.

Vogel, D. (2003). Tracing the roots of the contemporary political consumerist movement: Marketized political activism in the U.S. in the 1960s. In *Politics, products, and markets. Exploring political consumerism past and present*, Micheletti, M., Follesdal, A., and Stolle, D., eds. New Brunswick: Transaction Publishers, 83-100.

Whales as persons

Paola Cavalieri

I

In the morning of a Sunday of December 2005, in the open water east of the Farallones Islands, about 18 miles off the coast of San Francisco, a female humpback whale on her usual migratory route between the Northern California coast and Baja California became entangled in the nylon ropes that link crab pots. The whale, estimated to weigh 50 tons, was spotted by a person at 8:30 a.m. The combined weight was pulling the whale downward, forcing her to struggle mightily to keep her blow-hole out of the water. Soon, an environmental group was radioed for help. By 2:30 p.m., the rescuers had reached the whale and evaluated the situation. Team members realized the only way to save the endangered leviathan was to dive into the water and cut the ropes. It was a very risky maneuver, because the mere flip of a humpback's massive tail can kill a human being.
"My heart sank when I saw all the lines wrapped around her" the first diver in the water said, "I really didn't think we were going to be able to save her." At least 12 crab traps, weighing 90 pounds each, hung off the whale. Rope was wrapped at least four times around the tail, the back and the left front flipper, and there was a line in the whale's mouth. The crab pot lines were cinched so tight that the rope was digging into the whale's blubber and leaving visible cuts.
Four divers spent about an hour cutting the ropes with a special curved knife. The whale floated passively in the water the whole time, they said, giving off a strange kind of vibration. "When I was cutting the line going through the mouth, her eye was there winking at me, watching me, following me the whole time" one rescuer said. "It was an epic moment of my life - I will never be the same".
When the whale realized she was free, she began swimming around in joyous circles. She then came back to each and every diver, one at a time, and nudged them, pushed them gently around and flapped. "It felt to us like she was thanking us, knowing that she was free and that we had helped her", the rescue divers said, "it was the most incredibly beautiful experience of their lives"[1].
In recent centuries, whales have been savagely hunted to near extinction. Even today, despite decades of international restrictions, they are killed by the thousand every year. But, at present, most nations in the world are opposed to whale hunting, millions of people regard the killing of whales as inconsistent with current moral ideals, and a multifarious group of committed individuals wages a continuing war against whale hunters from a small number of countries. Is this new perspective justified?

II

A few years ago, legal scholars Anthony D'Amato and Sudhir Chopra dealt with this question in a dense essay, in which they argued that it is time to extend to whales the most fundamental of all human rights - the right to life[2]. In support they advanced a juridical argument connected with the broadening international consciousness which manifests itself in the history of the policies of the international institutions concerned with "whaling" - as, with an unpleasant

[1] Peter Fimrite, "Daring rescue of whale off Farallones", *San Francisco Chronicle*, December 14, 2005, http://sfgate.com/cgi-bin/article.cgi?file=/c/a/2005/12/14/HUMPY.TMP

[2] Anthony D'Amato and Sudhir K. Chopra, "Whales: Their Emerging Right to Life", American Journal of International Law 85 (1), 1991.

locution, whale hunt has come to be defined. Claiming that these policies moved through five stages - free resource, regulation, conservation, protection and preservation - and are now pointing to a sixth, entitlement, D'Amato and Chopra discuss the six stages employing purposive as well as descriptive materials, a reflection of the fact that customary international law is a synthesis of qualitative and quantitative elements. Given its role in the argument, their historical reconstruction is worth recounting in some detail.

According to D'Amato and Chopra, the free resource stage includes all recorded history up to World War I, when large-scale whale hunting, beginning in the eleventh century, had developed into a commercial industry leading to the near extinction of many species of whales. While the middle of the nineteenth century saw the biggest of all the whaling periods, favored by advances in technology, after the War the realization by the industries that their profits depended upon the availability of sizable numbers of whales gave rise to the regulation stage, culminating in 1931 when a Convention for the Regulation of Whaling was concluded. which set forth regulations covering all waters, and the private companies, motivated only by hope for the longevity of the industry, started negotiating new agreements concerning hunting areas and seasons.

Due to poor enforcement of these agreements, however, the extermination of the entire whale species soon surfaced as a major concern. Thus, the year 1948 saw the creation of the International Whaling Commission (IWC), meeting annually to define quotas and to identify protected species. Though for at least two decades the IWC's role was scarcely effective, when growing international awareness generated a new trend towards conservation, the Commission became the locus of the clash between opposing views. The first half of the 1970' witnessed serious conflicts between nonwhaling and whaling states, and when the U.N. call for a ten-year moratorium of commercial whaling was rejected by the IWC, serious questions were raised about the IWC role, and proposals were made calling for the U.N. to assume jurisdiction.

In what D'Amato and Chopra call the protection stage, the change in attitude was catalyzed by events outside the IWC. In 1977, CITES (Convention on International Trade in Endangered Species) characterized certain whales as endangered species. At the 1979 IWC meeting, Australia announced the intention to oppose whaling both domestically and internationally, stressing the high intelligence of whales and a growing community conviction of the immorality of whaling. By 1980 it became clear that moratorium was destined to be the topic of the decade. In 1982, when the U. N. Convention on the Law of the Sea was opened for signature, article 65 read: "States shall co-operate... and in the case of cetaceans shall in particular work through the appropriate international organizations for their conservation..." This article constitutes a clear legal gesture toward giving whales standing in international law beyond the limits of the IWC.

Starting from 1982, however, a wholly new form of protectionism opens the preservation stage. As dramatically reflected in the escalating "whale wars", the preservationist wants to ban all whaling, irrespective of whether a particular species is endangered. In 1982, the proposal of a phase-out of commercial whaling, seeking zero quotas for all commercial whaling by 1986, and subject to review thereafter, was approved by IWC. In efforts to appease the traditional whaling nations, the ban on "commercial" whaling was understood to exempt "aboriginal subsistence whaling" and "scientific whaling".

The moment when the five-year moratorium was revaluated by the IWC's meeting in Noordwijk, in 1990, paves the way for the entitlement stage. The Commission condemned as unnecessary the killing of whales "for research", and asked the involved countries to reconsider their policies. Meanwhile, virtually all commercial whaling activity ceased. But, preservation prefigures entitlement when the moratorium becomes permanent, at which point it is no longer definitionally a "moratorium".

It is at this passage in their historical overview that D'Amato and Chopra introduce their properly legal argument, which can be summarized as follows. The idea of entitlement clearly implies a major theoretical change: to claim that whales are "entitled" to life means to recognize this right as belonging to the whales themselves. But, though involving a radical philosophical shift, the entitlement stage actually represents simply an incremental advance in the series of the progressive stages in question. Set within the framework of international jurisprudence, what the above historical discussion has implicitly sketched is that trend in the component of customary international law which is called *opinio juris*. The development of international custom is a dynamic process: the seeds of a future conflict-resolving synthesis are present in the clash of thesis and antithesis constituting the claim conflicts among states. Thus, to anticipate a customary trend is to argue that, in a sense, it already exists. In the case of whales, the practice of states has moved through several stages that are best characterized as increases in international breadth of consciousness - and such combination of practice and consciousness is just what formally constitutes the material and psychological elements of general custom. Since what states do becomes what they legally ought to do, by virtue of a growing sense that what they do is right, proper and natural, the dawning sense of duty to the whales is evidence of a sense of obligation that constitutes the *opinio juris* component of binding customary international law. In this light, the attainment of the final stage - the entitlement of whales to life - in its inevitability has already been anticipated in the law.

The idea of having an entitlement includes a notion of a moral right that can inform existing law or push it in a certain direction. In a legal context, when a court accepts the moral claim of right and recognizes it as somehow subsisting in the law all along, though legal precedent was to the contrary, it is said that the court "articulates" the preexisting right. Along these lines, an international court could articulate a right to life of whales arising from the customary law practice of their preservation. This because whales' entitlement is already implicit in international law as resulting from progression through the previous stages, and from a sense that further development is morally legitimate.

III

Thus, according to D'Amato and Chopra's argument, the new perspective regarding whales is legally justified. But is the sense that such further development is *morally legitimate* warranted? In other words, is the new perspective also ethically sound? Should we grant whales a right to life? In concluding the presentation of the argument from *opinio juris*, D'Amato and Chopra notice that history has seen the continuous widening of the circle of rights holders, with a progression in ascribing fundamental rights to women, children, the mentally enfeebled and racial minorities. Such ascription, it should be added, has usually occurred through the inclusion in the number of "persons". And it is just the notion of person which can offer a clue to moral enquiry within this context.

Roughly speaking, ethics has as its object two sorts of theory of conduct. Morality in the broad sense is an all-inclusive theory of conduct, which includes precepts about the character traits to be fostered and the values to be pursued. Morality in the narrow sense, or social morality, consists instead of a system of constraints on conduct, usually expressed in terms of negative duties, whose task is to prevent harm to others - first and foremost, in the two main forms of the infliction of suffering and the taking of life[3]. In our philosophical landscape, the notion of person, usually contrasted with the notion of thing, has always played an important role

[3] Geoffrey J. Warnock, *The Object of Morality*, Methuen, London, 1971, p. 148; Peter F. Strawson, "Social Morality and Individual Ideal", *Philosophy: The Journal of the Royal Institute of Philosophy* 36 (Jan. 1968).

with reference to the protection from such harms[4]. For, while "person" is defined so that it is a descriptive term, whose determinative conditions of application have to do with the presence or absence of certain factual characteristics - most prominently rationality and self-consciousness - the assignment of descriptive content is guided by moral considerations[5]. And if, traditionally, to say of some being that it was a person meant to ascribe it a particular moral status, such as to prevent its use as a mere means to others' ends, the notion has gradually come to be especially tied to the question of the wrongness of killing. Accordingly, in present debates, to say of some being that it is a person is not only, as we have suggested, to ascribe it some rights, but most prominently, to grant it the right to life[6].

Is the concept of person coextensive with the concept of "member of the species *Homo sapiens*"? Arguably not. On the one hand, the notion of person being a creature of ethical theories, it may be pointed out that, historically, its theological use in connection with God has prevented it from simply becoming another term for human being. On the other, and more theoretically, in recent years an *ad hominem* argument directed at the paradigm of human equality has drawn philosophical attention to the inconsistency of denying a moral role to biological characteristics like race or sex while at the same time attributing a moral role to another biological characteristic such as species[7]. Against this background, an important strand of thinking in contemporary moral philosophy, arguing that the facts which are morally relevant *in themselves* are not biological facts, but rather psychological facts, has claimed that the concept of a person is the concept, not of a being belonging to a certain species, but of a being endowed with certain mental traits[8].

In particular, elaborating on Locke's basic idea that a person is a being that can consider itself as itself in different times and places,[9] many authors have argued that the mental trait which is central to personhood is not so much rationality, as rather the property of being aware of oneself as a distinct entity, existing in time and endowed with a past and a future - in other words, self-consciousness[10]. It is evident that what is at play here is is the connection between personhood and the prohibition of killing. For if a being is aware of oneself as a distinct entity, existing in time and endowed with a past and a future, it clearly has the possibility of conceiving of one's death as the discontinuance of one's existence, and of dreading, and being harmed by,

---

[4] Adolf Trendelenburg, "A Contribution to the History of the Word Person", *Monist*, July 1910. The best-known formulation of the person/ thing dichotomy can be found in Kant's moral philosophy; see Immanuel Kant, *Foundations of the Metaphysics of Morals*, trans. Lewis W. Beck, Upper Saddle River, N.J.: Prentice-Hall, 1997, p. 45.

[5] On this, see generally the special issue of *Etica & Animali* devoted to "Nonhuman Personhood" (vol. 9, 1998).

[6] Joel Feinberg, "Abortion", in Tom Regan, ed., *Matters of Life and Death*, 2nd ed., New York: Random House, 1986.

[7] Peter Singer, *Animal Liberation*, 2nd edn, New York: The New York Review of Books, 1990, p. 9.

[8] Paola Cavalieri, *The Animal Question*, Oxford University Press, New York 2001, pp. 117 ff.

[9] John Locke, *An Essay Concerning Human Understanding*, Cleveland: World Publishing Co., 1964, book 2, chap. 9, part 29, p. 211.

[10] Such perspective, though grown in the English-speaking world and detailedly developed in the context of contemporary bioetical discussions of the morality of abortion and euthanasia, has antecedents in continental philosophy as well. Leibniz, for example, connects personhood with consciousness of self and recollection of a former state (Gottfried Wilhelm Leibniz, *Epistula ad Wagnerum de vi activa corporis, de anima, de anima brutorum*. 1710), and even Kant, in spite of all his insistence on rationality, claims that it is the fact of being able to represent to themselves their own selves that elevates persons above all living beings (Immanuel Kant, *Anthropology from a Pragmatic point of View*, trans. Victor Lyle Dowdell, Southern Illinois University Press, Carbondale, Ill. 1978, book I, part I).

this discontinuance; and if the function of rights is to protect interests, the interest of such being in its continued existence ought to be protected by a right to life. In this context, then, our initial question becomes: Are whales self-conscious?

In spite of the difficulty of deciphering the minds of beings as evolutionarily distant from us as whales - suffice it to think of how alien the acoustic-aquatic cetacean cognitive environment is for beings like us, whose natural environment is visual-terrestrial - it can hardly be doubted that whale brains are impressive pieces of biological hardware, supporting a sophisticated type of awareness. Cognitive scientists, emphasizing the psychological mechanism of knowing one's self through interaction, argue that the brain creates the self through relationship. Over million of years, whale brains evolved through a similar process as those of humans - the need for complex societies and relationships. In the first-ever comprehensive analysis of its kind, a study guided by psychologist Lori Marino used computed tomography to investigate the pattern of encephalization in some fossil cetacean species in the past 47 million years, and analyzed these data along with those for some modern species[11]. Marino's conclusion is that the highly expanded brain size of cetaceans is, in a sense, convergently shared with humans, and that, while evolving along quite different paths, the brains of primates and cetaceans arrived at the same cognitive space.

Despite the fact that logistical problems with observing giant cetaceans in their removed habitats make it difficult to obtain in their case the same detailed evidence - regarding e.g. the capacity for mirror self-recognition or for verbal language apprehension - which is now available for their smaller relatives, the dolphins,[12] conclusions analogous to those stemming from Marino's study have recently been reached by scientists studying whale behavioral patterns. For example, according to biologist Hal Whitehead from the Cetacean Science Center at Dalhousie University, whales learn and live in ways that previously have only been identified as "human"[13].

If culture can be defined as behavior or information affecting behavior that is transmitted between individuals by non-genetic means - namely, social learning - there is ample evidence for culture and cultural transmission in whales. This holds in particular, though by no means exclusively, with reference to the specific patterns of communicative vocalization better known as whale songs, that are emitted at a much greater wavelength than human-produced sounds and whose transmission speed in the water is four times faster than the transmission speed in the air. Taking an ethnographic, as contrasted with an experimental, approach, biologists found evidence both for horizontal - or within generation - and for vertical - or parent-offspring - cultural processes. Innovations spread rapidly in social units, with many members of the group abandoning traditional feeding habits in favor of new habits introduced by creative individuals, and whale mothers create with their children close bonds which can stretch over decades, devoting time and energy to rearing them[14]. We have now a number of scientific descriptions of imitation and teaching, as well as of complex and stable vocal and behavioural cultures in many groups of whales - cultures which, representing an effect of genes affecting culture, or culture

---

[11] Lori Marino, Mark D. Uhen, Nicholas D. Pyenson and Bruno Frohlich, "Reconstructing cetacean brain evolution using computed tomography", *Anatomical Record (The New Anatomist)* 272B, 2003.

[12] Denise L. Herzing and Thomas I. White, "Dolphins and the Question of Personhood", *Etica & Animali*, Special issue: "Nonhuman Personhood" 9, 1998; Thomas I. White, *The Sea Peoples*, Oxford: Blackwell, forthcoming.

[13] Jonathan Dieli Colburn, "Listening to whales", interview with Hal Whitehead, *San Francisco Chronicle*, January 9, 2003, http://www.sfgate.com/cgi-bin/article.cgi?f=/chronicle/a/2003/01/09/MN184029.DTL

[14] Luke Rendell and Hal Whitehead, "Culture in whales and dolphins", *Behavioral and Brain Sciences*, 24 (2), 2001; M. T. Weinrich, M. R. Schilling and C. R. Belt, "Evidence for acquisition of a novel feeding behaviour: lobtail feeding in humpback whales, *Megaptera novaeangliae*", *Animal behaviour* 44, 1992; Whitehead, H., "Cultural Selection and Genetic Diversity in Matrilineal Whales", *Science*, 282, 1998.

affecting genes, have no parallel outside humans and had previously only been suggested for our species[15]. Indeed, cultural transmission is so crucial that menopause, previously thought a unique human characteristic, has now been detected in several whale species, arguably in connection with the importance of increasing the life span of older females who are the main source of information for the group.

Learning has a strong role in the development of vocal patterns, and cetacean vocalisation is an important aspect of their underlying social structure. Male humpback whales produce at any given time nearly identical songs which change yearly, while unknown means of learning enable them to keep singing in unison; and it can happen that, when some humpback whales migrate, they teach humpbacks in their new neighbourhood to start singing differently[16]. Female sperm whales display dialects - groups emit typical "codas" of several clicks partially overlapping, and individuals from different vocal clans jointly modify their codas into identical patterns, in a friendly vocal duet[17]. Since dialects will survive for several generations, and since whales in groups with different dialects tend to interact with each other quite often, cetaceans can be said to offer the only nonhuman example of multicultural societies where each individual has its own culture but is also interacting with individuals in a different culture[18].

Finally, various instances of whale behaviors directly testify to the presence of the backward, present, and forward looking attitudes forming the foundation upon which awareness of oneself as a distinct entity existing in time is mounted. A relevant backward looking attitude is revealed e.g. when hordes of whales returning to their original territory after long-distance trips first sing the old songs of the previous year, and then the new songs. The existence of a conscious self in the present, with the attendant ability to attribute mental states to others, is apparent for example in cases of gray whales doing acrobatic maneuvers to warn approaching vessels of their presence so as to avoid risks of serious damage. And undoubtedly, female killer whales' tutoring of their offspring in the dangerous activity of shallow water hunting offers evidence of the requisite forward- looking attitude in the form of a capacity for formulating and carrying out plans[19].

All considered, then, it can be claimed that the neurological and behavioral complexity of whales, as well as their elaborate communication skills, suggest that cetacean brains can produce not only a rich inner life, but also that capacity for self-consciousness that is deemed necessary for personhood[20]. If so, a consistent application of our moral standards would validate the present

[15] Luke Rendell and Hal Whitehead, "Culture in whales and dolphins", cit.

[16] Michael J. Noad, Douglas H. Cato, and M. M. Bryden, "Cultural Displacement and Replacement in the Songs of Australian Humpback Whales", *Nature* 408, 537 2000.

[17] L. S. Weilgart, "Vocalizations of the sperm whale (*Physeter macrocephalus*) off the Galapagos Islands as related to behavioral and circumstantial variables", Doctoral dissertation Dalhousie University, Halifax, Nova Scotia, Canada, 1990.

[18] Jonathan Dieli Colburn, "Listening to whales", interview with Hal Whitehead, cit.

[19] "Secrets Of Whales' Long-Distance Songs Unveiled", interview with Christopher Clark from Cornell University, March 24, 2005, *Spacedaily*, http://www.spacedaily.com/news/life-05t.html; "Gray Whale Migration Update", February 25, 1998, report by Etai Timna, Channel Islands National Marine Sanctuary, http://www.learner.org/jnorth/spring1998/critters/gwhale/Update022598.html; C. Guinet and J. Bouvier, "Development of intentional stranding hunting techniques in killer whale (*Orcinus orca*) calves at Crozet Archipelago", *Canadian Journal of Zoology*, 73, 1995.

[20] This is a conclusion which is also suggested, though in lesser detail, by other authors such e.g. Michael Tooley (*Abortion and Infanticide*, Oxford University Press, Oxford 1983, pp. 412), Peter Singer (*Practical Ethics*, 2nd, ed,, Cambridge University Press, Cambridge 1993, pp. 117-119), Harlan B. Miller ("Science, Ethics, and Moral Status", Between the Species, Vol. 10, n. 1-2 (1994 etc.

opinio juris, thus corroborating the reform at which it points - that is, the extension of the right to life to whales.

IV

In the light of all this, it seems plausible to conclude that in any conflict between whales and whalers the latter lack any moral entitlement to kill the whales, and that the international community is bound to enforce the protection of the whales both globally and nationally.

Before reaching this conclusion, however, an objection must be met. It is a distinctively legal, as contrasted with ethical, objection. For it might be claimed that, since the moment when D'Amato and Chopra's essay was published, in 1991, the trend regarding whaling has slightly changed. Though the great majority of states have responded enthusiastically to the anti-whaling movement, with the United States emerging at the forefront of the controversy as a supporter of the moratorium and unilaterally enacting pieces of legislation intended to augment the enforcement power of the IWC, a pro-whaling bloc not only keeps existing, but is becoming more vocal. Thanks to an "opt out" clause providing that a member nation in opposition to any IWC regulation needs only to file an objection to be considered exempt, Norway has lodged an official objection to the moratorium and has continued to whale commercially since 1993. Iceland has awarded itself a quota for "scientific" whaling, and after the moratorium entered into force, Japan has started a systematic program of "scientific research" directed by the government- linked Institute for Cetacean Research which, after the collection of data, directly markets whale flesh. Recently, moreover, the moratorium is facing one of its toughest challenges yet. At the 2004 IWC meeting in Italy, the Commission's chairman put forward a plan designed to hasten the adoption of a Revised Management Scheme (RMS), implying that the moratorium on commercial whaling should automatically end, with whaling restricted to coastal waters for five years and allowed anywhere after that. The resolution based on such proposal commited the IWC to proceed to completing a draft RMS for possible adoption at its next meeting, with Japan, Norway and Iceland threatening to leave the IWC if it did not adopt the RMS. And while at the 2005 meeting in South Korea member governments voted to uphold the moratorium, and Australia succeeded in passing a nonbinding resolution condemning "scientific" whaling and calling on Japan to halt the program, Japan is still actively endeavoring to reverse the trend, and to foster a return to the discredited pro-whaling policies of the past.

Does all this have an adverse impact on the argument from *opinio juris*, with the consequence of relegating the ethical case for whales' personhood and right to life to the abstract level of a merely theoretical aspiration? Arguably not. For it is not really unusual that ethically justified principles or rights backed by that sense of obligation that constitutes the *opinio juris* component of customary international law are not accepted by all nations. The conventions that are today collectively known as the "1949 Geneva Conventions" on the treatment of prisoners of war and of civilians in war, for example, have been subscribed to only by about 150 countries out of 192; and when the Universal Declaration of Human Rights was ratified through a proclamation by the U.N. General Assembly in 1948, eight nations abstained - including countries as different as Czechoslovakia, Saudi Arabia, South Africa and USSR[21]. But - quite apart from the fact that countries can, and often do, expand their moral consciousness - such lack of consensus, however regrettable, did not, and does not, prevent the involved nations from seeing the principles or rights they agreed to not only as ethically sound but also as binding customary international law, and from endeavouring to enforce them by various means, such as political pressures and economic sanctions. This holds in particular in a moment when the nation-state appears

[21] University of Minnesota, Human Rights Library, "Human Rights Education", http://www1.umn.edu/humanrts/education/4thR-F97/EleanorRoosevelt.htm

in the process of losing many of its prerogatives in favor of a globalized community[22]. Due to this process, which is clearly testified by the international recognition of the legitimacy of some forms of humanitarian intervention to stop genocides or to impose the respect of human rights within the boundaries of independent countries, the various stances and policies of the individual nation-states, far from being seen as the unchangeable outcomes of wholly autonomous entities, tend now to be considered as the proper objects of international moral censure and correction. In view both of the past record in the field of international jurisprudence and of the ongoing process of globalization, it can be argued that the fact that countries like Japan, Norway or Iceland keep opposing any granting of an entitlement to life to whales is a regrettable reality which cannot - and should not - jeopardize the emerging basic international consensus regarding the whales' status.

If this is so, then the overall argument so far developed stands. But if the new perspective regarding whales is justified, an important change is in order. We have mentioned that when the U.N. demand for a ten-year moratorium of commercial whaling was rejected by the IWC's Scientific Committee, questions were raised about the IWC role, and proposals were made calling for the U.N. to assume jurisdiction. This idea has become more relevant today. It seems plausible that, now that consciousness about whales has broadened, an institution which was initially created with the goal of regulating whales' exploitation can no longer be seen as the best organization to deal with their protection. As humanity as a whole comes to recognize the moral standing of whales, the time is ripe to remove human/cetecean relations from the hands of the former whaling nations. It would be fully in line with the present trend towards greater global governance in a variety of areas - trade and the environment, as well as peace and the protection of human rights - to create a new, *ad hoc* U.N. institution with the task of internationally declaring, and then elaborating in a series of covenants, the whales' right to life.

Among other things, the creation of such an institution would have the important side-effect of neutralizing the threat of withdrawal which the pro-whaling nations constantly use as a sort of blackmail towards the IWC. Admittedly, the problem of how to induce compliance would still remain. But this problem - which is common to all fields of international law - would not be altered for the worse in this new scenery. As it was in the past, when pro-whaling nations were prone to base their concrete decisions on their assessment of U.S. intentions, it will still be up to the anti-whaling countries to lead the battle, and to provide with their policies of suasion, pressure and sanctions the required enforcement mechanisms for whales' right to life.

[22] Peter Singer, *One World: The Ethics of Globalization*, Yale University Press, New Haven 2002.

Can food safety policy-making be both scientifically and democratically legitimated? If so, how?

Erik Millstone
Professor of Science Policy, SPRU – Science and Technology Policy Research, University of Sussex, Brighton BN1 9QE, United Kingdom, e.p.millstone@sussex.ac.uk

Abstract

This paper provides an analysis of the evolution of thinking about the role of scientific knowledge and expertise in food safety policy-making, and in risk policy-making more generally from the late 19th century to the present day. It highlights the defining characteristics of several models that have been used to represent and interpret the relations between policy-makers and expert scientific advisors. Both conceptual and empirical strengths and weaknesses of those models are identified, focusing in particular on the ways in which they deal with scientific uncertainties and social choices. By drawing on both empirical evidence and conceptual analysis, an account is provided of the conditions under which food safety policy-making could become both scientifically and democratically legitimate.

Keywords: food, policy-making, expertise, risk, legitimation

Introduction

The politics of food safety is a highly contestable and contested field. Not only are there disputes about the safety and acceptability of particular products and processes, there are also general disputes about how safety should be appraised and how food safety policy issues should be decided. Many acknowledge that scientific evidence and advice should play an important role, but there is far less agreement on what that role should be and how scientific and policy considerations should interact. There is, furthermore, a general philosophical problem about the extent to which policy-making can be either or both scientifically and democratically legitimate.

Ezrahi was not the first to argue that criteria of scientific and democratic legitimacy are distinct and incompatible, but he has articulated those arguments most vigorously (Ezrahi, 1990). Political issues are settled democratically if all eligible protagonists can contribute to decision-making and if the majority view prevails. On the other hand scientific issues are not (normally) settled in a majoritarian manner. Even if all citizens of particular national states agree that the earth is stationary at the centre of the universe, and that the sun, the planets and the stars all rotate around the centre of the earth, that does not make that cosmological theory true, either within those national states or more generally. Scientific truth is not decided democratically, while ethical and political issues are not decided experimentally. Consequently, some insist that scientific and democratic legitimation are not just different but irreconcilable.

Stages in the evolution of theories of science in policy

The idea that policy judgements can be based on uniquely reliable knowledge has a long history, going back at least to Plato. For much of the 19th century and the first half of the 20th century, however, the role of experts in policy-making was rarely thought of as either essential or problematic. Policies were legitimated with narratives insisting that all relevant considerations

and information had been fully taken into account – a rhetorical tactic that continues to have some currency.
The Second World War and the crises of the Cold War created the conditions in which scientific expertise and policy-makers became entangled as never before. Two important traditions influenced the ways in which the role of science in governance came to be understood in the mid-twentieth century; both had their roots in the late nineteenth century.

'Decisionist' models of science and policy

Weber and Durkheim argued that policy-making could be made more rational and scientific than it had been; but only partly, not entirely. Their 'decisionist' model presupposed a clear and strict division of labour between what Habermas termed: "...the objectively informed and technically schooled general staffs of the bureaucracy..." (including the experts) on the one hand and political leaders on the other (Habermas, 1971: 83). Weber recognised the superficial attraction of the idea of assigning responsibility for all aspects of policy-making to bureaucrats and technocrats, but insisted it was unrealistic because policies could never be decided solely by reference to facts or to scientific considerations. Although the choice of 'means' may be rationalised, the choice amongst the 'ends' of policy, and the underlying values, remain irredeemably subjective (Weber, 1958).
On Weber's model policy-making should comprise two separate set of deliberations, and correspondingly two distinct lines of accountability. Ministers should be responsible to elected representatives for their choice of policy goals, and through them to the electorate. Bureaucrats and experts, on the other hand, should be accountable to ministers for effectively pursuing the goals set from above, and to other experts for the knowledge and judgements that they bring to bear in the discharge of their responsibilities. A graphical representation of this model is provided in Figure 1.
That supposed division of labour came to be seen as problematic. In a relatively static pre-industrial society, political goal-setting might be indifferent to up-to-date scientific knowledge, but in a technologically dynamic society, those responsible for goal-setting may need a great deal of scientific and technical knowledge and/or advice about the potential benefits and risks arising from new science and technology. Otherwise policy-makers would not even know which areas of policy to develop.
Once scientific knowledge became pivotal to policy judgements about the products of technological innovations, the binary division of labour envisaged in the Weberian decisionist model broke down. If experts contribute to the deliberations on goal-setting, as well as selecting the mean by which those goals would be reached, two questions arise: how should scientists (and experts generally) best make their contribution, and what role remains for policy-makers?
Decisionism was, moreover, predicated on the assumption that the relevant experts were adequately knowledgeable, and unified, for their task. If the information available to the experts

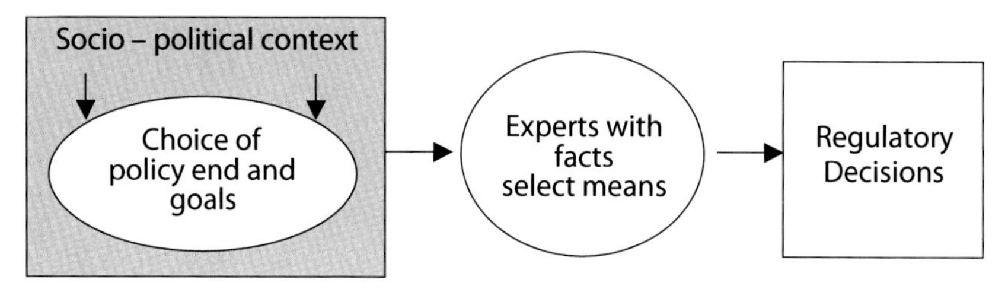

Figure 1. The Weberian decisionist model: 'Politics first, then experts'.

is incomplete, uncertain or equivocal, and/or if different groups of experts make conflicting judgements, then their deliberations will be inconclusive, so the science-based decisionist processes cannot reach closure without relying on arbitrary assumptions.

Technocratic models of science-based policy-making

In France, positivists developed a alternative vision of expertise and policy-making. Saint-Simon and Comte were not warning against technocracy but enthusiastically recommending it. They made optimistic assumptions about the progress, accuracy and adequacy of science and argued that public administration by impartial experts should replace governance by those with vested interests.
The 'technocratic' model of policy-making, as it is known, has often been encapsulated in the slogan that policy should be based on, and only on, 'sound science'. The assumption of technocracy, that scientific and technical considerations are not just necessary but also sufficient for policy decision-making, implies that policy-making should be delegated to experts, because only they possess relevant knowledge. Elected representatives' and government ministers' responsibilities are confined to recruiting the best experts. The conceptual structure of the technocratic model is represented graphically in the Figure 2.
Technocratic models are rarely articulated explicitly by policy-makers or by their expert advisors, but whenever policies are represented as if based on, and only on, sound science then an implicit appeal is being made to a technocratic model. The technocratic model assumes that the science and all the facts are entirely objective, socially and politically neutral and readily available. Technocratic rhetoric is therefore vulnerable to the argument that the evidential base and the understanding of experts is incomplete, unreliable or equivocal.

Risk governance post-1945

After 1945, in the context of an expanding regulatory state, policy-makers found themselves increasingly under pressure to make decisions about the acceptability of newly emerging technologies, or in response to newly emerging evidence of risks from technologies that were already in use, especially about health, safety and environmental consequences. Policy-makers were, however, poorly prepared for those challenges.
Governments often acknowledged that the protection of public and environment health required controls on the products and processes of, for example, the food and chemical industries. However, they had no basis for judging which products or processes needed to be restricted, nor the extent of those restrictions, so they turned to selected experts to provide scientific advice. The scientific community was also ill-prepared for this new role. Few scientists understood how to make themselves and their knowledge useful to policy-makers, and scientific knowledge of what was, and was not, safe was fragmentary and uncertain.
By the mid-1950s, scientists only had the ability to identify materials that caused severe harm which occurred either rapidly or frequently, or both. The disciplines of epidemiology, toxicology and ecology were rudimentary and the available evidence was fragmentary, incomplete and

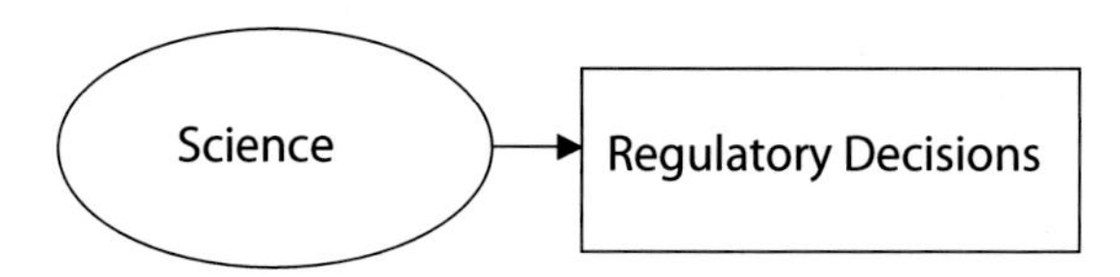

Figure 2. The Technocratic Model.

equivocal. If asked, many scientists then would have said that, apart from the causes of acute bacterial food poisoning and a handful of notorious toxicants such as lead and arsenic, either 'they did not know what was and was not safe' or 'there were risks everywhere', but that was not much help to policy-makers (Goldberg, 1985). A set of alliances then developed between, on the one hand, a fraction of the scientific community that found that it could gain an increasingly important role in, and influence over, policy-making processes, and on the other hand the officials and elected representatives responsible for administering those aspects of policy-making.

Those alliances often coalesced around the premise that thresholds of exposure to potentially hazardous materials could be readily identified, below which adverse effects would not occur. Scientists could help policy-makers by indicating what those threshold levels might be; policy-makers could then take responsibility for deciding the most appropriate means to (try to) ensure that those designated levels were not exceeded. Concepts such as an 'acceptable daily intake', 'recommended daily allowance', 'maximum residue level', 'maximum tolerated dose' and 'threshold limit value' were coined and regularly invoked, especially in the USA. As the US National Research Council explained: "The threshold hypothesis has been criticized as inadequate to account for some toxic effects, and it has not been accepted by [US] regulators as applicable to carcinogens, but it remains a cornerstone of other regulatory and public health assessments." (US NRC, 1994: 31).

In European countries (including the UK) prior to the mid-1990s, food safety policies less frequently portrayed in terms of quantitative targets, instead European policy-makers adopted a vocabulary of ensuring that products are 'safe' or 'acceptably safe'. One advantage to policy-makers of scientists identifying 'safe' levels below a supposed threshold was that it absolved them of responsibility for making explicit social and evaluative judgements about whether or not the risks were 'acceptable'. If there are no 'risks' then judgements of acceptability need not arise. It was, however, no coincidence that many of the officially selected expert advisors worked for, or with, the industries whose products they assessed. That was typically justified by arguing that relevant expertise was concentrated within those particular experts.

Under those conditions, instead of scientists working to an explicit policy agenda set for them by politicians, the policy-makers were portrayed as secondary to their experts or even entirely redundant. As Habermas suggested: "The dependence of the professional on the politician reversed itself. The...[politician] becomes the mere agent of a scientific intelligentsia, which, in concrete circumstances, elaborates the objective implications and requirements of available techniques and resources as well as of optimal strategies and rules of control..." (Habermas 1971: 63). Understandably, policy-makers did not routinely assert that they had become mere agents of a scientific intelligentsia, but they maintained their enthusiasm for invoking the rhetoric of 'sound science' to legitimate the decisions that were being taken. In effect, in those circumstances the decisionist model was inverted.

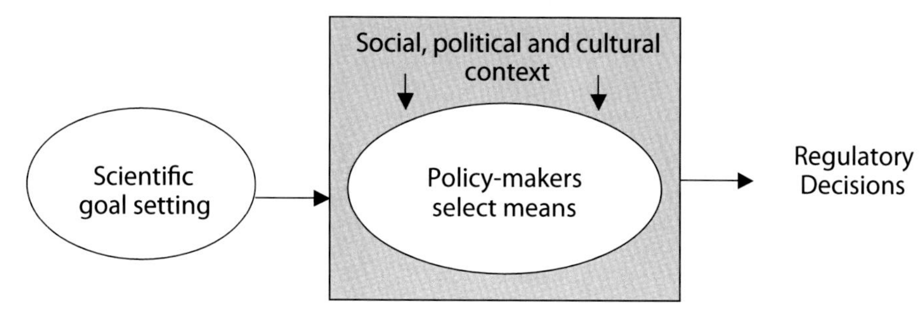

Figure 3. Inverted decisionism.

Inverted decisionism in the USA

By the late 1950s it had become increasingly difficult for US regulators to sustain technocratic representations of their policy-making processes. The US Federal government is inherently plural, composed as it is of the executive, the legislative and the judicial branches. Technocratic narratives were undermined by separate and contrary interventions of Congress and the courts. In 1958 Congress passed legislation that explicitly imposed a political framework upon the interpretation of science for policy. Under the provisions of the Delaney Amendment, Congress stipulated how incomplete and uncertain scientific evidence about possible cancer risks should be interpreted by federal government advisors and public officials (1958 Food Additive Amendments to the 1938 Federal Food, Drug and Cosmetic Act). Congress deliberately deprived the US Food and Drug Administration of a considerable portion of its scope to exercise discretion when responding to scientific uncertainties and evidence of possible carcinogenicity. The Delaney Amendment stipulated that if there was any evidence indicating that a chemical could induce cancer 'when ingested by man or animal' then it must not be permitted for use as a food additive. For such compounds, the only acceptable threshold was zero. Under those conditions, technocratic accounts of policy-making lost their plausibility because self-evidently policy decisions are products of both political and scientific judgements.

European approaches: Pre-March 1996

The situation in Europe was, however, very different. European governmental systems, especially in France, the UK, Spain and the Netherlands were monolithic rather than plural. The policy-making system in the German Federal Republic was potentially more pluralistic because the *Länder* were relatively autonomous. In all those countries there was, however, no freedom of information legislation. And science-based policy-making was routinely conducted behind closed doors, through opaque and un-accountable processes in which science and politics hybridised, while the process was misrepresented with technocratic narratives.

Since the European Commission was a product of the melding of European policy-making cultures, it too adopted technocratic rhetorics and institutional structures, though more often implicitly than explicitly. For example, the European Commission's Scientific Committee for Food (SCF) did not just report on what was known, and unknown, about the putative effects of food chemicals on consumer health; the SCF decided which compounds were acceptable and which were not, and which uses of which compounds were acceptable and which were not. Policies were represented as solely the product of scientific deliberations; policy decisions emerged from bodies that were officially labelled as purely scientific. To the extent that the Commission played an explicit role it was to rubber stamp the recommendations of the SCF. Discreetly, the Commission played a crucial role in selecting its preferred expert advisors.

The persistence of technocratic narratives into the 1980s and 1990s was especially conspicuous in UK and European BSE policy-making. Between 1986 and March 1996, UK ministers and European Commission officials routinely insisted that BSE policies were based on and only on sound science. In practice, but behind closed doors, the selections and interpretations of scientific evidence and advisors was exquisitely sensitive to the policy contexts in which experts developed and articulated their advice (van Zwanenberg and Millstone, 2005).

US developments

US policy-making regimes were confronted by several profound challenges. The conduct of the Vietnam War resulted in the introduction of a Freedom of Information regime, which was strengthened following the Watergate scandal. Disclosure entailed that the scientific uncertainties could no longer be concealed. At least some of the uncertainties became undeniable, and consequently technocratic narratives lost their residual plausibility. If the science is uncertain, science-based policy-making is open-ended and indecisive. An alternative variant of inverted decisionism was developed. Policy-making came to be represented as a two-stage process, science was portrayed as in the lead with policy-makers following and responding.

Policy-making could still be represented as 'science-based'; once scientists have decided what is, and would be, either safe or unsafe, there is a role for policy-makers to decide how best to implement the advice of the scientists, taking account of non-scientific factors such as the costs of alternative courses of actions, and the ways different groups of protagonists might respond to policy signals and/or regulatory requirements. They might also legitimately juxtapose anticipated benefits to anticipated risks (and uncertainties) and make judgements about the acceptability of such trade-offs.

This new narrative of policy legitimation was curious in one key respect. On this inverted model, it is the scientists who are portrayed as identifying the goals that should be reached, while policy-makers are confined to deciding which means are most appropriate, with which to reach the science-derived targets. It is not difficult to see why policy makers need to have regard, for example, to the costs of alternative courses of action and their likely effectiveness. What is harder to understand is how scientific deliberations can be sufficient to choose between alternative and competing policy goals. That conceptual anomaly did little, however, to inhibit the popularity and diffusion of inverted decisionist models and narratives.

Inverted decisionism was adopted not just rhetorically, but also as a design principle for the structure of policy-making institutions. Scientific advisory committees were established and asked to provide advice on policy matters. That advice could then be passed to, and received by, a separate part of the bureaucracy that would be responsible for conducting and completing the final stage of the policy-making process.

On this model, public policy makers can be held responsible, through normal processes of democratic political accountability (including for example the legislature and the courts) for obtaining and being guided by the advice of the scientific experts, and for their subsequent down-stream discretionary policy judgements. Expert scientific advisors, on the other hand, can be accountable, but only to their scientific peers. Non-experts on this model have no standing as judges of, or contributors to, the deliberations of the experts. In particular, that restriction is supposed to apply to the policy-makers, since on this model experts are presumed to provide scientific advice that is entirely independent of policy-makers and/or other interested stakeholders. It is the supposed neutrality and objectivity of science that is presumed to justify the role of scientific advice as providing the foundations upon which a legitimate policy regime could be constructed.

Problems with the plausibility and adequacy of the inverse decisionist model emerged soon after the model was first deployed, although those difficulties were more widely acknowledged by policy analysts than by policy-makers. The inverse decision model has, nonetheless, become the contemporary official orthodoxy. Science-based risk policy-making is routinely officially described as a two-stage process the first of which is called 'risk assessment' and the second of which is called 'risk management'. This revised model is represented graphically in Figure 4.

The 'Red Book' model

The inverted decisionist model was crucially articulated by the US National Research Council in a hugely influential book called *Risk Assessment in the Federal Government: Managing the Process*, which was bound in a red cover, and is often termed the 'Red Book' model (US NRC, 1983). A graphical representation of the Red Book model is given in Figure 5.

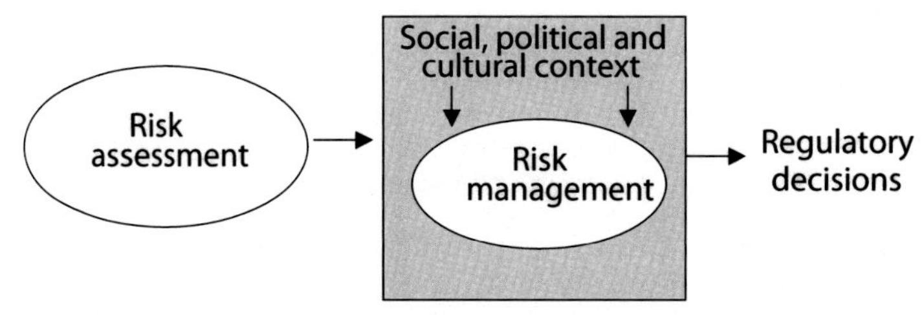

Figure 4. The revised inverted decisionist model.

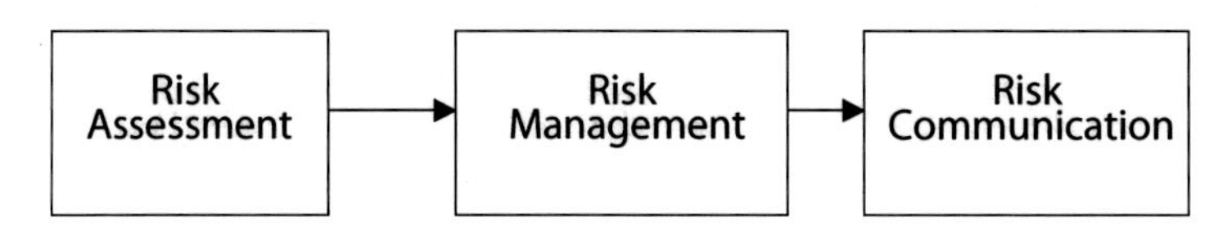

Figure 5. Red Book model.

A curious irony is that the Red Book came to be seen as setting out a science-first version of decisionism despite the fact that a careful reading of the text suggests that it adopted a more complex and nuanced model. The report was misinterpreted partly because the text is ambiguous, partly because key protagonists (including public officials, scientific advisors, and representatives of some powerful stakeholder groups) had an interest in representing scientific deliberations on risk as entirely independent of political considerations, and partly because of the binary contrast of the 'take home sound bite' provided by the distinction between 'risk assessment' and 'risk management'.

Although the Red Book argued strongly that the division of labour between scientific expert advisors and policy-makers needed to be far more carefully and clearly delineated than had previously been the custom, it did not argue that risk assessment could be a purely scientific enterprise. In ways that fundamentally undercut both decisionist and technocratic models, the Red Book emphasised that subjective value judgements are typically present, though often implicitly, in what are ostensibly represented as if they were purely scientific assessments of risks (US NRC, 1983: 28). Consequently the NRC recommended against an institutional separation of risk assessment from risk management because the report's authors saw risk assessment as an inherently hybrid activity (Jasanoff, 1990: 33).

In the 1990s the Red Book model spread from the USA to several multilateral bodies such as the OECD and the Codex Alimentarius Commission which, under the rules of the World Trade Organisation, sets global baseline standards for all internationally traded food and agricultural products.

European developments

The European Commission, along with the UK government and the governments of most other European countries (with very few exceptions), resisted the Red Book model and struggled to clung on to technocratic rhetoric until the BSE crisis of March 1996. One of the long-term effects of the BSE crisis in the UK and EU has been that it forced a re-conceptualisation and an institutional re-organisation of science-based risk policy-making, especially in relation to food safety. Several important European policy-making systems have been re-organised by reference to the 'Red Book model'. For example, as from 2004, EU food safety policy-making should be

decided at the Commission by the Directorate General for Health and Consumer Protection (DG-SANCO), which is responsible for 'risk management', but acting on the advice provided in scientific 'risk assessments' that a new separate body called the European Food Safety Authority should provide.
A similar division of labour has been institutionalised in Germany where in 2001 the Ministry of Agriculture was abolished and replaced by a new Ministry for Consumer Protection, Food and Agriculture (*Bundesministerium für Verbraucherschutz, Ernährung, und Landwirtschaft* or BMVEL). The BMVEL abolished the agency that had previous been responsible for providing both scientific and policy advice and reorganised it into two separate parts. The Federal Office for Consumer Protection and Food Safety (or *Bundesamt für Verbraucherschutz und Lebensmittel Sicherheit* - BVLS) is now defined as being responsible for risk management, while the Federal Institute for Risk Assessment (*Bundesinstitut für Risikobewertung* or BfR) is supposed to provide scientific risk assessments and advice, and not to make policy judgements.
In the UK, however, despite the popularity of the Red Book model in some policy arenas, the Food Standards Agency (FSA) is a hybrid body, responsible for both scientific advice and policy decision-making. The FSA has intermittently invoked some Red Book rhetoric, but without institutionalising a formal separation of scientific from policy deliberations that other European jurisdictions have thought necessary. One reason why this has occurred is that government ministers, reflecting on the political damage caused by the mishandling of BSE, have been keen to avoid taking responsibility for contestable and contested decisions. As the Secretary of State for Health said to the first Chairman of the FSA: "I will never hesitate to use you as my shield." That approach is in stark contrast with a prior insistence on the part of ministers that once the FSA was established, policy decisions would be taken by elected representatives rather than appointed officials.

Problematising decisionism

Numerous scholars have provided detailed evidence showing that the only reason why regulatory policy-making institutions were able to portray their policy-making processes with technocratic or decisionist models was because they contrived to construct representations of the scientific aspects of risk in narrow consensual terms, and by understating and concealing the uncertainties and concealing key non-scientific assumptions (Jasanoff, 1987; Funtowicz/Ravetz, 1993; Levidow *et al.*1997; Jasanoff and Wynne 1998; Millstone *et al.*1999).
To achieve that, policy-makers have carefully selected advisors from scientists who can be relied upon to provide advice that is broadly consistent with the prior policy objectives and commitments of ministers, and who are likely to acquiesce with technocratic or decisionist portrayals of decision-making (Erlichman, 1987). Jasanoff has argued that in the USA moreover in order to stabilize regulatory policy decisions, and enable them to withstand judicial and legislative scrutiny and challenges, institutions such as the FDA, EPA and OSHA have to 'naturalise' non-scientific judgements and to represent them as if they were essentially scientific (Jasanoff, 1990).
While much of the academic community has, in effect, ridiculed and abandoned both the technocratic and decisionist models as useful representations of how policies are actually made, policy-makers continue to pretend that science and policy have been, and are being, effectively separated from each other. Many scholars, by contrast, have argued that science and policy need to be more explicitly and effectively inter-related; their interactions are inevitable and so those interactions should be open and accountable rather than concealed and unaccountable. Those commentators have, in effect, adopted what is often referred to as a 'co-evolutionary' model of science in policy-making. It is co-evolutionary in the sense that scientific and non-scientific considerations are seen as mutually interacting rather than inhabiting entirely separate domains.

If the co-evolutionary model is more realistic than its predecessors, then institutions constructed in accordance with the Red Book model may face severe difficulties.

The co-evolutionary model of science and policy-making

Regulatory policy analysts and sociologists of science have documented ways in which ostensibly scientific representations of risk are profoundly influenced by contestable values and interests, and so have concluded that science has not provided, and can not provide, neutral and objective foundations to which policy-making can be anchored. As Jasanoff has remarked: "Although pleas for maintaining a strict separation between science and politics continue to run like a leitmotif through the policy literature, the artificiality of this...can no longer be doubted. Studies of scientific advisors leave in tatters the notion that it is possible, in practice, to restrict the advisory process to technical issues or that the subjective values of scientists are irrelevant to decisionmaking." (Jasanoff, 1990: 230). Accepting that premise entails abandoning both the technocratic and decisionist models (in both versions) and indicates an alternative approach. Figure 6 graphically represents this alternative 'co-evolutionary model'. This alternative is one that has become increasingly widely accepted by policy analysts, but only very rarely by policy-makers.

The key feature of this model is that it represents specific scientific deliberations as located in particular contexts, which have social, economic and policy dimensions. The model is predicated on the assumption that these contexts affect the content and direction of those deliberations. Consequently representations of risks are hybrid judgements constructed out of both scientific and non-scientific considerations, even if they may be presented as if they were purely scientific.

When scientists explore some aspect of the world, and develop representations of those aspects, they are firstly making some prior ontological assumptions about what kinds of objects and phenomena they are interested in, and about which are relevant and which should be ignored or discounted. Secondly they also are making some epistemological assumptions about what there is to be known, and what counts as relevant forms of knowledge and evidence. Thirdly they

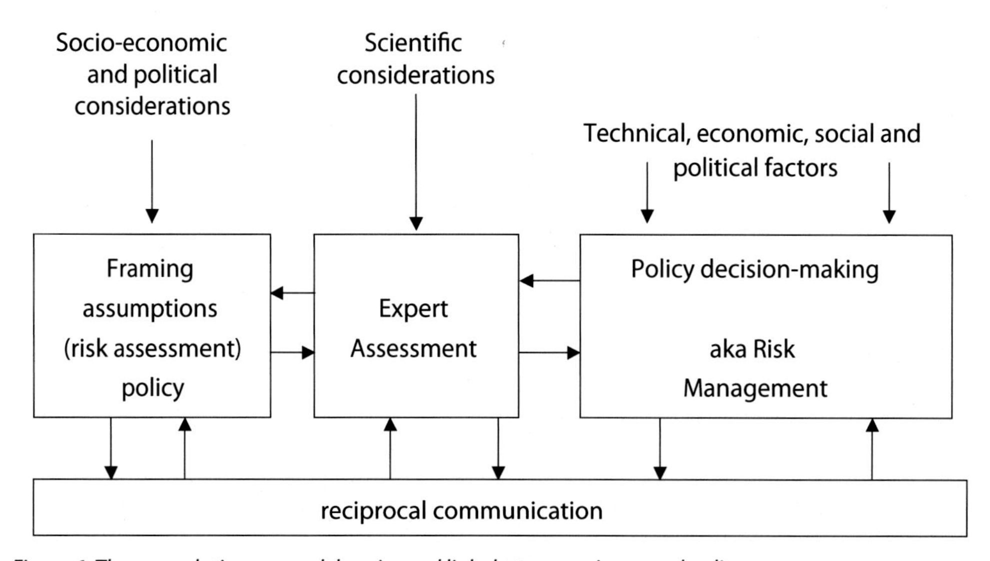

Figure 6. The co-evolutionary model: reciprocal links between science and policy.

are making methodological assumptions about what are the appropriate means for generating knowledge claims, and how those claims can be supported and/or challenged.

Elaborating the co-evolutionary model

Scholars such as Funtowicz/Ravetz, Wynne and Jasanoff who have advocated variants of the co-evolutionary model have gone beyond those generalities, and supplemented them by drawing attention to the ways in which historical, institutional, economic and cultural aspects of the contexts in which scientists operate (especially scientists chosen to provide official advice to public policy-makers) condition:

- the agendas that they address and those that they avoid.
- the types of questions that they seek to answer and those that they neglect or avoid.
- the types of evidence that they deem relevant and those that they discount or ignore.
- the ways in which the evidence is selected and interpreted.

When examining the role of scientific advisors in policy-making processes, co-evolutionary analysts have argued that the institutional and political contexts in which the advisors operate fundamentally influence what they refer to as the 'framing assumptions' of those experts. Analysts emphasize those 'framing assumptions' for four main reasons. Firstly they are very influential, secondly they have exercised their influence in an almost entirely invisible or unacknowledged way, thirdly they are readily contestable, and fourthly because the unacknowledged ways in which science and politics have been hybridised, and then misrepresented as if purely scientific, have been fundamentally implicated in regulatory policy failures, of which BSE and GM crops are just two of the most conspicuous examples (Millstone *et al.*, 2004).

The concept of a 'framing assumption' refers not just to underlying beliefs about the world but also to a broader set of considerations that condition the production of scientific claims and results. Scholars have identified many ways in which framing assumptions guide, explicitly or implicitly, the deliberations of expert scientific policy advisors. One important influence concerns the scope of those deliberations. They have shown how the deliberations of scientific advisors are crucially circumscribed by boundaries defined, in effect, by assumptions about the extent and limits of their agenda. In science-based risk debates, one of the main reasons why different groups of experts reach different conclusions is not because they reach conflicting interpretations of shared bodies of evidence but because they adopt differing framing assumptions about the categories of risks that they should address and those that they should discount or ignore. In other words, they reach different conclusions because they are answering different questions.

Illustrations of the co-evolutionary model

In the USA, risk assessments of the environmental impact of the commercial cultivation of genetically modified crops, in the mid- to late-1990s, only counted a change as an adverse effect if that environmental change harmed the commercial prospects of US farmers; all others were discounted. In Europe, the scope of risk assessments of the same crops extended to include effects on flora and fauna that could disrupt the non-agricultural environment. A few years later, the scope of scientific deliberations in both jurisdictions had widened in response to pressure from environmental and scientific groups, for example to include not just direct and short-term effects but also some indirect and long-term effects.[23]

Framing assumptions influence not just which questions are asked, but also which types of evidence are included or discounted and how the selected evidence is interpreted. For example, in many European countries where the introduction of GM crops has been controversial, one

[23] Those changes were embodied in the replacement of Directive 90/220 with Directive 2001/18.

of the controversial themes concerned the tricky issue of when a change that might arise as a consequence of growing GM crops should be counted as 'harm' and when it should be deemed merely to be an innocuous adaptation. For example, if a localised population of some variety of insects is diminished or eradicated, is that problematic if other populations remain undiminished, or only slightly diminished, in other locations?

In the UK, a set of controversial experiments known as the Farm Scale Trials of three GM crops were only conducted because some of the up-stream framing assumptions had been made explicit and had become the subject of public debate. Competing framing assumptions not only influenced the reasons for and against conducting those studies, but also influenced competing interpretations of the studies' findings. When scientists reported that the numbers of invertebrates was lower in the fields in which GM crops were cultivated, environmentalists interpreted that as demonstrating ecological harm, while some farmers adopted contrary interpretations, as providing a reduction in harm.

The safety and acceptability of food additives is regularly decided by reference to toxicological considerations, but other factors could also enter the frame. Emulsifiers are used to suspend oils and fats in aqueous solutions, and to suspend water droplets in fat products such as margarines. In recent decades the amounts of emulsifier used in our food supply has risen markedly, and so too has the fat content of our diets. The incidence of obesity is rising rapidly, and some have suggested that if the use of emulsifiers was more strictly controlled, that might make a significant contribution to combating the obesity epidemic. Emulsifiers, such as lecithin, are probably toxicologically innocuous but a decision to extend the scope of their assessments to include impacts on public health nutrition would be a policy decision (not a scientific judgement) about how much of which kinds of science were necessary and sufficient for an appraisal of the impact of emulsifiers on public health.

The only policy-making context in which the existence of up-stream framing assumptions has been explicitly acknowledged has been at the Codex Alimentarius Commission, which sets food safety standards for globally traded food products, under the auspices of the World Trade Organisation. The Codex Procedural Manual acknowledges that scientists' assessments of risks from food products are framed by prior up-stream framing assumptions that Codex calls 'risk assessment policy'.

The Codex Alimentarius Commission defines '**Risk Assessment Policy**' as follows:

- Determination of risk assessment policy should be included as a specific component of risk management.
- Risk assessment policy should be established by risk managers in advance of risk assessment, in consultation with risk assessors and all other interested parties. This procedure aims at ensuring that the risk assessment is systematic, complete, unbiased and transparent.
- The mandate given by risk managers to risk assessors should be as clear as possible.
- Where necessary, risk managers should ask risk assessors to evaluate the potential changes in risk resulting from different risk management options (Codex 2003, Appendix IV, paras. 13-16).

The Codex Alimentarius Commission and its constituent committees have been struggling to articulate risk assessment policies, but their efforts have been frustrated by several scientific committees' explicit refusals to accept the guidance provided, and by implicit deviations from that guidance. Nonetheless, the precedent set by Codex is likely to oblige Codex Member States to articulate their risk assessment policies, concerning which kinds of risks should be assessed by expert advisors, what kinds of evidence should be taken into account, and how uncertainties should he handled. If and when risk managers provide risk assessors with explicit risk assessment policy guidance that has been legitimated by democratically accountable procedures and when risk assessors explicitly follow that guidance, then important steps will have been taken towards satisfying the conditions under which science-based risk policy-making processes may become both scientifically and democratically legitimate.

Summary and conclusion

My contentions are that the co-evolutionary model provides a more accurate and adequate representation of how in practice politics, ethics and science frequently interact in food safety policy-making than any of its predecessors, secondly that there are up-stream as well as down-stream interactions. If policy-making processes were explicitly re-organised in terms of this model, and if risk managers took explicit and democratically accountable responsibility for risk assessment policy and for risk management decision-making, and if risk assessors explicitly acknowledged uncertainties as well as the ethical and political assumptions that guided their scientific assessments, then we may have policy-making processes that could reconcile both scientific and democratic legitimacy.

References

Codex Alimentarius Commission, Procedural Manual, Thirteenth edition, (see esp. Appendix IV. Working Principles for Risk Analysis for Application in the Framework of the Codex Alimentarius, paras 13-16, pp. 103-104) available at http://www.fao.org/documents/show_cdr.asp?url_file=/DOCREP/006/Y4971E/Y4971E00.HTM

Erlichman J (1987) 'Food watchdog denies conflict of interest', *Guardian*, 20 July 1987 p. 4

Ezrahi, Y (1990), *The descent of Icarus: science and the transformation of contemporary democracy*, Cambridge, Mass. : Harvard U.P.

Funtowicz S and Ravetz J (1993): 'Science for the Post-Normal Age', *Futures* Vol 25S, pp. 739-755

Goldberg, L (1985) 'Food toxicology: time for an agonizing reappraisal' in Gibson G and Walker R (eds.): *Food Toxicology: Real or Imaginary Problems?*, Taylor and Francis, London

Habermas, J (1971) 'The scientization of politics and public opinion', first published in *Technik und Wissenschaft als Ideologie*, Suhrkamp Verlag, 1968, and translated in English in *Toward a Rational Society*, Beacon Press, pp. 62-80

Jasanoff, S (1987) 'Contested Boundaries in Policy-Relevant Science', *Social Studies of Science*, Vol. 17, pp. 195-230

Jasanoff, S. and B. Wynne (1998) 'Science and decision-making', in S. Rayner and E. L. Malone (eds), *Human choices and climate change: Volume 1 – the societal framework*, Battelle Press, Ohio

Jasanoff, S (1990) *The Fifth Branch: Science Advisors as Policy-Makers*, Harvard University Press

Levidow, L, Carr, S, Wield, D, and von Schomberg, R (1997) 'European Biotechnology Regulation: Framing the Risk Assessment of a Herbicide-Tolerant Crop', *Science, Technology and Human Values*, Vol. 22, pp. 472-505

Millstone E, Brunner E, and Mayer S (1999) 'Beyond substantial equivalence', *Nature*, Vol. 401, 7 October 1999, pp. 525-526

Millstone E, van Zwanenberg P, Marris C, Levidow L and Torgersen H, (2004) *Science in trade disputes related to potential risk: comparative case studies*, European Science and Technology Observatory Technical Report Series, available at http://esto.jrc.es/detailshort.cfm?ID_report=1203

US NRC (1994) *Science and Judgment in Risk Assessment*, Commission of the Life Sciences, US National Research Council, Washington DC

Van Zwanenberg P and Millstone E (2005), *BSE: risk, science and governance*, Oxford University Press

Weber, M (1958) *Gesammelte Politischen Schriften*, 2nd ed. Tübingen

Part 1
Politics and ethics of transparency

Consumer concerns and ethical traceability: outline of a liberal argument

Volkert Beekman
Applied Philosophy Group, Wageningen University, Hollandseweg 1, 6706 KN Wageningen, The Netherlands, volkert.beekman@wur.nl

Abstract

Discourse about traceability in food chains focused on traceability as means towards the end of managing food-born health risks. The aim of traceability has been to ensure that consumers can trust that their food consumption is not risky in terms of health consequences. This discourse witnessed a call to broaden traceability and make it instrumental for the accommodation of consumer concerns about product and process characteristics of foods that are not related to health risks. This call envisions the development of ethical traceability. Then, the first question in need of scrutiny is whether ethical traceability might be a worthwhile objective. A justification discourse about ethical traceability needs to precede an application discourse. This paper presents a contribution to this justification of ethical traceability. The main line or argument is couched in terms of liberal distinctions, since the call for ethical traceability seems to be based on intuitions about consumer rights to informed choice on the food market. The paper suggests that two versions of ethical traceability find justification. First, ethical traceability as a tool to ensure that consumers are provided with foods that respect some threshold level of, e.g., animal welfare or fair trade as supported by an overlapping consensus of food values. Second, ethical traceability as a tool to facilitate that certain consumers are not provided with misinformation about product and process characteristics of foods that are relevant for reasonable, non-superficial values that are not supported by an overlapping consensus.

Keywords: consumer concerns, food ethics, informed choice, liberalism, traceability

Introduction

Regulatory and scientific discourse about traceability in food production chains hitherto predominantly focused on the development of traceability schemes as means towards the end of managing food-born health risks. The overall aim of developing traceability schemes within this context of risk management in production chains has been to ensure that consumers can trust that their consumption of food products as provided on the market is not risky in terms of health consequences. This regulatory and scientific discourse recently witnessed a call to broaden the established notion of traceability and thus to make it instrumental for the accommodation of consumer concerns about product and process characteristics of foods that are not related to health risks. This call envisions the development of something like ethical traceability. This notion of ethical traceability thus refers to the traceability of ethically relevant product and process characteristics of foods. Before engaging in the development of ethical traceability schemes, however, the first question in need of serious scrutiny is the question of why ethical traceability might be a worthwhile objective indeed. In other words, a justification discourse about ethical traceability needs to precede an application discourse. This paper presents a first and tentative contribution to the justification of ethical traceability. The main line or argument is couched in terms of some basic liberal distinctions, since the call for ethical traceability seems to be based first and foremost on intuitions about consumer rights to informed choice on the food market. The paper argues that the operational domain for ethical traceability should be positioned somewhere between needs like food safety that entail positive unconditional rights of consumers and positive unconditional duties of producers and regulators with respect to the provision and traceability of food products, and non-reasonable and/or superficial wants with respect to food that do not entail any rights and duties. The basic argument is that ethical traceability is operational within the domain of reasonable and non-superficial wants with respect to food, which entail conditional rights of consumers and conditional duties of producers and regulators.

Traceability of food safety

Rights of consumers

No doubt seems to exist that each and every consumer needs safe food. This need does not vary with the context of consumption and thus does not require any further qualification. This implies that consumers have a positive unconditional right to be provided with safe food products. This would hold true even if historically and geographically the right to safe food has not always and everywhere been recognised as such.

Duties of producers

If consumers have a positive unconditional right to be provided with safe food products, producers will have a positive unconditional duty to provide safe food products. This duty is only qualified by a non-moral clause that what counts as safe is always contingent upon available knowledge. This means that a duty to provide safe food products does not call upon producers to strive after a theoretically and practically unfeasible zero-risk. It does call upon producers to minimise food-born health risks. Furthermore, in a context of food production chains with more than two links the duty to provide safe food products implies a further duty to trace information about the safety of food products.

Duties of regulators

If needs inform rights and duties of different groups of actors in society, it will be the duty of regulators in a well-ordered liberal-democratic society to ensure that rights are respected by corresponding duties. This implies that with respect to consumers' right to safe food products, in a context of food production chains with more than two links, regulators have positive unconditional duties to secure the safety of food products by developing and enforcing traceability regulations.

Non-reasonable and/or superficial food values

If one moves from the domain of needs to the domain of wants, it should not be too difficult to see that not all wants or values with respect to food offer sound reasons to argue for rights and duties on part of different groups of actors. Two borderline cases should suffice to clarify this point. First, nobody will disagree that a cannibal does not have a valid claim to be provided with human meat. Some food values are simply beyond the realm of permissible values within well-ordered liberal-democratic societies because they impose infringements on the harm principle (Feinberg, 1987) or the justice principle (Rawls, 1971). It is of course contingent where the exact demarcation between reasonable and non-reasonable food values is to be drawn. Second, nobody will disagree that someone's preference for blackberries and disgust of blueberries does not provide sufficient reason to claim a right to be provided with blackberry pudding. Talking about rights and duties would loose any meaning, if it were not confined to values that transgress some level of superficiality. Again, the exact demarcation between superficial and non-superficial values is of course a matter of contingent discourse.

Reasonable and non-superficial food values

It seems that in contemporary European societies the following consumer concerns count as expressions of reasonable and non-superficial food values: 1) concerns about impacts on public health; 2) concerns about impacts of genetic modification; 3) concerns about impacts on animal welfare; 4) concerns about impacts on the natural environment; 5) concerns about impacts on international justice; and 6) concerns about the preservation of regional foods. It is important to acknowledge a relevant difference between the first two concerns and the last four concerns. Whereas it is perfectly possible to test products to ensure accuracy of information about food safety or genetically modified ingredients, such a proof of the pudding is not available for the latter concerns. All six concerns might be classified as substantive or end concerns to distinguish them from procedural or instrumental concerns about things like transparency, involvement, responsibility, trustworthiness and authenticity. Bluntly speaking, people only have procedural concerns because they are substantively concerned about public health, genetic modification, animal welfare, natural environment, international justice or regional foods. That is why procedural concerns may also be coined as instrumental concerns. Whereas the substantive concerns refer to reasonableness in terms of the harm and justice principles, the procedural concerns refer to reasonableness in terms of adequate information/knowledge and properly functioning markets. All six concerns share the characteristic that they go beyond merely personal interests like availability, convenience, price and taste. They are thus properly coined as moral concerns.

Rights of consumers

The fact of value pluralism within contemporary affluent societies implies that more than one set or hierarchy of reasonable and non-superficial food values exists or might exist. However, this does not yet seem enough to argue that consumers have a positive conditional right to be provided with food products that meet their values. It does seem reasonable to argue that consumers have a negative conditional right not to be misinformed about morally relevant product or process characteristics of foods. One should, however, not ignore the possibility of the existence of some overlapping consensus that holds it that, e.g., eggs, meat and milk from livestock production systems should adhere to some minimal standard of animal welfare. Then, focus changes from information to products and it does seem reasonable to argue that such cases call for a positive conditional right to be provided with certain foods. The condition is, of course, that these overlapping values are contingent.

Duties of producers

If consumers have a negative conditional right not to be misinformed about product and process characteristics that are relevant for some reasonable, non-superficial and non-overlapping food value, it will follow that producers have a corresponding negative conditional duty not to misinform consumers about these characteristics of their food products. This is actually a rather strong duty, since it would imply considerable changes to prevailing marketing practices.

Duties of regulators

Negative conditional rights and duties, on the one hand, call upon the government to develop regulatory assurances that these rights are respected by corresponding duties. The duty of regulators is, on the other hand, to facilitate the development of ethical traceability schemes as means towards the end of establishing market niches or segments for consumers and producers with similar food values.

Concerns about impacts of genetic modification

It might be wise to provide some further clarification of the above points by exploring three possible consumer concerns with respect to the much-debated yet ill-understood case of foods with genetically modified (GM) ingredients.

Health

First, it might be possible to have concerns about negative repercussions of foods with GM ingredients for public health. If these concerns are at least supported by enough scientific uncertainty about such risks to call upon a precautionary approach, the issue will become a case of needs with a negative unconditional right of consumers not to be provided with foods with GM ingredients until their safety has been established beyond reasonable doubt. Producers and regulators will then have the corresponding negative unconditional duties. If these concerns are not even supported by some scientific uncertainty, the issue will become a case of non-reasonable concerns with no rights for consumers and no duties for producers and regulators. This first possibility could thus be translated into cases of either needs or non-reasonable wants.

Justice

Second, it might be possible to have concerns about negative repercussions of foods with GM ingredients for either developing countries (intragenerational justice) or the natural environment (intergenerational justice). This is where 'old-school' Rawlsians (following 'A theory of justice' (1971)) and 'new-school' Rawlsians (following 'Political liberalism' (1993) will part. The first group of liberals would argue that it is enough to imagine a hypothetical overlapping consensus on the principles of intra- and intergenerational justice that could be reached, if the power of reasonable argument determined the outcome of socio-political discussions. The second group of liberals would argue that one needs an actual overlapping consensus to support positive conditional rights of consumers to be provided with foods without GM ingredients and to support the corresponding positive conditional duties of producers and regulators.

Naturalness

Third, it might be possible to have concerns about foods with GM ingredients because they interfere with either the order of nature or the creation of God. Such concerns are often voiced in the public debate but they cannot count on a supportive overlapping consensus. This is where anti-perfectionist Rawlsian liberals and perfectionist liberals (e.g. Raz, 1986) part. The difference between 'perfectionists' and 'anti-perfectionists' is that the latter argue that governments should somehow be neutral with respect to different visions of the good life, whereas the former argue that governments should be allowed to promote one or more specific visions of the good life. Anti-perfectionist liberals cannot see how producers and regulators in liberal-democratic societies could have positive conditional duties to ensure the preservation of the specific visions of the good that inform such concerns. They do, however, acknowledge the existence of negative conditional duties of producers and regulators that should ensure that consumers with these food values are not misinformed on labels that qualify certain foods as non-GM. Perfectionist liberals, on the contrary, do believe that producers and regulators have positive conditional duties to ensure opportunities to choose from a wide range of, e.g., regional foods. Otherwise consumers' freedom of choice would be an empty right.

Overlapping and non-overlapping food values

The above explorations might be summarised in more general terms by distinguishing two sub-categories of reasonable and non-superficial food values. If these values are supported by an overlapping consensus, ethical traceability should be developed as a public management tool to achieve the wanted changes in food production processes. It seems that concerns about animal welfare, the natural environment and international justice are based on overlapping, reasonable and non-superficial food values in present-day European societies. These concerns thus warrant regulatory enforcement of certain minimum standards of animal welfare, environment-friendliness and fair trade. Actually, this is not a case of informed choice but of reduced choice; the objective is to remove animal-unfriendly, environment-unfriendly and unfairly traded food products from the shelves of the supermarket. If these values, on the other hand, are not supported by an overlapping consensus, ethical traceability should be developed as a public-private governance tool to achieve the wanted changes in food information processes. The private side of this governance tool consists of voluntary positive labelling and tracing of food products with specific ethically relevant product and process characteristics. The public side of this governance tool consists of safeguarding consumers' negative conditional right not be misinformed by corporate marketing. Without this negative right there would be no point in positive labelling, since consumers would then not be able to judge the trustworthiness of the claims on food products.

Conclusion

The whole line or argument of this paper suggests that two versions of ethical traceability find justification. First, ethical traceability as a public management tool to ensure that consumers are provided with foods that respect some threshold level of animal welfare, sustainability or fair trade as supported by a contingent overlapping consensus of food values. Concerns about public health are excluded here, since food safety is not a want but a need and thus already covered by traceability schemes without the adjective 'ethical'. Concerns about genetic modification are excluded, since these concerns are not supported by an overlapping food value consensus. Concerns about regional foods are excluded, since it is fundamental to the nature of these concerns that they are only applied to specific food products. Second, ethical traceability as a public-private governance tool to facilitate that certain consumers are not provided with misinformation about product and process characteristics of foods that are relevant for reasonable, non-superficial values that are not supported by an overlapping consensus in contemporary European market democracies . It is important to notice that these justified versions of ethical traceability should both speak about product-information combinations, and that in both justified versions of ethical traceability improvements in tracing of information are instrumental to improvements in the provision of products with some substantive added value. These conclusions suggest focusing on the question of how the two distinguished ethical traceability regimes as management and governance tool might respectively look like. The two versions of ethical traceability might be further developed in terms of policy recommendations with respect to the development of ethical traceability schemes. Then, criteria and indicators need to be developed for a standard of labelled food products. Furthermore, contents and form of the information-demands and –supplies by consumers, retailers, food processing companies, farmers and supply companies need to be spelled out. Together these criteria, indicators, information-demands and –supplies specify ethical traceability schemes.

References

Feinberg, J. (1987). *Harm to others.* Oxford University Press, Oxford.

Rawls, J. (1971). *A theory of justice.* Oxford University Press, Oxford.

Rawls, J. (1993). *Political liberalism.* Columbia University Press, New York.

Raz, J. (1986). *The morality of freedom.* Clarendon Press, Oxford.

Ethical traceability

Christian Coff
Director of Research, Centre for Ethics and Law, Valkendorfsgade 30, 3rd floor, 1151 Copenhagen, Denmark, coff@ethiclaw.dk

Abstract

The paper gives a broad introduction to the notions of food traceability and ethical traceability. Present and potential implications for the agri-food sector are examined. Traceability is the ability to trace the history, application or location of an entity by means of recorded identifications. Today, food traceability is an instrument that serves many purposes: food safety, supply chain management, fraud control and prevention, valorisation of products/verification of product claims. It could also serve as a means for consumer information and communication with consumers. The term 'ethical traceability' is proposed as a tool for addressing specific ethical dimensions of food production practices. Ethical traceability is about keeping track of the ethical aspects of food production practices and the conditions under which the food is produced. It is a means to capture ethical values in the production history of food and represents in this sense a narrative and hermeneutic perspective on food ethics. Ethical traceability can serve as a tool for all actors in the food chain: suppliers, producers, processors, retailers and consumers.

Keywords: traceability, ethical traceability, communication, production history, informed choice

Traceability in the food chain

Traceability as a term only makes sense in modern societies where production and consumption have been separated. In a society where production and consumption occur in the same place and are carried out by the same people, or where trade is dominated by face-to-face transactions where buyer and seller can verify the quality, there is no need for verbalizing and formalizing traceability; it is there, it is inherent in the situation.

However, during the last two hundred years major changes have taken place in food production practices. Long-distance transportation means that buyer and seller do not meet face-to-face. In the early days of long-distance transportation one of the consequences was that food could be profoundly altered and decomposed during transit. Especially fresh produce was susceptible to deterioration. Hence, in the beginning of the 20th century the first record-keeping systems were developed in order to keep track of which grower delivered what, so that the grower could receive the proper price for his produce (USDA 2004:12).

Specialisation of food production practices means that food is increasingly processed outside the household and with the use of industrial and scientific techniques that are far from familiar to ordinary consumers. It also means that an increasing number of intermediaries like shippers, wholesalers, processors, repackers, brokers, importers and exporters are involved in the process. All this adds to the obscurity of how food is produced, how it is handled and from where it originates.

Furthermore, industrialization and scientific progress have changed production practices fundamentally. Also the food itself is being changed, for instance in the case of functional food. With the use of scientific methods new risks that were formerly undetectable have been identified in the food production chain.

The introduction of modern traceability schemes in the food sector occurred mainly as a response to food scandals, especially the outbreak of BSE (Bovine Spongiform Encephalopathy: mad-cow disease) in the 1980s in England, the find of dioxin in animal feed in Belgium in the late 1990s and the detection of pathogenic micro-organisms such as salmonella, listeria, clostridium and E-coli O157 and other contaminants in food. More recently fraud in the food chain has been exposed and covered extensively by the media. For instance in 2005 it was discovered in Germany that waste from slaughterhouses intended for pet food had been used for human food products.

Faults and frauds that happen during the production process endanger the wellbeing and life of 'innocent' consumers. In the long run such accidents may also damage producers: if a dioxin contamination is not detected before the food reaches the market, consumers are the first to be hit. But eventually such accidents strike back on the producers due to sanctions from the authorities and consumers and the negative media coverage. Hence, the detection of faults and frauds by the implementation of traceability schemes has become imperative. Since 2005 food traceability is required by the EU General Food Law Regulation (178/2002).

Thus traceability is introduced to deal with the growing complexity of the food chain based on mass production and global distribution and consumption. Traceability is used for record-keeping of products through all production stages, which makes it possible to trace a specific product at any time – and thus isolate contaminated or polluted elements and also expose fraud. The most internationally recognised definition of traceability belongs to the International Organisation for Standardization (ISO). ISO 9000:2000 refers to a set of quality management standards. To this set belongs ISO 8402. This standard defines traceability as "the ability to trace the history, application or location of an entity by means of recorded identifications". On a practical level the intention is to set up logistic record-keeping systems that make it possible to trace product-flow through all production stages and thus to locate the exact origin and flow of food products.

However, there are other reasons and purposes of traceability than the ones mentioned. Table 1 shows the different present applications of traceability as well as some more imaginative or visionary objectives of traceability. The many different applications can be gathered into five main categories.

Food safety initiated the use of modern traceability schemes in the food sector and has been implemented by standards such as HACCP (Hazard Analysis of Critical Control Points) and ISO. The need to be able to recall contaminated products motivated food producers to combine and incorporate traceability systems with supply chain management processes that were originally implemented to improve efficiencies (Farm Foundation 2004:8). The latter – efficiency management – is found at the third level in Table 1. It concerns how producers and processors manage the flow of the production process to secure efficient use of resources and labour forces. This fact reveals an interesting characteristic of traceability, namely that it serves a variety of purposes (and not a single purpose) at the same time.

The second level in Table 1 is likewise interwoven with the others. Record-keeping of the production history is used for surveillance, control and fraud prevention. The fourth level is linked to the second as it concerns valorisation and verification of quality claims and assurance schemes.

All these different functions of traceability are increasingly being used in the food chain. For instance the two largest retailer companies in the world, Wal-Mart and Carrefour, are increasingly demanding complete traceability from suppliers in order to achieve the goals mentioned above (Bantham and Duval 2004). The above table has also led many commentators to see traceability as a tool: as a tool for supply chain management/internal management of resources in the company, a tool for political government of the food chain, a tool for control and anti-fraud, a tool for verification of product attributes and liability and as a tool for valorisation of otherwise invisible food qualities (like respect for animal welfare).

Table 1. Concrete traceability functions in the food sector.

Objectives of food traceability
1. Food safety and risk assessment
• Safety: identification/mapping of breakdowns in food safety – allowing recall of contaminated products for the purpose of protecting public safety
• Risk assessment: mapping of foods, food ingredients or processing technologies that may have significance with respect to food safety (e.g. hygiene)
• Food residue surveillance: food sampling at appropriate points testing for residues, e.g. pesticides
2. Control and verification
• Surveillance and auditing of producer and retailer activities
• Avoidance of fraud and theft: control of products by chemical and molecular approaches (biological 'food-prints')
• Identification of responsible actors (but also claims of innocence!)
• Ingredients definition
• Avoidance of negative claims (e.g. 'may contain GMO traces')
3. Efficiency and management
• Management of the supply chain
• Secure efficient use of resources (cost minimisation)
• Minimisation of recall losses
4. Valorisation and quality assurance of products
• Marketing, health, ethical and other claims
• Authenticity: identity of the product (food authentication) and the producer
• Quality: final product standard assurance
• Label verification
5. Information and Communication
• Transparency of the production history
• Informed consumer choice
• Confidence and loyalty
• Recognition of consumer claims and concerns
• Allows for comparison of different products
• Public participation; consumer services and companies' 'care lines' for consumer feedback

Ethical traceability

The fifth level is at present the least developed. It concerns the communication of production practices and production histories in the food chain. At this level communication is not restricted to the actors within the food chain like producers, retailers and food authorities; it also includes the consumers. At this fifth level traceability is about visibility, it is about making the production history of food visible to the eyes of the consumers. Traceability as a means for information provision and as a communication strategy opens up new perspectives on the relationship between consumers and the actors in the food chain on the issue of food production practices. This kind of information and communication strategy could in future allow for consumers'

informed choice. Probably it could also contribute to a better understanding and communication of consumer claims on the actors in the food chain.
Traceability usually involves hundreds of inputs and processes, and setting up systems for the handling and communication of all these data requires highly developed and usually computerized systems. Most of the information in traceability systems would from a consumer perspective probably be irrelevant, as it concerns technical matters that are only of interest for the actors in the supply chain. But some facts relating to traceability do relate to consumer concerns about the production history of food.
Traceability is an ethical issue in many respects and as such the notion *ethical traceability* covers many areas and has many meanings. Food safety is obviously an ethical issue, since it aims at protecting consumers from food-borne diseases and pollution. Avoiding fraud in the food chain is likewise ethical by nature. The assurance of correct information to consumers and verification of assurance and labelling schemes are also ethical matters.
Ethical traceability is thus not suggested here as a term that covers a specific assurance scheme concerning fixed ethical issues. Rather, it is used more 'imprecisely' and in a more general sense about the record-keeping of ethical aspects of food production practices and the conditions under which the food is produced. It is conceived as a 'narrative' and 'historical' way of capturing and reflecting on ethical values in the food production chain. As a communication and information strategy ethical traceability can be used by both consumers and stakeholders in the food chain. Firstly, by producers and processors who wish to secure a minimum level of ethical behaviour among their suppliers. Ethical traceability concerns information on the ethics of the production history essential for the producer's ethical judgement of its suppliers (see Coff, 2006, for a description of the link between food ethics and the production history). Ethical traceability could in this way serve as an internal management of ethical issues, for instance as part of corporate social responsibility strategies in companies.
Secondly, ethical traceability can be conceived as the information necessary for consumers' ethical judgement of the production history of food. Ethical traceability would as such be central for consumers' informed choice: information on the ethical aspects of the production history is essential to ethical consumers who are concerned by the impact of production practices on for instance animal welfare, working conditions and environmental sustainability.
The many uses of traceability make it a battlefield. There is widespread agreement that the need for fully documented traceability systems within the food chain has never been stronger (Morrison 2003:459), but there is tremendous disagreement concerning the purpose of introducing traceability in the food sector. Disagreements about the role of food ethics in production practices are reflected and exposed in the opinions of how to make use of traceability. Some argue that traceability should be restricted to internal use in the food chain while others emphasise its potential for providing information to consumers and creating dialogue between consumers and the actors in the food chain.

Acknowledgement

This paper is part of the EU-funded project *Ethical Traceability and Informed Choice in Food Ethical Issues* (see www.food-ethics.net). Many of the ideas presented here have been developed and matured in discussions with the partners of the project.

References

Bantham, Amy and Jean-Louis Duval (2004). *Connecting Food Chain Information for Food Safety/Security and New Value*. John Deere, Food Origins.

Bingen, Jim and Busch, Lawrence (eds.) (2006). *Agricultural Standards. The Shape of the Global Food and Fiber System*. Springer, Dordrecht. Series: The International Library of Environmental, Agricultural and Food Ethics, Vol. 6.

Coff, Christian (2006). *The Taste for Ethics. An Ethic of Food Consumption*. Springer, Dordrecht. Series: The International Library of Environmental, Agricultural and Food Ethics, Vol. 7.

Danish Ministry of Foods, Agriculture and Fisheries (2004). *Sporbarhed i fødevarekæden*. [*Traceability in the food production chain*]. Danish only. Danish Ministry of Foods, Agriculture and Fisheries, Copenhagen.

EU General Food Law Regulation (178/2002).

European Commission (2006). *Special Eurobarometer: Risk Issues*. 238/Wave 64.1 – TNS Opinion and Social. Available at http://europa.eu.int/comm/food/food/resources/special-eurobarometer_riskissues20060206_en.pdf [date of consultation: 12.02.06]

Farm Foundation (2004). *Food Traceability and Assurance in the Global Food System*. Farm Foundation's traceability and Assurance Panel Report, Oak Brook.

Food Strategy Division and Food Standards Agency (2002). *Traceability in the Food Chain. A preliminary Study*. Food Strategy Division and Food Standards Agency, London.

GS1 (2006) *The Global Traceability Standard*. GS1. Available at www.ciesnet.com [date of consultation: 23.03.06].

Hermite, Maire-Angèle: La traçabilité des personnes et des choses. Précaution, pouvoirs et matrîse. In *Traçabilité et responsabilité*. Philippe Pedrot (ed.). Economica, Paris. 1-34

Korthals, Michiel (2004). *Before Dinner. Philosophy and Ethics of Food*. Springer, Dordrecht. Series: The International Library of Environmental, Agricultural and Food Ethics, Vol. 5.

Larsen, E. (2003). Traceability in fish processing. In Michèle Lees (ed.) *Food authenticity and traceability*. Woodhead Publishing Limited, Cambridge. 507-517.

Lind, David and Elizabeth Barham (2002). The Social Life of the Tortilla: Food, Cultural Politics, and contested commodification, *Agriculture and Human Values* 21:47-60.

Morrison C. (2003). Traceability in food processing: an introduction. In Michèle Lees (ed.) *Food authenticity and traceability*. Woodhead Publishing Limited, Cambridge. 457-472.

Napoli, Paolo (2003). Administrare et curare. Les origines gestionnaires de la traçabilité'. In *Traçabilité et responsabilité*. Philippe Pedrot (ed.). Economica, Paris. 45-71

Pedrot, Philippe (ed.) (2003). *Traçabilité et responsabilité*. Economica, Paris.

USDA (2004). *Traceability in the U.S. Food Supply: Economic Theory and Industry Studies*. United States Department of Agriculture, Agricultural Economic Report Number 830.

Towards value based autonomy in livestock farming?

Karel de Greef[1], Carolien de Lauwere[2], Frans Stafleu[3], Franck Meijboom[3], Sabine de Rooij[4], Frans Brom[3] and Jan Douwe van der Ploeg[4]
[1]Animal Sciences Group of Wageningen UR, PO Box 65, 8200 AB Lelystad, The Netherlands, Karel.degreef@wur.nl
[2]Agricultural Economics Institute, Den Haag/Wageningen, The Netherlands
[3]Ethics institute, Utrecht University, The Netherlands
[4]Rural Sociology Group, Wageningen University, The Netherlands

Abstract

Pig farmers experience a discrepancy between their view on 'doing good to animals' and the societal pressure to adapt their farming procedures. Legislation is regarded by them as making a schism between animal welfare and rational production, omitting the opportunities of optimizing welfare from a stockmanship's perspective. At the same time, it seems that the governmental intervention hardly reduces the societal unease. An explorative study revealed that a discrepancy of moral systems between urban and rural populations plays a larger role than lack of morality. A project has started which combines sociology, ethics and interactive design to study a solution to this discrepancy. It is hypothesized that providing some degree of professional autonomy to farmers is useful for building a system that restores the balance between being responsible and having the opportunity to act on basis of the own values. This may result in a system that has a better chance of being maintainable because it is founded on the values of the farmers. Interviews amongst a structured variety of farmers are held to collect their typical farmers' values. This data set is used to make the farmers' morality explicit. On basis of this, a code of conduct (directed to the own professional group) and a code of practice (directed to the rest of society) is developed, verified and applied on some actual (controversial) cases.

Keywords: farmers values; professional codes; professional autonomy

Introduction

Livestock farming is increasingly criticized by society. The welfare of animals and the quality of the environment is thought to be sacrificed by farmers for economical reasons (Waelbers *et al.*, 2002). This social criticism has a moral character: the quality of farmers' ethics is doubted. The reaction of many farmers is to take a defensive position. They tend to deny that farming contributes to environmental pollution and causes animal welfare problems (Te Velde, 2002; Silvasti, 2002). Such reactions are of a non-moral nature and do not respond to the heart of the criticism. European governments have chosen to meet the social concern with legislatory measures, imposing handling and husbandry procedures on farmers. However, these measures do not only affect the competitive (economic) position of the farmers, the underlying thoughts are also considered to be partially in conflict with the farmers view on doing good do animals. Increasingly, it is realised that solutions to problems like this cannot be solved by scientific research alone, neither by policy measures alone. The Dutch Organisation for Scientific Research has formulated a research program in which ethics, research and policy are addressed mutually. Within this program, a project has started that addresses the presented livestock farming problem in a multidisciplinary way, combining sociology, ethics with stakeholder interactivity. The starting hypothesis is that farmers have moral views that underlie their farming management and that moral empowerment of the farming community is a strategy that could reduce the

problem. The concept of professional autonomy is used as a vehicle to test this hypothesis. Codes of conduct and codes of practice will be formulated in order to explicate and discuss the moral views and beliefs of farmers into formalised procedures. The project is a result of an inventory study in which farmers values were made explicit and seemed to deviate to deviate from urban values (Stafleu *et al.*, 2004; De Greef *et al.* 2006).
Below, a transcript of the research proposal is presented, enriched with some early thoughts and experiences. The project is planned to be conducted from 2006 throughout 2008. The EurSafe community is invited to comment the line of thought. Connection to similar approaches elsewhere is appreciated.

The problem

Closer analysis of the causes of the disturbed relationship between livestock farmers and society shows a more pronounced interest of the public in the intrinsic value of nature and the individual animal on one hand (Ketelaar-de Lauwere *et al.*, 2000) and a farmers' ethics of productionism which conflicts with the intrinsic value view on animals on the other hand (Thompson, 1995; Silvasti, 2002). These farmers' ethics are for the most part implicit. Therefore, farmers do not have an adequate (i.e. moral) voice in the pluralistic societal discussion . This is a major problem in a democratic-pluralistic society in which stakeholders are expected to defend their own position (Brom, 1997). As a result the discussion is not even and farmers may be literally overruled in the end, leading to superimposed legislation and a loss of farmer autonomy. This limited social basis may result in a lack of farmers compliance with governmental directives.

Aims of research

The principal aim of the project is to contribute to the 'social empowerment' of farmers to participate better in the public debate and to create a better basis for society's regulatory system based on values and trust. At first, the focus is on the *moral* position of farmers by making the implicit farmers' ethics explicit by empirical research in different groups of farmers. After validation of the results in an interactive process with farmers, the results will be put into practice by developing tools which facilitate the moral discussion. Both a code of conduct will be developed, which primarily aims at the farmers themselves and expresses their values and responsibilities, and a code of practice which also incorporates the values of the rest of society and describes the farmers' social responsibilities for example for animal welfare and the environment. To test the applicability of these tools, the concept of ´professional autonomy´ will be evaluated for two practical cases. The hypothesis behind this is that value based (professional) autonomy is a feasible concept for farmers for a robust societal role and for restoration of trust between farmers and society.

State of the art

Values of farmers as input for farmers' ethics

The values of farmers have been subject of different studies. Stafleu *et al.* (2004) described for pig farmers that, besides economical values, ´free entrepreneurship´ and ´working with nature and animals´ are important. Schoon and te Grotenhuis (2000) distinguished the values production efficiency, sufficient (cheap) food, ecological sustainability and continuity of the farm. Thompson (1995) recognizes 'working hard' as separate value. All these studies have been done in small and/or distinctive groups. It is, however, not likely that all farmers have the same 'set' of values. It is known that different farming styles can be distinguished (Van der Ploeg, 2003; Commandeur, 2003; De Lauwere, 2006), which are related to different attitudes towards for example animal welfare and the environment. Gender differences are also known (De Rooij *et al.*, 1995) and differences between organic and conventional farming (Fairweather, 1999; Lund *et al.*, 2004).

Codes as tools for the moral empowerment of farmers

Ethics can be put into practice by the development of codes. Hogenhuis (1993) describes three kinds of codes: An aspirational code, an advising code and a disciplinary code. The aspirational code is aimed to the profession itself and identifies values, ideals and responsibilities which are directing the practice of a profession (Hogenhuis, 1993). Both a code of conduct and a code of practice will be developed for farmers. The code of conduct for the farmers can be interpreted as an aspirational code as in this code the values of farmers are leading. The code of practice may be an advising code or a disciplinary code. An advising code makes professionals aware of professional problems and the moral relevance of these problems, and a disciplinary code gives global standards or detailed codes of practice for professionals in a certain, morally sensitive situation and is used to assess the acts of the professional and to sanction trespassing (Hogenhuis, 1993). The code of practice will combine the farmers' ethics with demand from the rest of the society, and may serve as a professional standard as mentioned in the definition of professional autonomy. The codes, especially when developed and validated together with the farmers. will be more useful than common codes like 'good farming practice' because of the explicit foundation in the actors values.

Professional autonomy

In this study, it is hypothesized that value based autonomy is a feasible strategy to give farmers a 'voice' in the public debate. Professional autonomy is an example for this from the human heath care domain. It means that society gives a mandate to the professional and his profession. Delivering self control for the profession and freedom and independence for the professional within certain limits, given by a professional standard (Raad voor de Volksgezondheid & Zorg, 2000). Although health care and agriculture are very different practices, professional autonomy may be a useful concept to agriculture and farmers as well. Farmers do have special knowledge, not only originating from their education but also from their rich tradition, and they also describe their profession as being a 'vocation'. A keyword in getting a mandate from society is 'trust'. Patients do trust their doctors, but does the public trust 'their' farmers? Is there are enough basis for ascribing professional autonomy to farmers? This is subject of the proposed research.

Planning of the research, and methods

Part I: Sociological inventory of farmers values

Key question: What are the key values of farmers related to their profession?

In the first part of the research, the farmers' values are made explicit by an empirical research and are validated in an interactive process with farmers (focus groups). The identification of normative frameworks that govern, be it in an implicit or in a more explicit way, farmers' behaviour and attitudes towards nature and in particular towards their animals, will be based on the 'farming styles approach' developed in Wageningen (a.o. Van der Ploeg 2002). As it is expected that different groups of farmers have different values, a differentiation will be made between different sectors (pig farmers and dairy farmers), different farming styles, gender differences and between organic and conventional farming. In order to grasp relevant heterogeneity it is estimated that for each sector some 20 extensive (and probably iterated) interviews will be needed. A (stratified) survey among some 100 respondents is planned as well.

Part II: Ethical analysis

Key question: What does a specific farmers ethics look like?

The farmers' values from part I are the input for the ethical analysis which will consist mainly of conceptual analysis and development of morality. The 'positive morality' of the farmers will be examined on internal coherence and consistency and will be positioned in the public debate. Finally, it will be examined how it is positioned regarding the critical and public morality (c.f. Brom, 1997). The resulting explicit farmers' ethics (one or more!) will be validated in an interactive process with farmers. As a part of this, the degree to which the farmers' ethics matches the actual acting of farmer's will be assessed together with farmers by analyzing social controversial farming practices (such as castrating). The analysis will provide input for part III and IV

Part III: Integration of the results using code development

Key question: Can the farmers values be converted to codes, and what can be the role of codes and standards value based autonomy?

In part III, a code of conduct and a code of practice will be formulated, together with farmers and other relevant social actors The code of conduct is a 'set' of standards which matches the farmers' ethics and which helps the farmer to make relevant choices in his daily farming practice. The code of practice also is a set of standards, which is formulated together with farmers, but which is mainly imposed by society. The code of conduct is a mirror of the farmer's own actions and the code of practice the mirror of the extent to which his actions are socially accepted.

Part IV: Interactive evaluation on basis of two cases

Key question: Are the codes strong enough to acquire a professional autonomy?

As a validation of the previous parts of the study, the opportunities for acquirement of a more autonomous position for the farmers are evaluated using two practical cases: health management and welfare conditions. For both these cases, it will be studied whether acquirement of some degree of autonomy is feasible. Such a mandate implies that the farmers values (and codes) are accepted and trust is given by the stakeholder groups to the professional group. This study will be conducted in workshops, in which stakeholders such as veterinarians, and representatives from government and animal protection groups, will participate.

Originality and impact

The connection between the problem diagnosis (lack of explicit farmers' ethics) and the approach to solve the problem ('moral empowerment' resulting in 'social empowerment' through acquisition of some degree of autonomy) are new. The hypothesis that value based autonomy may be applicable to farmers provides a concept that combines and enriches old concepts like 'social responsibility' and 'licence to produce'.

The study is scientifically relevant as it combines social, ethical and interactive approaches in a highly relevant application. Furthermore, it is foreseen that it will make a contribution for agricultural ethics as it will add a 'ethics of production' concerning animals to the one Thompson (1995) has described for non-animal agriculture. Interesting ethical theoretical topics will be studied like the relationship between positive, critical and public morality in a pluralistic-democratic society and like the concept of professional autonomy which will be carried further than the human healthcare field. From an ethical and sociological point of view systematic research of differences between ethics of groups of farmers is interesting. The proposed research is truly multi-disciplinary as it is carried out by ethicists, sociologists and agricultural scientists. The main aim of the research contributes to solve an important policy problem i.e. the failing role of farmers in the public debate and the resulting non compliance with external rules using both social and ethical research in a practically relevant domain. The results (comprising both tools and analyses) are promptly valuable for both the actors/stakeholders themselves and to policy accounts. In this way, science and society profit mutually.

References

Brom, F.W.A.(1997). Onherstelbaar verbeterd. Biotechnologie bij dieren als moreel probleem. Thesis Utrecht University, the Netherlands.

Commandeur, M. (2003). Styles of Pig Farming. A Techno-Sociological Inquiry of Processes and Constructions in Twente and The Achterhoek. Thesis Wageningen University, The Netherlands.

De Greef, K.H., F.R. Stafleu and C.C. de Lauwere (2006). A simple value-distinction approach aids transparency in farm animal welfare debate. Journal of Agricultural and Environmental Ethics (2006) 19:57–66.

Fairweather, J. R. (1999). Understanding how farmers choose between organic and conventional production: results from New Zealand and policy implications. Agriculture and Human Values 16: 51-63.

Hogenhuis, C.T.(1993). Beroepscodes en morele verantwoordelijkheid in technische en natuurwetenschappelijke beroepen: een inventariserend onderzoek. Ministerie van Onderwijs en Wetenschappen.

Ketelaar-de Lauwere, C.C., H.J. Blokhuis, J.C. Dagevos, A.H. Ipema and J.A. Stegeman (2000). Changing human-animal relationships and their impact on livestock farming in 2040 (in Dutch with English summary). IMAGrapport 2000-6, Wageningen-UR, Wageningen.

Lauwere, C.C. de (2006). The role of agricultural entrepreneurship in Dutch agriculture of today. Accepted for publication in Agricultural Economics.

Lund Y., S. Hemlin, J.White (2004). Natural behavior, animal rights or making money- A study of Swedish organic farmers'view of animal issues. Journal of Agricultural and Environmental Ethics 17: 157-179.

Raad voor de Volksgezondheid & Zorg (2000). Professionals in de gezondheidszorg. Raad voor de Volksgezondheid en Zorg, Zoetermeer.

Rooij, S. de , E. Brouwer, R. van Broekhuizen (1995). Agrarische vrouwen en bedrijfsontwikkeling. Uitgave: Studies van Landbouw en Platteland 18/ Wetenschapswinkel rapport 116, Wageningen.

Schoon, B., R. te Grotenhuis (2000). Values of farmers, sustainability and agriculture policy. Journal of Agricultural and Environmental Ethics 12, 17-27.

Silvasti, T. (2002). The cultural model of "the good Farmer" and the environmental question in Finland. Agriculture and Human Values 20: 143-150.

Stafleu. F.R., C.C. de Lauwere, K.H. de Greef,. P. Sollie, S. Dudink (2004). Boerenethiek Eigen waarden als basis voor een 'nieuwe' ethiek. Een inventarisatie. Verkennende studie NWO.

Te Velde, H., N. Aarts, C. van Woerkom (2002). Deailing with ambivalence: Farmers'and consumers'perseption of animal welfare in lifestock breeding. Journal of Agricultural and Environmental Ethics 15: 203-219.

Thompson P.B.(1995). The Spirit of the Soil. Routlegde, London and New York.

Van der Ploeg, J.D. (2003). The virtual farmer. Past, present and future of the Dutch peasantry. Royal Van Gorcum B.V., Assen.

Waelbers K, F.R. Stafleu, F.W.A. Brom, (2002). Het ene dier is het andere niet! Verschillen in (morele) grondslagen in veterinairbeleid voor Landbouwhuisdieren en dieren die in natuurgebieden leven. *CBG reeks 14*, CBG Universiteit Utrecht.

Participatory methods and food policy: Different approaches for different purposes

Jon Fixdal
Project manager, Norwegian Board of Technology, Prinsensgate 18, N-0105 Oslo, Norway, jon.fixdal@teknologiradet.no

Abstract

During the last 10-20 years we have witnessed the development of several different methods for public participation in technology policy and assessment. Consensus conferences, scenario workshops, citizen panels, citizen summits, round tables and café-workshops are some examples. The methods share a common ambition in providing affected citizens an opportunity to deliberate and formulate policy advice on important issues. At the same time they differ in numerous ways, such as with respect to how the deliberations are organized, how they perceive "affected citizens", how topics are framed, and how the participants state their opinions. Hence, there is no "standard approach" to participatory processes. Food policy embraces numerous topics that could be – and has been – debated in participatory projects. Food safety, genetically modification of food, labelling, health and nutritional issues, and the ethics of food production are examples. But how can we know what participatory method to choose when we have a topic at hand? This is the key question for this presentation. It focuses on how the characteristics of a topic, the purpose of the participatory process and key characteristic of participatory processes needs to be pared for a successful participatory process.

Keywords: public participation, participatory methods, method selection

Introduction

Technologies are an integral part of almost all aspects of modern life. But while most technologies help solve problems they often create new ones two. Today we can travel faster and longer than ever before, but we are all familiar with the problems of local pollution and emissions of green house gasses. Similarly, information technologies make it possible to communicate more efficiently over grater distances, but problems related to surveillance and privacy are well known. Genetic modification of food can increase the nutritional value and enhance preservation of foods, but there is uncertainty about the long term health effects of eating such food. No doubt, there is a need for policy awareness and action on numerous issues related to the development and use of technologies.

Traditionally, experts have played a key role in assessing the possible advantages and challenges related to the use of technologies. They have also served as advisors to policy makers about threats, opportunities and policy options. The reason is, of course, that expert knowledge and the presumed disinterestedness of experts are believed to result in comprehensive and unbiased advice.

However, during the last 10-20 years a growing number of scholars have argued for increased public participation in science and technology policy making. In short, there are three main arguments for increased participation (see for example De Jong and Mentzel 2001; Laird 1993; Sclove 1992):

- Strengthening democracy: It is argued that expert based problem solving disenfranchises the public of its democratic right to control policy, and that it therefore is incompatible with

democratic ideals. The public should be involved in the definition of issues, in the assessment of consequences and in evaluating policy actions.
- Better informed decisions: This line of arguments highlights that ordinary citizens can have a local, personal knowledge that differs from the more specialized and generalized knowledge of experts, and that public participation thus can make decisions better informed. Also, many questions of technology policy have strong normative dimensions, and in assessing such question ordinary citizens can be just as expert as the academic expert.
- Enhancing legitimacy: It is argued that affected citizens have a possibility to participate and influence decisions affecting their own lives, policy decisions are likely to be more legitimate and socially acceptable.

Parallel to the (mostly academic) debate about the need for public participation, we have witnessed the growth of a series of different participatory methods. These are different conferences and workshops through which citizens affected by a policy issue can be informed about that issue and formulate advice to policy makers on what their main concerns are and how that issue ought to be handled.

However, the range of processes that can be (and has been) labelled 'participatory' poses the question: how do we choose between them? Attempts have been made to make categoryzations of types of participatory models and link them to proper types of policy issues (see, for example, Renn *et al.* 1993; Rowe and Frewer, 2000).

Such categorizations can provide some insights into the areas of availability of different participatory methods. However, I will argue that never can serve as guidelines for choosing method in a given situation. First, because the variety of participatory processes that can be altered in different ways can make different processes suitable for the same topic. Secondly, policy issues can be framed in so many ways that it is almost impossible to develop any typology of corresponding problem formulations and participatory processes. Therefore, as an alternative to the attempts to ling method to issue, I will argue that answering three sets of questions can be of crucial help. To show why this is so, and what the questions are, it is useful to take a closer look at three different types of participatory processes, and a selection of different topics related to food policy and safety.

Three types of participatory process

There exists, as mentioned, a variety of different processes that has been labelled participatory. It falls far beyond the scope of this paper to make an account of all of these. However, to illustrate the differences between some of these models, I shall present three distinctly different types. These are:

1. A "jury-structured" panel of non-organized citizens: one type of participatory processes are focused around a panel (or in some cases several panels) of 15-25 ordinary, non-organized citizens recruited through newspaper advertisement or personal invitation. The participants take part only as themselves and not as representatives from some organized interest group. Through their participation they are typically informed about a policy issue by a number of experts, and they discuss the topic thoroughly among themselves. The participants normally work together for several days. At the end of their work they formulate advice to policy makers on how the issue they have been discussing can and should be dealt with.

   Examples of this type of participatory processes include consensus conferences (see for example Joss and Durant, 1995), Citizen juries (Crosby 1995) and Planning cells (Renn *et al.* 1993).

   Participatory processes with this structure have been organized on a series of different topics, such as 'Technological animals' (Danish Board of Technology 1992), genetic modification of food (NENT 1996), and localization of a waste-treatment facility (Renn *et al.* 1996).

2. A second type of processes can be labelled "round tables". These are processes where representatives of interest groups with an interest in a policy issue or affected by a specific decision meet to discuss relevant issues. Often, the aim will be to try to reach agreement on how a topic should be dealt with or how a specific policy should be formulated. Alternatively, the participants can try to agree on which topics they do agree, and on which topics they cannot agree and for what reasons. The round table is thus a process where negotiation plays a key role, and where the round table symbolizes an egalitarian structure where all participants are meant to take part on equal footing.
   Examples of this type of participatory process include the Canadian "Round tables on the environment and economy" (NRTEE 1993) and Negotiated rule making (Fiorino 1995).
   Round tables have been used on numerous topics, such as developing guiding principles for sustainable forestry, discussion of the cultivation of crop plants with herbicide resistance, and for developing standards for occupational safety.
3. The third type of participatory process I will call "scenario workshops". These are workshops focused at developing future visions within some topic area, while also often trying to identify challenges – political, technological, social and economical – that may arise when trying to meet the visions. The key qualification for participants is that they have some knowledge that can help shed light on the topic, be it as member of an interest group, representative of an affected business, as an affected citizen or as an employee of a governmental agency. Discussion typically alternate between small groups and plenum.
   "Future workshops" have been organized on topics like sustainable urban living (Andersen *et al.* 1993) and the future of outdoor recreational life (Teknologirådet 2006).

These three types of participatory processes differ in numerous ways, including:
- how deliberations are organized: is it as a dialogue between a panel of citizens and a panel of experts, as negotiations between interest group members or as
- the "qualifications" of the participants: why are the participants qualified to take part in the process? Is it because they are and interested, non-organized citizen, a representative of an interest group, or a person with some special knowledge about a topic of future interest?
- how topics are framed: is it a clearly defined set of policy options, is it as a set of future visions within a topic area, or as a need for policy guidelines for a specific topic?

These differences show that there is no "standard approach" to participatory processes. When confronted with a specific policy issue, several aspects need to be considered when designing a participatory process. To illustrate this even further, we can take a closer look at some issues related to food policy and safety.

Some issues in food policy and safety

Issues related to food policy and safety are among the most frequently debated policy issues of today. The span of topics is wide ranging, influencing a broad variety of policy areas. Ordinary citizens, interest groups, governmental agencies and elected officials are engaged in numerous debates, such as:
- Genetic modification of food: what does it mean to genetically modify food? What can be achieved in terms of e.g. increased nutritional value and improved conservation? Are we tampering with nature? Is there any health risks related to eating genetically modified food?
- Food quality: How good is the quality of the food we eat? How does storage affect the quality?
- Labelling: Do we know enough about the food we buy? Can labelling be improved to tell us in an understandable way what we are going to eat? Does the labelling tell us what we really need to know? Which requirements should labelling meet to be considered satisfactory?

- The 'medicinification' of food: Newspapers and other media are every day full of recommendations and warnings about what you should and should not eat. How does the increased focus on the health effects of eating different types of food influence our understanding of eating? Are our meals starting to become arenas for our daily medicination? Are the social aspects of eating being suppressed, and how may that influence our health?

No doubt there is a variety of topics that could be debated in a participatory process. Many of these topics are obvious candidates for a participatory process, but how can we design a best possible process? I will argue that in answer this question, three partly related aspects of the process need to be considered:

1. What characterises the topic(s) that we want the participants to discuss?
2. What type of recommendations or assessments do we want to get out of the participatory process?
3. Who are the participants?

Let us look at these questions one by one.

What are the characteristics of the topic?

The many topics in food policy range from highly normative issues such as whether we are tampering with nature, to more practical challenges like how to ensure understandable labelling of food.

To discuss the first of these it will be essential to provide the participants a broad introduction to the topic, with contributions from persons holding different views. Participants will need time to develop their understanding, discuss various aspects of the topic and to formulate recommendations.

If we want to discuss requirements for satisfactory labelling, it will be necessary to inform participants about how labelling is done, which standards are used, etc., while at the same time allowing them to discuss improved or alternative ways of labelling. Finally, it will be necessary to provide sufficient time to develop their understanding.

What is the desired type of recommendations?

Recommendations can, as noted, take a variety of forms. But the key point is that *within* a given topic, like food safety, the topic can be framed in different ways depending on the type of recommendations organizers want. The topic does not by itself determine what the "proper" recommendations are. It is the organizers who need to decide what kind of recommendations they want. They can, for example, take the form of a broad discussion of the key challenges related to enhancing food safety, and the problems related to improper labelling. Alternatively, they can be presented as comments to a set of predefined scenarios for food production and the pros and cons of each of these. A third alternative can be recommendations in the form of a set of guidelines or criteria for safe food production.

Who are the participants?

Finally, it is crucial for the organizers to ask who they want to include in the participatory process. Many policy issues affect publics that can be constituted in different ways. It is, as noted by McGarity (1990) increasingly difficult to determine who speaks for the public interest on any given topic. Therefore, it is up to the organizers to choose which definition of the public to use. If we look at proper labelling of food, it can be perceived primarily as a topic affecting the broader public. If so, organizers would want to organize a participatory process with a selection of ordinary citizens. Alternatively, proper labelling of food can be perceived as a matter of debate between affected interests, like food producers, environmental organizations, interest groups for allergic persons and consumer organizations. If so, a process allowing negotiations between these interests would be necessary. Or, we can perceive the topic primarily as a practical challenge, and therefore want to include persons that can help develop a best possible system for labelling. This may include e.g. industrial designers, ordinary citizens (as consumers and "users" of labelling), manufacturers of labelling machines, employees in grocery stores, and representatives from governmental authorities.

The three above questions need to be considered together. To some extent, we can expect a certain "correlation" between the answers to each of them. For example, if the topic is the risks and advantages of genetic modification, it will not be unreasonable to conclude that it will be interesting to know how the broader public perceives these issues. At the same time a thorough discussion will require insights into a series of issues. Hence, a jury-style participatory process may seem fit. Thus, careful consideration of the three above questions will provide organizers of participatory process a good starting point for designing a proper process.

Nevertheless, the variety of possible problem formulations, definitions of the affected public, and types of recommendations, leaves it a challenge to organizers to design the process. This leads me to a crucial point: designing and organizing participatory processes is to a large extent a practical skill. The answers to the questions posed above cannot be answered solely by reading available literature. It is necessary to do it in practise. And, to make it even more complicated, cultural differences may make experiences from one country difficult to transfer to another. Therefore, learning by doing is crucial to develop the skills required to design and conduct a best possible participatory process.

Conclusions

Participatory processes can differ with respect to how the process is organized, how the affected public is defined, and how the topic is framed, the latter two easily illustrated by the many issues related to food safety and . These differences make it virtually impossible to develop any complete typology of possible topics and corresponding processes. However, careful consideration of three questions can be of great help when conducting a participatory project. These are: What are the characteristics of the topic? What is the desired type of recommendations? Who are the participants? In addition, practical experiences are crucial to develop the necessary skills to involve the public.

References

Andersen, I., O. Danielsen, M. Elle and L.D. Nielsen (1993). *Vision of urban ecology*. Danish board of technology, Copenhagen. 85p.

Crosby, N. (1995). Citizen juries: One solution for difficult environmental questions. In *Fairness and competence in citizen participation.* Renn, O., T. Webler and P. Wiedemann (eds). Kluwer, London.

Fiorino, D. (1995). Regulatory negotiation as a form of public participation. In *Fairness and competence in citizen participation.* Renn, O., T. Webler and P. Wiedemann (eds). Kluwer, London.

De Jong, M. and M. Mentzel (2001). Policy and Science: Options for democratisation in European countries. *Science and public policy*, 28: 403-412.

Joss, S. and J. Durant (1995). *Public participation in science. The role of consensus conferences in Europe.* Science museum, London. 110 p.

Laird, F. (1993). Participatory analysis, democracy and technological decision making. *Science, technology and human values*, 18: 341-361.

McGarity, T. O. (1990). Public participation in risk regulation. RISK – Issues in health and safety, 1: 103-109.

NENT – National committee for research ethics in science and technology (1997). *Fast salmon and technoburgers*. Final report from a consensus conference. NENT, Oslo.

NRTEE (1993). *Building consensus for a sustainable future – guiding principles.* National round table on the environment and the economy, Ottawa. 25 p.

Renn O., T. Webler and H. Kastenholz (1996). Procedural and substantial fairness in Landfill siting. *Risk: Health, Safety and Environment*. 7: 145-168.

Renn, O., T. Webler, H. Rakel, P. Dienel and B. B. Johnson (1993). Public participation in policy making. A three step procedure. *Policy sciences*, 26: 189-214.

Rowe, G. and L. J. Frewer (2000). Public participation methods: A framework for evaluation. *Science, technology and human values*, 25: 3-29.

Teknologinævnet (1992). *Technological animals: final report from a consensus conference.* Teknologinævnet, Copenhagen. 35 p. (Danish Board of Technology).

Teknologirådet (2006). *The future of outdoor recreational life* (Fremtidens friluftsliv). Teknologirådet, Oslo. 41 p. (Nowegian Board of Technology)

Farmers markets as a new arena for communication on food

Anne Moxnes Jervell, Hanne Torjusen and Eivind Jacobsen
SIFO (National Institute for Consumer Research), Sandakerveien 24 C, Bygg B, P.O. Box 4682, Nydalen, N-0405 Oslo, anne.jervell@sifo.no

Abstract

Consumers in supermarkets may have few sources of knowledge about how and where the food they buy is produced. The personal interaction is minimal and information needs are met by brand marketing, labelling and country of origin declarations. Food choice is in general governed more by habit and product placement than by ethical concerns. The new Farmers markets, introduced in Norway in 2003 as 'Bondens marked' reintroduces a direct link between farmers and modern consumers and introduces a new arena for communication about food. A central criteria is that the market vendors are also producers and have full information about how the products for sale are grown, reared or processed. Another central criteria at farmers markets is that products are local. This restricts the range of products on offer, but gives an opportunity for consumers to learn about local conditions for agriculture and to support local farmers. Farmers markets give preference and opportunities to local products and producers that lack access to shelf space in the dominating food retail system. Local also means fewer food miles and increased transparency. Based on interviews with more than 400 visitors at Bondens marked in different small town and city marketplaces we analyze the motivations, attitudes and shopping behaviour of customers in this arena. We also discuss how such new arenas for communication about how and where food is produced may influence consumer concerns about the ethics of food production.

Keywords: farmers markets, food marketing, consumer concerns

Introduction

It is a paradox that there is probably more information available about food now than ever before. Yet, the basics of where, how and by whom of production and distribution, are somehow mostly obscured from consumers. Consumers in supermarkets often have few sources of knowledge about how and where the food they buy is produced. Consumer information needs are supposedly met by brand marketing, other labelling and perhaps country of origin declarations. In private label products information about by whom and where products are produced is generally avoided. Consumers receive little information about how food is grown, reared, produced, transported, processed and handled, and animal products are often stripped off any reference to the animals they once where (Vialles 1994). New products are introduced by commercials rather than personal contact, and convenience and personal health rather than ethics and environment are the selling arguments. Put simply, the geographical and social distance between production and consumption has increased (Lieblein *et al.* 2001), and a lot of strong mediating commercial and bureaucratic institutions have gotten in between, blurring communication along the food chain (e.g. Goodman 1999). Therefore, food choice is in general governed more by habit, marketing and product placement than by ethical concerns (Dulsrud and Beckstrøm 2005).
Given this picture, one may even ask, as Christian Coff does, whether ethical consumer choices are possible at all (Coff 2005; see also Lang and Heasman 2004:280 or Warde 2002). Consumers lack the necessary knowledge (due to poor transparency in the food chain), autonomy (due to

commercially skewed marketing) and relevant alternatives (due to retailers' and manufactures' power and consumers' habituated practices) to be able to perform intelligent and ethically guided choices. Hence, the notion of political consumption as a way to exercise ethical power in purchasing situations is illusory, according to Coff (pp. 60-61). The necessary preconditions for most consumers to make ethically guided choices are simply not fulfilled, due to the way food production, manufacturing, distribution and marketing is organized in our societies.
This situation asks for alternative ways of communication and distribution along the food product chains. The introduction of 'new Farmers markets' expose consumers to such an alternative arena for food exchange and communication. This market model was introduced in Norway in 2003 as 'Bondens marked'(Jervell *et al.* 2003) and within 2 years spread to 19 different locations throughout the country. In a market dominated by 4 large retail chains Bondens marked reintroduces a direct link between farmers and modern consumers and introduces a new arena for communication about food. Our question in this paper is whether this new arena will influence consumer attention to and attitudes regarding the ethics, sustainability and safety of food production and thus influence also the more conventional food chains. The question is discussed using a combination of secondary sources: studies of farmers markets mainly in the US and England and material from a larger study including observations and customer interviews at Bondens marked (Svennerud *et al.* 2004, Svennerud and Jervell 2004).

The farmers' market concept

The new Farmers Market concept, introduced in the US in the 1970s, and spreading to Northern Europe in the last decennium, has been sucessful and has gained support from producers, consumers as well as the local communities (McGrath *et al.* 1993, Holloway and Kneafsey 2000, Latacz-Lohmann and Laughton 2000, Jervell 2001, Payne 2002, Jolly 2002). The core of the FM concept is the personal contact between producer and customer. "A Farmers' Market is one in which farmers, growers or producers from a defined local area are present in person to sell their own produce, direct to the public." (NAFM 2006). Since the market vendors are also producers they should have full information about how the products for sale are grown, reared or processed. The buyers right to know about the origins of the product is central, and local citizens are given an opportunity to be active participants in the local food chain through asking questions, requesting special products and through using (part of) their food budget to support local, often small-scale businesses in agriculture and food processing.
The spread of the model from the US to Europe can be seen as part of the 'quality turn' among consumers, where ".. quality conventions embedded in face-to-face interactions, trust, tradition and place support more differentiatied, localized and 'ecological' products and forms of economic organization" (Goodman 2004, p. 5). In the US the number of markets, the number of producers and consumers attending markets, and even the proportion of farmers relying on FM as their main marketing channel have grown (Payne 2002).
Local products, freshness and contact with growers/producers are attributes that are confirmed as basic reasons for customer patronage of Farmers markets in the US (Gallons 1997, Eastwood 1999). Local origin is at the core of the FM concept. This may contribute to the popularity of FM as around 75% of the surveyed respondents in US customer studies state a preference for local (state) produce, while a much smaller proportion are familiar with the labels for local products (Gallons 1997, Eastwood *et al.* 1999). The localness is ensured both by the rules that state maximum distance from the market (often with possibilities for exception) and by the rules that do not allow reselling and that demand that growers/producers should be present in the market in person.
Studies of the US FM indicate a shift in consumer interest and market focus over time, but also that markets differ according to region, urban/rural location and a number of other characteristics

(Podoll, 2000). FM serving different neighbourhoods, in different locations and at different times of the week may develop their own distinct profiles: In the Californian city of Santa Monica there are different markets in the city at different locations: of the two Saturday markets one is primarily organic (60% organic produce) while the other offers reasonably priced fresh produce in a neighbourhood in lack of fresh food supermarkets (City of Santa Monica 2002).

Customers at the Norwegian Bondens marked

As part of a larger study of the introduction of farmers markets in Norway observations and customer interviews were conducted in five different location in the first season (Svennerud *et al.* 2004). A two-page questionnaire included questions about the importance of knowing how food was produced, that food was produced locally, about shopping behaviour in the market, and shopping behaviour regarding organic food, shopping at farm shops and about whether BM contributed to increased knowledge about farming.

Some of the findings from the customer surveys support the hypotheses that exposure to farmers markets may change attitudes and behaviour. In Bergen where markets were held seldom and producers came from far away the importance of local production was seen as less important . Customers in Trondheim, where markets were held regularly every week rated that food was produced locally as more important compared to customers in Oslo and Bergen.

Customers at BM were interested to know how the food was produced, and those who had visited the marked before more so than those who were there for the first time (Svennerud *et al.* 2004,10). This may indicate some kind of influence or learning process through being at BM and interacting with the people there. Similar findings have been reported from CSAs (Community Supported Agriculture), where the reasons for joining in the first place (such as availability of fresh produce) gradually were changed or supplemented by other kinds of motivation for continuing to buy food in this way (such as being part of a community of interest, or supporting local farmers) . First-time customers placed less importance on how their food was produced compared to those who had visited markets before. In general, customers who dropped in by chance considered how food is produced to be less important than those who intentionally came to the market.

Customers were asked how often they purchased organic produce (never, seldom, sometimes, often). Those who often bought organic products regarded knowing how food is produced as more important than customers in other groups. Customers who buy from farm shops view both how and where food is produced as more important than people who do not purchase at such outlets. Younger people were less concerned about where their food was produced, while those with least education were most concerned about buying food produced locally. Customers who regularly bought organic products were more likely to consider the markets as a factor bringing them closer to agriculture, compared to those who never bought organic food.

Communication about food at Bondens marked

That products are local restricts the range of products on offer, but gives an opportunity for consumers to learn about local conditions for agriculture and to communicate with local farmers. Farmers markets give preference and opportunities to local products and to small producers that lack access to shelf space in the dominating food retail system. Local means fewer food miles and thereby increased transparency. Consumers can more easily evaluate and confirm the product stories offered by the vendors, often also by visiting them at the site of production. Many of the vendors at Bondens marked combine sale at farmers markets with sales from farm shops.

The information and personal contact with producers is part of the quality offered to consumers at the market. Some observers have commented that BM is a context where it is all about "the truth on a plate"[24]. There is no room for cheating. The stories about food are authentic and real – they are about the facts behind the product. Such contact opens opportunities to distinguish and market products on the basis of local and individual producer qualities and to communicate more complex messages about food production. Do producers use this opportunity to develop their products and communication?

One lesson learned from the first season of Bondens marked in Norway is that the consumer interest in organically grown food is larger than the offer. While 50% of the consumers sometimes or often buy organic products, only 30% of the producers had organic products for sale. In Oslo, as many as every fourth customer, reported purchasing organic products often. Many consumers comment on this, but only a small proportion of surveyed vendors, were influenced by participation to convert to organic farming (Svennerud and Jervell 2004). In the US organic produce – or products produced without pesticides etc. have a larger proportion of the Farmers market, but varying according to place and customer base.

There is also in Norway evidence that environmental concerns and ethics may be successfully included as sales arguments in a Farmers market setting. Among the two price winning vendors at the Norwegian farmers market in 2005 is a farmer who sells meat from pigs that live a free-range life in the farm forest. Both animal welfare and superior taste are used in sales promotion (BM, 2006). The animal may come more to the forefront in a BM context, as compared to in a supermarket, where information is likely to be available, but not in a personalised and specific way. Direct references to the actual place where the animals have grazed, the animals, and the persons who have reared them and made food products from their meat, are found among several BM participants. "I relate my products directly to the nature where the animals are raised", says one producer and BM participant. She explains that she uses large photos at her stand at BM, showing the mountains and grazing landscape of Lyngen, pictures of the goats, as well as pictures of herself with the animals and when preparing the meat products.

Discussion

As a source of knowledge and by offering alternative choices Bondens marked may be one of several small steps towards meeting the preconditions for ethically guided choices. Norwegian consumers have only been exposed to Bondens marked for three seasons, and while information of the markets existence may have reached a large proportion of the population in some of the locations, the market in the largest city (Oslo) is located in a part of the city where most of the customers come from the close-by residential area. As an alternative marketing channel for food the phenomena is marginal, even in the US where the phenomena has grown in importance over more than 30 years.

Despite this, the symbolic value of the phenomena may be larger, influencing the way consumers think and talk about food, and also the actions of producers and other actors in the value chain. Exposure and success of products in Bondens marked has already contributed to give niche products access to retail chains. Animal welfare and localness of products are quality attributes communicated at Bondens marked that are seldom promoted in the conventional retail stores. Increased consumer interest in and concern about where and how the products they buy are produced is one of the factors that can influence future food trade and production.

[24] Commented by Tore Stubberud at a seminar for "Bondens Marked" participants 1. April 2006. Also referred in his book "Herregårdsmat", where he attributes this saying to his co-writer, the chef Eyvind Hellstrøm.

References

Coff, C. (2005). *Smag for etik. På sporet efter fødevareetikken.* København: Museum Tusculanums Forlag.

Eastwood D.B, J.R. Brooker and M D Gray, (1999). Location and other market attributes affecting Farmer's market patronage: The case of Tenessee. *Journal of Food Distribution Research*, 31,1, 63-72.

Dulsrud, Arne and Jan Roar Beckstrøm (2005). *Å sette pris på hylleplassen.* Fagrapport nr.2-2005, SIFO, Nydalen

Gallons, J., U.C. Toensmeyer, J.R. Bacon, and C.L. German. (1997). An analysis of consumer characteristics concerning direct marketing of fresh produce in Delaware: A case study. *Journal of Food Distribution Research.* 28:98–106.

Goodman, David (1999): Agro-Food Studies in the 'Age of Ecology': Nature, Corporeality, Bio-Politics. *Sociologia Ruralis* 39, 1, 17-38

Goodman.. D. (2004) Rural Europe Redux? Reflections on Alternative Agro-Food Networks and Paradigm Change. *Sociologia Ruralis* 44, 3-16

Jervell, A. M. (2001). Farmers' Market – direkte kontakt mellom produsent og forbruker. *Landbruksøkonomisk forum,* Nr. 2/2001, s 5-18.

Lang, T. and Heasman, M. (2004). *Food Wars. The Global Battle for Mouths, Minds and Markets,* London: Earthscan.:280

Latacz-Lohmann, U. and R. Laughton (2000). Farmers' Markets in the UK – A study of farmers' perceptions. Farm *Management*, 10: 579-588.

Lieblein, G., Francis, C. A., Torjusen, H. (2001). Future interconnections among ecological farmers, processors, marketers, and consumers in Hedmark County, Norway: creating shared vision. *Human Ecology Review* 8 (1): 61-72.

Mcgrath, M. A., Sherry, J. F., And Heisley, D. D 1993. An Ethnographic Study of an Urban Periodic Marketplace – Lessons from The Midville Farmers Market. *Journal of Retailing* V69[N3], 280-319. 1993.

NAFM (2006). National Association of Farmers' Markets Criteria. http://www.farmersmarkets.net/ April 2006

Payne (2002). *U.S. Farmers Markets – 2000. A study of emerging trends.* Oregeon State University Extension Service/USDA

Steven, Anderson, Alison and Meethan, Kevin (2002). *The Changing Consumer: Markets and Meanings,* London: Routledge.

Svennerud, M., A.M. Jervell 2004. *Bondens marked som salgskanal. Resultater fra en undersøkelse blant deltakere i 2003. NILF-Notat 18: 2004. Oslo*

Svennerud, M., A.M. Jervell og B. Øyen 2004. *Kunder ved Bondens marked 2003.* NILF-Notat 17:2004. Oslo

Vialles, N. (1994). *Animal to edible.* Cambridge. Cambridge University Press.

Warde, A. (2002). Setting the scene: changing conceptions of consumption, i Miles *et al.* (eds): *The Changing Consumer: Markets and Meanings,* London: Routledge.

Consumer information about farm animal welfare: A study of national differences

Unni Kjærnes and Randi Lavik
The National Institute for Consumer Research, Oslo, PO Box 4682 Nydalen, 0405 Oslo, Norway, Unni.kjarnes@sifo.no

Abstract

This paper will focus on a key issue is current debates on efforts to improve farm animal welfare, namely consumer information. Labelling programmes have emerged as a central measure. Public opinion surveys were conducted in seven European countries (Hungary, Italy, France, The Netherlands, Great Britain, Norway and Sweden) in September 2005. The survey has been conducted as part of Welfare Quality (FOOD-CT-2004-506508), a research project co-financed by the European Commission within the 6th Framework Programme. A question on what kinds of information the respondents wanted gave very varied answers across the countries, both in terms of general wishes for more information and the types of information that they wanted. While Hungarians were not sufficiently informed about farm animal welfare, the Dutch seemed more disinterested to that question. The type of information requested also varied across the countries; while the Italian and the French respondents were in favour of a simple welfare assurance mark, the Norwegian and the Swedish were much less interested in such a mark, the other countries being in between. The paper will explore possible explanations to this variation. It will draw upon a theoretical understanding of food consumption as a matter of institutionalisation, where this institutionalisation is influenced organisationally and normatively by conditions on the consumer side as well as the by character of the food provisioning system, regulatory arrangements, etc. The paper will draw mainly upon the survey data, limiting the possible types of information to characteristics of consumption, but including also interaction in the market, trust in institutional actors, consumer agency, etc.

Keywords: animal welfare, consumption, comparative analyses, institutionalisation, political consumerism

Extended abstract

Farm animal welfare is an emerging issue on the public and increasingly also political agenda in Europe. Attention is directed towards unacceptable ways of handling animals in modern agriculture (Roux and Miele, 2005). Concerns have first of all been expressed via mobilisation through animal protectionist and animal rights' organisations, pushing for stricter regulations. Increasingly, however, animal welfare has become an issue in the food market, subject to differentiating labelling programmes introduced by producers and retailers, as a separate issue or as an aspect of food quality. Also, people (or at least some) express their concerns and activism via their role as purchasers of food (Blokhuis *et al.*, 2003). As important, other actors often frame people's involvement and responsibilities as a matter of choice in the market (Hughes, 1995). Markets and morals thus seem to be combined in ways where people's individual responsibility as buyers emerges as a crucial point. Individual choice is a theoretical model that assumes a demand reflecting particular preference structures and a differentiated market supply that matches this demand (Bennett, 1997). Information is seen as central for promoting an issue

like animal friendliness. This paper will problematise the links between information demand, responsibility and this individualised understanding of the consumer role.
The approach is comparative and institutional. A major point is that practices and opinions are analysed as macro level, national characteristics rather than individual psychological constructs and ethical priorities. At this macro level of analysis we regard the "consumer" role as emerging within particular discursive and institutional settings. Buying food is an everyday routine practice, organised and normatively regulated in highly diverse ways. The markets that people buy from are shaped very differently, the societal divisions of responsibilities for key food issues, like animal welfare, can be very varied, the political understanding and role of "food consumers" can be quite diverse, and the household organisation of food related activities is far from uniform (Kjærnes, Poppe and Lavik, 2005).
Public opinion surveys were conducted in seven European countries (Hungary, Italy, France, The Netherlands, Great Britain, Norway and Sweden) in September 2005. The work is part of Welfare Quality (FOOD-CT-2004-506508), a research project co-financed by the European Commission within the 6th Framework Programme. A question on what kinds of information the respondents wanted gave very varied answers across the countries, both in terms of general wishes for more information and the types of information that they wanted. While few Hungarians and Italians felt sufficiently informed about farm animal welfare, the Dutch seemed much less interested. The paper explores three possible explanations to this variation; linking information demand to consumer choice, political consumerism, and general scepticism and distrust in food institutions. The method is ordinary regression analysis, where information demand (an index) is the dependent variable and where a series of explanatory variables are introduced in a step-wise procedure. The explanatory variables include 1. gender and age, 2. place of shopping, 3. trust in different actors, 4. consumer activism, 5. attitudes towards animal welfare and human-animal relations, and 6. opinions on the importance of farm animal welfare.
The analyses show, first, that people across Europe invariably display very positive attitudes towards the welfare of farm animals. This general interest is also reflected in demand for information. Demands are, however, quite non-differentiated and do not follow general patterns in consumption practices. Second, political consumerism or consumer activism (see eg. Micheletti, 2003) seems to be important for involvement in animal welfare issues as consumers in Northern Europe, much less so in the south. But in spite of that, the activism is not very critical – or particularly altruistic or "ethical". People with ethical concerns do want more information, but they are not many. Third, trust is important for interrelations between food consumers and other actors in the food system. But the impact of trust is not very large as long as people are not engaged in animal welfare through their daily consumption practices.
Public opinion about farm animal welfare remains largely just that, a matter of opinion, a good cause that most people find it easy to agree with. Information demands generally also reflect this positive attitude. There may be a number of explanations to that. The societal distribution of responsibility for farm animal welfare may direct attention towards very different arenas, such as authority regulation. Market differentiation based on varying levels of treatment of animals is in several of the countries limited and even opposed, like in Scandinavia. Moreover, we see that many people are interested, but we also know that that does not necessarily imply that people are worried about the conditions in their own country. And if they are, they may judge other forms of action more efficient – or see their own voice as a consumer as having little influence. Later analyses will allow us to go deeper into such questions.
Does this mean that animal welfare is irrelevant as a consumer issue? Or that animal welfare concerns can only become part of food consumption practices as representing explicit activism? To us, that would be wrong conclusions. Consumption and conditions for consumption may certainly become subject to turbulence and activism. So also in relation to farm animal welfare. Information, perhaps quite detailed, may in that respect represent an important condition for

involvement and mobilisation as well as signifying openness around production standards. Still, food consumption is generally highly routinised, embedded in tacit, taken for granted norms and expectations and organisational structures. When the Dutch, as judged by demand, have the highest proportions of animal friendly products, including for example all eggs sold in supermarkets, but seem to have so little interest in more information, one reason may be that animal welfare has become institutionalised as part of routine food purchases. Routine, unquestioned purchases are rarely associated with complex information needs. Alternatively, as is often the case in Southern Europe, very encompassing concepts of food quality have developed, generally framed by the region of origin, and animal welfare seems to be included in that.

References

Bennett, R. M. (1997). Farm animal welfare and food policy, *Food Policy, 22*(4), 281-288.

Blokhuis, H. J., Jones, R. B., Geers, R., Miele, M. and Veissier, I. (2003). Measuring and monitoring animal welfare: Transparency in the food product quality chain, *Animal Welfare, 12*, 445-455.

Hughes, D. (1995). Animal welfare: the consumer and the food industry, *British Food Journal, 97*(10), 3-7.

Kjærnes, U., Poppe, C. and Lavik, R. (2005). *Trust, Distrust and Food Consumption. A Study in Six European Countries. Project Report No 15.* Oslo: The National Institute for Consumer Research.

Micheletti, M. (2003). *Political Virtue and Shopping. Individuals, Consumerism, and Collective Action.* New York and Houndmills: Palgrave Macmillan.

Roux, J. and Miele, M. (2005). *Farm Animal Welfare Concerns. Consumers, Retailers and Producers. Welfare Quality Reports No.1.* Cardiff: Cardiff University, School of City and Regional Planning.

The presence of animal welfare-friendly bodies: An organised or disorganised achievement in the food supply chain

Emma Roe and Marc Higgin
Cardiff School of City and Regional Planning, Cardiff University CF10 3WA. UK

Abstract

This paper explores the market for food products derived from cattle, chickens and pigs that are considered to have had a welfare-friendlier life. Welfare-friendly claims hold considerable ambiguity in meaning since there is no precise definition of what better 'animal welfare' means in practice. However, despite this ambiguity there are numbers of animals that are being made into food products which carry labelling that suggests higher animal welfare, and in addition many animals or parts of animals which experience the same living standards but which don't ever get labelled to suggest welfare-friendliness. Through the development of an 'economy of qualities' (Callon *et al.* 2002) within the food market there have been a number of private initiatives by major retailers, farmers' cooperatives, independent standard bodies, manufacturing brands within the UK which has supported the development of a market for 'welfare-friendly' food products. How do these organisations work together to realise the economic potential through product labelling or corporate branding of meat/dairy or egg products from welfare-friendly production practices? Or in other words, by what mechanisms do some bodies or body-products of animals attain, retain or lose power as welfare-friendly as they move through the different organised spaces of the supply chain?

Keywords: animal welfare, agro-food, network, market, quality assurance, schemes, retailer, brands

Introduction

There is a growing market for products that carry animal welfare-friendly packaging description. In this paper we discuss how a welfare-friendly product is achieved through critically thinking about the process by which a thing, an animal body, becomes labelled as a welfare-friendly product. We work with a socio-material (Latour 1993, Whatmore 2002, Murdoch 2006) and socio-technical (Stassart and Whatmore 2003) network approach to understand how relations are assembled around the body; the processes of transformation that occur to the body, as society mobilises it, and circulates it, through socio-material networks; and how its welfare-friendliness becomes a socio-technical achievement. As it moves through these networks different practices and different socio-technical devices instil different qualities to the body. For example, the beef steer through the stunning and slaughtering processes in the abattoir becomes a dead animal's body. The process of evisceration (through the practices of removing the animals' skin, head, bodily organs and spinal cord) transforms the dead animal's body into that of a meat carcass. The practical inspection of the Meat and Livestock Commission gives the meat carcass a stamp, a mark that facilitates its movement into the human food chain as meat. The event of an animal's body becoming meat carcass is one of many stages of transformation in the agro-food network. Importantly, it is material qualities of the animal's body that affords the potential for the movement of flesh as meat between different actors in the spaces between food's production, retailing and consumption.

We focus upon three devices where the practices that surround these devices become key in the assembling of a product carrying a welfare-friendly slogan or packet-description on the supermarket shelf. The first device is the quality assurance scheme and specifically those farms that are members of a quality assurance scheme. Second is the retailers brand, specifically how variety in brand values reflects differing engagements with concerns about farm animal welfare. Third is the process of category-segmentation, specifically the strategic position of welfare-friendly claims (a higher value product attribute) within a product category that often includes retailer own-brand and manufacturer brand products.

For the consumer, animal welfare as an attribute of meat is always a mediated presence. It is the different modes of this mediation that this paper pursues through studying key sites of transformation for the animal becoming welfare-friendly meat.

"all quality is achieved at the end of a process of qualification, and all qualification aims to establish a constellation of characteristics stabilized at least for a while, which are attached to the product and transform it temporarily into a tradable good in the market" Callon (2002:199)

Animal welfare, as an implicit (i.e. not identifiable through sensory perceptions) quality of meat, can therefore only emerge through a series of qualifications. This has to be reliably assessed so assurance can be given that the animals experienced a better quality life. In our analysis we discuss the complexity by which animal welfare, as a quality, is co-constructed by multiple, not-necessarily-economic actors. We consider how they work together to assemble the marketing potential through product labelling or corporate branding of meat/dairy or egg products from welfare-friendly production practices?

What we will show is that we are witnessing an assemblage of properties, devices and practices that are facilitating this process.

A socio-material, socio-technical approach

The study of the market for animal welfare-friendly products would be severely limited if the animal's body was to be taken out of the analytical process, since it is the living animal, the dead body of the animal as meat, and the stories of the represented animal on product packaging that is a significant figure in this network. Therefore, this study of what lies in the spaces between a farm animal's experience within a particular production system, and the eating practices of a reflexive consumer who has chosen a welfare-friendlier meat product will trace the body of the animal becoming welfare-friendly product as a socio-technical achievement. This approach sympathises with the body of work that seeks to break down the modernist binary division between studies of nature and studies of society, by taking a hybrid geographical approach (Whatmore 2002). Also, it aims to study as closely, the activities of the material, fleshy, bodies of the animal (alive, dead or re-presented as living through welfare-claims), as the practices of society that circulate and mobilise the materiality of the animal's body or body-parts or body-product and the stabilising activities of socio-technical devices.

In this section we will reflect briefly on what are the benefits from taking a socio-material networked approach to agro-food studies, as opposed to the supply-chain or value-chain analytical approach that often dominates studies of retailing activities. The supply-chain approach (see Fearne *et al.* 2001) does not facilitate a detailed analysis of how different retailing cultures exist which complexify the process of product manufacture between farm and retailer; it focuses on a rather analytically, limiting uni-directional, functional, logistical movement of the product down the supply chain. The values-chain literature has merits in its ability to identify how values are built-up through the chain and thus takes a wider view through including a greater number of actors that are significant to the manufacture of a product. The major criticism of both approaches is that they do not work closely with, nor appreciate the materiality of the thing, which in the case of food is a plant or animal that places its own limits and constraints on human activities in agro-food networks. They do not ask questions about how the plant or animal becomes enrolled in the agro-food network; nor ask which processes and practices are crucial to their transformation into a packaged product on the supermarket shelf, which may occur outside of the process of adding value. Agro-food networks have particular characteristics which are contingent to working with processes of life and death, even with substantial manipulation by humans into rates of growth, selective-breeding for body conformation etc.

A body of work by Callon *et al.* (2002), which studies both the function and the organisation of markets does attend more closely to the socio-technical activities and the reflexive activity of the market. This work is useful to this study because even though it does not engage fully with living processes it does outline clearly the process of transformation and the role of qualification in organising the supply and demand of products. What this study hopes to add to this work is recognition for the bodies of sentient animals which are central to the development of a market for welfare-friendly foodstuffs within the context of socio-technical devices in the market and the reflexive organisational activities of the market.

Socio-technical devices: Quality assurance schemes

The food safety act of 1990 in the UK accelerated the development of farm assurance schemes (FAS). FAS provided a way for large retailers to gain some control over the supply chain, or to at least be able to monitor it through traceability and auditing (Lindgreen and Hingley 2003). Farm assurance when communicated to consumers becomes a Quality assurance schemes (QAS). QAS was a tool for putting in place many of the desired characteristics of a product that are not discernible by visual inspection (FAWC 2005). Suppliers could provide that information through their membership of a scheme. QAS exist for different production systems. Assured Food standards is the largest QAS in the UK, owned by the industry, and covers all major sectors of the livestock and dairy industry in the UK. Its standards for animal welfare are above the minimum national legal requirements. Their growth in membership is due to the pressure from the large multiple retailers (notably not the food service industry) who are in large, nearly unanimous numbers, choosing to buy only from assured farms. This QAS is communicated to consumers through the use of a logo. Products carrying this logo carried welfare-friendly claims (Roe *et al.*2005). This is in a context where this industry standard has been criticized for trumpeting its animal welfare credentials – mainly vis-à-vis imported products from countries with lower legal requirements on animal welfare.

As well as industry standards, other organizations, in particular NGOs have realized the potential of setting their own standards in order to develop alternative production systems, using their brand image and the trust felt by consumers to position themselves as mediators of quality. RSPCA set up the Freedom Food standard in the early 1990s with the express goal of improving the lives of as many farm animals as possible; as expected products where found with welfare-friendly claims carrying this logo. The standards are based on the FAWCs 5 freedoms and aim to introduce higher levels of animal welfare than the 'base' schemes as represented by AFS. Additionally, there are a number of organic production quality assurance schemes. The Soil Association is the largest organic certification body and the largest organic charity body in the UK. Animal welfare features prominently both in the standards and the marketing of the Soil Association brand:

"*No system of farming has higher animal welfare standards than organic farms working to Soil Association standards.*" (Certification News No. 53 Spring/Summer 2005).

As the furore that followed this claim by the Soil Association testifies, animal welfare standards and their certification are not simply 'objective, value neutral science' (Hanataki *et al.pers. comm.*). Of note from the empirical evidence is that the number of products found that carried both the logo and welfare-friendly claims was disproportionate to the number of farms that are a member of the scheme. The AFS scheme logo was found on only a very small number of products

carrying welfare-claims, which surprisingly was also true of the Freedom Food logo. However, in contrast the organic logo was found on a number of products carrying welfare-claims.
The farm on which an animal lives is assessed for membership of an assurance scheme. The validating of the farm by the socio-technical device of a scheme inspection allows the animal's body to become sold as meat in selected markets. Yet, the communication tool of the QAS in the form of logos is used on packaging in a selection of instances where they also accompany welfare-friendly claims. Thus though the device of a QAS is appreciated as the mechanism on farm for increasing welfare of animals, its use as a communication tool to consumers appears intermittent due to the power of the retailers to decide on packaging design and information.

Socio-technical devices: Retailer brand values
The 'brand' is becoming the dominant 'mediation' between the public and food and animal science when eating livestock-based products. Significantly, the corporate social responsibility aspirations of the major retailers now includes animal welfare, and thus animal welfare is becoming part of the package of values included in the Corporate Brand (Friedberg 2004). Animal welfare, is a dimension of farming and food quality standards, is usually included as part of the list of things which one places trust in when one purchases meat (Allaire 2005). The trust in the brand displaces a need to have trust in the meat. This shift in responsibility for safety and quality of the food sector has coincided with the inexorable growth of the market share the major retailers hold (Thankappan and Flynn 2006). Along with this growth and increasingly fierce inter-firm competition, has come increased scrutiny of their practices and their products; they have become more accountable on issues such as environmental impact, worker welfare and animal welfare (Hanataki *et al.pers. comm*):
"the supermarkets occupy a powerful position in Britain's food supply, but one that has subjected them to intense scrutiny and criticism, and the perilous, ever-present possibility of a tarnished brand" (Freidberg 2004:521).
Animal welfare as an ethical concern is defined through the province of the media (the voice of the people) and civil society movements like non-governmental organisations.
It is these civil movements that have demanded supermarkets to respond to calls for ethical sourcing (Friedberg 2004). Not all retailers are promoting their animal welfare concerns and action as strongly as others. This reflects different target consumers that the retailer brand is orientated towards. Thus depending upon where the body of animal is eventually sold may affect whether it gains a market credential as being welfare-friendly through the fickleness of a retailer's brand values.

Socio-technical devices: Category segmentation.
The corporate retailers segment and differentiate product lines based on a complex process of category segmentation. Category segmentation describes how a category has a range of products with different values and accompanying packaging strategy that orientates itself in various ways between concern for the price, care for the self placed towards a personal health and personal political/ethical angle, through to care for others including for example 'humans' by fair trade, 'animals' by animal welfare and 'plants (or more generally 'the environment') by sustainable farming practice like organic. These differentiating factors are found on products within different quality bands simply put as 'good', 'better' and 'best'.

Acknowledgement

The present study is part of the Welfare Quality research project which has been co-financed by the European Commission, within the 6th Framework Programme, contract No. FOOD-CT-2004-

506508. The text represents the authors' views and does not necessarily represent a position of the Commission who not be liable for the use made of such information.

References

Allaire, G (2005). Quality in economics: a cognitive perspective. Pp.66-92 In M Harvey, A McMeekin and A Warde (eds) *Qualities of Food*, Manchester University Press, Manchester.

Callon, M. Méadel, C. and Rabeharisoa, V. (2002). The Economy of qualities *Economy and Society* 31 (2) 194-217

Farm Animal Welfare Council (FAWC) (2005) *Report on the welfare implications of Farm Assurance Schemes* Farm Animal Welfare Council June 2005, London

Fearne, A., Hornibrook S., Dedman, S (2001). The management of perceived risk in the food supply chain. *International food and agribusiness. Management Review* 4. 19-36.

Friedberg, S. (2004) The ethical complex of corporate food power. Environment and Planning D: Society and Space, 22: 513-531

Hanataki, M. Bain, C, Busch, L. (*pers. comm.*) Differentiated Standardization, Standardized Differentiation: The complexity of the Global Agrifood system. RAISE, Michigan State University

Lindgreen, A. and Hingley, M. (2003). The impact of food safety and animal welfare policies on supply chain management. The case of the Tesco meat supply chain. *British Food Journal* 10(6) pp. 328-349

Latour, B (1993). *We have never been modern*. Harvester-Weatsheaf, Hemel-Hempstead.

Murdoch, J. (2006). *Post-structuralist geography* Sage, London

Roe, E, Murdoch, J, Marsden. T (2005) The retail of welfare-friendly products : A comparative assessment of the nature of the market for welfare-friendly products in six european countries. In A. Butterworth (ed) *'Science and Society improving Animal Welfare' Welfare Quality conference proceedings* 17/18 November 2005, Brussels

Soil Association

Stassart, P. and Whatmore, S. (2003). 'Metabolising risk: food scares and the re-assemblage of Belgian beef', *Environment and Planning A, vol. 34. no. 11.*

Thankappan, S. and Flynn, A. (2006). Exploring the UK red meat supply chain. *BRASS Working Paper Series no 32.* BRASS centre, Cardiff.

Whatmore, S. 2002 *Hybrid Geographies* Sage, London

Private regulatory approaches and the challenge of pesticide use

Nicolien van der Grijp
Institute for Environmental Studies (IVM), De Boelelaan 1087, 1081 HV Amsterdam, The Netherlands, nicolien.van.der.grijp@ivm.vu.nl

Abstract

During the past decades, private strategic actors have developed regulatory initiatives that aim to remedy the negative impacts of the conventional agrifood system. Against this background, this paper describes the evolution of regulatory initiatives by non-state actors that are relevant from the perspective of pesticide use, while highlighting the role of the key strategic actors. It distinguishes three stages in the evolution of regulatory initiatives by non-state actors, including the conceptualisation of new approaches, their institutionalisation in regulatory formats, and the harmonisation of regulatory initiatives. The main conclusion of the paper is that many of these initiatives have potential benefits, as they provide a practical elaboration of concepts of safe and sustainable agriculture. Moreover, several of them include mechanisms for stimulating participation, and opportunities for learning and upgrading. Finally, it is being argued that under certain conditions private regulatory initiatives can be considered forms of 'democracy in practice', as they provide a flexible and reflexive means to respond to challenges in terms of environmental and health problems in agriculture.

Keywords: private regulation, pesticides, governance, framing, participation

The evolution of private regulatory initiatives

During the past decades, private strategic actors have developed several regulatory initiatives that aim to remedy the negative impacts of the conventional agrifood system (e.g., van der Grijp and Den Hond, 1999). In the early years, the main impetus for these approaches was based on a concern about environmental and social issues. Farmers' organisations, alternative food companies and scientists played a dominant role in them. In the 1990s, however, under the influence of the emerging concepts of sustainable development and corporate social responsibility, the conventional food and retailing industry became increasingly involved in regulatory initiatives with environmental and social objectives that were often framed as quality assurance schemes.

The emphasis as regards content started to shift from environmental and social issues to food safety issues around the turn of the century. Initially, this shift was prompted by the urge felt by the food and retailing industry to deal more rigorously with the traditional risks of chemical and bacterial contamination in the face of increased international trade. After the 9/11 attacks in New York and Washington, the food safety focus was further strengthened out of fear for biological terrorism. Very recently, several non-state actors have begun to develop a new strand of regulatory initiatives that promote healthy eating habits in the struggle against obesity and food related diseases, *inter alia* in the hope to curb the exploding costs of healthcare.

Against this background, this paper aims to describe the evolution of regulatory initiatives by non-state actors that are relevant from the perspective of pesticide use, while highlighting the role of the key strategic actors. It distinguishes three stages in the evolution of regulatory initiatives by non-state actors, including the conceptualisation of new approaches, their institutionalisation in regulatory formats, and the harmonisation of regulatory initiatives.

The paper is structured as follows. Section 1.2 is about the stage of conceptualisation and concerns the development of alternative approaches to the chemical-based approach to agriculture, and the subsequent efforts to identify the distinguishing features of these approaches and to define them in appropriate concepts. Section 1.3 depicts the stage of institutionalisation that can be characterized by an increased use of a regulatory discourse leading to the development of regulatory outputs such as criteria, guidelines, or standards that further define the alternative approach. Section 1.4 deals with the stage of harmonisation that is aimed at a convergence of existing regulatory approaches by the development of overarching systems that make content and procedures more uniform. Section 1.5 compares the regulatory approaches by non-state actors in terms of substance and procedure.

The stage of conceptualisation

The basis for modern conventional farming has been established in the early 20th century, when new developments in chemical and biological sciences and increased mechanisation created the conditions for a major transformation towards increased farm productivity by the scaling up of production practices (e.g. Hough, 1998; Pretty, 2005). In response to this rapid takeover, several production methods and approaches have been developed that can be considered an expression of a critical attitude towards the direction that conventional agriculture had taken. In the first place, the anthroposophist Rudolf Steiner gave a series of eight lectures in Silesia in Germany in 1924, which marked the beginning of the biodynamic agriculture movement (Steiner, 1993). The biodynamic production method puts the emphasis on the condition of the soil, and focuses on the use of certain preparations for fertilisation purposes and the positive effects of cosmic forces on crop production.
In the second place, the concept of organic agriculture was developed by farmers in the UK and the USA, more or less in tandem with that of biodynamic agriculture (e.g., Northbourne, 1940; Balfour, 1943; J.I. Rodale, 1945). Similarly, it focuses on a healthy soil as the basis of sound agricultural production systems. In this respect, organic agriculture relies on ecosystem management and favours agronomic, biological, and mechanical methods, as opposed to using external agricultural inputs, such as synthetic fertilisers and pesticides.
In the third place, the concept of integrated production was developed by scientists from the discipline of biology (Boller, 2005). Although initially their focus was on the biological control of pests, they became soon interested in the development and application of various broader concepts that aimed to offer solutions for dealing with pests and diseases in agriculture. These concepts are the so-called integrated approaches, including integrated pest management (IPM), integrated crop management (ICM), and ultimately integrated production as the method incorporating the highest level of integration (Steiner, 1977).
In the fourth place, the concept of integrated farming was launched by farmers, under the influence of the Brundlandt report and the Earth Summit in Rio de Janeiro. It is in fact an interpretation of sustainable development in the agricultural context and is seen as an approach to sustainable farming that would be realistic and achievable for the majority of farmers.

The stage of regulation

In the late 1920s, the first standards for biodynamic agriculture were formulated by national biodynamic associations that also started to use the Demeter logo which still symbolises the movement today. In the late 1960s, the British farmers organisation Soil Association in the UK was the first entity that developed standards for organic production. Soon organic farmers associations in other countries followed. These early initiatives were driven by the need of organic farmers in a certain region to have a common definition of organic that could provide assurance to consumers and prevent fraudulent claims and unfair competition.

Switzerland was the pioneer with standard setting for integrated production, with a group of fruit producers establishing standards for apples in 1978 (EUREP, 1998). Several farmers associations in other countries followed the Swiss example, but it was only in the late 1980s that production under integrated production became a more broadly institutionalised undertaking. In that period several ambitious programs were set up and accompanying certification schemes were developed. From a scientific perspective, the International Organization for Biological Control of Noxious Animals and Plants (IOBC) became increasingly involved in the elaboration of integrated production by formulating a framework of principles and general guidelines (El Titi *et al.*, 1993). In addition to this framework and its revisions, the IOBC published crop specific guidelines for all major crops in Europe. As distinct from integrated production, the concept of integrated farming also gathered a wider appeal in the 1990s, as farmers organisations in several European countries started to elaborate the criteria for this production method. Major examples included the labelling scheme set up by LEAF (Linking Environment And Farming) in the UK and the Farre Charter in France.

Food safety schemes, many of them based on integrated approaches, emerged in the late 1990s, after several food scandals were thought to damage consumer confidence in the agrifood industry (e.g., Lang and Heasman, 2004; Reijnders, 2004). In reaction to these scandals, the industry developed food safety schemes with regulatory features, fuelled by the realization that the industry lacked control over the previous links in its supply chains and that the commercial risks due to incidents would be difficult to insure. This counted relatively strongly for the large retailers with thousands of products on offer and on top of that their special responsibility for the retailer home brands which constitute a considerable amount of their turnover. For them, the task of managing food safety within their own quality departments was simply too big.

The stage of harmonisation

The proliferation of standards led to a call for harmonisation at the international level. The earliest initiative in this direction was taken by the pioneers of the organic movement. IFOAM, or the International Federation of Organic Agriculture Movements, was the first non-state actor that formulated international standards that should form a bottom line and common ground for national initiatives. It published the first version of the IFOAM basic standards for organic production and processing in 1980. These standards have been revised many times since. Since 1997, Demeter International is a member of IFOAM and as such committed to incorporate the standards for organic agriculture in its own system.

During the 1990s, several initiatives were launched by coalitions of leading companies in the agrifood sector, representing what Fulponi (2006) calls 'an industry grassroots harmonisation effort' and indicating that the industry was starting to take a global approach to managing the food system. One of the larger initiatives by the industry was the foundation of the Euro-Retailer Produce Working Group (EUREP) in 1996 by a group of 13 large European retailers which aimed to publish sector-oriented protocols of standards for good agricultural practice (GAP).

Meanwhile, the multinational food industry developed its own strategy. Unilever sought cooperation with Groupe Danone and Nestlé, and in May 2002, the three of them officially launched the Sustainable Agriculture Initiative Platform (SAI Platform). The SAI Platform aimed to jointly develop sustainable agriculture principles and standards through the assessment of practices and experiences, but it did not aim to set up a certification system. It claimed putting the priority on creating the right mindset for the implementation of sustainable agricultural practices and to cooperate with producers in order to find solutions, instead of applying a top down approach.

Similarly in the early 2000s, several European farmers organisations promoting integrated farming joined forces in the European Initiative for Sustainable Development in Agriculture (EISA). In order to further its objectives, EISA published a Common Codex for Integrated Farming in 2001 that lists principles and suggestions for agricultural practice.
With regard to food safety, the Hazard Analysis and Critical Control Point (HACCP) system was initially seen as the ideal basis for a global food safety standard and was as such recognised by the Codex Alimentarius Commission. However, the food and retailing industry did not consider the HACCP system as a complete answer to their food safety concerns as the system provides a methodology for risk analysis but does not provide a solution for risk management.
Consequently, the retailing industry initiated the Global Food Safety Initiative (GFSI) in the context of the retailer-led business organization CIES – The Food Business Forum. The GFSI originally intended to develop one common food safety standard but did not succeed to develop one standard that could replace all existing standards. Instead, it developed a benchmark system to be able to assess existing schemes against 'a global set of voluntarily but universally accepted standards for food safety, quality and security.' However, as the GFSI is being dominated by the world's largest retailers, other stakeholders in the agrifood industry felt less inclined to affiliate with the initiative. ISO then picked up the signal that there was a market opportunity for an international food safety standard which would make other standards superfluous. The ISO 22000 standard was published in 2005 after agreement was reached between all participating countries. The future success of the ISO 22000 standard is expected to depend on its acceptance and implementation by public authorities and business.

Comparison of regulatory approaches by non-state actors

A comparison of the content of the larger international programs for safe and sustainable agriculture shows that most of them claim to elaborate the concept of sustainable development but that they come up with diverging interpretations. In their book Food wars, Lang and Heasman explain the Food Wars thesis and three competing food system paradigms. They argue that two possible science-informed visions for the future are emerging, both seeking to transform the productionist paradigm that became dominant after World War II. One is what they call the life sciences integrated paradigm and the other the ecologically integrated paradigm. When comparing the programs of private actors with these paradigms, it seems obvious that the systems for organic and biodynamic agriculture represent the so-called ecologically integrated paradigm, that has as core assumption the recognition of mutual dependencies, symbiotic relationships and more subtle forms of manipulation, and that furthermore aims to preserve ecological diversity.
It is, however, more difficult to classify the other initiatives. There are several arguments to see them as a part of the life sciences integrated paradigm because they are focused on intensive agriculture, and possibly allow the use of GMOs. However, some of them also have the potential to develop in the direction of the ecologically integrated paradigm, under the condition that the environmental content of the program will be strengthened and a choice will be made for rigorous pesticide reduction strategies.
As this paper has shown, non-state actors have been an important driving force behind the development of programs containing standards for safe and sustainable agriculture and have important stakes in them. However, despite this stakeholder involvement, the majority of the larger initiatives cannot be qualified as multi-stakeholder partnerships, because their organisational structures tend to favour one type of stakeholder by allowing it a dominant position in decision-making. At the same time, however, several of these initiatives include mechanisms for stimulating participation, although, within boundaries, and opportunities for learning and upgrading. In this respect, it can be argued that under certain conditions private

regulatory initiatives can be considered forms of 'democracy in practice', as they provide a flexible and reflexive means to respond to challenges in terms of environmental and health problems.

References

Balfour, E.B. (1943). The Living Soil. Faber and Faber, London.

Boller, E.F. (2005). From chemical pest control to integrated production. A historical review. Written for the occasion of the 50th anniversary of IOBC. Available on http://www.iobc-global.org. 23 p.

El Titi, A., Boller, E.F. and Gendrier, J.P. (1993). Integrated production: Principles and technical guidelines. IOBCwprs Bull. 16 (1) 1993. Available on http://www.iobc-global.org. 96 p.

EUREP (1998). Integrated production of fruit and vegetables. EUREPGAP c/o FoodPlus, Cologne.

Fulponi, L. (2006). Private voluntary standards in the food system: the perspective of major food retailers in OECD countries. In: Food Policy 31: 1-13.

Hough, P. (1998). The global politics of pesticides. Forging consensus from conflicting interests. Earthscan, London.

Lang, T. and M. Heasman (2004). Food Wars: The global battle for mouths, minds and markets. Earthscan, London and Sterling. 365 p.

Northbourne, W.E.C.J. (1940). Look to the Land. Dent, London.

Potter Gates, J. (1988). Tracing the evolution of organic/sustainable agriculture. A selected and annotated bibiliography, US Department of Agriculture. Available on http://www.nal.usda.gov/afsic.

Pretty, J. (ed.) (2005). The pesticide detox: towards a more sustainable agriculture. Earthscan, London. 240 p.

Reijnders, L. (2004). Food safety, environmental improvement and economic efficiency in The Netherlands. In: British Food Journal, 106: 388-405.

Rodale, J.I. (1945). Pay Dirt: Farming and Gardening with Composts. Devin-Adair Company, New York.

Steiner, H. (ed.) (1977). Vers la production agricole intégrée. IOBCwprs Bull. 1977/4. Available on http://www.iobc-global.org. 153 pp.

Steiner, R. (1993). Spiritual foundations for the renewal of agriculture. A course of lectures. Bio-Dynamic Farming and Gardening Association, Kimberton. 310 pp.

Van der Grijp, N.M. and den Hond, F. den (1999). Green Supply Chain Initiatives in the European Food and Retailing Industry. R-99/07. Institute for Environmental Studies, Amsterdam. Available on http://www.vu.nl/ivm.

Organic values and animal production

Henk Verhoog and Jan de Wit
Louis Bolk Institute, Hoofdstraat 24, 3972 LA Driebergen, The Netherlands, h.verhoog@louisbolk.nl

Abstract

Food crises in connection with conventional animal production have led to an increased demand for animal products of organic origin. This applies to pigs and poultry in particular. As a consequence we see a sharp rise in the production of organic pigs and poultry, mainly by new converters. The products are processed and traded by large companies and sold in supermarkets. Prices are kept as low as possible. In a densely populated country like the Netherlands this leads to large-scale intensive organic animal production systems. These developments raise the question whether this still is 'organic'. Empirical research has shown that different views on 'organic' exist within the organic sector. For those who see organic as a market niche only, there is no problem as long as producers stick to legal rules. But many organic producers and consumers think that values related to ecological principles (working with closed cycles, regional production), animal welfare and fairness determine whether it is organic or not, and not the legal rules. These values have recently been formalised by IFOAM (International Federation of Organic Agriculture Movements) into four ethical principles: the principle of health, ecology, fairness and care. The intensification of organic animal production does clash with at least some of these principles. This raises the question whether the European Regulation should be adapted or not. It could be done at the level of rules (land-based production instead of input-production) or by incorporating ethical principles and ethical procedures within European Regulation.

Keywords: organic values, animal production, ifoam, european regulation

Developments in organic meat production

Until 1995 organic poultry and pigs in the Netherlands were mainly kept in relatively small production units, often integrated on mixed farms (with dairy and/or arable production). From 1995 onwards, veterinary rules became tighter, hindering the small-scale production of pigs and poultry. Simultaneously marketing possibilities for organic meat and eggs (especially abroad) grew and processing and retail companies increased their product specifications. In 2003, more than 50% of the Dutch pig production had already been concentrated on specialized farms with 100 sows plus belonging fattening pigs; and poultry production with more than 13.000 laying hens. Since then, the total number of laying hens has doubled and the main pork-processing industries stopped collecting small numbers of pigs. Production in the Netherlands is now dominated by relatively large-scale, specialised farms, mainly owned by recent converters with hardly any land of their own (on average around 7.5 ha). They sell a major part of their manure production (which is around 540 kg N per ha).

This development could take place because no standards had been formulated concerning the origin of pig and poultry feed, in contrast with the standards for organic dairy production, where at least 50% of the feed should be home-produced. As a result, most of the feed is from organic origin (>80%), but only a minor part (<10%) is produced at the pig and poultry farms themselves. Moreover, most of the animal feed concentrates (>70%) originates from abroad, with a growing percentage of concentrate feed coming from distant areas such as Latin America and the Far East. If all animal feed were to be produced within the Netherlands, more than the total present area of organic arable production would be required (Prins, 2005). Through all this,

the structure of the organic sector has become highly comparable with that of conventional agriculture in the Netherlands.
This development has negative side effects, such as:
- an increasing conflict with the consumer image of organic products and production;
- environmental problems as a result of inefficient nutrient utilization by high concentration of animals kept loose (high phosphate loads and a contribution to the acidification of natural areas due to ammonia evaporation);
- higher energy consumption for feed production due to transport. Bos (2005) calculated that a pig ration with 100% inland-produced feed requires 25% less energy compared to a current ration with only 15% inland feed;
- few and highly standardized human-livestock interactions. Large animal numbers limit the possibilities for adequate individual animal care, which is likely to be essential in order to improve animal health status in organic production without extensive use of medicines and preventive measures harming animal health (such as debeaking hens to prevent feather picking and cannibalism).

Different views of organic agriculture

Many pioneers in the field of organic farming are concerned about these developments in organic animal production systems. They feel that the side effects just mentioned go against the organic values and principles, although they are not forbidden by any rules. They believe that the rules laid down by European Regulation are minimum rules, and that one should not forget the organic philosophy which is behind these legal rules. But obviously not everybody, especially the new converter, has the same idea of what 'organic' means.
Different views on 'organic' have been distinguished in the literature:
- In a study on the meaning of 'naturalness', Verhoog *et al.* (2003) distinguished the non-chemical view (use of natural substances), the agro-ecological view (natural in the meaning of self-organisation) and the integrity view (referring to the characteristic nature of living beings).
- Meeusen *et al.* (2005) mention the calculating chain, the traditional chain, the unique chain and the responsible chain as different 'world-views' (chains) of organic stakeholders.
- Alroe and Noe (2006) speak about different perspectives: organic agriculture as a protest movement (protest against conventional agriculture), as a logo-poietic system (value-driven) and as a market niche.

The elements of the intensification of organic animal production in the Netherlands as described earlier, typically fit into the market niche approach (calculating chain) within a global market, thus leading to a structure that is very much the same as in conventional farming. Related to this, organic farming is mainly seen as a production system in which only the inputs are taken into account: without synthetic substances or GMO's (non-chemical view). There is nothing in the rules about whether the feed should be home- or regionally produced or not, as it should be according to the agro-ecological view.
These developments could have consequences for the image of organic agriculture, and thus for the success in the market-niche. This success is partly dependent on this image of 'organic': when organic can hardly be distinguished from conventional, why should consumers pay more for the organic products? Moreover, when the consequences become more widely known, it may have a negative impact not only on the market oriented calculating chain itself, but also on the value-driven part of the organic movement ('responsible chain'). This part of the organic movement, as well as consumers and producers in the 'protest movement', have a view of organic agriculture in which so-called public goods like a good environment and good animal welfare are also important, besides producing healthy food. To prevent a splitting up of the organic movement

along different views on organic agriculture, something must be done about this situation, in which the rules are not adequately expressing the intentions of the common values.

Organic values (principles)

But what are theses values, which are at stake in this case? Organic agriculture has been a value-driven movement from its very beginning. The founders emphasized different values, but these values have in 2005 been integrated into four ethical Principles by the International Federation of Organic Agriculture Movements (IFOAM):

- The Principle of Health: organic agriculture should sustain and enhance the health of soil, plant, animal, human and planet as one and indivisible.
- The Principle of Ecology: organic agriculture should be based on living ecological systems and cycles, work with them, emulate them and help sustain them.
- The Principle of Fairness: organic agriculture should be built on relationships that ensure fairness with regard to the common environment and life opportunities.
- The Principle of Care: organic agriculture should be managed in a precautionary and responsible manner to protect the health and well being of current and future generations and the environment.

In the Preamble it is said that these Principles are the roots from which organic agriculture grows. The ethical principles serve to inspire the organic movement in its full diversity and they are presented with a vision of their worldwide adoption.

Looking at the consequences of pig and poultry production in the Netherlands, we can conclude that they conflict with several of these Principles, first of all the Principle of Ecology. Part of this Principle is the idea of closed production cycles and the conservation and creation of agricultural- and biodiversity. When production-cycles get too big and the food miles too high it becomes increasingly difficult to keep the cycle closed (with respect to resource and energy use), as is shown in the intensification of organic animal production. And allocating responsibility also becomes more difficult when the cycles are too big. To feel responsible for the environment and the people in the organic chain, as implied by the Principle of Care, it is important that there is transparency and that all stakeholders are involved. One could also say that this development is not sustainable. The organic value of sustainability is ideally characterized by the fact that its three components (people, planet and profit) are fully integrated.

The aim of keeping production cycles as closed as possible can also be seen as part of the value of localness (nearness, regionality). This value is not explicitly mentioned in the four IFOAM Principles, but it clearly turned up in an empirical research project with focus groups in five European countries. This research is part of a European research project called Organic Revision (Padel, 2005). The outcome of this research shows that several important values are 'not covered in detail by the new IFOAM principles': proximity (localness); environmental and bio-diversity conservation; animal welfare and animal health; the whole systems or holistic approach; professionalism and careful processing.

As to proximity, Padel says: "for many it appeared that close proximity between production and consumption was seen as a natural progression from other organic values...". The dimensions of proximity and regional trade mentioned by focus groups are: farm income, traceability and trust, communication, food miles and product quality. Before the formulation of the four IFOAM Principles 'to foster local and regional production and distribution' was one of the 'Principle aims of organic production and processing'. At present it is somewhat cryptically mentioned as part of the Principle of Ecology: "Organic management must be adapted to local conditions, ecology, culture and scale".

Besides the Principle of Ecology and Care also the Fairness Principle could be mentioned. Compared to organic farmers that stick to land-based animal production systems, one can argue that input-based systems create an unfair economic advantage for the latter.

IFOAM ethical principles and EU Regulation

The situation sketched comes down to a conflict between the minimum rules of official standards (European Regulation: EEC 2092/91 and 1804/99) and the traditional organic values, which have been formalised by IFOAM as four ethical Principles, which have been accepted after a worldwide participatory process. In European Regulation, decision-making follows very different routes (political lobbying, etc.). The ultimate aim of European Regulation is to provide free trade within the European Community and private regulations are seen as hindering this development. One can argue that it is due to this regulation that the market-niche approach mentioned before could gain strength. The Regulation says that member states of the EU may not prohibit or restrict the access of organic products to their internal markets on grounds relating to the presentation of it as organic when the requirements laid down in the Regulation are met. We understand that this applies to ethical grounds as well.

So how to proceed in this situation? One way is to let the ethical dialogue take place within a private organisation such as IFOAM, and implement the results within the procedures that lead to Regulation at a European level. IFOAM already has got a special position as a voice from the private organic sector. The IFOAM-EU-Regional Group meets once a year with representatives of the European Commission. Other possibilities of influencing the decision-making process are limited to the drafting process, the advisory committees, and the parliament that comments on the drafts.

An ethical dialogue is supposed to be 'power free', and should in principle be open to all citizens, or in our case all people involved in the organic movement. It is no surprise therefore that the initiative to formulate ethical principles was taken by the private organisation IFOAM, in a deliberative, participatory process involving many people all over the world. It is a worldwide organisation, intending to be an open global platform for all organic stakeholders. IFOAM does not only represent the organic movement in several international institutions, it also wants to be leading in the field of new perspectives and views. The IFOAM Directory (2005) says that it wants to be 'attractive for every important actor in the sector, respecting and accepting different points of view in order to develop a commonly accepted agenda for the sake of the overall goal', which is the worldwide adoption of ecologically sound systems that are based on the Principles of Organic Agriculture. IFOAM looks as the ideal forum for developing and implementing structures in which a power free ethical dialogue can take place between all stakeholders. So why does the European Commission not delegate the ethical dialogue to an organisation such as IFOAM? Ideally all legal regulation should be a proper reflection of the values held by people in a particular community.

Another solution is trying to implement ethical or regulatory principles within the European Regulation itself. At the moment part of the research-program of the Organic Revision Project mentioned before, is to see how organic values can be incorporated in the ongoing revision process. By adding rules with respect to the Principle of Ecology (such as land-based production) one could counteract further intensification of organic agriculture. At another, more fundamental level, one could think of more general procedures of how decision-making processes at a regulatory (European) level can stimulate more participatory procedures, involving all stakeholders. But at present there still is a clash between the principles of a 'discourse ethic' (power free ethical dialogue), and the principles underlying political lobbying, which is at the basis of influencing European decision-making. Attempts to include a structured dialogue about ethical principles in political decision-making procedures will be very difficult (Verhoog, 2003).

References

Alrøe, H.F. and Noe, E. (2006). *What makes organic agriculture move – protest, meaning or market? A polyocular approach*. Paper presented at the XXI ESRS Congress, Hungary, August 22-25, 2005

Bos, J.F.F.P. (2006). *Intersectorale samenwerking in de biologische landbouw: mengvoergrondstoffen met binnen- of buitenlandse oorsprong; effect op energieverbruik van mengvoerproductie*. Report 114, Plant Research International, Wageningen.

IFOAM (2005). *Organic agriculture worldwide*. Directory of the member organizations and associates. Bonn.

Meeusen M. *et al.* (2005). *Zicht op dierlijke biologische ketens* (with English summary). Report 2.05.01 LEI-WUR, Den Haag.

Padel, S. (2005). *Focus groups of value concepts of organic producers and other stakeholders*. Report D21, EEC 2092/91.

Prins, U. (2005). *Verzelfstandiging van de biologische landbouw op het gebied van mest, voer en stro*. Report LV57, Louis Bolk Institute, Driebergen.

Verhoog, H. *et al.* (2003). The role of the concept of the natural (naturalness) in organic farming. *Journal of Agricultural and Environmental Ethics*, 16: 29-49.

Verhoog, H. (2003). Über die Eingeschränktheit der Ethikdebatten zur Gentechnologie. In: C. Hiss (Ed.), *Der GENaue Blick*, Ökom Verlag, München. 82-96.

Part 2
Foundational issues in philosophy and ethics

Biotechnology, disagreement and the limitations of public debate

Bernice Bovenkerk
University of Melbourne, Currently linked to the Ethics Institute, University of Utrecht, Heidelberglaan 2, 3584 CS Utrecht, The Netherlands, bernicebovenkerk@yahoo.com.au

Abstract

In discussions about the merits of biotechnology we are faced with a diversity of opinions and attitudes, giving rise to persistent disagreements. I distinguish six different sources of disagreement, which correspond to different levels of 'fundamentality'; the more basic the values and worldviews that specific arguments appeal to, the more fundamental the disagreement is, and the more likely it is to become intractable. This could be one reason why the biotechnology debate tends to revolve around more pragmatic issues, such as the safety of GMOs and the need for labelling, while other issues, especially those of a more metaphysical nature, remain in the background. Appeals to more fundamental concerns are often left out of public debate, because it is argued that in a pluralistic society we will never be able to reach agreement on these. However, I want to argue that limiting public debate about biotechnology to less fundamental levels of disagreement, does injustice to the biotechnology controversy, because it misrepresents the actual way in which participants in the debate reach their positions on biotechnology. To a certain extent all the different categories of disagreement are based on moral viewpoints – as they all have evaluative aspects – and on a deeper level they can be traced to specific worldviews and metaphysical views. We would be abstracting from the wider reasons that move people, if we limited the debate to less fundamental issues. More importantly, framing the biotechnology debate solely in less fundamental or pragmatic terms is inherently biased, because it tends to favour pro-biotechnology standpoints.

Keywords: biotechnology, pluralism, public debate

Introduction

In discussions about the merits of modern biotechnology[25] we are faced with a diversity of opinions and attitudes, giving rise to persistent disagreements. Advocates of agricultural biotechnology are of the opinion that GM crops have the potential to alleviate world hunger and ease pressures on the environment (Meyer, 2001), whereas critics believe the exact opposite, namely, that biotechnology will increase the gap between rich and poor and will add to already existing environmental problems (Hindmarsh, 1991; Kloppenburg, 1996; Anderson, 2000). In the medical field, proponents argue that biotechnology gives us invaluable information about the building blocks of life and can help fight disease, whereas opponents argue that many of the claims researchers make are exaggerated; they question the validity of their scientific paradigm (Ho, 1999; Wills, 2002) and point to the costs of using biotechnological techniques, especially

[25] The Cartagena Protocol on Biosafety defines modern biotechnology as 'the application of: a) in vitro nucleic acid techniques, including recombinant deoxyribonucleic acid (DNA) and direct injection of nucleic acid into cells or organelles, or b) fusion of cells beyond the taxonomic family, that overcome natural physiological reproductive or recombination barriers and that are not techniques used in traditional breeding and selection' (Cartagena Protocol on Biosafety, article 3 (i), 2000). When people refer to modern biotechnology they often imply genetic engineering, which is the horizontal transfer of genes between different organisms, often belonging to different species, and which is made possible by recombinant DNA technology.

the costs to non-human animals (Jochemsen, 2000). In general, proponents argue that people have been changing organisms to suit their own needs for centuries and that there is nothing new about genetic engineering (Fedoroff and Brown, 2004). Opponents, on the other hand, argue that the speed with which changes are taking place in organisms through genetic engineering, and the possible consequences of this technology, set it apart from earlier interventions in nature (Anderson, 2000). Furthermore, it has been argued by opponents of biotechnology that crossing the species barrier is unnatural, that changing the genetic make-up of organisms to suit human needs attests of human *hubris*, that it violates animals' integrity, and turns animals and plants into mere instruments (Brom, 1997; Verhoog, 1992). As this quick foray into 'the biotechnology debate'[26] shows, this debate is a complex one, in which disagreements have many different dimensions and operate on many different levels. How can we make decisions in the face of such persistent and multi-layered disagreement? The way that seems to be favoured is by concentrating on the more pragmatic issues, such as labelling and risk reduction, whilst side-stepping the more fundamental questions. It is often argued that since we will never be able to reach agreement on the more fundamental issues, it is best to simply leave them out of the debate and focus instead on issues with (apparently) less 'moral baggage' (Ackerman, 1989). In this contribution, I want to argue that this limitation of the debate does injustice to the biotechnology controversy, because it misrepresents the actual way in which participants in the debate reach their positions on biotechnology. Moreover, I want to suggest that framing the biotechnology debate solely in pragmatic terms is biased, because it tends to favour pro-biotechnology standpoints.

Sources of disagreement

Even though many of these specific issues are unique to biotechnology, the diversity of opinions encountered in the biotechnology debate is a reflection of the diversity of opinions found in society in general. In modern Western societies we are faced with moral, religious, and cultural diversity. The terms 'cultural' or 'social pluralism' are often used to refer to the plurality of values and worldviews stemming from this diversity (Lawrence, 1996).The existence of social pluralism can have several different sources. People can disagree about something because they lack all the relevant facts bearing on the case, because they base their viewpoint on different sources of evidence, or because they interpret the facts differently. We could label this 'factual disagreement'. In the biotechnology debate, factual disagreements exist, for example, about whether GM crops lead to more or less herbicide and pesticide use. At first sight, this type of disagreement appears to be easily solved by pointing to scientific evidence. However, disagreement also exists on the level of scientific theory. Biologists inspired by an ecological worldview approach genetics from a more holistic perspective than molecular biologists, for example. This could be termed 'scientific disagreement'. Another example is the question whether or not the current methods used for testing the safety of GMOs are adequate (Carman, 2004). People do not only disagree about factual knowledge and the methods of knowledge generation, but their disagreement may already begin at the level of definitions of the core terms used in the debate. Our choice of definitions determines how the debate is framed. For example, core concepts such as 'substantial equivalence' and the 'precautionary principle' are open to several different interpretations, and there is disagreement about whether to define GMOs as 'inventions' or 'discoveries'. All of these could be called 'definitional disagreements'. Of course, how one decides to define certain core notions in a debate and one's initial position on the topic are typically interrelated: one can define terms to suit one's own interests or

[26] When I refer to 'the' biotechnology debate, I of course realize that there is no one such all-encompassing debate. Rather, we encounter countless smaller debates about specific issues involving biotechnology.

values. On a deeper level, such definitional disagreements are not only about how we describe certain core terms, but also about power. The dominant discourse in society tends to be that of powerful groups; when these groups choose to employ certain definitions they are determining the terms of discussion (Kleinman and Kloppenburg, 1991). Because of the bias that can be inherent in definitional disagreements, they can sometimes be redefined in terms of 'interest-based disagreements': disagreements that are based on some sort of material or economic interest. Both parties in the biotechnology debate accuse each other of self-interested motives, even though they appear to advance arguments in the name of the greater good. An example is the claim that countries that enforce labelling rules for GM products are not doing so for the sake of the consumer's freedom of choice, but in order to throw up trade barriers. Another type of disagreement can be traced to conflicts between different, often fundamental, values. We could term disagreements based on this type of argument 'value disagreements'. For example, the question whether animal testing should be allowed depends on conflicting views about the proper moral status accorded to animals. Value disagreements can be based on conflicts between specific worldviews. For example, proponents of biotechnology generally tend to have an optimistic view about human progress through science and technology, while opponents tend to have a more pessimistic view, based on past technological failures. This type of disagreement could be termed 'metaphysical disagreements'. Another example of a metaphysical disagreement is the one between anthropocentric and non-anthropocentric worldviews. Our metaphysical views influence our moral viewpoints; however, they will not determine them, as two people with the same metaphysical views can base different moral viewpoints on these. Nonetheless, metaphysical views inform and influence our moral viewpoints by structuring the field of possibilities. In this sense, metaphysical disagreements operate on a more fundamental level than value disagreements.

Levels of fundamentality

Of course, it could be maintained that many of the disagreements we encounter in the biotechnology debate are simply due to conflicts of interest – for example between the interests of farmers to improved harvests and of consumers to safe food. However, many of the conflicts of interest in this debate can also be traced to opposing values or worldviews; they are interrelated in the sense that one's values and worldviews are influenced by one's more material interests and vice versa. Similarly, factual, definitional, and scientific disagreements have value aspects, because they involve evaluative judgments, for instance in the experimental design of scientific research or in the interpretation of test results. While the different types of disagreement classified here often merge together in real-world conflicts, they nonetheless remain analytically separable and are, therefore, of considerable heuristic value in narrowing down the more fundamental sources of disagreement. It can then become clear that, for instance, a conflict about the risk of introducing a specific GM crop into the environment is not based only on a scientific disagreement, but also on an underlying value disagreement about the evaluative steps in the type of risk analysis employed. When we dig deeper, it might also become clear that the conflict is based on what risks we are willing to impose on our environment, and this in turn could be based on a disagreement about the status of non-human nature. To a certain extent all of the categories of disagreement I have distinguished are based on moral disagreements: the viewpoints expressed all have evaluative aspects, and on a deeper level they can often be traced to specific worldviews and metaphysical views. Our opinions of the merits of biotechnology are therefore, ultimately, embedded in our views of the proper place of human beings in the world; of course, for many this does not only relate to our relationship to non-human entities, but also to God.

It appears, then, that the different sources of disagreement I have identified correspond to different levels of 'fundamentality'. 'Fundamental' as I use the term here refers to the structure of our personal system of values and worldviews; the more basic the values and worldviews that specific arguments appeal to, or derive from, the more fundamental the disagreement becomes. Moreover, the different levels are interrelated; the less fundamental ones are often based on, or embedded in, the more fundamental ones. We could picture the different levels of fundamentality in a similar vein to a proposal by Simon Blackburn (1998), who suggests that our moral reactions to certain behaviour could be thought of in terms of 'a staircase of practical and emotional ascent'. In our case, it might be easier to think of it in terms of a staircase from which we descend, starting with simple factual disagreements that can either be easily solved or cannot, for example because the disagreement rests on the next step of scientific disagreement. The latter in turn could rest on evaluative differences, such as what the proper aim of the research that generated the disputed facts should be, and it is, therefore, based on the next step of value disagreements. These value disagreements could in turn be based on the next step of diverging metaphysical worldviews, for example a holistic versus an atomistic one.

Intractability

The foregoing categorization might help to explain why the debate tends to revolve around less fundamental, more pragmatic, issues, such as the safety of GMOs and the need for labelling of genetically modified products, while other issues, especially those of a moral or metaphysical nature, remain in the background. I have found this to be the case in researching literature and media coverage about biotechnology and especially in the way that governments deal with legislation issues (Hain *et al.*, 2002; Hindmarsh and Lawrence, 2001). Part of my PhD-research consists of making a comparative analysis between the Netherlands and Australia about the way in which the biotechnology controversy is publicly debated. In Australia there tends to be a lot of focus on risks and labelling, and cost-benefit analyses are made in order to deal with these issues. Action groups who oppose certain biotechnological procedures and applications, such as GenEthics, even adopt this more pragmatic language as well. In the Netherlands there seems to be more opportunity to discuss more fundamental moral or metaphysical issues. Action groups, such as the Dutch Society for the Protection of Animals, do not shy away from using more value-laden language either. However, from interviews with members of the Dutch Committee for Animal Biotechnology it becomes clear that even though in Committee discussions there is room for moral issues, the more fundamental the disagreements become, the less likely they are to be discussed.[27] Objections that refer to more fundamental moral or metaphysical views, that are officially made against Committee recommendations, by the aforementioned and other organizations, are - in the view of these organizations and some Committee members - not addressed satisfactorily. It appears, then, that even though more fundamental 'steps' of the descending staircase play a role in the debate, they are often not voiced. The likely reason for this is that when disagreements are of a more fundamental nature it becomes harder to generalize the different viewpoints, as they appeal to deep-rooted values that are perceived to be subjective. This is not to say that these values cannot be defended by rational argument, but merely that there will be more disagreement already about the premises of such arguments. In general, one would expect, therefore, that the more fundamental the values and worldviews that are involved in a disagreement, the more likely this disagreement is to become intractable. One could argue that intractable disagreements can never be rationally resolved and that we

[27] The Committee for Animal Biotechnology has a double function. Firstly, it advices the Minister for Agriculture on license applications for biotechnological procedures with animals, and secondly, it stimulates public debate about animal biotechnology. See Committee for Animal Biotechnology (1998).

should therefore leave those disagreements out of the debate. However, when we manage to get the reasons for our disagreement out in the open, we will gain a deeper understanding of what divides us and if we cannot come to agree, this will at least give us a basis to 'agree to disagree' (Sunstein, 1997). Gaining an understanding of the motivation behind our opponents' points of view might help us to respect each other and this will reduce the chance of an escalation of conflict. Otherwise, all we can hope for is 'tolerating' other people's viewpoints, which means that we remain indifferent about what moves them.

Limited debate

If my observation is correct that moral and even metaphysical views are involved in most disagreements about biotechnology, this raises questions about the way in which the biotechnology issue is currently being debated. The focus on pragmatic aspects does not do justice to the wider reasons why people take a stance for or against biotechnology. A person's position, even on such a practical issue as labelling of GMO's, cannot be seen apart from her broader set of assumptions about biotechnology, about our place in the world, about the moral status of nature, and even about the proper role of food in our lives. In my opinion, we would be abstracting from the 'real' reasons that move people, if we limited the debate to less fundamental issues. The debate then does not reflect the actual sources of the disagreements and obscures the reasons behind arguments. How can participants in a debate have any hope of reaching mutual understanding and respect, if not actual agreement, if they are only discussing 'the tip of the iceberg'?
Furthermore, limiting the debate to more pragmatic issues, and to those issues that can be dealt with by way of making a cost/benefit analysis, tends to create a bias in favour of proponents of biotechnology. More fundamental moral and metaphysical views can often not be expressed solely in pragmatic terms. As became clear in the introduction, opponents of biotechnology appear to have more deontological concerns about limits we should not transgress or actions that are wrong in principle rather than just because of their consequences. Deontological concerns, however, do not lend themselves for the 'calculation' of a cost/benefit analysis, because they cannot so easily be quantified. Someone who does not find biotechnology problematic in principle, is likely to have less qualms about casting the debate in pragmatic terms than someone who opposes it on intrinsic grounds. After all, the question in such a 'pragmatic debate' is no longer whether biotechnology is desirable at all, but rather, how we can regulate it in order to lower associated risks and enhance people's autonomy. Deontological arguments, on the other hand, tend to address the desirability of biotechnology in itself. Framing the biotechnology debate solely in less fundamental or pragmatic terms, in other words, is inherently biased, and therefore unfair, because it tends to favour pro-biotechnology standpoints.

References

Ackerman, B. (1989). Why Dialogue? *The Journal of Philosophy*, 86 (1): 5-22.

Anderson, L. (2000). *Genetic Engineering, Food, and Our Environment. A Brief Guide*. Scribe Publications, Melbourne. 191 p.

Blackburn, S. (1998). *Ruling Passions*. Clarendon Press, Oxford.

Brom, F.W.A. (1997) *Onherstelbaar Verbeterd. Biotechnologie bij dieren als een moreel probleem*. University of Utrecht, Utrecht. 305 p.

Carman, J. (2004). Is GM food safe to eat? In: *Recoding Nature: Critical Perspectives on Genetic Engineering*. Hindmarsh, R. and Lawrence, G. (eds). UNSW Press, Sydney: 82-93.

Cartagena Protocol on Biosafety. (2000). Article 3 (i).

Committee for Animal Biotechnology. (1998) Annual Report.

Fedoroff, N. and Brown, N.M. (2004). *Mendel in the Kitchen. A scientist's view if genetically modified foods.* Joseph Henry Press, Washington D.C. 370 p.

Hain, M., Cocklin, C. and Gibbs, D. (2002). Regulating Biosciences: the Gene Technology Act 2000. *Environmental and Planning Law Journal*, 19 (3): 163-179.

Hindmarsh, R. (1991). The Flawed "Sustainable" Promise of Genetic Engineering. *The Ecologist*, 21 (5): 196-205.

Hindmarsh, R. and Lawrence, G. (2001). *Altered Genes II the future?* Scribe Publications, Melbourne. 253 p.

Ho, M.W. (1999). *Genetic Engineering. Dream or nightmare?* Gateway, Dublin. 385 p.

Jochemsen, H. (2000). *Toetsen en Begrenzen. Een ethische en politieke beoordeling van de moderne biotechnologie.* ChristenUnie, Amsterdam. 261 p.

Kleinman, D.L. and Kloppenburg, J. Jr. (1991). Aiming for the Discursive High Ground: Monsanto and the biotechnology controversy. *Sociological Forum*, 6 (3): 427-447.

Kloppenburg, J. Jr. and Burrows, B. (1996). Biotechnology to the Rescue? Twelve Reasons Why Biotechnology is Incompatible with Sustainable Agriculture. *The Ecologist*, 26 (2): 61-67.

Lawrence, M. (1996). Pluralism, Liberalism, and the Role of Overriding Values. *Pacific Philosophical Quarterly*, 77: 335-350.

Meyer, G. (2001). *Fighting Poverty with Biotechnology? Report from a Copenhagen Workshop on Biotechnology and the Third World.* University of Copenhagen. Centre for Bioethics and Risk Assessment, Copenhagen. 31 p.

Sunstein, C. (1997). Deliberation, Democracy and Disagreement. In: *Justice and Democracy: Cross-Cultural Perspectives.* Bontekoe, R. and Stepaniants, M. (eds.). University of Hawaii Press, Honolulu: 93-117.

Verhoog, H. (1992). The Concept of Intrinsic Value and Transgenic Animals. *Journal of Agricultural and Environmental Ethics*: 147-160.

Wills, P.R. (2002). Biological Complexity and Genetic Engineering. *Conference Proceedings of the Environment, Community and Culture Conference*, Brisbane.

Cynicism and corporate integrity: Obstacles and possibilities for moral communication

Erik de Bakker
LEI, Wageningen University and Research Centre, Burgemeester Patijnlaan 19, P.O. Box 29703, 2502 LS The Hague, The Netherlands, erik.debakker@wur.nl

Abstract

Paying thorough attention to cynical action and integrity could result in a less naive approach to ethics and moral communication. In this article the issues of integrity and cynicism are discussed on a theoretical and a more practical level. In the first part, Habermas' approach of communicative action (1981) is confronted with Sloterdijk's concept of cynical reason (1983). In the second part, the focus will be on the constraints and possibilities for moral communication within a business context. Discussing the corporate integrity approach of Kaptein and Wempe will do this. This approach can be considered as a valuable and insightful contribution to the question of how to deal with (dilemmas of) conflicting interests, open discussion, fairness, and strategic decision-making in the context of stakeholder dialogue. However, it is concluded that Kaptein and Wempe seem to overstretch the concept of corporate integrity by their inclination to make it an all-purpose remedy for corporate dilemmas.

Keywords: social theory, business ethics, corporate responsibility, integrity, cynicism

Introduction

Moral communication in the context of free market competition is often called into question. A genuine discussion about morality and ethics seems difficult if not impossible to realise in such a context. One could argue that the economic interests at stake in the market contradict with the ethical prerequisite of non-strategic communication, which is characterized by open moral discussion and sincere involvement of the participants. Therefore, one could conclude that corporations are not in a position to freely deliberate about moral issues. A cynic might add that moral statements of corporations are merely window-dressing.
This article concentrates on the key issue of integrity in moral communication. The following questions will be dealt with:

- Is the presupposition of non-strategic communication, which is thought to be necessary for moral deliberation, a realistic assumption?
- Is most if not all moral communication in societal practice not cursed with power struggles and strategies?
- Is moral communication possible in an economic market context in spite of the constraints and dilemmas of corporations?
- Can the problem of integrity, cynicism and hidden agendas be solved by the development and improvement of corporate integrity?

The first two questions will be discussed by confronting Habermas' approach of communicative action (1981) with Peter Sloterdijk's concept of cynical reason (1983). In the second part of the article, the focus will be on the last two questions. The issue of moral communication, integrity and cynicism will be addressed on a more practical level by discussing the corporate integrity model of Kaptein and Wempe (2002). Finally, the article will consider whether this model offers a fruitful perspective to tackle cynicism.

Communicative action and cynical behaviour

Habermas' theory of communicative action

In his theory of communicative action (1981) Habermas tries to develop a philosophical and sociological account of the conditions that are necessary for moral communication on a rational basis. An important component of his analysis is the distinction between communicative integration at the level of the lifeworld and functional integration as brought about by economic and political subsystems that are steered through money and power. While social coordination in the lifeworld is structured by processes of communication that aim at reaching agreement about validity claims, social coordination in the semi-independent subsystems of economic markets and bureaucratic administrations rests on the strategic pursuit of private interests. In the view of Habermas reasonable discussions about norms and morality are only possible under the conditions of communicative rationality that characterize the lifeworld.

Although Habermas provides us with an insightful linguistic and philosophical understanding of the different goals and claims connected with communicative and strategic rationality, his duality between the lifeworld and economic and political subsystems is problematic. If one tries to apply this duality empirically, it will seem almost impossible to draw a clear-cut line between a communicative lifeworld and strategically oriented subsystems.

A second objection that could be raised against Habermas concerns his presupposition of free discourse and authenticity that ideally underlies communicative reason. An important question in the context of moral communication is to what extent can be counted on the integrity of the parties involved. The presupposition of integrity, which is necessary for a rational discourse that is free from power, is probably one of the most difficult problems in the theory of communicative action. Furthermore, Habermas' theory seems to have a blind spot for the effects of cynicism on rational discourse, in particular the use of hidden agendas.

Sloterdijk's concept of the cynical reason

According to Sloterdijk cynicism nowadays has become ubiquitous and goes together with discretion and resignation. He characterizes the cynical reason of our modern age as '*the enlightened wrong consciousness*' (1983: 37). The modern cynic has made himself immune against rational critique. He is aware of morally 'correct' behaviour but still acts morally 'wrong'. Cynicism rests on the belief that 'reality is hard' and that one has to consider all realistic means that are possible to pursue a private goal. Therefore, cynicism can be defined as all

Table 1. Open and hidden agendas in connection with cynical action orientations.

Action orientations	Open agendas		Hidden agendas	
	at one side	at all sides	at one side	at all sides
Strategic context in which self-interest dominates	Unilateral strategic action	Mutual strategic action	Unilateral cynical action	Mutual cynical action
Persuasion context oriented at intersubjective agreement	Unilateral persuasive action	Mutual persuasive action		

Source: De Bakker (2001: 50)

'morally incorrect' orientations, attitudes and actions that are (implicitly or explicitly) justified by referring to the way things are in (hard) reality (De Bakker, 2001: 48).

To classify different modes of social action the existence of open and hidden agendas is a useful criterion. This criterion can serve as a method to analyse cynical action in different social contexts and can shed light on the key role of authenticity in social action.

The scheme above makes clear that cynicism can be unilateral (one person uses a hidden agenda) or mutual (everyone uses hidden agendas). In the latter case it can lead to situations in which social actors are keeping up appearances about their integrity in front of a wider public, while behind the scenes all kind of hidden strategic manoeuvres are taking place. The scheme of cynical action orientations presented above also makes clear that the problem of a lack of integrity is both possible in a persuasion (communicative) context and in a strategic context.

Although the analysis of Sloterdijk can be viewed as an insightful contribution to a less naïve philosophy that is aware of concealed power struggles and strategies, it does not offer much guidance on how to deal with integrity on a more practical level of moral communication. The following sections will address a more practical approach to integrity in the field of business ethics: the concept of corporate integrity of Kaptein and Wempe.

Social responsibility and corporate integrity

Integrity as "wholeness" and the corporation as a moral entity

The everyday use of the term integrity (in a moral sense) is often featured by a negative view. People normally think of integrity in terms of lapses or deficiencies: someone who proofs to be impeachable or corrupt. Instead of concentrating on this negative view on integrity, Kaptein and Wempe focus on a more positive understanding of integrity. They point at the following three characteristics (Kaptein and Wempe, 2002: 90-93):

1. A person of integrity is an autonomous thinker; he or she has a stable set of values and norms; he or she stands and strives for something and is true to his or her ideals.
2. A state of internal integration; a person of integrity is capable of integrating his or her values, deeds, and their effects in his or her life in a natural way.
3. A state of external integration: a person of integrity is integrated into his or her environment and sensitive to social issues, and willing to account for himself or herself.

Wempe and Kaptein consider the corporation as an autonomous moral entity that bears responsibility as a whole (Kaptein and Wempe, 2002: 145-152). They argue that within each corporation an organisational structure and culture exists that can be distinguished from the individuals who work within the corporation. Because of the identifiable culture and structure that underlie corporate practices, it is possible to view the corporation as a moral entity and to judge the actions and intentions of a corporation in moral terms.

Corporate integrity relates to the corporate efforts that are localized in the corporate culture and structure and refers to the coherence between corporate efforts, conduct, and consequences (Kaptein and Wempe, 2002: 152-154). Just like a person of integrity, a corporation of integrity has to strive for wholeness or consistency between (1) different values, norms and ideals that motivate action; (2) words and deeds; and (3) corporate practices and the outside world (societal expectations and demands). However, the pursuit of corporate integrity is hindered or complicated by dilemmas that go together with moral communication in a business environment. These can be described as the entangled hands dilemma, the many hands dilemma, and the dirty hands dilemma.

- The entangled hands dilemma refers to different roles employees in a corporation may have to handle. It may be the case that these different roles are not well coordinated and

the corporation can be damaged. Examples of the entangled hands dilemma are improper use of corporate assets or the personal acceptance of promotional gifts.
- The many hands dilemma is visible in the relations that employees within the corporation maintain in order to fulfil the responsibilities of the corporation. The many hands dilemma points at the problem that the internal specialisation and division of labour, which characterize modern corporations, often leads to the dilution of responsibility.
- The dirty hands dilemma is visible in the relations maintained with the various stakeholders on behalf of the corporation. A corporation has to take the claims of different stakeholders into account and has to find a balance that ensures its continuity. In the process of balancing the different interests and expectations it is almost impossible for the corporation to avoid dirty hands.

Within the scope of this article, the many hands dilemma is most significant. This dilemma refers to the problem of moral communication with external stakeholders and the possible problem of conflicting interests and constraining demands. Dirty hands are unavoidable and must be made but in the view of Kaptein and Wempe this can be done in a responsible way, if the corporation reflects on its wider social obligations towards stakeholders.

The corporate integrity approach

Kaptein and Wempe distinguish five behavioural principles for managing stakeholder relations with integrity: (i) openness, (ii) empathy, (iii) fairness, (iv), solidarity, and (v) reliability (2002: 237 ff.). These behavioural principles operate as criteria to assess the corporation's relations and behaviour. They argue that these behavioural principles will be more easy to fulfil, if the corporate culture and structure embody certain ethical qualities: clarity, consistency, achievability, supportability, visibility, discussability, and sanctionability. As Table 2 shows, each of these corporate ethical qualities relates to the three fundamental dilemmas and focus on the organizational context that should be taken into account, if one aims to comprehend the possibilities and constraints of employees in moral communication and responsible behaviour.

Corporate integrity: a remedy against cynicism?

Kaptein and Wempe show convincingly that normative principles and organizational ethical qualities are always interrelated. Therefore, the theory of corporate integrity makes (the lack of) efforts and intentions of moral agents to a certain degree (sociological) understandable and avoids the pitfalls of simple moral judgements and accusations. For sure, although Kaptein and Wempe do not explicitly talk about the problem of cynicism in modern societies, the concept of corporate integrity offers a practical perspective to deal with (cynical) obstacles and to fulfil corporate social responsibility. In particular, the ethical qualities with respect to the dilemmas of entangled and many hands can be seen as remedies against internal organizational cynicism and window dressing. If corporations feel responsible to realize these ethical qualities (in an open and transparent manner), corporate integrity can be a good medicine against cynical behaviour on the shop floor or employee's indifference and it will provide better possibilities for moral communication.

However, a critical note has to be made about the inclination of Kaptein and Wempe to make the concept of corporate integrity an all-purpose remedy for corporate dilemmas, including the probably most difficult dilemma of dirty hands. This can be explained by their broad characterization of integrity. According to this broad characterization integrity also means external integration with the outside world. A corporation has to strive for reasonable coherence with its environment by fulfilling social roles and tasks and 'making a contribution to the

Table 2. Ethics qualities of corporations.

Ethics qualities	Organizational dimension		
	Entangled hands - responsibilities with regards to the organization	Many hands - responsibilities within the organization	Dirty hands - responsibilities on behalf of the organization
Clarity	Employees have clarity about how they should treat the assets of the organisation	Employees have clarity about their job-related responsibilities	Employees have clarity about what stakeholders expect of them
Consistency	Referents make enough effort to treat the assets of the organization with care	Referents make enough effort to fulfil their functional responsibilities	Referents make enough effort to satisfy stakeholder expectations
Achievability	Expectations on how the corporate assets should be treated are feasible	Employees have the means to fulfil their job-related responsibilities	Created stakeholder expectations can be satisfied
Supportability	The organization fosters support for the careful treatment of the corporate assets	The organization fosters support for an adequate coordination between employees	The organization fosters support for satisfying stakeholder interests
Visibility	Consequences of how the corporate assets are treated are visible	Consequences of how job-related responsibilities are fulfilled are visible	Consequences of stakeholder expectations are satisfied are visible
Discussability	Dilemmas, problems, and criticisms regarding how the corporate assets are treated can be discussed	Dilemmas, problems, and criticisms regarding how job-related responsibilities are fulfilled can be discussed	Dilemmas, problems, and criticisms regarding how stakeholder expectations are satisfied can be discussed
Sanctionability	Staff will be sanctioned, if assets are deliberately mishandled	Staff will be sanctioned, if functional responsibilities are deliberately neglected	Staff will be sanctioned, if stakeholder expectations are deliberately ignored

Source: Kaptein and Wempe (2002: 254).

smooth functioning of the whole' (Kaptein and Wempe, 2002: 92). The question arises whether a quintessential trait of integrity is not exactly the opposite: not integrating with the outside world but remaining true to what one sincerely beliefs in spite of social pressure from the environment. The suggestion that integrity in this context means 'acting in accordance with the *legitimate* expectations of those around us' (Kaptein and Wempe, 2002: 92) only seems to shift the problem. What are legitimate expectations and who defines them?In this context it should also be considered that an important cause of dirty hands dilemmas might be cynical or opportunistic behaviour of other economic rivals or stakeholders. Perhaps the truth is that a corporation of (great) integrity can face unsolvable ethical dilemmas in its relationships with other rivals or stakeholders just because of its (true) integrity, dilemmas that can only be solved in a morally incorrect way by acting expediently and using hidden agendas. Integrity is of great value but can also involve painful experiences.

References

Bakker, de, E. (2001). *De cynische verkleuring van legitimiteit en acceptatie (The Cynical Discoloration of Legitimacy and Acceptance).* Aksant, Amsterdam.

Habermas, J. (1981), *Theorie des kommunikativen Handelns (The Theory of Communicative Action).* 2 Vol., Suhrkamp, Frankfurt am Main.

Habermas. J. (1996), *Between Facts and Norms. Contributions to a Discourse Theory of Law and Democracy.* (translated by W. Rehg), MIT Press, Massuchusetts.

Kaptein, M. and J. Wempe (2002), *The Balanced Company. A Theory of Corporate Integrity.* Oxford University Press, Oxford.

Sloterdijk, P. (1983), *Kritik der zynischen Vernunft (Critique of Cynical Reason).* 2 Vol., Suhrkamp, Frankfurt am Main.

Protecting future generations through submajority rule

Kristian Skagen Ekeli
Department of Philosophy, NTNU – Norwegian University of Science and Technology, 7491 Trondheim, Norway, ekeli@hf.ntnu.no

Abstract

Democratic decisions not only affect present people, but also near and distant future generations. Despite the fact that voters and their elected representatives have the power to make decisions that can have a serious impact on the living conditions of future people, succeeding generations do not have the opportunity to influence present political decision-making processes. Against this background, the purpose of this paper is to propose and consider two new constitutional devices, the aim of which is to give minorities of legislators a political tool to represent and protect the interests of future generations. The common denominator of the proposed reforms is that they represent examples of submajority rules that grant defined minorities of legislators certain procedural rights. The first device empowers a minority of at least 1/3 of the legislators to demand that the final enactment of a law proposal should be delayed until a new election has been held, if they believe that the law in question can inflict serious harm upon posterity. The second implies that a minority of at least 1/3 of the legislature can require a referendum on a bill that can have a serious adverse impact on the living conditions of future people. I will argue that these submajority rules can serve as useful means to encourage more future-oriented public deliberations and decisions. Although the proposals face some important problems, it is argued that they can be defended on the basis of central ideals in recent theory of deliberative democracy.

Keywords: deliberative democracy, future generations, political representation, public deliberation, submajority rules

Deliberative democracy[28]

Before proceeding, I will briefly outline some central ideas in deliberative democracy that play an important role in my arguments for the suggested reforms. From the point of view of deliberative democracy, democratic decisions can only be regarded as just or ethically justifiable if they result from a process of thorough and reasoned public deliberation – where all affected parties or their representatives have had the opportunity to participate. This implies that importance is attached to the process of public deliberation that takes place among the decision-makers *before* the issue in question is decided through voting. The point and value of a thorough process of deliberation is to improve the basis of information and enhance the level of reflection among the participants. The hope is that good reasons and inputs of new information will have a bearing on people's preferences and the way they vote. Thus, public deliberation is assumed to have a transformative effect, in the sense that the initial or pre-deliberative preferences of the participants will undergo a change that can lead to more rational and impartial decisions (see also Goodin 2003 and Ekeli 2005).

[28] I would like to thank Espen Gamlund, Brian Garvey, Sverre Grepperud and John O'Neill for valuable comments and discussions.

The proposed constitutional reforms

The common denominator of the proposed constitutional devices is that they represent examples of what can be termed *submajority rules*, that is, 'a voting rule that authorizes (i) a predefined numerical minority within a designated voting group (ii) to change the status quo (not merely to prevent change) (iii) regardless of the distribution of other votes' (Vermeule 2005:76). The 'status quo' referred to in condition (ii) is typically procedural rather than substantive. According to Adrian Vermeule, submajority rules are rarely or never used directly for final substantive decisions, such as the passage or defeat of legislation. Rather, they are used for procedural matters. This also applies to the submajority rules that I propose, which can be regarded as procedural rights ascribed to minorities of legislators.

1. The right of minorities to demand delays

A minority of at least 1/3 of the legislators should be granted the right to demand that the final enactment of a law proposal should be delayed until a new election has been held, if they believe that the law in question can inflict serious harms or risks upon posterity.

2. The right of minorities to demand referendums

A minority of at least 1/3 of the legislators should be empowered to require a referendum on law proposals that can have a serious adverse impact on the life-conditions of posterity. In this way, a minority can place an issue to the people for final approval. With regard to this proposal it is important that the electorate get sufficient time to gather relevant information as well as to consider and discuss the bill. Therefore, there should be a time interval of at least one year from the minority demands the referendum until it is held. On the other hand, in order to avoid a too time consuming process, I believe that the popular vote should be held within two years after the minority required the referendum.

At this point, it is important to make two clarifications with regard to the proposed reforms. In the first place, a minority of legislators should only be allowed to demand a delay or a referendum, if they present a *prima facie* case for the assumption that the law proposal in question can inflict serious harm or risk upon posterity. Thereafter, the burden of proof should shift to those legislators who reject the minority's harm scenario. Secondly, since conflicts about the reliability of competing harm scenarios involve issues of constitutional law (i.e. such conflicts concern the distribution of powers between minorities and majorities as specified in the constitution), I believe that they should be resolved by a special constitutional court or by some similar body such as the state's supreme court. In cases where controversies arise, the legislators who want to prevent a delay or a referendum (for instance 10% of the representatives in the parliament) should be allowed to initiate legal proceedings. But, as pointed out above, in such cases the onus of proof should rest with those who reject the minority's harm scenario after the minority has presented their *prima facie* case.

The case for the proposed submajority rules

The proposed constitutional procedural devices can, in several ways, indirectly affect decision outcomes, because decisions on procedural matters (i.e. on how decisions should be made) can have effects on outcomes. In what follows, I will argue that the reason for this is primarily that these submajority rules will have important effects with regard to agenda-setting, public deliberation and exchange of information.

Agenda-setting

Agenda-setting will affect decision outcomes for the simple reason that 'nothing can emerge as the output of a democratic process unless someone has first put it on the agenda' (Goodin 2003:163). The proposed submajority rules will affect the distribution of agenda-setting power, because they will to some extent have the effect of distributing the competence to control the agenda away from majorities to minorities. The reason for this is in part that minorities are empowered to determine how long an issue should be on the agenda (which applies both to delays and referendums), and to decide how the issue should be placed on the voting agenda (which applies to the right to require referendums and to some extent to delays). In this way, minorities are allowed to force the majority to pay more attention to certain issues affecting posterity. The suggested submajority rules can also give minorities of legislators the opportunity to increase the political visibility and the public awareness of the issues in question. I will return to this below.

Public deliberation and exchange of information

Another important aim of the proposed constitutional devices is to improve the process of deliberation and decision-making about issues that can have a serious impact on the living conditions of posterity. The purpose is to improve the basis of information and enhance the level of reflection among legislators and voters. Since both submajority rules allow minorities to slow down the process of collective decision-making, they can promote a more thorough and well-informed process of public deliberation about certain issues or law proposals. First of all, to the extent that minorities use their right to demand delays or referendums in order to protect future needs, this will ensure that both the electorate and politicians have the opportunity to consider the proposals in question more closely before a decision is made by simple majority vote either in the legislature or in a referendum. Secondly, in this process they will have time to gather new information which can in turn affect the subsequent process of deliberation, agenda-setting and decision-making. For example, new information can have effects on agenda-setting, because '[l]egislative majorities set their agendas in light of the information known to them and the information known to relevant publics; by changing the latter, submajorities may force a new agenda upon the majority' (Vermeule 2005:83). Thirdly, if minorities requires a delay or a referendum, decision makers will have more time to come up with, discuss and consider alternative courses of action, which might have more desirable consequences with regard to future generations than the bills that were initially introduced. In this way, the proposed devices can lead to a process of public discussion and deliberation that might lessen the problem of bounded rationality – the problem that our knowledge, imaginations and reasoning abilities are limited and fallible. According to James Fearon, public discussion might lessen the impact of bounded rationality for two reasons. 'First, it might be "additively" valuable in that you might think of some possibility that hadn't occurred to me, and vice versa. Second, it might be "multiplicatively" valuable in that in the course of discussion we might think of possibilities or problems that would not have occurred to either of us by ourselves' (Fearon 1998:49-50).

Devices such as delays and referendums can, as indicated above, also induce more public awareness and engage citizens more directly when it comes to issues affecting future generations. First, if minorities use their right to require delays, this will ensure that the people have the opportunity to consider the law proposal more closely during election campaigns. Besides, citizens would be given the chance to determine the composition of the legislative assembly before the law proposal can be enacted through majority vote. Second, a direct democratic device such as the referendum can provide a useful institutional mechanism for engaging citizens more directly in public deliberations about environmental and technological issues that can have a serious impact on future people.

At this point, an interesting question related to both the proposed devices emerges. This is whether, or to what extent, delays and referendums will encourage present citizens to take the needs of posterity into account when they cast their votes. Although it is difficult to answer this question, it is likely that these devices will promote more public debate and awareness about issues that affect future generations. Moreover, there is considerable evidence that people's preferences and perspectives change in response to additional information – particularly during political campaigns.[29] If these assumptions are correct, delays and referendums can make the interests and needs of future generations more 'imaginatively present' in the minds of the voters – despite the fact that future people cannot be 'communicatively present' in public deliberations.[30] This might in turn encourage voters to behave in a more principled and impartial fashion, that is, act on the basis of an altruistic rather than self-interested stance. Of course, more public debate and inputs of information about issues affecting posterity cannot guarantee that the electorate will, in fact, take the needs of succeeding generations into account when they cast their votes. It is, however, an important precondition for achieving this end.
Finally, I will mention another important way that the proposed devices can affect decision outcomes. Given the costs related to delays and referendums (e.g. both devices are time consuming), majorities would be encouraged to put forward more future-oriented law proposals that they believe are acceptable also for minorities who are concerned for the welfare of posterity. In this way, these submajority rules may even prevent proposals that have adverse future effects from coming on the voting agenda, because they can create strategic incentives to avoid such proposals.

Problems facing the proposed reforms

So far I have presented several arguments for the suggested constitutional reforms. In what follows, I will briefly mention three problems facing the proposed submajority rules.

Strategic abuses

The first and perhaps most important problem confronting the suggested devices is that they can be abused by minorities for strategic or egoistic reasons. It might be tempting for a minority in the parliament – who are not concerned for the well-being of future generations – to require for instance a delay in the hope that a bill they dislike or oppose would never be passed. To what extent such strategic abuses of the suggested devices are likely to occur in the real word is very difficult to predict. Nevertheless, the problem of strategic abuses must be taken seriously when assessing the desirability of the proposed reforms.

The burden of proof

As I pointed out above, the suggested reforms imply that a minority of legislators who wants to require a delay or a referendum must present a *prima facie* case for the assumption that the bill under consideration can inflict serious harm or risk of harm on future generations. Thereafter, the onus of proof should shift to those legislators who reject the minority's harm scenario. Some might object that this distribution of the burden of proof is unfair, and that the onus of proof should rest solely with the minority who wants to require a delay or a referendum. Something can be said for this line of criticism. But if the burden of proof solely rests with the minority, the practical value of the proposed devices would probably be undermined, because it is often very difficult to foresee how present decisions will affect near and distant future people in view of

[29] See, for example, Gerber and Jackson (1993).

[30] The distinction between 'communicative presence' and 'imaginative presence' is discussed in Goodin 2003, ch. 9 and 10.

uncertainty and ignorance of the future (see also Ekeli 2004). Since future generations cannot influence political decisions, I believe that minorities of legislators should be empowered to demand a delay or a referendum provided that they can present a *prima facie* case for their decision.

Disagreement about what best serves the interests of posterity

Some might object that the proposed reforms are problematic because minorities and majorities of legislators will, in a number of cases, disagree about what best serves the interests of posterity. For example, a majority might want to pass a bill that would imply increased use of nuclear energy with the justification that this would reduce CO2 emissions and the likelihood of future harms related to global warming. On the other hand, a minority might claim that such a policy would expose future people to grave risks. I do agree that such disagreements are likely to arise, but I do not think that such disputes pose a serious problem for the suggested reforms. The proposed submajority rules will not close off debates about what best serves the interests of posterity. They only give minorities a political tool to increase the political visibility and to encourage a more thorough and well-informed process of public deliberation about issues subject to such disagreements, before a final decision is made through majority rule in the legislative assembly or in a referendum.

Concluding remarks

The aim of this paper has been to throw light on how submajority rules can be used in order to represent and protect future interests. I have argued that a good case can be made for the proposed reforms in view of central ideas and ideals in recent theory of deliberative democracy. Although I have primarily focused on the arguments in support of the suggested submajority rules, I do believe that they face some important problems that deserve more attention than they have been given in the present paper. Anyway, the present contribution is intended as a beginning, not the end, of a discussion about whether future interests ought to be protected by means of submajority rules.

References

Ekeli, K. S. (2004). 'Environmental Risks, Uncertainty and Intergenerational Ethics', *Environmental Values*, 13:421-448.

Ekeli, K. S. (2005). 'Giving a Voice to Posterity – Deliberative Democracy and Representation of Future People', *Journal of Agricultural and Environmental Ethics*, 18:429-450.

Fearon, J. (1998). 'Deliberation as Discussion'. In *Deliberative Democracy*, J. Elster (ed.). Cambridge: Cambridge University Press, pp. 44-68.

Gerber, E. R. and Jackson, J.E. (1993). 'Endogenous Preferences and the Study of Institutions', *American Political Science Review*, 87(3):639-656.

Goodin, R. E. (2003). *Reflective Democracy*. Oxford: Oxford University Press.

Vermeule, A. (2005). 'Submajority Rules: Forcing Accountability upon Majorities', *Journal of Political Philosophy*, 13:74-98.

Scientific advice and the ideal of certainty

Anna Paldam Folker and Peter Sandøe
Danish Centre for Bioethics and Risk Assessment, The Royal Veterinary and Agricultural University, Denmark, Rolighedsvej 25C, 1958 Frederiksberg C, Denmark, apf@kvl.dk

Abstract

The role of scientific adviser to the general public is potentially tricky in that it involves a number of possible tensions between the requirements of science and the requirements of scientific advice. One such tension concerns the management and presentation of scientific uncertainty. While uncertainty is at the heart of science, scientific advice seems to call for certainty and unambiguous recommendations. In the first part of the paper we illustrate this tension by presenting the findings of an interview study of Danish nutritional scientists. The study shows that nutritional scientists feel their roles as 'public advisers' and 'scientists' differ in that the former requires an appearance of certainty that relegates scientific uncertainty to the background. In the second part of the paper we question the premise that certainty is a virtue in scientific advice. The role of the scientific adviser to the public is to offer practical guidance - something which does not necessarily require certainty from the adviser. We argue that it is possible for scientific experts to convey scientific uncertainty while at the same time to deliver clear advice but that scientists collectively should make an effort to avoid conflicting recommendations.

Keywords: appearance of certainty, scientific advice, scientific consensus, scientific uncertainty

Introduction

Scientific advice-giving to the general public[31] is often thought to involve a complex choice in which scientific uncertainty is balanced against the provision of definite answers. This balance might seem to compromise the scientific basis of the advice.
This is not only an issue of effective communication however, i.e. of adequate ways to present scientific results to the public. It also raises larger issues regarding the ethics of scientific advice-giving and the proper role of the scientific adviser.
To provide an empirical foundation for discussion of these issues, we conducted an interview study among scientists in the field of nutrition. This area was chosen because it is characterised by public attention and by various actors with different agendas that continuously influence each other (e.g. rival experts, the media, politicians, NGOs and the food industry). Research on nutritional questions tends, in other words, to have a wide impact. Because of the keen public attention, it is often a short distance from research to scientific advice. Thus the nutritional scientists themselves are frequently required to play a dual role: to be scientists and public advisers at the same time.

[31] A prototypical scientific advice in this context is a recommendation by scientific experts that is based on scientific studies and aimed at policy makers, officials, the relevant industry or ordinary citizens. It often involves an attempt to synthesise the knowledge available in a given field at a particular time. Scientific advice-giving can be explicitly normative, but it need not be.

The interview study

Data and methods

The interview study was conducted in Denmark in 2004-2005. To obtain a common focus the interviews were centred on a specific case: the Danish variant of what was, in fact, during the 1990s, an international controversy over the relationship between sugar and obesity. The study involved 16 semi-structured interviews, half of them with prominent scientists within the field of nutrition – typically men in middle age originally trained as medical doctors. Apart from their involvement in the case these scientists were selected on the basis of their function as public advisers on nutritional issues. The other eight interviewees – two dieticians, a politician, and five central administrators in governmental departments dealing with health and nutrition – were selected as key informants regarding the case and the field of nutritional policy broadly conceived.[32] [33]

Findings

Analysis of the interview study showed that scientists do perceive a tension between the requirements of science and the requirements of scientific advice. On the one hand, science is seen as marked by doubt and uncertainty. Within their disciplines scientists are required to critically asses their work and the work of others. It is generally accepted that results are usually qualified and may well be subsequently revised. Scientific advice-giving, on the other hand, seems to involve a requirement to conceal doubt and uncertainty in order to steer action – often on the basis of incomplete knowledge. In the study scientific advice-giving is described as a matter of taking a step from scientific doubt to an acceptance that a certain course of action is right.

> Scientist: Even if we have this doubt that drives research forward as I just told you about with these four or perhaps even more ways to go regarding the role of sugar in terms of public health and the development of obesity. Then there is a another element, and that is if the researcher steps out of his role as a scientist in order to give advice to the public either by means of communication to the media or advisory committees to the public authorities on various issues, then you have to leap – almost as described by Kierkegaard – qualitatively you have to take a leap out of the doubt you have as a researcher into the decision... the choice to believe that this is the truth after all, and that this belief has such a strength that you choose to act on it. This leap... some people are willing to make it easily while others are more worried. And some refuse to make it at all. They won't risk – again in the terminology of Kierkegaard – the seventy thousand fathoms of water. But you *have* to leap because you live in the present. You are here and now.

As the quote illustrates, the step from the state of uncertainty (science) to acceptance that a certain line of action is right may not be an easy one for scientists to make. It is a step in which the scientist must try to handle the doubt and uncertainties inherent in scientific material. In the interview study the scientists describe the sense in which these conditions of science are swept aside in their role as public advisers – they must in the words of the scientist quoted above "leap out of the doubt".

---

[32] The interviews have been recorded, transcribed, and analysed according to the principles of Template Analysis as described by King (2004).

[33] Apart from the interviews data sources include relevant articles within nutritional science, public debates, commentaries in news papers and journals, and administrative as well as juridical statements and exchanges.

In the attempt to live up to their role as scientific advisers, experts may appear more certain than is warranted by the scientific findings. This is described by one of the scientists:

> Scientist: Many people have such a simplistic conception that something is right and something is wrong.
>
> Interviewer: yes...?
>
> Scientist: But that is not the way it is in the results we achieve. In our results something indicates one direction while other things indicate other directions. But very often we are required to report clearly what is right and what is wrong. Often that is not possible. And... well... at the same time you are also...at the same time you perhaps also signal that you are more certain than actually the case, because sometimes you have to... Consider the analogy between a doctor and his patient. If the question is whether the patient should have an operation, sometimes you end up saying that an operation is the right thing to do – even though you know in your heart that you are uncertain about it.

Another scientist also speaks of public advice-giving and the expectation of certainty:

> Scientist: They ask you to read a crystal ball... but it is really difficult to do this in a balanced way.

An impression of certainty can be given in various ways. Individually it may be a matter of keeping doubt to oneself as mentioned in the passages above – presenting firm conclusions with conviction and authority. Collectively it may be a matter of approaching the public via a scientific consensus that does not convey the underlying scientific controversies and disputes. In the interview study, scientists emphasised the importance of scientific consensus in public. As one said:

> Scientist: We cannot afford to communicate in disagreement, when it is about... when the issue is what the experts know. I mean they are supposed to have the most knowledge about a given field. What kind of advice are they able to give? They *have* to provide a consensus. Minority views are no good... they make the advice useless.

The level of collective advice thus introduces a further complicating factor. While the individual adviser has to deal with scientific uncertainty, as we have seen, collective advice-giving also may involve the management of divergent views among the experts.

Discussion

According to the interview study, scientific advisers in the field of nutrition feel there is an expectation that they will give the appearance of certainty. This corresponds with a more general observation at the theoretical level that scientific advisers are under pressure to present an appearance of certainty (Kaiser, 2005; Rip, 1985; Rip, 1996). They are said to be faced with expectations that often require them to adopt and defend firm conclusions even when uncertainties exist. This push in the direction of certainty is also behind a certain 'professional ideal' of consensus that scientists as public advisers have been said to pay tribute to (Rip, 1985). According to this ideal, scientists should never disagree about the facts in public. They should rather refrain from speaking out until they have reached a solid consensus among themselves – either by means of a genuine agreement, or if necessary by suppressing minority views.

According to Rip (1985, 1996) this pressure to appear certain persists because a number of interests are thereby served. The air of certainty may, for instance, be useful to experts because it bolsters their own standing and that of their area of expertise – something which may be very important for research funding and public impact. It can also be seen as a defence against critics of scientific power and authority. Moreover, the appearance of certainty may be useful to those officials, decision makers and parts of the industry that invoke science to legitimise or back up their claims, or to undermine the claims of their opponents.[34] The association of scientific advice and certainty may also be reinforced by journalistic practices that tend to downplay or even omit qualifications and uncertainties when reporting the claims of scientists.

Although these suggestions contain some truth, they may not capture the whole picture. To appear certain could also be seen as a response to a certain consideration over the public interest. For the public it may often be hard to assess scientific uncertainty, i.e. to judge the relative weight and importance of the various kinds of scientific uncertainty. As compared with scientists, ordinary citizens often do not have the same informed opportunity to assess scientific doubt and uncertainty. Something similar occurs when rival experts disagree in public. In this case it may be impossible for the public to assess whether there are rational grounds for believing one expert rather than the other (Goldman, 2001). They only have recourse to such things as performance in the media, personal charisma, and the like.

At this point, scientific advice to the general public differs substantially from individual counselling. In individual counselling (e.g. in the relation between doctor and patient) it is possible for the expert to explain doubts and uncertainties and thus relate the provision of information to the particular circumstances at hand. Scientific advice to the general public does not come with the same opportunities. There is no automatic opportunity to follow-up on, or explain, uncertainties. One way or another uncertainty must be managed at the onset.

Hence, insofar as scientific advisers are genuinely interested in the successful provision of advice, it is perfectly understandable that they attempt to avoid situations in which the public (for rational reasons of their own) are not able to take their advice – either because they do not have an adequate understanding of the issues to make a proper assessment of any scientific uncertainties, or because there are too many plausible yet conflicting scientific advisers speaking at the same time.

The question, however, is whether an appearance of certainty is the best way to meet these difficulties. To answer this question it might be a good idea to consider what is truly valuable about scientific advice. This cannot be the provision of certain, indisputable knowledge, since this is often difficult or impossible even for experts to give.[35] What experts can provide are qualified judgements regarding their domains of expertise and critical comment on these judgements. This is no small feat – indeed, such information is often indispensable in that the relevant information is otherwise unavailable to non-experts. The value of scientific advice, we argue, lies in the provision of the best judgement of the adviser, i.e. a judgement that proceeds to an unambiguous recommendation that is clearly based on an assessment of the available knowledge within the field *including* any potential uncertainties. This is not a matter of deciding on people's behalf. On the contrary, recipients of such advice have to exercise autonomy and judgement in choosing to either accept the recommendation or act on a different assessment. In this way there need not be a tension between clarity of guidance, on the one hand, and scientific doubt and uncertainty, on the other. It does not follow that scientific advice should be construed as the provision of certainty. Rather, it may be possible for scientific experts to

[34] Note however that scientific uncertainty – just as well as scientific certainty – can be used strategically in the service of political or industrial interests.

[35] Though not always. There are indeed areas in which the scientific knowledge base is overwhelmingly well established, i.e. secure.

convey scientific uncertainty while at the same time delivering clear advice, i.e. to steer a middle course in which the assessment of uncertainty is neither downplayed nor totally left to people's own management.

A further complicating factor, however, is the relationship of one scientist to other scientists. Scientific experts giving their best judgement may disagree; and if the public is presented with rival experts that contradict each other, in their best judgements, the value of advice-giving (clarity of guidance) will be compromised – it will not be rational for the general public to use such advice to make up their own minds about what do to. For this reason, attempts to secure scientific consensus might still be worthwhile. Such consensus can be seen as a way of ensuring that differences of opinion have, at least minimally, been debated before the public. Thus it is the goal here, not to suppress minority views or relevant scientific uncertainties, but to reach an agreement of a higher order: a common recommendation that nevertheless reveals some disagreements.

Conclusion

Scientific advice-giving to the public often involves a tension between fulfilling the expectation of useful guidance and being true to the science on which the advice is based. In this paper we have argued that there are things to retain and things to discard in current thinking about this tension. On the one hand, there is no need to see a conflict between clarity of guidance and scientific uncertainty. The provision of clear advice need not be tied to an appearance of certainty. On the other hand, there is good reason for scientists to try to obtain a scientific consensus – not in order to suppress minority views, but because it may be too difficult for recipients to make up their own minds on the basis of conflicting advice.

References

Goldman, A.I. (2001). Experts: Which Ones Should You Trust? *Philosophy and Phenomenological Research*, vol. LXIII, No. 1: 85 – 110.

Kaiser, M. (2005). Certainty and uncertainty in science. In: *Biotechnology-Ethics. An introduction.* Landeweerd, L., Houdebine, L., and Termeulen, R. (eds). Firenze: IAAS – EDAP. 135 - 146.

King, N. (2004). Using Templates in the Thematic Analysis of Text. In: *Essential Guide to Qualitative Methods in Organizational Research.* Cassell, C. and Symon, G. (eds). Sage Publications. 256 – 270.

Rip, A. (1985). Experts in public arenas. In: *Regulating Industrial Risks.* Otway, H. and Peltu, M. (eds). Butterworth and Co (Publishers) Ltd. 95 – 110.

Rip, A. (1996). Expert Advice and Pragmatic Rationality. In: *The Sociology of the Sciences, vol. II.* Nowotny, H. and Taschwer, K. (eds). Cheltenham: Edward Elgar Publishing. 294 – 310.

Controlling biodiversity? Ethical analysis of the case of swine fever and wild boar in Denmark

Christian Gamborg and Peter Sandøe
Danish Centre for Bioethics and Risk Assessment (CeBRA), Royal Veterinary and Agricultural Universty, Rolighedsvej 25, DK-1958 Frederiksberg C, Denmark, chg@kvl.dk

Abstract

Biodiversity has high priority these years – not many would argue against a concern for biodiversity. However, our prioritising of biodiversity is dependent upon the level of nature we are addressing. Big mammals are lacking in Danish nature. They have been here before, why not reintroduce them or allow for natural migration to promote biodiversity and wildness? But a "richer" nature is not the only value at stake. Denmark also has a position as the world's largest exporter of pork. Wild boar carries a risk of acting as a vector for infecting pigs in livestock production with classical swine fever (CSF). At face value it seems like a classical conflict between considerable economic interests and profound conservation concerns. To get to a full understanding and open up for a more comprehensive discussion we need to address the embedded ethical issues. What degree of control are we ready to exert and how do we really value nature? The paper argues that besides the distinction and potential conflict between nature as a value and the common good, there is also an interesting distinction to be made between "micro" nature – including the virus that cause classical swine fever – and "macro" nature – medium sized flora and fauna. Predominantly, micro nature is something we want to control, whereas we want macro nature to be let loose. This dichotomy of 'nature' has to be addressed.

Keywords: agriculture, biodiversity, food export, risk assessment, values

Introduction

Should wild boars be reintroduced to Denmark to promote biodiversity? Through years of (systematic) selective breeding the wild pig has been transformed more or less into a "bacon and tinned food pig" (Clausen, 1968). According to Danish agriculture the Danish society is based on production systems in agriculture, utilising the majority of the land and if there is going to be additional wild animals in Denmark they should be fenced off (Andersen, 1996). Classic swine fever (CSF) is of increasing concern in Europe as wild boar appears to play a significant epidemiological role posing a risk of transmitting CSF virus to domestic pigs (Artois *et al.*, 2002) OIE (Office International des Epizooties) lists it as one of the highly contagious List A diseases. In 1997 more than 11 million pigs had to be destroyed in the Netherlands because of CSF, estimated to represent a net economic loss of 2 billion euros (Meuwissen *et al.*, 1999). Other countries in Europe have also experienced large-scale culling due to CSF between 1991 and 2001, making the disease one of the most feared of domestic animals (Laddomada, 2000). Free-ranging wild boars are perceived by agriculture as a ticking bomb under the world leading Danish export of pork with a value of 3.5 billion euros. But being a leading food exporter has a price.
Until not so long ago, wild animals were something to be shot and eaten or heavily fought. As a result many wildlife species became extinct or were driven away. Over-utilisation and loss of proper habitats took the beaver away 2,000 years ago. Wild boar was also an important game animal and was almost extirpated in the 17th century and completely gone in the beginning of the 19th century. Many other wildlife species which were not seen as directly useful or even

more so, as pest or direct threats to farming (or even to humans), such as wolf, or as a threat to other game animals, such as several birds of prey after pheasant was introduced as a game animal to Denmark or a threat to fishery such as Comorant, were brought down in the 19th and early 20th centuries (Olsen, 1997; Alex-Hansen, 1946). Similar developments have been experienced throughout Europe.

Now, things have changed. There is a pronounced "back to nature" trend, implying that our view of nature has changed from predominantly nature as a resource in a narrow sense, something to be moulded and tamed to our wishes or a direct threat to humans, to a view emphasising nature protection and biodiversity preservation (Gamborg and Sandøe, 2004). After the use of judicial and biological instruments such as protection of species acts and habitat conservation, the most recent development is comprehensive ecological restoration, including species reintroduction (Goulding *et al.*, 2003; Leaper *et al.*, 1999; Nolet and Rosell, 1998;).

In Denmark nature protection organisations have put forward the suggestion that free-ranging wild boar should be reintroduced as part of a national biodiversity strategy. In addition, some animals from the neighbouring country Germany, where wild boar in some parts is becoming abundant because of a series of mild winters and the increasing use of maize on the fields, may increasingly migrate to Denmark.

It would seem that a "richer" nature with a higher abundance of plant and animal species and more variety in structures would be considered unconditionally a common good. However, not everybody shares the view that the return of extinct species – such as wild boar – is in fact to be promoted. This is partly because somebody has to provide land and that somebody has to carry the risk of more damages – and diseases, such as classical swine fever. To open up for a more comprehensive discussion we need to ask: What degree of control are we ready to exert and how do we really value nature?

Wild boar in Denmark

There is no free-ranging wild boar in Denmark, anymore. Wild boar is only found in Denmark in fenced off populations in parks where time and again some of the animals escape. But in 1995 Danish agriculture became aware of a population of 23 free-ranging wild boars originating from fenced populations in the southern part of Denmark, close to the border of Germany. The question was: should this population be allowed to stay or should it be brought down to avoid the risk of infection to domestic pigs?

From a biological point of view it was argued that big mammals were lacking in the forests, and a new national forest strategy stressed the need for re-establishing the natural fauna to promote a more dynamic forest ecosystem. From a moral point of view nature conservationists argued that wild boar was part of authentic Danish nature and that wild boar ought to part of the more versatile flora and fauna, according to not only Danish nature politics but also to signal that Denmark is serious about the Convention on Biodiversity when we expect poor developing countries to preserve animals such as elephants.

However, from an economic point of view, Danish agriculture and pig industry strongly objected to free-ranging wild boar, seeing it as a severe pest animal which could cause farmers the loss of a great deal of money, even leading to personal tragedies by spreading CSF from wild boar to domestic pig.

At the end of 1995, the Minister of Environment decided to order all free-ranging wild boars in Denmark shot, apparently after substantial pressure from Danish pig industry, and despite that the minister publicly expressed his regret. However, the decision caused a media furore and two months later the decision was revoked. However, near the end of 1996, a new decision was taken: bring down the wild boars with all means possible (Steinar, 1997). All wild boars were shot. Tests later revealed that none of them were carriers of any transmittable diseases. Danish agriculture

expressed its willingness to pay shooting prices of approximately 250 euros per wild boar. Since then, there have been no reported sightings of free-ranging wild boars in Denmark.
In 2005, a comprehensive risk analysis of free-ranging wild boar was concluded (Alban *et al.*, 2005). The remit of the group performing the analysis was to assess the risk of getting CSF in Denmark (where the last incident was in 1933) as a result of free-ranging wild boar compared to existing risks. According to the assessment, the risk of CSF among domestic pig may increase a little if free-ranging wild boar is to be found in Denmark again. As long as there is no swine fever north of the Kieler kanal, the probability of transmission of the disease through migrating wild boar is insignificant. Another mode of transmission of CSF could be through the deposition of tourists' meat waste in the southern part of Denmark where – if there was a free-ranging population of wild boar – wild boar would eat the infected meat. According to the risk analysis, however, the most likely way that CSF may be transmitted to domestic pigs, is through animal transports, import of breeding animals or semen, illegal import of meat and hunters going hunting abroad (ibid.) – i.e. not related to wild boar.

Controlling biodiversity – ethical issues at stake

On the face of it the Danish case could be seen entirely as a classical conflict between nature conservation and economic interests with two major groupings, one valuing production and one valuing nature. But there is more at stake, which may change the power relation and which may change the way we look upon conflicts like these. As a whole, society has acquired a broader view on the value of nature, which is of a broader, non-use kind. So no-one denies the value of wild nature. It is more a question of placing different weights on different values, and here, it seems, one is still influenced or committed by one's place and function in society (e.g. urban dwellers versus farmers).
However, despite this more encompassing attitude to nature, ambivalence exists. Some parts of nature – such as swine fever – must still be combated with all means possible, whereas other parts deserve our concern and protection. In the case of the wild boar, we may answer these questions by distinguishing two types of nature: One type could be called "micro" nature, and the other "macro" nature. There is a difference as to how and how much we want to control the type of nature in question.
"Micro" nature signifies the organisms and processes which are not visible to the naked eye, hereby including the virus that causes CSF. Traditionally, we examine this type of nature through comprehensive risk assessments, i.e. evaluating hazards, probabilities and its impact on other parts of nature, like bigger mammals. Consequently, management of "micro" nature is essentially risk management, i.e. identifying disease carriers, mode of transmission, and hence potential culling of such carriers and strict control of the immediate environment – and communicating the measures taken, entering a dialogue with stakeholders. No intrinsic value seems to be attached to nature at this level.
"Macro" nature is the "big" nature – medium sized flora and fauna – the kind of nature which eight out of ten urban dwellers in Denmark want to visit, experience and interact with. In this case, it is wild boar as a nature and landscape element which is the focal point. No-one denies that more biodiversity here is of positive value. Assessing this type of nature is more a question of examining the impact of other (e.g. human) actions on flora and fauna through biological studies, socio-economic surveys of attitudes and preferences and practical considerations. Management, including damage control, has its focus on catching, culling and fencing off areas.
Clearly, there is a schism here. From the lay person perspective, most of us expect that there is a high degree of control of the "micro" nature, that is to say that potential damages and diseases are managed. One the other hand, there is an outspoken wish for letting the "macro" nature run loose creating a greater variety of species, getting a more dynamic nature. From the scientific

and managerial environment there seems to be a much more risk averse attitude when it comes to "micro" nature, contrasting the position on "macro" nature where there seems to be a higher degree of accepting that we don't really know how things are going to develop.
The problem is that the connection between these two types of nature, "micro" and "macro", is not perceived. "Micro" nature is discussed within specific academic disciplines, such as etiology and epidemiology whereas discussions on "macro" nature take place within ecology, geography, landscape management and so on. A common discussion of nature, and its management, is absent. If we begin to see "micro" and "macro" nature as connected we could be better at making decisions on where to draw the line in controlling nature, making choices about the control of biodiversity and management in general, and better equipped for entering fruitful discussions.

Conclusion

The current debate about wild boar and classical swine fever seems primarily perceived as a traditional conflict between economic interests and nature conservation. But there is another layer in the debate, asking how do we in reality look upon nature, and to how large a degree do we want to control it? Predominantly, "micro" nature is something we want to control, whereas we want "macro" nature to be let loose. This dichotomy of 'nature' has to be addressed. To address this layer of ethical issues and underlying value questions contributions are needed from the "micro" nature as well as the "macro" nature arenas. And moreover, it should ideally be done in a way which does not see the two discussions as separate from each other.

References

Alban, L., *et al.* (2005). *Classical swine fever and wild boar in Denmark: A risk analysis.* Danmarks Fødevareforskning. ISBN: 87-91587-01-8. 118 pp.

Andersen, V. (ed.) Miljø- og Energiministerens konference med organisationerne om natursyn. Miljø- og Energiministeriet, Copenhagen.

Alex-Hansen, B. (1946). Danmarks vildsvin. *Dansk Jagttidende,* 63(9): 121-122, 136-137.

Artois, M., Depner, K.R., Guberti, V., Hars, J., Rossi, S. and Rutili, D. (2002). Classical swine fever (hog cholera) in wild boar in Europe. *Rev. sci. tech. Off. int. Epiz.*, 21(2): 287-303.

Clausen, H. (1968). Fra vildsvin til bacon- og konservesgris. *Tidsskrift for Landøkonomi*, 155: 143-153.

Gamborg, C. and Sandøe, P. (2004). Beavers and biodiversity: the ethics of ecological restoration. In: *Philosophy and Biodiversity*. Oksanen, M. (ed.). Cambridge University Press, New York:. 217-236.

Goulding, M.J., Roper, T.J., Smith, G.C. and Baker, S. (2003). Presence of free-living wild boar *Sus scrofa* in southern England. *Wildlife Biology*, 9, Suppl.: 15-35.

Laddomada, A. (2000). Incidence and control of CSF in wild boar in Europe. *Veterinary Microbiology* 73(2-3): 121-30.

Leaper, R., Massei, G., Gorman, M.L. and Aspinall, R. (1999). The feasibility of reintroducing Wild Boar (*Sus scrofa*) to Scotland. *Mammal Review*, 29: 239-259.

Meuwissen, M.P.M., Horst, H.M., Huirne, R.B.M. and Dijkhuizen, A.A. (1999). A model to estimate the financial consequences of classical swine fever outbreaks: principles and outcomes. *Preventive Veterinary Medicine*, 42: 249-270.

Nolet, B. and Rosell, F. (1998). Comeback of the beaver Castor fiber: An overview of old and new conservation problems. *Biological Conservation* 5: 165-173.

Olsen, S. (1997). De fortabte dyr vender tilbage. *Kaskelot,* (115) : 2-31.

Steinar, M. (1997). Kan vi være os selv bekendt? *Jæger* 10/97: 96-97.

The moral basis of vegetarianism

Espen Gamlund
Department of Philosophy, Classics, History of Art and Ideas, University of Oslo, P.O. Box 1020 Blindern, N-0325 Oslo, Norway, espen.gamlund@ifikk.uio.no

Abstract

During the last decades there has been an ongoing ethical debate about vegetarianism. This debate has attracted several philosophers, theologicians and others. The purpose of this paper is to provide an account of the moral basis of vegetarianism. Generally speaking, there are two dominant approaches to the moral basis of vegetarianism: Peter Singer's animal welfare ethics, and Tom Regan's animal rights ethics. According to Singer, we have an obligation to become vegetarians because factory farming inflicts suffering on sentient animals. Moreover, this suffering is avoidable because meat is not necessary to provide us with a healthy diet. According to Regan, we have an obligation to become vegetarians because animals are experiencing subjects of a life with an inherent moral status value, and they have a moral right not to be caused unavoidable harm. I will offer a third approach that I will term a recognition theory of moral status. This theory implies that the way we conceive of ourselves as human beings constitutes the normative basis for recognizing the moral status of animals. I will argue that the commitment to become vegetarians can be defended in view of this theory.

Keywords: vegetarianism, moral status, recognition theory of moral status, self-conception, respect and concern

Arguments for vegetarianism[36]

In the contemporary debate, a number of different arguments or justifications have been put forward in favor of vegetarianism. In *Deep Vegetarianism*, Michael Allen Fox lists ten possible grounds for accepting a vegetarian lifestyle[37]:

1. health;
2. the moral status of animals;
3. impartiality or disinterested moral concern;
4. environmental concerns;
5. the manipulation of nature;
6. world hunger and social injustice;
7. interconnected forms of oppression;
8. interspecies kinship and compassion;
9. universal nonviolence;
10. spiritual and religious arguments.

As pointed out by Fox, these arguments should be regarded as converging and mutually reinforcing. By themselves, these arguments may have little weight, whereas together they

[36] I am indebted to Kristian Skagen Ekeli for valuable comments and suggestions to earlier drafts of this paper.

[37] Fox (1999), p. 61. I believe this list takes into account the five arguments for vegetarianism offered by William O. Stephens (1994), (i) the argument from distributive justice, (ii) the argument from environmental harm, (iii) the feminist argument from sexual politics, (iv) the argument from moral consideration of animals, and (v) the prudential argument from health.

make a strong case for vegetarianism. In this paper, I will focus on arguments from the moral status of animals.

The moral status of animals

The argument from the moral status of animals is perhaps the most common argument within the philosophical discourse on vegetarianism. Several different positions have been developed during the last three decades, two of which are worth examining briefly: Peter Singer's animal welfare ethics, and Tom Regan's animal rights ethics.[38]

Singer's animal welfare ethics

Singer and Regan both agree that we have a moral obligation to become vegetarians, but they disagree concerning why we have this obligation. Singer was perhaps the first philosopher to introduce the topic of animal considerability and vegetarianism. In his book *Animal Liberation*, Singer argues that we have an obligation to become vegetarians. We should become vegetarians because today's factory farming inflicts suffering on sentient animals. This argument proceeds from the following two premises. (1) All sentient beings have moral status. The ability of having conscious feelings of pleasure or pain, i.e., sentience, is taken to be the morally relevant property for the ascription of moral status. (2) Humans do not need meat for a healthy diet, i.e., meat is a trivial need. Proceeding from these premises, Singer argues that factory farming is morally wrong because it inflicts unnecessary suffering on sentient animals.

Some have objected that Singer's position does not give a very strong protection of the interests of animals. This is because Singer is a utilitarian. According to the principle of utility, one should choose the course of action which increases the overall welfare of all affected parties, i.e., all sentient beings – humans and nonhumans. This implies that it is ethically permissible to produce and eat meat if this can be done without inflicting pain and suffering on the affected animals, and if this increases the sum of utility or welfare of the affected parties. Regan rejects this utilitarian line of argument.

Regan's animal rights ethics

Regan has developed his animal rights ethics in opposition to Singer's animal welfare ethics. As I said, Regan agrees with Singer on the conclusion that we ought to become vegetarians, but he does not accept the utilitarian basis of Singer's position. In his book *The Case for Animal Rights*, Regan develops a deontological ethic inspired by Immanuel Kant. Regan argues that there are good reasons for ascribing inherent moral status value to all living beings who can be "subject of a life", which presupposes self-consciousness. This includes all normal mammals of one year or more, and also birds. According to Regan's "principle of respect" all moral subjects who have inherent moral status value should be treated as ends in themselves, never merely as means.[39] This seems to rule out the above mentioned implications of Singer's utilitarian position. In view of the principle of respect, it is problematic to use animals as means for food production, even if they were killed painlessly. Regan assumes that subjects of a life have a *prima facie* moral right not to be caused avoidable harm. And, since humans can avoid eating meat without being made worse off, we have a moral obligation to become vegetarians.

Regan's position appears to offer a better protection of animals than Singer's utilitarianism. However, Regan's defence of vegetarianism rests on the controversial philosophical assumption that animals can be right-holders. In my opinion, this controversial assumption is unnecessary in order to come up with a cogent justification of the moral status of animals and vegetarianism. While

[38] See Peter Singer (1975 & 80) and Tom Regan (1980 & 83)

[39] See Regan (1983), pp. 248-250.

I accept the principle of respect, I think that the assumption that animals can be right-holders should be rejected – primarily because this represents a contested basis for vegetarianism.

A recognition theory of moral status

So far I have looked at two alternative justifications for vegetarianism. I now wish to consider a third approach that can be regarded as a recognition theory of moral status.[40]

The question of moral status

It can be argued that moral respect and concern for humans implies respect for human freedom and self-determination, as well as a concern for human welfare. Besides, as human beings we are in possession of an equal inherent moral status value and dignity which demands moral respect of a particular kind. This dignity and moral status is attached to our personhood and capacity for agency, rationality, reasoning, etc. Obviously, nonhuman beings lack these abilities. In so far as these abilities are considered to be necessary conditions for the ascription of moral status, then the absence of these abilities will be a sufficient condition for denying moral status to them. As a result, animals are often regarded merely as means to our ends, such as food, clothing, cosmetics, etc. In what follows, I will develop a response to this anthropocentric position.[41] The central idea of the proposed recognition theory of moral status is that the way we conceive of ourselves as human beings constitutes the normative basis for either the rejection or the approval of ascribing moral status to animals. Suppose we form and develop certain kinds of attitudes or dispositions. If we assume, for the sake of argument, that these attitudes express respect and concern for something, then I believe we can make the further assumption that those things, toward which these attitudes are taken in a direct manner, are ascribed moral status. As regards the notions of moral status value, I suppose one can say that in so far as a thing is ascribed a value in itself by a person, it has *intrinsic value*. But in so far as there exist an agreement on the thing's value among a group of persons, and this belief is institutionalised in the group, the thing is ascribed an *inherent value* as a moral status value.[42] This inherent value is then ascribed to these beings in their own right.[43] I believe that this can provide a basis for extending our attitudes of direct moral respect and concern from humans to nonhumans.

Descriptive similarities and normative relevance

A precondition for ascribing moral status to animals is to develop what I will term a wide self-conception, in contrast to a narrow self-conception. The *narrow self-conception* implies that one focuses on whatever separates oneself from animals. For example, only humans can be responsible moral agents with the ability to make free and informed choices. Using the terminology of Aristotle, this narrow self-conception makes it likely that one will only identify with the rational (*lógos*) part of one's soul, which specifically belongs to human beings. And, in so far as this is the case, one will perhaps regard these differences as morally relevant for differential treatment of humans and animals when it comes to respect and concern. This seems to be a prominent way of understanding oneself as humans today, and I believe that this influences the way in which animals are treated in current food industry.

---

[40] This recognition theory of moral status is inspired by the endorsement theory of Christine Korsgaard (1996), as well as the recognition theory of moral validity proposed by Jon Wetlesen (1999).

[41] Singer and Regan's positions are also developed as responses to this argument.

[42] I borrow the term "moral status value" from Wetlesen (1999).

[43] I have developed this line of thought in more detail in Gamlund (2006).

The *wide self-conception*, on the other hand, implies that one has a much wider conception of oneself as a human being, where emphasis is placed on those abilities one has in common with animals. Turning again to Aristotle, we can infer that apart from the rational part of the soul, which belongs specifically to humans, one has certain abilities in common with animals, such as consciousness, sentience and striving. If one adopts a wide conception of oneself, these abilities are considered to be morally relevant properties for the ascription of moral status. Inherent value is ascribed to those beings who have these abilities. This includes self-conscious and conscious animals, sentient animals, and other non-conscious beings with a striving for self-preservation. This argument rests on two central assumptions. First of all, a descriptive assumption regarding the existence of striving, perception, suffering etc. Secondly, a normative assumption about the moral relevance of these properties. My point is that one will recognize certain factual properties as normatively relevant grounds for the ascription of moral status in so far as these properties correspond with one's conception of oneself as a human being.

A wide self-conception of oneself is a fundamental precondition for granting moral status to animals, and for treating them with direct moral respect and concern. This does not necessarily lead to the assumption that humans and animals have an equal moral status value. Rather, like myself, I suppose that many people would endorse a gradual position in view of these considerations.[44] Anyway, in order to have a direct moral concern and respect for animals, one must identify with the abilities and properties that these beings have in common with us. If this attitude is adopted, it is reasonable to apply the golden rule to actions and omissions affecting animals.[45] For example, this implies that one should not inflict pain and suffering on animals which one would not want others to inflict on oneself, if one had been in their situation. In my view, there are also good reasons to accept the principle of respect if one develops these attitudes.

If one accepts the preceding line of reasoning, what implications will this have for the issue of vegetarianism? Some might argue that vegetarianism does not follow from the outlined argument. Something can be said in favour of this position. One can object that it is morally permissible to eat animals provided that they are killed painlessly. Moreover, one can accept my line of reasoning without reaching the considered judgement that one should become a vegetarian, since there is a morally relevant difference between the moral status of persons and animals. So, if animals used for food production enjoy good life conditions, and are killed painlessly at the end of their lives, a case can be made for treating humans and animals differently.

In response to these objections I would emphasize two points. First, to have direct moral respect and concern for animals would rule out that we regard them solely as means to our ends. Given that animals are ascribed moral status they should be treated as ends in themselves, not merely as means. In my opinion, this makes it morally problematic to raise animals (under good conditions) for the purpose of killing them (painlessly), and eating them. Second, although I assume that there is a morally relevant difference between persons and animals, it can be argued that meat is a non-vital human need, and that it is unnecessary to use animals solely as means for food production in order to satisfy vital human needs. In other words, in our present world human non-vital interests and needs are given priority over the interests and vital needs of nonhumans.

If one acknowledges the moral status of animals, one should renounce one's non-vital needs and interests if the pursuit of these interests presupposes the suffering, or even death, of animals. In fact, if one were to fulfil those needs and interests, it would presumably imply recognising the mere instrumental value of animals. It seems to me that one cannot maintain the view that meat-eating is morally permissible provided that animals are killed painlessly, while at the same

[44] A gradual position is developed by Wetlesen (1999).

[45] I am indebted to Jon Wetlesen for this suggestion.

time granting these animals moral status of the kind I have suggested. In light of the preceding considerations, one should become a vegetarian.

Concluding remarks

Central to the proposed recognition theory of moral status is that our self-conception constitutes the normative basis for recognizing the moral status of animals. In so far as one has a wide conception of oneself, one should acknowledge that there are certain similarities between oneself and animals, such as a capacity for suffering. I have argued that this recognition theory can provide a basis for vegetarianism. In this paper, I have only briefly outlined an argument for this conclusion, and I will develop this on another occasion.

References

Fox, M. A. (1999). *Deep Vegetarianism*. Philadelphia: Temple University Press.

Gamlund, E. (2006). "Who Has a Moral Status in the Environment? A Spinozistic Answer," *The Trumpeter*, forthcoming.

Korsgaard, C. M. (1996). *The Sources of Normativity*. Cambridge: Cambridge University Press.

Regan, T. (1980). "Utilitarianism, Vegetarianism, and Animal Rights," *Philosophy and Public Affairs*, Vol. 9, No. 4, pp. 305-324.

Regan, T. (1983). *The Case for Animal Rights*. Berkeley: University of California Press.

Singer, P. (1975). *Animal Liberation, A New Ethic for Our Treatment of Animals*. New York: Avon Books.

Singer, P. (1980). "Utilitarianism and Vegetarianism," *Philosophy and Public Affairs*, Vol. 9, No. 4, pp. 325-337.

Stephens, W. O. (1994). "Five Arguments for Vegetarianism," *Philosophy in the Contemporary World*, Volume 1, No. 4, pp. 25-39.

Wetlesen, J. (1999). "The Moral Status of Beings Who are not Persons; A Casuistic Argument," *Environmental Values* 8/3, pp. 287-323.

Problems of principlism

Laurens Landeweerd
University of Maastricht, Faculty of Health and Science, Dept. of Health Ethics and Philosophy, PO box 616, 6200 MD Maastricht, l.landeweerd@zw.unimaas.nl

Abstract

Principlism, or principle based ethics, was one of the first approaches in bioethics. An important problem in principlism is that the interpretation of highly generalized principles to a certain practice with all its particularities, is not self-evident. Even with the idea of a reflective equilibrium and a specification of these principles, one still needs certain criteria that indicate how a principle is to be interpreted in practice. Part of the issue lies in the very idea of a method of ethics. However, to reduce ethics to mere practical regularities would also imply a problem of justification. So, the move away from theory has not solved this foundational problem either. Especially in regard to the ongoing GM-controversy it will be necessary to develop a democratic alternative for principlism in food ethics. Maybe ethics can only partly be made explicit, in the form of rules, guidelines or principles, with the sole purpose of clarity or uniformity. But then, what will be left to justify our beliefs?

Keywords: regulism, regularism, principlism, pragmatism, pluralism, food ethics, applied ethics, GM-food

Introduction

Food ethics as a separate field is a relatively new branch in applied ethics. With the development of novel techniques, especially in genetics, food industry is rapidly stepping into a new age. We encounter problems with regard to safety and precaution, solving famine in development countries, the development sustainable ways to produce food, and animal ethics. Since these developments change the face of food production in agriculture and fishery in an unprecedented way, it necessitates ethical reflection on whether all aspects of these developments are desirable, as well as on how new technologies can be implemented in agricultural practice in a responsible way. This not only requires a reflection on the practical issues at hand, such as the risks involved in the development and implementation of novel techniques, ways of dealing with these through precautionary measures, accounting for questions of animal suffering, or questions of nature preservation. It also requires reflection on how to reflect on these issues.

The standard ways of dealing with ethical questions in other fields of bioethics are also applied to the area of food ethics. Since food ethics does not concern people directly it may seem obvious not to refer to the existing schools in bioethics as applied in medicine. However, indirectly, people are always involved in questions of agriculture and food. First of all, there are the farmers and fishers working in the industry, second of all, there are the consumers, thirdly, we have the bio-engineers and geneticists who are providing for novel developments in the field, and fourthly, we have the realm of politics. An overemphasis on issues like sustainable development might blur other ethical issues at hand. Differing values on food consumption necessitate an approach that acknowledges the pluralistic nature of the field. To impose a certain scientifically founded view on the production and consumption of food would not only work counterproductive, it would actually be unethical. However, to develop an ethic that accounts for this pluralistic nature of the field is more difficult than it seems.

The disadvantages of principlism in food ethics

In 2001, transgenic DNA was found in several traditional maize stems in Mexico. When this fact was made public, politic and public outcry followed instantaneously, causing a debate on the negative sides of the production of GM food that has not quieted down until now. Although scientists often motivate their own work as being beneficial for the 'common good', the fact that these GM products were developed in a partly commercial environment and the fact that scientists misrepresented many aspects of the development of GM food did not contribute to the acceptance of GM food (Scott, 2003), which pre-structured the debate in a direction in which scientists were suspected to have a double agenda, commercial as well as scientific.
The experience with GM food in, amongst others, the UK showed that one cannot dismiss with a dialogue with the public a priori (Durant and Lindsey, 2000). Because the public was ignored in the implementation of GM food in society, the public came to distrust the novel technique, and set back the development of GM-food for commercial purposes by a decade.
The main problem that flowed from the controversy about GM maize was that it created a high degree of mistrust towards the academia. Scientists were regarded as suspect. The inability of science to communicate with the public was a main influence on the further development of GM food.
Even in a crisis situation, one cannot dismiss with open dialogue. In 2002, in Zambia, 300.000 tons of food aid was offered from the USA to combat famine, but unfortunately, it was offered in the form of GM food (Adcock, 2005). The aid was refused, referring to the food as poison, since the Brittish Medical Association had voiced its concerns about GM food. If the West didn't want to offer GM food to their own subjects yet, that third world countries should not be used as a guinea pig, especially not in a crisis situation where one is already forced to take on any help needed.
The Zambian government spoke in a discourse of post- or anti-colonialism, whilst USA implicitly still embraced the colonial discourse; the rich west helping the poor south. In the academic debate as well as in society at large, one is stepping away from the 'benevolent Maecenas'-approaches to food aid to the Third World. This approach, although sometimes solving direct needs, did not contribute to the further development of structural changes in the countries in need. A dialogue with local governments on an equal level appeared more asked for than an imperialistic approach, that only helped the needy in case of direct emergency.
As in any science, in food engineering or biotechnology, one would be tempted to only praise the benefits of one's own field and disregard any comments coming from the layman, or the non-professional. However, the goals a scientist sets himself within the walls of his or her laboratory often differ widely from the values held in society. A scientist, for example, may feel he is contributing to the common good by designing a type of rice that can grow much faster, so it can be harvested within the time the rain season lasts in a relatively dry country. However, sentiments in that country itself may be adverse to using types of rice which genes have been 'tampered with'. Furthermore, some may feel that the rich west should not meddle with internal affairs in such a way.
How can one take into account the autonomy of groups and governments as well as individuals? And in which cases should one give precedence to the one, and in which to the others? It may seem that an account of autonomy as it is presented in principlist approaches of ethics can solve this problem. However, there are several problems in principlist approaches of ethics. The first versions of principlism in bioethics did not provide for a method of interpretation to arrive at the right conclusion. Other versions, such as specified principlism or the method of reflective equilibrium did address this problem. But in these versions, there is also still a problem of interpretation, since it is not stated how to specify or how to balance principles towards a

reflective equilibrium. So how can we involve stakeholders and the public in scientific decision making without crashing the development novel technologies?

Steering between rule and practice

Any applied ethics, food ethics included, forces us to steer between what I, by mouth of the philosopher Robert B. Brandom, (Brandom, 1994; Brandom, 2001) would call two reductionisms in ethics: the reduction of ethical contents to rules and principles that are loose-standing from the practical reality, and the reduction of ethical content to the regularities of behaviour, to the mere practice as it is already conducted. Brandom defines these two reductionism respectively as regulism and regualrism.

Regulist approaches to ethical questions suffer from a reductionism since the ethical content of a rule cannot be brought to bear upon a practice without adding another rule that is necessary to guide the way in which a principle is applied to that practice. This however, necessitates yet another rule of interpretation. The regress in this case is infinite.

In practice, regulism is a symptom of ethical approaches that run barren because of their dogmatic insistence on their truth value, without regard for the practice on which they need to be applied. One often encounters such barrenness in either dogmatic ethical standpoints, or in bureaucratic systems.

Regularism is a form of reductionism that leads from relativism to nihilism. A regularist would state that since one cannot formulate rules to guide our behaviour, we should restrict ourselves to the regularities of behaviour we already perform. This type of pragmatist reduction to reality as it is fails to give any justificatory explanation for our actions at all. It does not guide our actions, it only embraces our factual behaviour as it is.

Regularism is a symptom of ethical relativism, when it's thought through to the extreme. It can be found in quietist attitudes towards moral behaviour as well as in neo-pragmatic approaches to ethics. It attempts to account for the diversity in practice, and respect it as it is. I also tries to refrain from taking a position in which only the self-held truths are valued. It tries to embrace other views on the basis of an inbuilt relativism. However, in the end, there is no basis for any position to be justified. All is mere practice, there is no ground justification behind any practice.

To be able to develop a position that accounts for the diversity of the practice, without retreating in a mere descriptivist position, from which one can neither endorse, nor dismiss with different positions in normative practice, it is necessary to steer in between this Scylla and Charybdis of rule-reductionism and regularity-reductionism (of behaviour), one will need to give a proper place to the different opinions held by the stakeholders (farmers, consumers, politicians, scientists), without reducing all questions of ethics to questions of either democratic decision-making or merely anthropologically interesting footnotes. To be able to do that, one must steer in between an ultimately self-serving universalism, and an ethical relativism.

Principlism is a regulist approach to ethics. It states a certain rule without formulating in what way this rule is to be applied to a certain practice. Even if it states a method of interpretation – such a specification or balancing (in finding a reflective equilibrium between rule or principle and practice, one will still find that there is no further method of application of how to specify, or how to balance. Whose autonomy, for example, should be respected, and under what circumstances? How does one apply a principle of respect for autonomy on concrete situations? Where is the limit between informing the public and influencing their opinion? Adding another rule that would state how to specify a principle, or how to balance it to other principles and the concrete practice, would only bring with it a similar problem. It is an infinite problem.

To turn to neo-pragmatism or pluralism for an answer would also be problematic. The more one allows for the practice of the field of application itself, the less one is able to define what justification for the ethical judgements that arise from this practice. Abstaining from rational argumentation would render any ethical discussion void. Furthermore, if one looks at the pluralistic nature of practice, conflicting beliefs and values will never allow for a single ethical guideline to be reached.

An ethic like an edifying philosophy proposed by Richard Rorty (Rorty, 1979; Rorty, 1989) seems to suffer from a specific problem: "Rorty recommends the transition to this type under the aspect of fruitfulness. Edifying philosophy, he says, will lead to "new, better, more interesting, more fruitful ways of speaking" (M. 360). I am afraid, however, that abstinence from argument and fruitfulness go together badly. More precisely: it is exactly the exclusion of argument which seems to me to render the remaining practice of edifying philosophy regrettably unfruitful." And "[...] all processes of refutation, falsification, of argumentative discussion and clarifying dispute are eliminated. Hence it is to be expected that this results not, as Rorty says, in "keeping a conversation going" (M. 378), but rather in the opposite: in the silting up of these conversations, in which nothing is really at stake, in peaceable conversational murmurs, in discursive entropy." (Welsch, 2005). So, in a neo-pragmatist ethic one cannot even hold to some position or another oneself, since any ethical belief one were to hold, or any argumentation one would adhere to, to support ones beliefs, would need to be dismissed with, thus ultimately silencing any ethical debate as such.

Conclusion: Towards a dialogical ethics?

It appears principled ethics stands in the way of a dialogical approach, since, in its old form, it is structured a s a deductive ethics. So instead of coming to an ethical judgment through dialogue, one comes to an ethical judgment through derivation. This is a handicap also suffered from in the new form of principle based ethics. A turn to the societal practice itself would not solve the problem of interpretation, since it would not solve the problem of conflicting beliefs and value systems. It seems we need democratization on the level of stakeholder identification, in order for a dialogue to take place that allows for all dimensions of food production and distribution. In such a dialogue, one should remain open to changes in ones own belie system, without having an over-ironic position towards it.

In food ethics, one can observe a predominance of questions of Safety and precaution in the implementation of new production technologies. These questions flow forth from a debate on what is 'sound science' and what is 'sound technology'. However important these themes are, there are other ethical questions that are relevant, that don't appear within a mere scientific perspective. Questions of tradition, (post)colonialism, culture, and even ethnicity are at least equally important, though they appear to be non-food –issues. This, rather than the level of the ethical judgments themselves, is the level on which pluralism should be given its proper place.

References

Brandom, Robert B. 1994. Making it Explicit. Cambridge: Harvard University Press.

Brandom, Robert B. 2000 Articulating Reasons: An Introduction to Inferentialism. Cambridge: Harvard University Press.

Durant, J., Lindsey, N. (2000) "The Great GM Food debate - a survey of media coverage in the first half of 1999" Report 138 of the Parliamentary Office of

Science and Technology.

Kinderlerer, J. Mexico and genetically modified Maize. In: Landeweerd, L, ter Meulen, Houdebine, L. M. (2005) BioTechnology ethics; an introduction. Florence, Pontecorboli.

Adcock, M. Zambia Food aid scenario. In: Landeweerd, L, ter Meulen, Houdebine, L. M. (2005) BioTechnology ethics; an introduction. Florence, Pontecorboli.

Landeweerd, L Principlism and the Problem of a Deductive Ethics. In: Landeweerd, L, ter Meulen, Houdebine, L. M. (2005) BioTechnology ethics; an introduction. Florence, Pontecorboli.

Rorty, R. (1979) Philosophy and the Mirror of Nature. Princeton, NJ: Princeton University.

Rorty, R. (1989) Contingency, Irony, and Solidarity. Cambridge: Cambridge University Press.

Scott, D. Science and the Consequences of Mistrust: Lessons from Recent GM Controversies. In: Journal o Agricultural and Environmental Ethics. 2003; 16(6): pp. 569-582.

Welsch, W. (2004) "Richard Rorty: Philosophy beyond Argument and Truth?" From: The Pragmatic Turn in Philosophy, eds. William Egginton and Mike Sandbothe, Albany: State University of New York, pp. 163-185.

Passive nature and safe food?

Svein Anders Noer Lie
University of Tromsø, Dep. of Philosophy, 9037 Tromsø, Norway, sveinn@sv.uit.no

Abstract

Knowledge produced within science becomes more and more indispensably important in food-production. In a fundamental sense I see this as highly problematic. There is a difference in how science and the food-eater see nature that is both a difference in content and level of understanding. In general the debate about food ethics and food safety takes it for granted that science delivers knowledge about cause and effect while the public or professional ethicists make judgements about the favourability of the product. In this paper I use much space on explaining why I believe that the scientific knowledge is defined by an already fixed opinion of the ethical aspect of the product – an opinion that cannot be suspended. The levels of ontology and ethics are confused first and foremost because the ontology of science demands that these to levels should to be separated and not confused. This becomes especially apparent in debates about products originating from biological sciences, and I give some examples of how this leads out in absurd opinions about such products. It is crucial to understand the ontology of science, which I call "the anti-telos maxim", to understand the discussions on bio-ethical and food safety questions.

Keywords: teleology, ontology of science, is and ought, risk-evaluations

Introduction

Food-production is, of course, much older than modern science, and has up till recent times been an integrated activity where knowledge, ethics, religion and culture have converged in the manifested products and the praxis of production. When science now becomes steadily more crucial as underpinning for the production of food, this activity will be increasingly more *modernised* in a Habermasian sense. The discussion of ethical and social questions that halters behind the introduction of new agri/aqua-cultural methods is a clear example of this. More specifically we are experiencing the tensions between a scientific concept of nature and other concepts of nature.

In this paper I particularly look at the concept of risk (and safety) to reveal the inconsistencies between *the scientific ontology of nature* that is a condition for the whole invention of new methods and other ontologies that have to come into play when the products are introduced "into the world". It is important to understand the full range of the ontology of science, which I call the anti-telos maxim of nature, to see why these inconsistencies are more than the result of the lay-peoples' failure to adjust to reality. I use modern biotechnology as an example.

Telos and anti-telos

For Aristotle the question about the nature of nature was given an answer through his *teleology*. Every being had its own *telos*, its own "inner end" to realise. The realisation of this inner end was regarded as a *natural movement*. For Aristotle, nature had a direction, which humans or other beings could cross. Galileo Galilei challenged the teleology of Aristotle by showing that the stone was "searching" to the ground not because of an inner want/shortage (steresis), but because of an *outside* force - a force that could be described through mathematics, and whose principle could be formulated as a law of nature. Operating with *inert matter* as a starting point

formed a need for these "active forces" to originate from outside the being itself. For the world does not stand completely still either. Change exists, and has particular modes. In this situation the powers of nature where "afforded" natural laws. Laws (i.e. powers) from outside govern the behaviour of the inert matter. Hence, natural beings as such could not be said to have a Nature. As there only exists states that either govern or is governed – beings of nature have to be called heteronomous. Subsequently, from Galileo and on the being of beings were thought of as *passive* (Ellis, 2002, e.g. p. 62,). (Something that, if more closely examined, of course is impossible. It is a claim that nevertheless has strongly survived in an "intermediate layer" as a maxim. But I will not go further into this here).
This is the core situation of the deep dualism that was only waiting to be systematically exploited by someone like e.g. Descartes or Hume. Further on Descartes and others tried to widen the extension of this basic view of nature, developed within the scope of astronomy, into biology, but it was Darwin who first convincingly uncovered specifically how biology, evolution and living natural kinds could fit the notion of anti-telos by making it plausible that biological nature had no purpose or end of its own. The equivalence in biology to the "falling stone" in physics could be found in the view that comprehends the evolving organism as a *passive* result of "inner determinism" (genetic coding + blind mutations) and outer pressure (selection), most consequently worked out by Jaques Monod in "*Le Hasard et la Nècessitè*" (1970).
The theory of evolution is engaged in answering the question about how selection works. For biology as well as natural science in general, it is as I have pointed out outside the field of vision to search for "right nature", "wrong nature" etc. Biology has no criterion to claim that there should not be "a desert in the Amazon Jungle", and it will never find such a criterion. To make it quite clear: Biology as a science is a part of the dualism that results in the delineation between: facts and norms, object and subject, passive and active, nature and man, natural and artificial, is and ought etc. - a dualism that makes up the foundation of modernity in general and in science in particular (Latour 1996). On the other hand: the problem with the general scheme of modernism is easily revealed in biology and is in particular illustrated within the science of molecular biology.
Apparently the object of molecular biology does not behave like a totally passive being, but its behaviour is not easily fitted into the category reserved for humans either, namely the category of intentionality. When an organism "is silencing" a gene that the molecular biologist tried to fit into the genome, this is caused neither by "a natural law",[46] nor an intentional act. Even though this must be evident to everybody, the gen-tech-industry and the whole prospecting of future economy, grown out of the *science* of molecular biology, goes on by leaning on the deterministic assurances made by the anti-telos maxim.
I have tried to put forward a short version of the full range of what I regard as the ontology of science. The further enterprise in this paper is to see how much influence this anti-telos maxim have or should have on the *role* of science and *accordingly* on the role of the "normative discourse" as well. This is crucial to uncover if we assume that the field of food-production will be increasingly more dependent on science-based knowledge.

The value of anti-telos

The requirements of *anti-telos* make it impossible to ascribe any sort of moral value to the "real" reality. Something principally passive cannot be attributed any conduct as such, and much less

[46] As I have made clear, I do not support the notion "Natural Law", although I use it here to make a point. In my opinion it is not the law/regularity that have causal powers, rather it is the dispositions of beings. The ontology of dispositions as it is particularly worked out by Stephen Mumford, and partly by Brian Ellis, is what makes it possible to see realistic alternatives to the anti-telos maxim.

then any moral conduct or value. What is important to understand here is that the ontology of modern science makes it a priori impossible to do harm to nature, as such. To science everything is natural except the supernatural. To take an example: for a molecular biologist it is against reality and the very essence of science to regard genetic engineering as problematic in itself. The difference between a hen that has two wings, two legs, a head with a beak, living in an hierarchy with a roster on top, and a "hen" that is just a growing and egg-laying "meatball", without legs, head and wings, is totally without significance to *science*. Very few have elaborated the *normative* status of this "non-normative" position and realized the full range of the official ontology of science, maybe because any step in any other direction seems to be towards teleology, and therefore steps towards what is paradigmatically considered unscientific.

The notion of anti-telos also has, as I have indicated already, a strong influence on the defining of *the scope and being* of the normative aspects of human life. Today, when we say that natural science should give descriptions, not "prescriptions" the whole set-up of argument and thinking is dependent on a *specific* picture of nature that views it as passive. Accordingly, the whole effort "to have an ethical debate" is *as such* always already embedded in this view on *nature*. The cards have already been distributed by the ontological viewpoint. This is, however, not at all obvious, as the so-called ethicist understands his/her position to be *independent*, also from science and the scientific enterprise as such. What is important here is to recognize the fact that the biological scientific product (e.g. a genetically modified organism - GMO) is loaded with a non-normative normativity that creates problems for any serious debate that in addition always seems to halter behind the implementation of the product. I will now look closer at the role of science in the debate of genetic engineering (GE), so as to understand the general situation described above further and deeper.

The consequences of discussing only the consequences

Anyone who takes a closer look at the GE-debate will find it somewhat impenetrable. It looks as if the positions taken, do not communicate well with each other. The "un-reflected" lay-people concept of nature, employing the concept of naturalness, is "pedagogically rejected" by the scientist, (Meyer and Sandøe 2001) but is defended by the "environmentalist" with the argument of intrinsic value (Verhoog, 2003). The argument of intrinsic value, on the other hand, has no base in the ontology of science and therefore in "reality" The argument from naturalness is an argument that does not communicate with the scientist and the entrepreneurs of technology. The issue is therefore forced into a narrower discourse that *seems to be* the last common ground for both proponents and opponents, namely "risk-evaluations" (Wynne 2001). So the debate about GMOs and gene-technology in general is about risks. This is a debate that apparently conforms to the scientist's worldview. Why is this peculiar?

Generally, discussing risks in this manner seems to be the same as asking if the biology *works as planned*. In principle, it is the same as asking if the engine of a car works as planned. To answer this question seems reasonable for the scientist, although the word "works as planed" (i.e. have functions) is an anthropomorphic way of interpreting causality. This view allows the spectator to regard the genetic modification as *per se* "substantially equivalent" to "natural" genetic processes (E.g. mutations) This also, of course, is in line with the anti-telos view on nature, which means that it would be sufficient to investigate the biological functions that are intended by the altered genetic modifications. No wonder that the natural scientist feels at home when he is making risk-evaluations (Meyer and Sandøe, 2001). But although the scientist succeeds in narrowing down the range of the question to risk-evaluations, he has not dealt with the inevitable question in this subject-matter.

The question "does it work?" needs a reference that values specific kinds of nature, (like defining the function of a car requires definite purposes). But nature does not simply function, and, furthermore: according to the ontology of modern science it has no function, because it has no *telos* or nature – what so ever. Science rejects by its very essence the existence of the distinction between "right and wrong nature". The next and superfluous question is therefore: Work? – In *relation* to what? Or, "is safe/hazardous" - in relation to what? There has to be some particular and definable *value* that is endangered[47]. How can a scientist then participate in a debate he/she should reject as a debate built on the premise of an illusion?

My anticipation is: The science of genetics has already "neutrally" stated that the change in genetic dispositions is as insignificant as moving a grain of sand from one place to another (anti-telos). At the same time "the scientist" *somehow* recognizes that there is an ethical aspect hidden in the area under discussion. But blinded by the success in narrowing down the debate to a "risk-question", he overlooks the fact that a risk-judgement necessarily presupposes an evaluation of what is natural or at least normal – concepts that totally transcend the scope of the natural science. What follows here is that "the scientist" in the first place starts out with a normative non-normativness which is invisible to her and therefore ends up in making "neutral" evaluations on a basis that totally contradicts the official ontology (*anti-telos*) of natural science that in the end underpins his alleged neutrality. The "ethicist", of course, never comes upon the core of the question and ends up where he starts in his "eunuch-ethics ". This is a kind of triple inconsistency that makes the debate around modern agriculture into a long series of cover-up projects.

A very clear example of this confusion is illustrated by what professor in mathematics Norman Levitt says about the GMO-debate in Europe (Stangroom, 2005). He suggests that the European resistance against GMO`s can be deconstructed through looking at their particular cultural "dogmas of purity and danger". Levitt uses the anthropologist Mary Douglass to analyse the situation. This could have been a plain disagreement between two different "cultural interpretations" but is obviously not. As a scientist Levitt seems to be in a situation that enables him to disavow the "European arguments against GMO" by just calling attention to the fact that "this is a cultural interpretation" - and by that indicating that there exists an "uncultured interpretation" of GE. This is possible to anticipate because according to the molecular biological anti-telos view on "genes, life and nature in general", there is no nature of nature, and therefore nothing that could be "unclean" or pure. According to the anti-telos view on nature this is a clear example of a category-mistake. This means that every critique of GE would be "cultured", while the support of GE would not. In this case, as viewed by Levitt, we would be able to say that "the Europeans" have interpreted the technology while "the Americans" have managed to avoid that. To explain why the scientist speaking as a scientist can get away with such an inference, we have to recognize the significance of the anti-telos maxim of natural science.

For the scientist, risk-evaluations might seem to be a realistic and "rational" approach to GE , but it is not. Moreover, this also means that the debate in fact is homeless. Neither lay people nor the scientist can really participate under these circumstances, but for different reasons.

The GMO-object looks like an authentic *hybrid*, as Bruno Latour would have called it. But is it? A GMO carries every mark of an "artistic fact", but it is not. For a being to become an artefact it has to have a specific being or a function - that is changed by man. But if we presuppose that nature is totally passive we cannot at the same time change the nature of something. If a "gene is a grain of sand" I will by definition be prevented from making an artefact when I move a gene from a bacterium to a maize-plant. That is why Latour's concept of hybrids becomes dubious within his own philosophy when he at the same time insists on keeping himself to the *anti-telos*

[47] The anti-telos solution to this problem is normally to say that is and ought have been confused. If one takes into consideration the ontology of dispositions one will see that this solution is highly relative. But even if this value is "only cultural", this cultural value has to anticipate some "nature", either socially, biologically or even transcendentally defined.

maxim of science.[48] According to the anti-telos maxim everything is nature, and thus nothing or everything is natural. There could be no hybrids, when there is nothing natural in nature. In the end there could be no safe or unsafe food either.

References

Beck U. (1992). *Risk Society* Sage.
Cavel S. (1979). *The Claim of Reason* Oxford University Press.
Ellis B. (2002). *The Philosophy of Nature* Acumen.
Johanessen F.R. *Unpublished manuscript*. University of Tromsø.
Kvilhaug T. (1997, 1998). *Naturens realitet i teorier...* Agora
Latour B. (1996). *Vi har aldri vært moderne.* Spartacus.
Meyer G., Sandøe P. (Juni 2001). *Oplysning og dialog om ..*Suppl. til Gen-etik i praksis
Moss L. (2003). *What Genes can't do*. MIT press.
Mumford Stephen. (1998). *Dispositions.* Oxford University Press.
Reiss M.J and Straughan R. (1996). *Improving Nature.* Cambridge University Press
Stangroom J. (Ed.) (2005). *What Scientists Think.* Routledge.
Verhoog H- (July 2003). *Naturalness and the genetic modification of animals.* Trends in Biotechnology.

[48] Cf. p. 186 – Norwegian edition.

The human and social sciences in interdisciplinary biotechnology research: Trojan horses or useful idiots?

Bjørn Myskja and Reidun Heggem
Department of Philosophy, NTNU Trondheim, NO-7491 Trondheim, Norway,
bjorn.myskja@hf.ntnu.no

Abstract

Projects where social scientists and ethicists participate in interdisciplinary biotechnology research has increased recently. The intention has been to ensure a case-based critical and informed view on the project and the technology in general. Biotechnologists may have different reasons for accepting and inviting this external participation: (1) a genuine concern for the ethical and social acceptability of the project, (2) a belief that even if they find the project acceptable, there are reasonable grounds for disagreement, or (3) a strategic intention to accommodate public opinion, as a way to legitimize this application of the technology. The first two objectives are in keeping with the motivations usually given by the external participants and the political bodies that support interdisciplinary approaches to controversial technologies. The third reason implies that social scientists and ethicists are mere means to ensure public acceptance. The external participants become "useful idiots", just to legitimize the positions of the core scientists and may draw the attention away from social and ethical issues that can better be analysed independently of particular applications of the technology. On the other hand, the social scientists and ethicists may serve as Trojan horses eroding the trust in products due to myopic negative analyses, as argued by some biotechnology proponents. The presentation explores possible advantages and disadvantages of interdisciplinary research, and argues for a combined approach where social scientists and ethicists alternate between participation in interdisciplinary research and doing independent analyses to ensure technological insight as well as critical distance.

Keywords: interdisciplinary research, ethics, social science, biotechnology, public opinion

Introduction

Research in the ethical, political and societal impact of technology used to be conducted as meta-disciplines, studying the preconditions and implications of the technologies from the outside. This kind of studies of ethical and social aspects of technology has become increasingly important due to growing public scepticism towards the impacts of biotechnology on our health and environment. An indication of this is the standard set by the American Human Genome research programme that five percent of the budget should go to so-called ELSA-research. A primary concern used to be how the technology affected human lives, and how we should regulate the use of the technology to ensure that humanity could benefit without facing risk of serious harm. These have been common research fields in relation to medicine, modern biotechnology and the growing field of nanotechnology. This perspective from the outside ensured a critical approach, perhaps even overly critical in some cases, leading some biotechnology proponents to express scepticism towards this research as hostile to technology (Miller and Conko 2000). At the same time, there is little doubt that proper insight in the risks and uncertainties as well as other ethically significant matters of biotechnology required a solid background in the relevant sciences, lending support to rejection of meta-research as poorly informed. This lack of scientific background has been partly remedied by biologists and biotechnologists doing analyses of ethical and social aspects of their own technology. However, these scientists often lacked systematic knowledge and understanding of crucial theoretical and methodological aspects of human and social sciences, reducing the value of their contribution. It is possible, at least theoretically, to develop research competence in both biotechnology and ethics or social science, but in practice it is difficult to keep ahead in two different fields of knowledge. The solution was obvious: interdisciplinary research integrating meta-perspectives in particular research projects. But then we risk either that the representatives of the humanities perspectives become useful idiots serving the interests of technology proponents or Trojan horses that undermine the biotechnology from the inside.

Integrated projects

A way to solve the problem of lack of simultaneous competence in biotechnology and in the meta-disciplines is to integrate the ELSA-research in concrete biotechnology projects. This is not only a solution to the problem of competence, but enables the researchers to explore other significant aspects of the technology. In ethics, it is generally agreed that Aristotle is right that "actions are about particular cases" (Aristotle 1985, 1107a28-30). This is illustrated by the assessment of risks and uncertainties of the release of genetically modified organisms that depend on the properties of the organism, the nature of the modification, the particular environment where it is released and the knowledge and activities of the affected parties. Likewise, people's attitudes and the political regulations will depend on the particular characteristics of the project and the judgements can differ significantly from the ones given for more general questions about biotechnology, for example in the Eurobarometer survey (Gaskell *et al.* 2003). A different reason for doing this kind of integrated projects can be drawn from the insights derived in the field of science and technology studies, where the focus is the dynamics of scientific developments (Nydal 2002, 141ff.). Under such perspectives, the ELSA-research is not meta-research as it is traditionally understood, but an essential part of the research project itself. Still, we will regard these aspects of an integrated project as meta-research, because the nature of the questions asked in a number of such projects are closer to traditional bioethical approaches.

The growing awareness of the dynamic relation between science, technology and society has resulted in an increasing number of interdisciplinary projects, and an inclusion of the ethical dimension in research within the EU Seventh Framework Programme. Integrated ELSA-research in biotechnology projects is initiated not only by researchers from the humanities or from political authorities, but also by the technologists or natural scientists themselves. Some scientists obviously must agree that this kind of meta-research is necessary. But they do not necessarily regard it as necessary for the same reasons as those held by the ethicists and social scientists themselves. The reasons for including ELSA-research in biotechnology projects may include: (1) a genuine concern for the ethical and social acceptability of the project, (2) a belief that even if they find the project acceptable, there are reasonable grounds for disagreement, or (3) a strategic intention to accommodate public opinion, as a way to legitimize this application of the technology.

The first reason is expressive of a genuine concern for the values involved in and the impacts of applications of the technology, and we know that many scientists share these concerns. After all, biotechnologists did once impose a five-year voluntary moratorium on parts of their own research for ethical reasons (Barinaga 2000). In addition, many of the major issues in the biotechnology debate were first raised by biotechnologists. The second reason is one that is typical for people who are open-minded, and is expressed paradigmatically in Karl Popper's falisificationism. Good science is science that lends itself open to counter-arguments, and inviting researchers with perspectives that may in the end put an stop to development of the results of their own research is one way to live up to this ideal. The third reason is a strategic way to accommodate public opinion that the scientists themselves reject as ill-founded. Their intention is to use the meta-research for their own ends, which is gaining acceptance for application of the technology regardless of the results. We can say that they use the ELSA-research as mere means, and not as an end in itself. Before we take a closer look at the negative aspects of this instrumentalisation of ethics and social research in biotechnology, we will mention the positive potential in this strategic inclusion of meta-perspectives.

Caught by the argument

When someone uses strategic argumentation, we often regard that purely negatively as an instance of dishonesty. Still, in Machiavellian politics it is generally regarded as a necessary evil. The same is the case with people using the rhetoric of ethics for furthering egoistic purposes, such as including bioethics merely to accommodate public opinion. It is better to fight an honest cause we tend to think, by demonstrating through arguments why we should not be worried about the ethical and social impacts of biotechnology. We can however see a productive potential in this kind of hypocrisy, as Elster (1995) has pointed out. Strategic argumentation and action have a civilising force, altering the way we think about the issue. Thus, biotechnologists who argue about the need for interdisciplinary research on their project might be influenced by their own arguments, realising the strength of the more open position that many of their colleagues take to this kind of research. The cooperation with ethicists and scientists, discussing their own project, may also contribute to a change of heart. In this way, even a strategic inclusion of ELSA-research can be turned to a genuine interdisciplinarity. Still, this is only a possibility, more likely, the strategic approach will continue.

Useful idiots

When ethicists and social researchers participate in biotechnology research, the main goal of the research will be development of a particular product, for example a genetically modified plant. Regardless of the reasons for the inclusion of these meta-perspectives, the results are meant to have an impact on the regulation on the development and use of this plant. The scientists and regulators will evaluate the conclusions of the meta-research in relation to how it affects the main goal. If the conclusions are negative, they tend to regard it as predictable, since ethics often have this limiting function. When we find what is expected, we tend to hold the results to be less than interesting. However, if it is possible to read the results in a more positive vein, the interest is growing for two reasons. First, it is unexpected, and second and more important, it will contribute to an acceptance of the product. It is probable that biotechnologists developing a particular plant wish to continue their research and in the long run develop a product that is useful and commercially interesting. Thus, they will tend to highlight the aspects of the meta-research that contribute to such developments.

We can take as example a recent EU project; "Sustainable production of transgenic strawberry plants. Ethical consequences and potential effects on producers, environment and consumers". The European Commission funded this as a cooperative project (2000-2004) between five different groups of biologists, philosophers and sociologists in Norway, Netherlands and Germany. The sociological part of the study was to carry out national surveys in Norway, Denmark and England to investigate how people responded to this genetically modified strawberry and to detect what kind of ethical concerns they had. This future strawberry, which was supposed to be resistant to grey mould, was claimed to have environmental as well as economic benefits. The environmental benefits were connected to the possible decrease in use of pesticides as a result of the resistance of the grey mould.

The sociologists found that consumers in general were negative and sceptical to genetic modification, due to possible negative effects on health and the environment. When they focussed on this specific strawberry that might lead to a decrease in the use of pesticides, it seemed that it became more morally acceptable because it reduced a major problem in agriculture. These results were very well received by the biotechnologists. They neglected the fact that this was results from a survey of attitudes, not concrete action in shops. It is well known that there is a big step between what you say and what you do. Even if many answered that they certainly or probably would buy genetically modified strawberries, it does not mean that in the actual situation they would do that. It might as well be the attitudes towards genetic engineering in general that is the decisive argument on the action level.
We also have to mention that the positive attitudes towards this strawberry might be a result of the fact that strawberries are a small part of the consumers' food intake. On that ground the respondents feel that they easily can control and perhaps avoid genetically modified strawberries. If the product at stake was a product that is customary in our food tradition or a product that is used in many semimanufactured foods, then we might have a different result. The biotechnologists neglected these conditions, and presented the survey as an indication of a more positive attitude among the public to genetically modified products.
These kinds of presentations have an impact on other aspects of agricultural biotechnology, as well. If there can be found products that are favourably regarded by ethicists and well received by the public, they will be used as a foot in the door for the general acceptance of other biotechnology products. Even if the meta-research emphasise that the validity of their research is restricted to this particular product, they are unable to stop biotechnology proponents from making a wider use of their results. This is a mirror image of the tendency of biotechnology opponents to use the least acceptable examples of the technology as a general argument against every application of it. The problem for the social scientists or ethicists is that the fine print will disappear. All the conditions they make prior to the conclusion will not be understood or will be glossed over because they are too difficult and complicated or they are phrased in the language of a particular discipline or they make the exiting news less interesting for the biotechnologists. Only the positive aspects of the conclusion will be discussed, and the researchers have become useful idiots for the biotechnologists. Even biotechnologists who invite ethicists and social scientists for ethically sound reasons may use their colleagues in this way because they fail to understand the particularities of the research traditions they themselves don't belong to.

Trojan horses

Some biotechnologists are generally sceptical to the results of bioethics and social studies of science, regarding it as expressing conservative, risk aversive and technology sceptical views, based on insufficient knowledge. Thus, they put obstacles in the way of the development of necessary and beneficial products, and unjustly undermine the trustworthiness of science. They may be right. Parts of the ethics debate have been overly risk aversive, as is expressed in some very restrictive versions of the Precautionary Principle. If ethicists and social scientists with such views are participating in interdisciplinary research they may very well serve as Trojan horses for the anti-biotech lobby, willingly or not. Neither biotechnology research nor the general society is well served with integrating such researcher with negative perspectives on the technology in concrete projects. They will take up the time of the biotechnologists on fruitless debates and fail to give a balanced assessment of the value of the research and products.

Narrow and wide perspectives

The strength of participating in integrated projects lies in the possibility to gain correctives from the biotechnologists themselves in presentations of the research in surveys, interviews ands analyses, in the focus on a particular case, and in the increased knowledge the biotechnologists get on the societal and ethical significance of their own research. The weakness lies in the danger of becoming useful idiots or Trojan horses, but also in the narrow focus on the research project. This may draw attention from the larger picture in which this project is a minor part. The bioethical and sociological analyses of a genetically modified strawberry have no way to evaluate how biotechnology contributes to solving the hunger problem or to a more or less global justice by technology dissemination or patent restrictions. These wider perspectives are arguably the most important questions we can discuss in relation to the impact of biotechnology, and a wider perspective is a significant part of the biotechnology debate. Thus, we suggest that a combination of integrated projects and independent outside perspectives is necessary to ensure a sound analysis of biotechnology by the human and social sciences.

References

Aristotle (1985) *Nicomachean Ethics.* Hackett, Indianapolis.

Barinaga, M. (2000) Asilomar Revisited: Lesson for Today. *Science* 287 (5458), 1584-1585.

Elster, J. (1995) The Strategic Uses of Argument. In: *Barriers to Conflict Resolution*. Mnookin, R., Ross, L. and Arrow, K. J. (eds). Norton, New York. 236-257

Gaskell, G., Allum, N. and Stares, S. (2003) The Europeans and Biotechnology in 2002.: http://www.blauen-institut.ch/Tx/tP/tpG/0723EuBarometer.pdf

Miller, H.I. and Conko, G. (2000) Genetically modified fear and the international regulation of biotechnology. In: *Rethinking Risk and the Precautionary Principle*. Morris, J. (ed). Butterworth-Heinemann, Oxford. 84-104.

Nydal, R. (2002) *I vitenskapens tid. Introduksjon til vitenskapsfilosofi*. Spartacus, Oslo.

Moral co-responsibility in food production and consumption

Hilde W. Nagell
The National Committees for Research Ethics, Prinsensgt.18, P.O Box 522 Sentrum, N-0105 Oslo, Norway, hilde.nagell@etikkom.no

Abstract

This paper asks: When are we co-responsible for harmful outcomes? Individual responsibility depends on such notions as causation, choice, agency and the ability to do otherwise. However, when each one of us is only marginally attached to the harmful act, and the causal link between individual behaviour and the non-preferred outcome is vague at best, our common sense notions of responsibility tells us that we are not individually responsible. Call this the limited individual responsibility position. This paper will challenge this position, and establish a wider notion of responsibility based on the idea that we are sometimes responsible for how our actions together with others can have affects which concern society as a whole. The paper examines and critically questions three common excuses for avoiding responsibility in cases of collective action: "If I don't do it, someone else will", "I had no choice", and "I didn't know". I will relate the discussion to the production and consumption of food.

Introduction

How are we to address and assign moral responsibility[49] when many people are involved in an outcome? In modern production processes, actions are almost by definition collective and may have multiple unintended consequences. At the same time, the area for which an individual producer can exercise control has narrowed due to a higher degree of specialisation, differentiations of roles and more complex production processes. Consumers are normally thought to have extensive power in terms of their demands for goods. Increased information and labelling of products, and a larger selection of goods, may shift responsibility away from producers and over to consumers. My ambition here is modest. The point I want to make is not that we should ascribe considerable responsibility to individuals in cases of collective action. I will argue, however, that excuses derived from common notions of individual responsibility are not transferable when confronted with situations of collective action.

Individual responsibility and collective actions

When different actors contribute in many ways to a production process, it is difficult to identify who is morally responsible for outcomes. Within the realm of political governance, Dennis Thomsen has called this the problem of many hands (Thomson 1980:905, 2005:11). Christopher Kutz has also made an attempt to restore the concept of individual responsibility, and re-define it in terms of part-taking in collective action (Kutz 2000). These theoretical aspirations have inspired this paper, but before I can start my discussion, some preliminaries need comments. First, it is not entirely obvious why we should bother about responsibility at all. A consequentialist could argue that what is important to us are results- or state of affairs- not who produced them. However, morality is also about doing the right thing given the circumstances. Sometimes it is possible to change or affect these circumstances by acting together, in ways that will produce better state-of-affairs. My argument is that focusing on responsibility is important because

[49] Moral responsibility is here understood as being worthy of reactive attitudes such as praise or blame.

otherwise there is a narrow definition of individual responsibility which gives no incentive to engage in such collective enterprises.
Second, an important question immediately suggests itself: morally co-responsible for what? Opinions greatly differ when it comes to what constitutes harmful acts. For the purpose of this discussion, it suffices to state that the production, transformation and distribution of food sometimes involve harmful acts, acts that have severe negative effects on values such as human well-being, animal welfare, or sustainable development. [50]
Third, there is an important distinction between taking responsibility and being responsible. People in top positions know that it is part of their job to take responsibility. They may be required to leave office or to apologise in public for harms caused by the organisation they represent. However, this is not the same as saying that they are morally responsible. They may for instance not have been expected to have control or information over that particular action. Moral responsibility, therefore, is not coextensive with formal responsibility.

The limited individual responsibility position

A host of criteria have been discussed and re-butted in the philosophical literature on responsibility.[51] It is not the object of this paper to go into these difficult and challenging aspects of individual responsibility. My point of departure is a widely shared definition, what I will call *the limited individual responsibility position*: an individual is responsible if a) his or her actions are a cause of the outcome and b) his or her actions are not done in ignorance or under compulsion.[52]
A strong interpretation of the first criterion would be that you are responsible for an outcome if it would not have happened but for your act. A weaker interpretation would be to say that you are responsible if your act was at least a causal factor in the outcome (Thompson 2005:18). Underlying this first criterion is the *individual difference principle*: I am only accountable for harm if something I did made a difference to the outcome (Kutz 2000:3).[53] The second criterion is based on the idea that you are only responsible if you could have acted otherwise. Behind this is *the control principle:* I am only accountable for harm over which I have control. And I can only exercise control if I am properly informed and not under cohesion (Kutz 2000:3).
This position forms a coherent and strong definition of individual moral responsibility. However, my claim is that the definition leaves out situations where it would be wrong to say that individuals are not morally responsible, but where at least one criterion referred to above is being violated.

"If I don't do it, someone else will"

Consider a case where farmer A produces apples using a pesticide particularly harmful to the environment. Many factors could of course be relevant to whether this is considered to be a harmful act (development prospects in his country, consequences of alternative production methods, how he treats his workers etc). Let's assume, however, that this is all things considered a harmful act. A common excuse for avoiding responsibility is what Thomson calls "the excuse

[50] The Food and Agricultural Organization of the United Nations (FOU) has formulated a number of ethical issues central to food and agriculture, see Ethical issues in food and Agriculture, Rome 2001, FAOs ethics series, 1, ISBN 9251045593 x9601/E
http://www.fao.org/documents/show_cdr.asp?url_file=/docrep/003/x9601e/x9601e00.htm

[51] For an overview of positions, see for instance Fisher and Ravizza 1993 and 2000.

[52] For reasons of limitation, this definition does not address all relevant aspects responsibility, such as the distinction between intentional and non-intentional actions, or between acts and omissions.

[53] Here I will simply assume that moral responsibility is compatible with causal determination.

from alternative cause". The usual expression would be: "If I don't do it, someone else will". This excuse may come up in situations were my actions alone have a causal effect, but If I refrain from acting, someone else will act instead of me, producing exactly the same bad outcome.
The farmer could find support from this excuse: He only produces what the market wants. If there is a demand for his polluting fruit, that's what he will produce. Furthermore, being a small farmer, his production process does not have much effect on the total level of pollution caused by using pesticides. If he stops using the pesticide, however, some other farmer will expectedly sell more apples, and he will loose competition. According to the limited individual responsibility position, the farmer is not responsible.
What should be taken into account in this case is whether the individual farmer with low costs to him-self, together with other farmers, can take realistic steps to stop the use of that pesticide in their production. If that is the case, I believe that the apple farmer is responsible if he continues using the pesticide. In situations like this, it is natural to call for government intervention trough laws and regulations. However, detailed government regulations of this kind are often neither possible nor desirable. Furthermore, the logic of a free market economy resists strong government regulations almost by definition because these interventions create unfair rules of competition in the market. Third, regulations are often given in the form of "soft law" and not "hard law", making room for discretion and multiple interpretations, so the farmer will in any case not be relieved from his responsibility.
How does this look from a consumers' perspective? What apples I choose to buy will arguably have no causal effect on the outcome, since it is likely to come about anyway. The apple-farmer will not go away if I don't buy his apples, and it will not make him change his use of pesticides. I am only one of many hundreds if not thousands of consumers, so how can one possibly say that I am in any way co-responsible for the pollution he causes? According to the limited individual responsibility position, the first requirement for responsibility is not fulfilled. Each individual's actions in isolation are not a cause of the outcome. An interesting question is of course how weak that causal link can be before this excuse applies. Instances of weak causal effect of individual actions give rise to what Kutz calls I-WE problems (Kutz 2000:5). Where the point of insignificant causal effect is to be drawn is partly a question of how willing we are to regard our own acts as part of what others do.
The consumer is less responsible than the farmer because he has another position in the causal chain. The power of consumers is often emphasised, however, and producers tend to argue that they only produce what the market demands. If more people would be willing to pay extra for pesticide-free apples, that's what they would get. Consumer power is in practice limited in many ways, as I will argue in the next section. However, it does not remove the fact that if I buy a product, I contribute in the causal chain by being on the receiving end and signal a demand for the product, and I contribute by supporting the producer with my money.
Here it could be useful to draw a parallel to criminal law: the receiver of stolen goods is guilty on par with the thief. If fully informed and not under cohesion, you buy stolen goods, there is no excuse that someone else would have bought it anyway if you didn't. Neither is it an excuse that you only bough a small fraction of the total stolen property, so your purchase has no causal

effect on the incentives to steel.[54] Again the ability to co-operate with others is morally relevant: you are responsible if, with small costs to yourself, you have the alternative to act together with others, and thereby to reduce the amount of harm.

I didn't know

Industrial farming has received much criticism for inflicting unnecessary pain on the animals. Peter Singer, not surprisingly, has strong opinions on this:

> We should see the purchase and consumption of factory-farm products, whether by an individual or by an institution like a university, as a violation of the most basic ethical standards of how we should treat animals and the environment (Singer 2006).

Elisabeth Costello, the heroine in J.M Coetzes book by the same name, compares industrial farming and the way animals are treated with the horrors of the concentration camps:

> Let me say it openly: we are surrounded by an enterprise of degradation, cruelty and killing which rivals anything that the Third Reich was capable of, indeed dwarfs it, in that ours is in enterprise without end, self-regenerating, bringing rabbits, rats, poultry, live-stock ceaselessly into the world for the purpose of killing them (Coetzee 2004:65).

If, like me, you do not think it is wrong to bring animals into the world for the purpose of eating them, you may want to change the example. This is not the point, however. What is interesting is not whether this highly speculative comparison can in any way be defended, but whether not wanting to know about harmful acts makes a difference in terms of assigning moral responsibility. Costello adds:

> It was and is inconceivable that people who *did not know* (in that special sense) about the camps can be fully human (Coetzee 2004:65).

Degree of information is not a fixed entity, but something at least partly open to choice. People around the death camps in Germany said they did not know what was going on.[55] If someone did not know, but could have known with little effort to himself, and choose not to make that effort, the excuse from ignorance does not apply. Needless to say, this does not imply that knowing in itself is sufficient for establishing responsibility. However, in cases where an act is causal to an outcome in one of the ways described above, and there is no cohesion involved in the choice between alternatives, not knowing is not a good excuse if you are yourself responsible for the ignorance.

The excuse from ignorance is not valid when responsible for the ignorance, and it is not valid if the person should have realised that misfortunes of this kind were likely (culpable ignorance).[56] There are different ways in which you may be responsible for your own ignorance. In corporations or government institutions, a practice or routine may intentionally secure ignorance, as when production routines are such that the top management has no chance of knowing what is going

[54] We might think of situations were buying stolen goods is morally defensible, for instance when a government institution buy important cultural objects which belongs to them but has been looted from archaeological sites or museums. What could defend to buy in this situation is that if the object in question is bought by someone else, it will not be accessible to a broader public, and is a real danger that the object will be damaged or destroyed. A variant of this excuse is to say that I did it in order to prevent something worse from happening. A commander in war could argue that although he opposes the war, he stay on in order to prevent an escalation (Thomson 2005:21). However, this is an altogether different argument from the one discussed here. If it is true that it produces the best outcome to stay in the war, it cannot be a harmful act, all things considered. For a discussion, see Nagell 2005:xx

[55] I will not go into the difficult question of collective guilt.

[56] For a definition of culpable ignorance, see Hacking 1986.

on in the lower layers of their sub-contracting chain, making them instruments of their own ignorance.
I admit that this is complicated. Lack of information is often less of a problem then over-load of information. As consumers, we are increasingly bombarded with detailed information, not only telling us whether the product is organic or conventionally grown, but everything from the name of the farm, and the exact amount of energy used in the production process. If the level of information you have is relevant to the degree that you are responsibility for your consumer choices, more information should place more responsibility in your shoulders. This seems unreasonable, and should under any circumstances not discharge producers from their responsibility.

I had no choice

Excuses from compulsion come in many forms. They obviously have some relevance in situations where a choice is made under physical pressure (gun pointed at your head) or under the treat of severe sanctions. Less obvious cases are those where "no explicit order is given, but subordinates believe that a superior expects them to pursue a morally dubious course of action." (Thomson 2005:27). However, when nothing can be done by individual actors to correct harmful actions, each of us still has a special responsibility to call attention to these actions. Responsibility then, also in part requires taking into consideration what other actors do. Not blowing the whistle trigger off responsibility. This is not reflected in the limited definition of individual responsibility.
A variant of this excuse is what I will call the excuse from no available alternatives, as when consumers argue that their choice is restricted by what they find on the shelves: If there are no alternatives available to me, I could argue that I have no choice but to buy the product even though I know that by doing so I will contribute to a harmful act. This excuse is based on the assumption that I do not choose the range of alternatives within which I make some decision. However, this is not always really the case. We need to ask what alternatives were realistically accessible. If, with little effort to myself, I could widen the range of alternatives and thereby my own control, I am (co-) responsible if I fully informed continue to buy the polluting apples.

Conclusion

The limited individual responsibility position requires that we are not responsible for anything else then our own direct actions. However, I have argued that in some situations we are not only responsible for what we do alone, but also for we do when we act together.
I have argued that the scope of your responsibility should be extended to include situations where it is possible to coordinate your activity with others in order to avoid harm. Secondly, even when under direct orders from superiors or under conditions where it is not possible to do otherwise, you are responsible for calling attention to the harms involved. From a wider interpretation of individual responsibility you are required to take the harm of others into account, inform, and be prepared to make an effort to promote good ethical standards.

References

Coetzee, J.M (2004): Elisabeth Costello- eight lessons. London:Vintage.
Fischer, John Martin and Mark Ravizza (ed.)(1993): "Moral Responsibility." New York: Cornell University Press.
Fischer, John Martin and Mark Ravizza (2000): "Responsibility and Control." Cambridge: Cambridge University Press, Cambridge Studies in philosophy and law.

Hacking, Ian (1986): Culpable Ignorance of Interference Effects." in MacLean, Douglas (ed) (1986): "Values at risk" Maryland: Rowman & Littlefield Publishers, pp.136-155.

Kutz, Christopher *et al.* (2000): "Complicity- Ethics and Law for a Collective Age." Cambridge Studies in Philosophy and Law.

Singer, Peter (2006): "Factory Farming: A moral Issue", The Minnesota Daily, March 22.

Thompson, Dennis F: "Restoring Responsibility: Ethics in Government, Business, and Healthcare", Cambridge University Press, 2005, 360pp.

Williams, Bernard (1988): "Consequentialism and Integrity". In Scheffler, Samuel (ed): Consequentialism and its critics. Oxford: Oxford readings in philosophy, pp.20-51.

Genetically modified foods between ethics and public policy

Assya Pascalev
Department of Philosophy, 226 Locke Hall, Howard University, 2441 6th Street, NW, Washington DC 20024, USA, apascal2003@yahoo.com

Abstract

The paper explores the question: What ethical principles should inform a morally sound policy on genetically modified foods (hereafter GMF)? It focuses on the debate in the USA and on the widely held doctrine of political liberalism and its principle of liberal neutrality. A policy based on liberal neutrality requires policy makers not to favor any particular position and to remain neutral between those who support and those who oppose GMF. This paper advances the view that political liberalism cannot provide an adequate framework for policy decisions concerning GMF because political neutrality cannot protect adequately the interests, rights and values of those who oppose GMF on religious, moral or metaphysical grounds. Such individuals are left with few alternatives and have to pay a huge price to live in conformity with their values. Liberal neutrality ultimately favors the status quo, which erodes the personal integrity and freedom of choice of those who oppose GMF. A positive action to protect those who object GMF is morally required from policy makers and is grounded in the values of personal integrity, freedom of choice and in the principle of autonomy. They provide normative justification for rejecting GMF and, minimally, for positive societal actions and policy measures including not only mandatory labeling but also subsidies for organic farming and non-GMF. To allow for a genuine co-existence of diverse views and ways of life, policies should be set to guarantee that those opposing GMF have access to non-GMF, which access is not too burdensome and costly.

Keywords: ethics, public policy, political liberalism, neutrality, integrity

Introduction

In this paper, I explore the question: What constitutes a democratically acceptable policy on GMF and what ethical principles should inform such a policy? I focus my attention on the current situation and the state of the debate in the USA, where over 50% of the foods are genetically modified. Yet, producers are not required to label GMF and there has been little public debate on the acceptability and merits of GMF (Hopkins, 2001, Brown, 2001, and Hart, 2002). Below, I explore the normative basis for developing a democratically acceptable public policy concerning GMF and while at the center of my attention is the USA, my conclusion is more general and has implications for public policy beyond the American context. I consider a recent proposal by Robert Streiffer and Thomas Hedemann (2005) that policy decisions concerning GMF should remain silent on the merits of the ethical arguments for or against GMF. Instead, policy makers should develop policies acceptable to all stakeholders, including the opponents and proponents of GMF. Streiffer and Hedemann hold that such policies support mandatory labeling but not a ban on GMF (2005, p. 192). The authors ground their conclusion in the widely held doctrine of political liberalism and its principle of liberal neutrality. I argue, however, that political liberalism and neutrality cannot provide an adequate framework for policy decisions concerning the permissibility of GMF in the USA. Given the peculiar circumstances in the USA characterized by mass proliferation of GMF in the absence of consumer information and alternatives, political neutrality cannot adequately protect the interests, rights and values of citizens who oppose GMF on religious, moral or metaphysical grounds. Such individuals are left with very few alternatives and have to pay a huge price to live in conformity with their values. Since the balance has already been shifted silently in favor of GMF, liberal neutrality would *de facto* amount to endorsing the status quo, which erodes the personal integrity and freedom of choice of those who oppose GMF. I conclude that not neutrality, but a positive action protecting the rights of those who reject GMF is morally required from policy makers. This conclusion is grounded in the moral values of personal integrity and free choice, and the principle of autonomy. They provide normative justification for rejecting GMF and for positive societal actions and proactive policy measures, which include not only mandatory labeling but also subsidies for organic farming and non-GMF.

The Ethics and politics of GMF

The debate about the moral permissibility of GMF traditionally has centered around two types of concerns: intrinsic concerns calling for a moral evaluation of the very process by which GMF are produced, and extrinsic concerns related to their safety, environmental impact and social consequences. Those who oppose GMF on extrinsic grounds claime that GMF have harmful consequence, and those opposing GMF on intrinsic grounds appeal to moral problems inherent in the very practice of producing GMF.[57] With a few recent exceptions, the intrinsic objections to GMF evoke the unnaturalness of GMF, playing God, hubris, disrespect for nature and violating species sanctity, which appeals appear based on religious, non-scientific principles. Because of that, the intrinsic objections have been rejected by prominent ethicists as lacking rational, theoretical merit, and thus no moral and political validity. On the other hand, the extrinsic objections are based on empirical claims that lack full factual support. Many ethicists such as G. Comstock (2002a) and B. Rollin (1995) argue that the objections to GMF are not viable since they are not grounded on respectable premises. In a recent article, Robert Streiffer and Thomas Hedemann (2005) offer a reevaluation of the merits and political import of the intrinsic objections. They acknowledge the significance of the objections and argue that even if they fail to advance ethically sound arguments against GMF, the objections should be taken seriously in political sense and policy-making. This is because political liberalism requires that public policy be based on values that are acceptable to a "wide range of moral and philosophical worldviews likely to persist in a just and open society (Estlund, 1998, pp. 252-253)." Since modern democratic societies are characterized by diverse and reasonable moral views, or "comprehensive doctrines", which are often incompatible, the policies of the state must be constrained by the principle of neutrality, if they are to be legitimate. The principle of legitimacy demands that political arguments in a pluralistic society are independent of any particular system of values and policies are justified on neutral grounds (Streiffer and Hedemann, 2005, p. 197). This requires that the justification for the policy [including a policy on GMF – A.P.] be consistent with all reasonable comprehensive doctrines, and that the policy not be aimed at making it more difficult for some to live according to their reasonable comprehensive doctrines (2005, p. 1987)." When Streiffer and Hedemann apply the principles of political liberalism to the particular case of GMF, they reach the following conclusions: (1) "The typical intrinsic objections rely on ethical claims that do not fall into an overlapping consensus" and, therefore, they must not be used to justify a policy on

[57] For a discussion of the extrinsic and intrinsic objections to GMF, see Comstock, 2000a. It should be noted that the "intrinsic/extrinsic" distinction is somewhat crude: "an extrinsic objection might nonetheless raise the same moral issues as an intrinsic objection... For example, someone might believe that even though genetically engineering plants does not in itself constitute playing God, its widespread practice will lead to genetically engineering human beings, which would constitute playing God. Strictly speaking, this person holds an extrinsic objection, but clearly the moral component of this extrinsic objection reduces to that of an intrinsic objection " (Streiffer and Hedemann, 2005, p. 193-4).

GMF. (2) The rejection of the intrinsic objections is also based on ethical claims, which do not "fall into an overlapping consensus" and cannot be used to justify a public policy in favor of GMF either. According to Streiffer and Hedemann, this means that the intrinsic objections cannot be ignored in policy debates but they cannot inform the debate either. The policy must remain silent on the merits of the intrinsic objections, which can play only limited role in determining the content of the policy (2005, p. 194).

Intrinsic objections to GMF and their moral and political import

The political import of the intrinsic objections to GMF derives from the fact that such objections are part of legitimate, reasonable value systems of individuals who choose to avoid GMF on religious, moral or philosophical grounds. Streiffer and Hedemann note that the state has an obligation to respect these individuals and their value systems and not to set policies that are designed to violate them. This principle is called "neutrality of intent" (2005, p. 197). The authors note that most arguments against the intrinsic objections advanced by prominent ethicists appeal to principles that do violate the neutrality of intent and, thus, are illegitimate. The intrinsic objections cannot be dismissed in public policy on the grounds that they are unsound, faith-based, or inchoate. The most important question, then, becomes: How should intrinsic objections be taken into account in setting public policy in a pluralistic, democratic society? Streiffer and Hedemann seek to answer the question within the framework of political liberalism. They maintain that political liberalism offers a strong argument in favor of mandatory labeling of GMF because allowing unlabelled GMF on the market "does interfere with people's ability to live according to their reasonable comprehensive doctrines" (2005, p. 199). However, they reject the possibility of banning GMF as a matter of public policy because the ban would violate the neutrality of intent by violating the views of those who support GMF. Streiffer and Hedemann argue that, according to political liberalism, respecting the intrinsic objections does not require that the sate take any positive action to enable those who oppose GMF to purchase non-GMF. Even if a policy allowing GMF on the market makes it more difficult for some people to find non-GMF, as long as they are informed which foods are GMF, the state has fulfilled its duty of neutrality and has given citizens the information necessary to make choices consistent with their beliefs.

I welcome the renewed interest in the intrinsic objections to GMF and the recognition of their significance, but I disagree with by Streiffer and Hedemann on several counts. Firstly, I find their account of the various intrinsic objections incomplete. In recent years, there have been notable efforts to advance novel intrinsic objections such as the integrity-based objections to GMF (A. Pascalev, 2004). These objection do not appeal to the same moral principles and intuitions, which Streffier and Henemann attribute to the intrinsic objections, and which they ultimately reject as unsound from an ethical stand point. Rather, the integrity-based objections appeal to the concept of integrity, the importance of personal integrity to moral agents and ague for the philosophical relevance of food choice to personal integrity. The integrity objections draw attention to the profound philosophical and symbolic value of food and food choice in the lives of individuals and communities and to their connection to the individual's conception of the good. I submit that the integrity-based intrinsic objections do not suffer the defects of the intrinsic objections identified by Streiffer and Hedemann, and, therefore, I reject their conclusion that the intrinsic objections as a whole lack ethical merits and are not convincing. However, my purpose here is not so much to present and defend the integrity-based intrinsic objections, or to argue for the theoretical merits of the integrity objection in general. Instead, and this is my second point of disagreement with Streiffer and Hedemann, my goal is to show that their account of what constitutes an acceptable democratic policy on GMF within the framework of political liberalism fails to offer an adequate answer. For the sake of the argument, I will set aside the first issue about the theoretical soundness of the intrinsic objections and will adopt Streiffer and Hedemann's weaker thesis that the objections have political importance because they are components in the reasonable comprehensive doctrines of a large number of citizens who reject GMF on moral, religious or philosophical grounds.

Political liberalism, neutrality and food choice

Streiffer and Hedemann conclude that the intrinsic objections must be respected by the state in regulating GMF. The policies must include mandatory labeling which is necessary and sufficient to protect the interests and values of those opposing GMF and to allow them to live in accordance with their fundamental beliefs. To require more would violate neutrality and disrespect the beliefs of those who endorse GMF on the basis of incompatible but equally reasonable comprehensive doctrines. The final decision should be based on a majoritarian voting procedure and, ultimately, the proper policy would depend on the number of people who believe that GMF are morally objectionable, and on how important this belief is to them. Another factor in liberal policy decisions is "the extent to which the intrinsic objections could be accommodated without compromising other important political values (2005, p. 207-8)." In the case of GMF, the compromise would require mandatory labeling to give those who oppose GMF "a reasonable chance of avoiding it themselves, thereby affording them a greater chance at living according to their particular reasonable comprehensive doctrine (2005, p. 208)."

To evaluate the conclusion of Streiffer and Hedeman and the policies they recommend, we need to place them in the concrete social, economic and political context, in which they are to operate. Currently in the USA, there is a mass proliferation of GMF, which have over 50% of the food market. Yet, there is little or no consumer information which foods are GMF and there are no readily available alternatives to GMF. Organic farming has a relatively small share of the market and the prices are too high for citizens of lower and middle class income. Due to lobbying and economic power, the producers of GMF have silently shifted the balance in favor of GMF. Those who oppose GMF have very few options and have to pay a huge price to live in conformity with their values. In these circumstances, it becomes extremely burdensome and costly for the individual to avoid GMF. The introduction of GMF on the market in the absence of a public debate on the merits of GMF and without consumer access to relevant information (including labeling), lack of oversight and independent safety testing in the USA violate important forms of personal integrity and the individual freedom of choice of those who oppose GMF on moral, religious or metaphysical grounds. Such individuals are left with very few alternatives and have to pay a huge price to live in accordance with their values. The moral value of personal integrity and free choice and the principle of autonomy provide normative justification for rejecting GMF and, minimally, for positive societal actions and policy measures to protect the interests of those who refuse to consume GMF. Mandatory labeling can offer important information but the information has little value if, once informed, the individual has no opportunity to act on the information and to access non-GMF. In the US context, liberal neutrality would *de facto* amount to endorsing the status quo, which favors those who support GMF and erodes the personal integrity and freedom of choice of those who oppose them. Therefore, not neutrality, but a positive action protecting the rights of those who reject GMF is required from policy makers. Such an action must include not only mandatory labels but also subsidies for organic farming and non-GMF. To allow a genuine co-existence of diverse views and ways of life, the policies should be set to guarantee that those opposing GMF have access to non-GMF, which access is not too burdensome and costly.

References

Batalion, N. (2000). "Fifty Harmful Effects of Genetically Modified Foods" [on line]. Available on http://www.cqd.com/50harm.htm[date of consultation: 22/04/06].

Bovenkerk, B., Brom, F.W.A. and Van den Bergh B.J. (2002). "Brave New Birds: The Use of 'Animal Integrity' in Animal Ethics." *Hastings Center Report* 32 (1), 16-22.

Brown, K. (2001). "Seeds of Concern." *Scientific American* April, p. 51-57.

Comstock, G. (2001a). *Vexing Nature? On the Ethical Case against Agricultural Biotechnology.* Kluwer Academic Publishers, Boston.

Comstock, G. (2001b). "Ethics and Genetically Modified Foods." *SCOPE GM Food Controversy Forum,* 2.

Cox, D., La Caze, M. and Levine, M. (2001). "Integrity." In: *The Stanford Encyclopedia of Philosophy.* E. N. Zalta (ed.) [on line]. Available on http://plato.stanford.edu/archives/sum2001/entries/integrity/ [date of consultation: 22/04/06].

Estlund, D. (1998). "The Insularity of the Reasonable: Why Political Liberalism Must Admit the Truth." Ethics 108, 252-275.

Hart, K. (2002). *Eating in the Dark.* Pantheon Books, New York.

Homiak, M. (2003). "Moral Character." In: *The Stanford Encyclopedia of Philosophy.* Edward N. Zalta (ed.) [on line]. Available on http://plato.stanford.edu/archives/spr2003/entries/moral-character
[date of consultation: 22/04/06].

Hopkins, K. (2001). "The Risks on the Table." *Scientific American* 4, 61.

Horsch, R. (2001). Interview with Sasha Nemecek, "Does the World Need GMF? Yes!," *Scientific American* 4, 62-63.

McFall, L. (1987). "Integrity." *Ethics* 98, 5-20.

Pascalev, A. (2003). "You Are What You Eat: Genetically Modified Foods, Integrity and Society." *Journal of Agricultural and Environmental Ethics* 16 (6), 583-594.

Rollin, B. (1995). *The Frankenstein Syndrome: Ethical and Social Issues in the Genetic Engineering of Animals.* Cambridge University Press, Cambridge.

Streiffer, R. and Hedemann, T. (2005).The Political Import of Intrinsic Objections to Genetically Engineered Food." *Journal of Agricultural and Environmental Ethics* 18 (2), 191-210.

Thomson, P. (1997). "Food Biotechnology Challenge to Cultural Integrity and Individual Consent." *Hastings Center Report* 27 (4), 35-38.

Part 3
Trust in food

Food safety in Europe: New policy and shifting responsibilities

Lotte Holm and Bente Halkier
Department of Human Nutrition, Rolighedsvej 30, 1958 Frederiksberg C, Denmark, loho@kvl.dk

Abstract

Following a series of scandals the European Union launched an ambitious restructuring of food safety regulation in the last decade. Central to this are the increased emphasis on the responsibility of market actors and the assignment of a new public role for consumers. How is this achieved? And how is the new policy implemented in different national contexts? The paper presents results from a series of institutional analyses which were part of the comparative project "Consumer trust in food. A European Study of the Social and Institutional Conditions for the Production of Trust" conducted by six research teams and coordinated by the authors. Political and institutional change over the last decades in EU, Denmark, Germany, Italy, Norway, Portugal and UK was analysed building on documentary analysis of legal and policy documents, reports, and statistical information from public authorities, market actors and organisations in the food sector, supplemented by media-texts and open-ended, qualitative interviews with key informants. It emerges that, though legislation and core political principles appear to be shared by EU and the six countries, the actual implementation of policy and institutional principles vary considerably, depending on the particular national conditions of, and relations between, state, market and civil society.

Keywords: regulation, food safety, institutions, europe

Introduction

The new food safety policy in EU was a reaction to the breakdown of public confidence in EU policies and institutions which followed the BSE crises in 1996, which was preceded and closely linked to the BSE crisis in the United Kingdom (Millstone and Zwanenberg 2002). Here, the political and institutional response to the crisis was formed in a manner, which reflected British conditions for restoring public confidence in the food system Recent changes in the British food chain have given more power to retailers and promoted the development of integrated production systems, which are effective in ensuring of food quality, protecting brands and thereby the markets of dominant supermarket chains. The new British food safety control system built on the procedures and systems which were created by the market actors, such as in-house quality assurance schemes which include systems of HACCP ('Hazard Analysis and Critical Control Points') (Warde, Harvey and Wales, forthcoming). Secondly, the BSE crisis in UK, which followed a series of minor food safety crises, destroyed public confidence in the public food safety system and governmental authorities in this field, calling for severe changes of public administration systems. Independence of ministerial systems and clear separation of scientific advice from the management of risks became the key characteristics of the British reorganisation of food safety policies.
The EU food safety policy was heavily influenced by the changes in the United Kingdom. It was based on an involvement of private actors in the food chain, who were given main responsibility for food safety, and on a separation of risk assessment from the politics of risk management. The question is, how such a policy fits with the realities of other European countries where market and regulatory conditions may be quite different and political controversy around food safety issues either lacking or moving in other directions than in the United Kingdom and EU?

This paper is based on a series of institutional analyses which come out of a comparative research project called "Consumer trust in Food. A European Study of the Social and Institutional Conditions for the Production of Trust (TRUSTINFOOD 2002-04)". The aim of the project was to describe and interpret variations and changes in ordinary consumers' trust in food in the light of institutional conditions in the food sector. Research teams from six countries participated: Denmark, Germany, Italy, Norway, Portugal and UK. The project included surveys among the populations of the participating countries as well as country-studies of institutional conditions and change in the same six countries and at the EU level(Bergeaud-Blackler 2004;Domingues, Graca, and Almeida 2004;Ferretti and Magaudda 2004;Lenz 2004;Nielsen and Møhl 2004;Terragni 2004;Wales 2004).

The new regulation

In January 2002 the new European Food Regulation was adopted (Regulation (EC) 178/2002). The regulation covers all stages of food/feed production and distribution establishes the basic principle that the primary responsibility for ensuring the safety of the food, rests with the food business. However, public authorities have an important role too: "To complement and support this principle, there must be adequate and effective controls organised by the competent authorities of the Member States" (http://europa.eu.int/comm/food/food/foodlaw/index_en.htm).

Risk assessment, risk management and risk communication provide the basis for food law. Transparency of legislation and effective public consultation are seen as essential elements of building greater consumer confidence. Better communication about food safety, including full transparency of scientific opinions, are considered of key importance.

In all the six countries included in the Trustinfood project the new regulation has been implemented. This has meant that complex and scattered bodies of law are replaced by one single Food law in most countries. In all countries it is now official policy that principles for food safety inspection have changed. The adoption of HACCP principles which are suggested in the regulation implies that post-factum product control is substituted by process control. This takes the form of public authorities' control with private self-inspection systems. However, in Italy and Germany legislation has to be implemented at Länder/regional level even if it is accepted at federal/national level and national/Federal authorities have no power to ensure proper implementation, leading to variation as to the extent of implementation of EU regulation (Ferretti and Magaudda, fortcoming).

Institutional reorganisation

The BSE crisis had revealed severe mismanagement within the scientific committees of EU and poor coordination and collaboration between EU directories (Bergeaud-Blackler, 2004). The new risk policy was therefore to cover the whole food production chain from plough to plate and to separate responsibility for risk analysis from that of risk management, i.e. it was to separate legislation from scientific consultation and from food safety inspection. The European Food Safety Authority (EFSA) was set up an independent authority with main responsibility for risk assessment and risk management was placed in the General Directorate for Health and Consumer Protection (DG Sanco. This Directorate must ensure effective control systems in member states and evaluate compliance with EU standards.

Food safety systems in many member states have historically been organised in a manner which differentiated between specific types of foods (meat, fish, plant products) and specific phases of the production chain (primary production in agriculture and fisheries, industrial production and the retail sector), and sometime on a distinction between food for the home-market and food for export. As a result, the body of regulations and the number of institutions involved in food safety policy were numerous, and it was often considered unclear who was in fact responsible for which aspect of food safety.

As a consequence, in all six countries it has been on the political agenda to reduce the number of institutions responsible for food safety and to clarify the division of esponsibilities. However, the rate of success in this endeavour varies between the countries. In most countries new institutions were built with the overall responsibility for food safety in all areas of the food sector and for the whole food chain. In the United Kingdom, this new institution – the Food Standards Agency, was set up as a non-ministerial government office, directly linked to the Prime Minister, in the other countries similar institutions were placed under the authority of ministries. In Denmark and in Germany (on the federal level) responsibility for food safety was concentrated in only one ministry; in Norway, Italy and Portugal several ministries are involved.

Clear separation of consumer and producer interests was a priority in the new EU policy as was a stronger representation of consumer interests and protection. Only in the United Kingdom this separation was ensured through the setting up of the Food Standards Agency as independent of the old Ministry of Agriculture, Food and Fisheries, and placed outside normal departmental structures under the direct responsibility of the Minister of State (Warde, Harvey and Wales, forthcoming). In none of the other countries included in our analysis this separation was ensured. In Norway, Germany and Portugal – and Denmark at first - responsibility for food safety was placed under ministries which were also responsible for agricultural policies. (In Denmark the food control system was transferred to a new Ministry of Family and Consumer Affairs in 2004).

The BSE crisis raised concerns about the independence of science. Consequently, a clear separation of scientific analysis and advice from the political management of risk came to the front as a way to restore public confidence in food system.Measures to ensure this have been taken in EU and in all six countries by basing risk policy on the separation of three steps: assessment, management and communication. To varying degrees these steps are the responsibilities of separate institutions in the national regulatory systems.

Integral to the new EU risk policy was also the intention to separate responsibility for legislation from that of inspection. In EU legislation is the responsibility of the Council of Ministers whereas responsibility for enforcement lies with DG Sanco. Similar separations are established in Norway and the United Kingdom. In Denmark legislation and food inspection are not separated and Germany and Italy the situation is different by tradition. Here legislation is the responsibility of the national political and ministerial system, whereas enforcement such as food inspection is the responsibility of regional and local authorities. In Portugal, by tradition there isn't a separation between legislation and enforcement for food safety, and the responsibilities are shared by several central, regional and local public authorities, which creates problems of overlapping of responsibilities on food control.

A new role of consumers

During the BSE crisis in EU the idea of a harmonious relationship between the consumer and market was destroyed. Objectives had to be prioritised: consumer welfare or market effeciency? The outcome of the BSE crisis was increased priority to consumer interests. Thus 'the consumer' became a central food policy actor. Traceability and labelling was seen as a tool for consumer control over all steps of the food chain, that is, as a means of direct control by consumers on market actors. Consumers were not to be protected from market excess but seen as an omnipresent and unpredictable figure, which needed to be taken into account.

This shift can be found in several ways in some of the six countries. It is reflected in the re-naming of existing institutions or in the naming of new institutions, and in formulations of

objectives of new and old institutions. However, in Italy and Portugal e.g. a strong prioritization of consumer interest within food policy has not appeared yet.
In some countries this new prioritising of consumer interests also show in new policies regarding the workings of the administrative system. In the United Kingdom e.g., consumers are the only market actors who are given a legitimate role and place in the new Food Standards Agency (Warde, Harvey and Wales fortcoming). In Portugal, during the BSE crisis, it was the first time that scientific experts and representatives of the food market and the consumers were called upon for consultancy in such a political process (Domingues, fortcoming). This was a new hitherto unknown practice in the Portuguese parliamentary democracy, which began only in 1974 after forty years of dictatorship. In Germany some Länder started to draw upon consumers for consultation (Lenz, 2004: p 79). In the Nordic countries the long tradition of incorporating consumer organisations as 'concerned parties' in legislative processes and public administration continued. In Italy consumers are still mainly seen as receivers of education and information about food, but some observers in the food market argue, that there is an increasing awareness of consumer interests and policy which is going to change the way public authorities approach consumers (Ferretti and Magaudda 2004) p. 86 independent actors able to and responsible for making relevant choices in the food market. In order for consumers to be able to live up to their new status as competent and independent actors, able to and responsible for making relevnt choices in the food market, full information about food products was necessary. This was the aim of legislation about labelling and traceability. The shift in view of consumers was by no means uncontroversial and have led to debate and conflict in several countries. Practices with regard to the publication of results from food inspection vary considerably. In Italy and Portugal e.g. such results are never published, wheresas in Denmark a more radical publication policy is adopted – the 'Smiley system' (Nielsen, forthcoming) which readily informs consumers on the spot about recent food inspections of any premise selling food products or meals.

Harmonisation and varying market conditions

Changes in regulatory practices are not uniform across the EU and the six countries.. Changes converge with respect to the regulatory framing and diverge with respect to institutional reform. On a formal level the new EU law is implemented in all countries changing the focus of food inspection from one of end product control to process control. However, there are huge variations with respect to how this new legislation is handled in practice in different national contexts. The new EU law is a good example of *regulation* (Majone 1990) in the sense that it includes several of the characteristics of the regulatory type of governance. It builds on capacities already existing in the regulated field, spreads responsibility to a broader range of social actors and blurs traditional distinctions between public authorities and private actors. However, it is unlikely that this regulatory type of governance will change balances between public and private actors in the same manner in all countries.
Food markets vary across Europe, and it is a question to what extent the new division of responsibility fits food market structures in different national contexts. The changes in the food market which have led to integrated production chains have been visible not only in the United Kingdom, but in other countries as well. This is the case not only in Northern and Central Europe, but also in the South. Creations of integrated production systems have appeared in Italy as well as in Portugal. However, in these countries large parts of the food production system still consists of small and family owned businesses (Domingues, Graca, and Almeida 2004;Ferretti and Magaudda 2004) and in southern Italy a considerable proportion of the food market belongs to the hidden economy. Under such conditions, food safety policies, which are designed on the basis of large-scale integrated production chains such as the British, cannot be fully implemented. It is therefore to be expected that there will be big national and regional

variations in the future EU food safety system. Further, sector differences are inevitable since also in Northern European countries there are areas of the food sector which are characterised by small businesses, some of which are not well integrated in ordinary regulatory systems.
It is, however, also to be expected, that the Single European Market and the new food safety policy will promote change within the food production systems of member states. In Italy, e.g. it is expected, that EU regulation will promote improved integration between various levels of food production and distribution. The demands of traceability will necessitate technical and administrative innovation and incentives for small producers to come together in cooperatives and associations in order to better face the economic burdens implied by the new methodologies for ensuring food safety (Ferretti and Magaudda 2004). It is thus expected, that EU food safety policy will promote increased harmonisation of not only regulatory practices but also of market structures.

References

Bergeaud-Blackler 2004, *Consumer Trust in Food. Institutional Report: European Union* www.trustinfood.org.

Domingues, M., Graca, P., and Almeida, M. d. 2004, *Institutional Country Report: Portugal*, www.trustinfood.org.

Domingues, M. (forthcoming). Politicisation and Institutional Unclarity: the case of the Portuguese Food Agency

Ferretti, M. P. and Magaudda, P. 2004, *Institutional Country Report: Italy*, www.trustinfood.org.

Ferretti, M.P. and Magaudda, P. (forthcoming). The slow pace of institutional change in the Italian food system

Lenz, T. 2004, Consumer first? Shifting responsibilities in the German food system in the light of European integration and the BSE crisis, www.trustinfood.org.

Majone, G. 1994, "The Rise of the Regulatory State in Europe", *West European Politics*, vol. 17, no. 77, p. 101.

Millstone, E. and Zwanenberg, P. v. 2002, "The Evolution of Food Safety Policy-making Institutions in the UK, EU and Codex Alimentarius", *Social Policy and Administration*, vol. 36, no. 6, pp. 593-609.

Nielsen, A. (forthcoming). Contesting Competence: Change in the Danish Food Safety System

Nielsen, A. and Møhl, T. 2004, *Trust in Food - Country Report: Denmark*, www.trustinfood.org.

Terragni, L. 2004a, *Institutional Country Report: Norway*, www.trustinfood.org.

Terragni, L. Norway never had a BSE crisis. To be published in: Changing Responsibilities for Food in Europe (Halkier, B.and Holm, L.eds) . 2004b.

Wales, C. 2004, *Country report: United Kingdom*, www.trustinfood.org.

Warde, A., Harvey, M. and Wales, C (forthcoming). Recuperating from BSE: the shifting UK institutional basis for trust in food.

Politicising consumer trust in food: A socio-institutional explanation to variations in trust

Unni Kjærnes, Alan Warde and Mark Harvey
The National Institute for Consumer Research, PO Box 4682 Nydalen, 0405 Oslo, Norway, Unni.kjarnes@sifo.no

Abstract

The topic for this paper is the role and foundations for trust in food in modern societies. Trust in food is often analysed as a matter of individual risk perception and trustworthy risk communication. Generally overlooked by individual level explanations are observations of rapid macro level shifts and large and consistent cross-national variations. Such variations and shifts are not easily explained within these perspectives, which focus mainly on individual strategies and communication efforts. The paper will argue that trust is social and relational, meaning that we should seek to understand the dynamics of 'who trust whom in regard to what'. By this expression we do not suggest rational decision-making processes. The approach assumes that trust in food is influenced by long-term cultural factors as well as responses to the performance of central institutional actors. We are first of all looking for aggregate effects in terms of variations in institutionalisation processes, what we call 'triangular affairs' between provisioning systems, regulations as well as consumption. We therefore concentrate on macro-level interrelations between three poles: the food market, public regulation, and the consumers. The paper will present some findings from a comparative study in six European countries, Trust in Food, including public opinion surveys and institutional studies in Denmark, Italy, Germany, Great Britain, Norway, and Portugal. It will focus particularly on some key issues related to trust in various institutional actors. While there is relative consensus across Europe that civil society actors can be trusted most, followed by public authorities, and market actors and politicians the least, there is considerable national variation in the overall levels of institutional trust and also in the degrees of differentiation across the various actors.

Keywords: trust, food, consumption, institutionalisation

The background

Consumer distrust in food has emerged as a pressing issue on the political agenda over the last decade or so. Many have tried to understand this, with a variety of approaches and explanations, most of them concentrating on what is happening to consumers. But if you look at pan-European public opinion poll data, there are systematic variations in levels of trust which cannot be attributed either to universal distrust among consumers, or to their inability to understand or evaluate risk. We have to look elsewhere than individualistic explanations. There is obviously something distinctive about each country; but it cannot be some sort of national character, since levels of trust also vary significantly over time. This is our main interest. What is it in our modes of food consumption and their social and institutional environment that sustains trust in food in contemporary Europe? What is it that nurtures scepticism and distrust among food consumers and provokes intermittent crises? The paper is based on a study of opinions, organisational structures and strategies in six European countries based on a research project conducted between 2002 and 2004.

Trust is subject to much philosophical dispute. Following from the questions that we posed and the problematisation of 'the consumer' and consumer choice, we explore trust not only as social, but also relational. This means that instead of trying to contextualise individual opinions of trust, we study trust and distrust as emergent properties of on-going relationships between social actors. For example, in the case of food, relations people have *vis à vis* a producer or a retailer will be different from those they have towards government bodies. Mostly people want access to food which is safe, nutritious, gives fair value for money, is consistent, tasty, etc. Trust could be seen as an expectation, a practical confidence, that other relevant actors in the food system will behave in such a way as to ensure that some or all of these objectives are met. The main point is that it is *actors* who can be trusted or mistrusted. Trust is thus a strand, or a dimension, of a relationship between actors. Trust depends on who they are, what they do, and the interaction involved. We consider three sets of actors particularly important because they are directly involved – market actors who supply food, state agents responsible for regulation and governance, and consumers and their representatives. Above all, it is the extent to which consumers find these various actors trustworthy which concerns us. These are, theoretically, the key institutional formations from which public trust in food emanates. We have suggested that the state, market provisioners, and consumers may be differently institutionalised in different countries. When we talk of the institutional basis for trust in food, therefore, we consider that the core of any explanation requires an analysis of the relationships between these three 'poles', a triad of relationships, or, what we name 'triangular affairs'. All three poles may vary from country to country, but to understand trust, we need to understand the relationships between them. We use both survey and institutional data as inputs to this analysis. In doing this, we see consumers and consumption in a particular way, where 'the institutionalisation of consumption' is a key concept, a concept that directs attention towards distinctive patterns of how consumption is organised and normatively founded.

Food nations

Whether by big transnational food companies with worldwide brands or globally renowned retailers bringing an ever-expanding variety of counterseasonal produce from every corner of the world, food transcends national boundaries. So do panics and scandals. Avian flu knows no absolute barriers, geographical or biological. BSE provokes fear in Europe or Japan. And political responses and frameworks have become increasingly transnational. Yet, remarkably, we find in our data that food nations and cultures survive. Deep differences and divergent historical trajectories persist. Discovering that trust and problems of trust are not the same all over Europe has led us to appreciate the significance of the national and societal basis of trust. This is not to argue, of course, that convergence and processes of globalisation have no impact on the way different national societies develop. But, whether in terms of public opinion, or the way that food is provided and consumed, or how states respond to change and crises, societal patterns are powerfully shaped within national boundaries.

Starting out with public opinions, country variations are often different from what might initially have been expected. Great Britain, which has experienced the gravest and most disruptive scandals over food in recent years, appears as a country where people display considerable enthusiasm regarding food safety and optimism for a range of other food issues as well. It is less surprising that Danes and Norwegians are generally trusting, but they are so to a lesser degree than Britons. Germany, on the other hand, is marked by widespread public distrust and contention within the food system. Italy, which is renowned for a strong focus on fresh, high quality food, is marked by the highest and most consistent levels of distrust regarding all aspects of food and the food system. Finally, and also quite surprising considering the findings in Italy, the Portuguese are considerably less distrustful towards their food, but they are highly sceptical towards authorities and supermarkets and pessimistic regarding ongoing changes.

Consumers also differ significantly in what they do in practice from country to country. To some extent this should not be surprising, but, especially in the light of much recent sociological and cultural discussion, nonetheless needs emphasising. If the Spanish sustain a norm of eating their main evening meal at 10 pm and Norwegians their lunch at 11 am, there is a logic to these coordinating routines within countries which is uncompelling between countries. While tourists adapt, locals still meet and eat together, in households or wider social networks, within a national social space. Whether to have one or two main meals a day, how often and where to shop, the amount of home food preparation, the use of microwave ovens, and a raft of other aspects of what people do mark differences in behaviours that typify consumers nationally. How much money to spend on food, or whether food safety or quality is the number one issue, express national particularities. For that reason, we speak of an institutionalisation of the consumer and consumer behaviours and culture as societal features of different countries. It is not as if these patterns never change – indeed, we find that they are shifting considerably – but they change in different ways in different countries.

Turning to how food is produced and delivered to consumers, there is no uniform or common 'supermarket phenomenon' across Europe. Certainly, there is evidence in all countries of the growing importance of large outlet food retailing owned by big chains. But there the similarity stops. Indeed, taking the two countries most dominated by supermarkets, Norway and the United Kingdom, the differences are most extreme, whether in the variety of food, its geographical sourcing, its cultural diversity, its orchestration of food quality, or the extent to which the retailers as such control the whole food chain. Germany divides between price discounting retail chains and smaller traditional green or quality-oriented outlets. Portugal has experienced a foreign invasion of retail chains, whilst Italy's Co-op is just about holding its ground against French supermarkets. More widely, however, the role of farmers and primary producers, often with important political organisations and electoral clout, or of home-country manufacturers, have varying weights and differing profiles from country to country. The provisioning systems of different countries are still primarily national realities.

At the governmental level, since the early 1990s there has been a huge upheaval in the institutional ways of dealing with food, its safety and its quality. Here the issue of national versus transnational is certainly complex within Europe, and there is more than an appearance of common models being adopted across the European Community. In response especially to BSE, the British were the quickest and most radical in their establishment of a Food Standards Agency (FSA), with independence from government and a more prominent institutional role for consumer organisations. The European Commission re-organised its government of food related issues under a new DG Health and Consumer Protection (DGSanco) and the European Food Safety Authority (EFSA). It seems as if these quick responses, and the way they were done have had the effects of restoring confidence in markets, and the legitimacy of authorities when it comes to food safety. But the Nordic states differ quite significantly between them in the way they regulate food or 'empower' the consumer, and not always in ways that support the current literatures on varieties of states and markets. Germany responded to the trust crisis – at the state institutional level – in quite a distinctive way, by strengthening – and at the same time re-directing - a producer orientation with the 'greening' of the agricultural ministry. It was an 'agricultural turn'. The Italian response was a further regionalisation of responsibilities, while the Portuguese had difficulties making any definitive decisions at all. In each country, national governmental structures have responded to crisis and change in distinctive ways, within existing institutional and political constraints. The result, overall, is certainly not a diminution in differences between national ways of governing food or according consumers role and representation, and in many ways, differences have sharpened. There is a paradoxical European reality of increasingly common models and frameworks *and* national divergence.

A socio-institutional explanation to variations in trust

In treating trust as an emergent aspect of these societal relationships, three aspects are especially important: the societal division of responsibilities between actors; the matches and mismatches between the norms and expectations of the various actors, especially in relation to their practices and performances; and the societal 'configurations' of these relationships, or how the relationships fit together.

The first aspect involving the division of responsibilities for key food issues is crucial for understanding trust in terms of social relationships between actors. There is no one particular type of division of responsibilities that we can say will produce trust or distrust. The relationship between state and market actors differs according to their respective responsibilities as they have been institutionalised in different countries. In the area of food new roles of the state as well as of the market are emerging, for example by making direct monitoring and inspection a market responsibility, while public authorities take on a more indirect role as auditor. What we see is a process of re-regulation rather than generally declining role of the state. Shifting responsibilities may also establish new boundaries between the public and the private. Responsibilities may be redistributed between market actors, the state and private households. Policies which centre upon 'the consumer' as a private and individual figure tend towards the privatisation and individualisation of responsibilities and thus potentially the de-politicisation of food issues. However, the very fact of their being consumer policies implies the politicising of consumption. Handling consumer distrust then pulls food consumption out of the private and into the public sphere.

For this type of reason, one of the most striking conclusions from this type of analysis is that scandals happen differently in different societies, not only in terms of magnitude of impact, but equally in terms of institutional response or consumer protest. A food crisis may even reinforce trust if its resolution is achieved through meeting consumers' pre-existing norms and expectations of those it holds responsible for handling crises.

A crisis about food may be the occasion in which tensions and mismatches between actors' norms and expectations become manifest, discussed publicly and in the media. But it is longer term changes in consumer behaviour, in the way food is provisioned, or the way the state is governed, that often constitute the underlying dynamics of trust and distrust. After all, BSE was a scandal that broke at a particular time, and in relation to a particular threat to human health, but it was predicated on developments in agricultural technologies and production methods that had being taking place on an historical scale.

Finally, as the third aspect of a socio-institutional explanation, we turn to the way relationships fit together amongst the cast of actors involved in food provisioning and consumption. Given that we are seeking explanations at the societal level and looking at national configurations, we have privileged three main actors: consumers with their shared national characteristics; the market food provisioners; and the state actors (regulatory bodies, experts, etc.). The national configurations between these three relational poles were described in terms of 'triangular affairs'. Between the societally institutionalised consumers, national provisioning systems, and the state endowed with different modes of governance and organisation of responsibilities, there are shared and nationally typical relationships between consumers and provisioners; consumers and state; and state and provisioners.

How the relationships fit together in the overall configuration is more important than the strength of trust embodied in any one of the relationships taken separately. Even though Norway and the United Kingdom are so different in terms of each of the three principal relationships, consumers manifest high levels of trust in food because of the concordance between norms and expectations running between them. Simplifying, what really matters is that everyone expects everyone else to be doing what they are doing with respect to everyone else in the triangular affair. What is actually expected of whom in Norway and Britain, and how responsibilities are organised, how and what different actors do, matters much less than consistency and conformity of norms and expectations between our three polar actors. Governments, retailers and manufacturers must do what they are expected to do – and what they are doing must be acceptable to people as consumers and citizens.

Likewise for distrust: actors differ between each other both with respect to what they expect of each other, and in the norms of what they do in relation to each other. Germany displays relatively high levels of distrust, and, using the configurational approach, we argue that conflicts of expectations and norms arise not only between consumers and the other two poles, but also in what the state expects of consumers and market actors. Its support of green and local food against the mass standardised foods of discount retailers aggravates the tensions. Italy presents quite a different configurational basis for even higher levels of distrust: a distrust of state regulation, and more importantly of rules imported from the European Community, meets a consumer torn between modernising lifestyles and conserving tradition, conceptualised as the local, the 'typical'. In this analysis, distrust is much more than an absence of harmony and concordance between actors and their norms and expectations. There are positive conflicts, reflecting the historical capacity of food to generate controversy, now fuelled by the rapid changes in consumption, regulation and provision of food.

Why increasing predictability cannot do the job alone when we aim to establish trust in the agri-food sector

Franck L.B. Meijboom
Ethiek Instituut, Utrecht University, Heidelberglaan 2, NL-3584 CS Utrecht,
f.l.b.meijboom@ethics.uu.nl

Abstract

In spite of all attention to the safety and quality of food, public trust in food is still subject to discussion. A common way to address the public hesitance to trust the parties within the food sector is a strong focus on increasing predictability. This mainly results in measures that aim to increase transparency and provide information. The argument that underlies this focus is that the real problem of trust is a lack of predictability and a need for possibilities of a trustor to anticipate on what will happen when one trusts. Without denying the relevance of explicating routines and patterns in the sector, I argue in this paper that predictability is not a sufficient condition if the aim is to build public trust in the agri-food sector. For this claim I present three arguments. First, the current food sector faces us with the problem that patterns and routines are not simply elements that are present and only have to be revealed with the help of sufficient information. This has implications for the scope of transparency, which is elaborated in my second argument. Transparency cannot be restricted to mere openness. It has to include more fundamental assumptions that underlie actions and policy. Finally, trustful expectations are not only based upon predictable patterns, but also on anthropological or moral assumptions. These assumptions enable us to trust even when predictable patterns fail or are conflicting with each other. Hence they are an essential addition to the emphasis on trust and predictability.

Keywords: trust, food, transparency, normative expectations

Trust and predictability

Much time and effort have been spent to make food healthier and safer since the last decades. Nevertheless, Slovic's claim that "many in the public have become more, rather than less, concerned about risk" (1999) also holds for the food sector. In spite of all improvements at the level of risk analysis, it has been argued repeatedly that there is a need for rebuilding and maintaining public trust in the agri-food sector. The FAO stated that the food safety system must be able to "manage risks and inspire trust" (2003) and the European Commission emphasised that there is a "need to develop trust" (Byrne, 2004). This tension between the improvements with regard to food safety and quality and the public hesitance to trust food can be considered as an argument for the claim that trust is something different than deliberately taking risks as has been argued by scholars as Coleman (1990). Different authors (e.g., Lahno, 2002; Lagerspetz, 1998) argue that, although trust is relevant in risk related situations, it is not a matter of taking risks based upon a risk-benefit calculation.
However, when we grant this critique on this risk approach – and I think there are good reasons to do so[58] – the question arises of what is an adequate way to develop trust in the food sector if trust is something different than taking risks.

[58] It is beyond the scope of this paper to elaborate on this critique at full length. In the literature on trust this critique has been formulated in different ways (cf. Baier, 1994; Lagerspetz 1998; Lahno, 2002).

A common way to address this situation focuses on increasing predictability, in practice, on increasing transparency and providing information. The argument that underlies this focus is that the real problem of trust is a lack of predictability and a need for structures on which a trustor can anticipate. Consequently, it is important to provide information about the trustee and to increase transparency to reveal the patterns that underlie the behaviour of the other party since it enables the trustor to predict and anticipate. This focus on information on underlying standard procedures and routines, either regarding the object of trust (e.g., the product), or the subject of trust (the trustee) can easily be recognised in regulatory frameworks. However, making a situation more predictable by providing information or increasing transparency is a necessary, yet not a sufficient condition if the aim is to build public trust in the agri-food sector. For this claim I have three arguments. First, predictable patterns are not always obvious in the current sector. It is not something that is implicitly present and can be revealed with the help of more information or transparency. The search for predictable patterns is complicated. This is (a) the result of a lack of predictable patterns and (b) because it is not always clear which of the available patterns is applicable. Second, increasing transparency in order to make the other agent or the situation more predictable can only be an accurate tool under certain conditions. Transparency has to be conceived in a broader way that includes more than risk-related issues. Finally, predictability is mostly referred to in terms of expectations "that" someone will act in the favourable way. However, we also have trustful expectations that are not based on regular patterns, but on anthropological and moral assumptions.

The problem of revealing patterns

Trust needs a certain level of predictability. This is what is meant by "anticipatory trust" (Sztompka, 1999) or "predictive trust" (Hollis, 1998). This type of trust is based upon the expectation that the other party will act according to normal patterns and routines. If clear patterns and routines are available, it is easier to predict how the trustee will react and what to expect. For instance, if you have bought a product for many years, you will expect that its safety remain unchanged the next time you buy the product. Hence you rely on this being so even though there is always a risk that this may be the first time the product is unsafe. This does not only hold for the object of trust, but also for the subject of trust. Predictable patterns and experiences contribute to mutual knowledge about beliefs and expectations. Hence, as a trustor, you are in a better position to formulate expectations about each other when you are confronted with uncertainty. Information about the patterns and routines that underlie the behaviour of a trustee is relevant in order to develop trust.

Nevertheless, providing information in order to reveal the patterns that lie beneath the actions of the other party is complicated in the food sector. First, we regularly are confronted with new situations in which it is not easy to predict what we can expect. An example is biotechnology. Since this technology is a relatively new in food we lack the predictability and familiarity that can serve as a first basis for trust. Partly, because, we are confronted with new benefits, unknown carry-over effects and long-term effects, which may lead to unforeseen consequences. Partly, since we lack a history in which trustees have shown their reliability. Hence there is not a clear pattern or history that only has to be explicated or revealed in order to show the trustor that his reliance is warranted.

Second, though predictable patterns are sometimes available, they can be conflicting. Hence it is still difficult to predict whom we may expect what. The introduction of food products with a health claim illustrates this. The relation between food and health is not new, yet lowering blood cholesterol with the help of a dairy product is new. Normally you do not drink milk for lowering elevated levels of cholesterol, just like you do not take liquid drugs when you are thirsty. Both for drugs and for dairy products we have clear patterns and traditions that provide predictability that explicates what we can expect regarding issues as safety and justice. This helps to trust with regard to food and pharmaceutical products. However, since a food product with a real health claim can be categorised in both groups, there is not one unambiguous pattern available upon which one can formulate trustful expectations. Trust is either based on patterns that are familiar in the food sector, although it has a health claim, or it is based on the patterns of the pharmaceutical domain, although it is a food product. Thus, the introduction of such a diary product complicates pattern-based-trust. In this situation more information or increased transparency contributes to the explication of patterns of behaviour and action, however, it does not address the fundamental problem of which of the conflicting patterns is the most applicable and serve as warrant for one's reliance.

Hence, predictability and familiarity are often no longer strong enough to serve as reason for trust: a consumer has to find other reasons that can function as basis for trust.

Transparency: more than openness

My second argument concerns transparency. Improving transparency is as an important tool to deal with the lack of predictable patterns or the conflict between existing patterns. However, its usefulness is subject to conditions. Transparency is sometimes equated to openness or making all available information public. This, however, neither establishes trust nor solves the problems related to the lack of patterns. O'Neill (2002), correctly, criticises the relation between trust, openness and information. She observes "we are flooded with information about government departments and government policies, about public opinion and public debate (...)." Additionally, she argues that "if making more information about more public policies and institutions and professionals more widely and freely available is the key to building trust, we must be well on the high road towards an ever more trusting society (...) trust should surely be within our grasp." This last conclusion is obviously only rhetorical. In spite of the available information the level of public trust in the food sector did not increase in a proportional way.

This illustrates that transparency defined in terms of sheer openness is only adequate to a limited extent as tool to build trust. For instance, when meetings of a Food Safety Agency are open to attend by the public this does not automatically build trust. The public character of a meeting offers the opportunity to check the agency's competence and goodwill. However, to check this we need more than just openness about patterns. One should not only show the decisions that have been made or the measures that are being proposed. The arguments and the theoretical models that underlie the decisions should be communicated too (Rasmussen and Jensen, 2005). Such an extended object of transparency entails that there has to be explicit attention to arguments and evaluative judgments that underlie a position or decision. This implies that both normative assumptions and factual premises have to be explicated and subjected to discussion. The explication of these fundamental issues helps to trust when predictable patterns fail or are not compatible with each other.

Since trust is relying on the other party's competence and goodwill towards you or the object of trust, a broader concept of transparency is relevant with respect to both competence and goodwill. With regard to competence, an extended object of transparency is relevant since then it will become clear that the trustee is able to deal with the normative deliberation that is necessary to deal with issues as the safety of GM-food or the prevention of Avian Influenza. Moreover, communication on the evaluative judgements that one has taken gives the trustor an indication of one's goodwill.

In short, transparency as a tool to reveal routines and predictable patterns is a too marginal definition to address issues of trust in the food sector. When patterns are conflicting or lacking, transparency has to include clarity with regard to assumptions that underlie the decisions and policy of the trustee.

Anthropological and normative assumptions

For my third argument I return to the patterns and routines. Without doubt, information with regard to predictability is useful in cases of uncertainty; especially when patterns have a profound dimension, e.g., when they are the result of conventional norms. When I buy my food, I trust that my Euro notes are accepted. This is not based upon experience only, but on conventions too. When the supermarket operates within the European monetary system, I trust them to accept my money even when I do not have any experience with this retailer. Within such conventions one has stronger reasons to trust.

Nevertheless, even within these conventions a serious problem still exists. Routines or conventions do not necessarily informs us about the actual behaviour of the trustee. Since humans have the freedom to act, we have the freedom to leave patterns and act against predictability. This means that we are confronted with new uncertainties: Even if there are predictable patterns, one is still uncertain whether another will act according to the expectations. We can only rely on patterns if we have indications that the other party is reliable. Thus, establishing trust in the food sector cannot be limited to providing information on patterns and procedures. It equally should show whether the trustee will follow these patterns and routines, what incentives does he/she have for responding to my trust?

The answer to this question is related to anthropological and normative assumptions. Let's start with the anthropological assumptions. In many accounts of trust we can recognise anthropological premises. For instance, in theories with a rational choice background it is expected that individuals confronted with uncertainty act in line with their interests and strive to maximize utility. This assumption does not only hold on a descriptive level. It is not only *likely* that agents act rationally, they also *should* do so. This claim is linked to a prudential rather than a normative "should". Within rational choice accounts one does not only expect *that* an agent will act in line with his interests, but it is also expected *of* him. This illustrates that trust include both the expectation "that" someone will act in a favourable way and an expectation "of" the other agent (Hollis, 1998). This stronger claim enables us to trust when we lack predictable patterns, since we can still expect of the other party to act in the favourable way if it is in his interests.

Furthermore, normative assumptions enable us to formulate expectations that are beyond the level of predictability. For instance, I do not only expect *that* the government ensure an adequate food safety system, I also expect it *of* them even if it were in their interest to do otherwise. I believe that they have a moral duty to do so and that I am entitled to expect this. This assumption of entitlement can be defended in different ways. First, one can argue that if one's trust is (implicitly) accepted by another a specific performance is required. In line with the moral dimension of making and keeping promises (Scanlon, 1998), trusting implies certain obligations to respond in the expected way. Second, the acceptance of trust by the trustee is not a necessary condition for someone's moral expectations. My belief that an adequate food safety system is something that can be expected of the government is not depending on whether my expectations are explicitly accepted. I consider myself entitled to expect this. This belief can be justified in different theoretical ways, e.g., on utilitarian grounds, from Kantian perspectives, or from contractualist accounts (e.g., Scanlon, 1998). This last approach seems promising to me, since I do not consider it hopeful to look for trusting expectations that can be morally justified independent from a trusting relation.[59] Only from a relational perspective it is possible to differentiate between various types of obligations connected with the various types of relationships (Misztal, 1996). This enables us to formulate what one may reasonably

[59] This does not imply that an expectation based upon trust cannot be justified from a third-person perspective.

expect of one another, which can serve as reasons to trust even when predictability-based-trust is complicated.

This illustrates that attention to anthropological and normative assumptions is not a luxurious dimension in the analysis of trust, but are a necessary addition to the current attention to revealing patterns and routines by providing information and enhancing transparency.

References

Byrne, D., (2004). "The Regulation of Food Safety and the Use of Traceability/ Tracing in the EU and USA" Available on: http://europa.eu.int/comm/dgs/health_consumer/library/speeches/speech168_en.pdf [19 April 2006]

Coleman, J.S. (1990). *Foundations of Social Theory*, Belknapp, Cambridge. 1014p.

FAO. (2003). *Expert Consultation on Food Safety: Science and Ethics*, Rome, FAO.41p.

Hollis, M. (1998). *Trust within reason*, Cambridge UP, Cambridge. 170p.

Jones, K. (1996). 'Trust as an affective attitude', *Ethics*, 107, p.4-25

Lagerspetz, O. (1998). *Trust: The Tacit Demand*, Kluwer, Dordrecht. 177p.

Lahno, B. (2002). *Der Befriff des Vetrauens*, Mentis, Paderborn. 461p.

Misztal, B.A. (1996), *Trust in modern societies*, Polity, Cambridge. 308p.

O'Neill, O. (2002). *A Question of Trust*, Cambridge UP, Cambridge. 100p.

Rasmussen B, Jensen, K.K.. (2005). *The hidden values. Transparency in decision-making processes dealing with hazardous activities*, Schultz, Copenhagen, 47p.

Scanlon, T.M. (1998). *What we owe to each other*, Belknap, Cambridge.432p.

Slovic, P. (1999). 'Trust, emotion, sex, politics, and science: Surveying the risk-assessment battlefield', *Risk Analysis*, 19/4, p.689-701

Sztompka, P.. (1999). *Trust: a sociological theory*, Cambridge UP, Cambridge, 226p.

Part 4
Ethics and safety in food discourse

GMO controversy in Japan over test growing of PEPC transgenic rice plants

Mamoru Fukamizu, Seiko Yoshida, Kohji Ishihara
Creative Research Initiative "Sousei," Hokkaido University, North 21, West 10, 001-0021 Sapporo, Japan, tech-ethics@mail.cris.hokudai.ac.jp

Abstract

With the increase of GM crops in the world, many GM crops have been imported to Japan. To deal with this situation, Japanese regulatory ministries have set up a system of safety assessment and labeling. As for the growing of GMOs, Japan has reached the stage of test growing in research institutions, with a view to beginning commercial growing in the near future. However, the test growing of GMOs has raised a new controversy of GMO in Japan. We will discuss the ethical problems of GMOs growth with reference to the controversial case in Hokkaido prefecture, Japan's biggest food supplier. Moreover, we will try to apply Brian Norton's convergence hypothesis proposed in environmental ethics to agricultural ethics and suggest a concept of risk communication helpful to solve the problem and indicate the future vision of agriculture.

Keywords: GMO, risk communication, convergence hypothesis, agricultural ethics

The problem of the food self-sufficiency rate and GMO policy in Japan

To understand the background of GMO controversy in Japan, it is necessary to know about the problem of the food self-sufficiency rate. Japanese food self-sufficiency rate has declined to 40% (calorie based) in the last 40 years. Grain self-sufficiency rate has decreased to only 27% (MAFF 2005a). Japanese import dependency of foods has been brought about by social factors, such as decrease in agricultural population, change in eating habits, and culling out of domestic foods by imported and cheaper foods. The decrease in farm houses, especially, is one of the most severe problems. The number of Japanese farm houses has decreased by more than 10% in the last 5 years (ibid.), which has been caused by concentration of young population in cities, retirement from farming and abandonment of farming. Moreover, the high cost of new entry into farming prevents newcomers into farming. The downward trend of workforce in farming will reduce Japanese self-sufficiency rate even more.
GMO controversy in Japan has been developed against the backdrop of Japanese food supply system that is dependant on import. Since commercial growing of GMOs has not begun in Japan, all GM foods available in Japan come from abroad. The Ministry of Health, Labour and Welfare decided to impose mandatory safety review on all GM crops and established labeling rules on April 1st, 2001. As of Feb. 14th, 2006, 75 GM crops, which include the following 7 species —potato, soybean, sugar beet, corn, rape seed, cotton and alfalfa—, have been approved (MHLW, 2006). Japanese labeling regulation stipulates that foods, which are composed by materials that are within the top three by weight by GM crops as well as more than 5% GM crops, must be labeled. Conversely, it does not request to label foods that include less than 5% of GM crops or does not meet the above weight condition for GM crops. Moreover, it is not requested to label 1) unintended mixture of not over 5% GM crops during the course of distribution, 2) processed foods such as soy sauce and cooking oil produced by GM crops if recombinant DNA and proteins produced by recombinant DNA are eliminated (MHLW, 2001). Under such an insufficient labeling

system, there remains the ethical problem that Japanese consumers' right to choose will not be sufficiently guaranteed even if food companies obey the labeling regulation.

Inefficient risk communication on test growing of transgenic rice plant

Let us turn to the controversial case of the growing of GMO in Hokkaido. In May 2003, National Agricultural Research Center for Hokkaido Region announced to execute a test growing of PEPC transgenic rice plants (hereinafter referred to as PEPC rice plants) in the open agricultural field at the center. At the same time, the center announced to hold a prior explanatory meeting for the public. (PEPC rice plant refers to the rice plant to which PEPC genes of corns is introduced. In Japan, besides PEPC rice plant, some kinds of GM rice plants with functions such as cool weather resistance and pathogen resistance have been and are being planted in the research agricultural fields. Moreover, GM potato, rape, corn and soy are under research and development.)

Up to this time, PEPC rice plant had been under research and development in closed green house and isolated fields. It was the fist time for them to test PEPC rice plant in the open field. The center held prior explanatory meeting in order to win understanding of stakeholders and the public. However, the participants of the meeting felt that they were not informed sufficiently. The participants were especially unsatisfied with the information given on the isolation distance necessary for avoiding crossing by GM rice plants. Researchers gave a "scientifically" based explanation using available data, and did not give any answer to reasonable questions, such as, "What might happen if a stronger wind blows than that has been expected in the experiment?" The meeting resulted in making organic farmers and consumer's organizations, such as co-op, to form solidarity as opponent groups rather than winning their understanding (Yoshida 2006). This meeting confirmed that one-way risk communication without considering people's concern will fail. While it has been emphasized the inappropriateness of one-way risk-communication (NRC, 1989), they failed to make efforts to conduct interactive risk communication. This failure to conduct risk communication in the early stage made the possibility of efficient risk communication and adequate discussion difficult.

In Japan, the problem of GM rice plant has a background which escalates the conflict. Since rice is Japanese staple food, Japanese are very interested in its safety, brand, and eating quality. The policy concerning rice constitutes the bedrock of Japanese agricultural policy and rice plants are one of most important crops for Japanese farmers. Considering this background, agricultural ethics concerning GM rice plants should take into account the following two factors:

1. The possibility of derogation of the brand value of Hokkaido. Hokkaido has a brand value as a good producing region in Japan. If the commercial growing of GMOs began in Hokkaido, it is possible that Hokkaido brand value will derogate. Hokkaido prefecture made some efforts to keep the brand value after the beginning of the commercial growing of GMOs, by introducing an ordinance which regulates the growing of GMOs (see the last section of this paper). However, even if they succeed in preventing crossing, consumers might hesitate to buy agricultural products made in Hokkaido.
2. The significance of developing GM rice. The development of GM rice plants with functions such as cool weather resistance, high-photosynthesis, and pathogen resistance aims at increasing yield point. However, rice consumption in Japan is declining more and more because of the diversification of Japanese dietary life. Japanese rice consumption per capita has declined to half in the last 40 years (MAFF, 2005b). Considering the downward trend of rice consumption, the purpose of the development of GM rice is arguable. (As for the purpose of stable production, GM technology may contribute to achieve this purpose. However, it is necessary to evaluate the benefit of GM technology for this purpose more cautiously.)

Agricultural ethics should take into account impact assessment based on economic and social factors, aside from safety assessment of food products. We should not forget that the political regulation of GMO has direct influence on agricultural regions. Considering the broad influence on regions, the regulation should be executed based on disclosure of information and a certain degree of agreements among stakeholders. Agricultural ethics concerning GMO should be discussed not only from the theoretical perspectives, such as consumer autonomy, intrinsic value of plants and ecosystems, but also from the practical perspectives, such as efficiency of risk communication and the future vision of agriculture.

Risk communication toward convergence

The failure in risk communication at the prior explanatory meeting of the test growing of PEPC rice plant came from not only the one-sidedness of communication, but also from the content of risk message. At the meeting, researchers in GMO gave information on safety assessment concerning human health and ecosystem, necessity of developing GMO to deal with expected demand of foods due to the growing of world population in the future, and benefit of GM technology. Opponents of GMO also admit these issues are important. However, if we analyze their statements carefully, we can find that their concerns are more related to the problems such as the fear of mastery of seeds by major biotech companies, the imbalance of the distribution of risk and benefits, the crossing by growing of GMOs and the labeling. The discordance between the information given at the explanatory meeting and the information wanted resulted from the discordance of their concerns. While researchers took into consideration only the risk to human health and environment, the public has also concerns with the impact on the region, daily diet life, and subsistence farmers.

As pointed out in the literature of risk communication (Fiorino, 1989), the public's concerns with risks are more related to social and political consequences than those of researchers. Taking into consideration such situations, risk communications on GMO should not be limited to scientific knowledge. To broaden the coverage of risk communication to social issues does not mean to ignore the demonstration of such scientific knowledge. Rather it would contribute to broaden the coverage of communication and achieve a compromise. To achieve a compromise, we should start from a shared understanding among proponents and opponents of GMO.

Here we would like to draw on Brian Norton's "convergence hypothesis" (Norton, 2003; 2005). He points out that opposite anthropocentric and non-anthropocentric environmentalists could "approve many, perhaps all, of the same policies" (Norton, 2005, p. 508). Of course we should take into account that this hypothesis is proposed concerning the opposition of environmental ethics. However, if we note that one of the reasons of opposition concerning GMO is the difference in sense of values and difference in the stance toward agriculture, Norton's convergence hypothesis, which could indicate a way of dealing with severe opposition, seems to be applicable to the GMO controversy.

It would be helpful to note Norton's expression: "provided these antithetical theories are properly formulated" (ibid.) in considering application of the convergence hypothesis to GM controversy. It is true that the claims of both sides are in a striking contrast. However, it is also true that they have never sat at a table to discuss the issues. This means that they do not understand each other's claims. Their claims have not been even formulated. By sitting at the same table and discussing in person, opponents and proponents would be able to understand their claims respectively.

We emphasized the importance of risk communication not only because we hope that this will be effective in the conciliation between opposites, but also because we think that agricultural politics and risk communication on GMO will influence regional agriculture in the future, the future role of biotechnology in agriculture and our dietary life as well as that of future generations. For example, in order to achieve sustainable development in agriculture, there need to be sufficient number of farmers. This means farming should be attractive to young

people. However, future views of Japanese agriculture envisioned by both proponents and opponents of GMO are not so promising. While the proponents intend large-scale farming based on biotechnology which may drive out small-scale farmers, the opponents propose low input agriculture with low profitability that is labor-intensive.
Then what kind of convergence could be expected? It is noteworthy that both proponents and opponents criticize the current trend of agriculture and are concerned with the future of agriculture. In this respect, sustainable agriculture could be a candidate of the goal for convergence. The point is to avoid either-or discussion and carry on positive discussion on the future vision of agriculture, considering the legitimate rights of stakeholders and the fair distribution of risk and benefits among them. Risk communication in Japan has often been one-sided, where policy makers want to inform the public of "almost decided" policies. We have learned that this type of risk communication makes the situation worse for proponents of technology as well as opponents of GMO.

Future challenges to risk communication in Hokkaido

Some local governments in Japan have set up and are preparing guidelines to regulate the growing of GMOs (Sassa 2006). (Japanese government has set up regulation for GMO in 2003 following *Cartagena Protocol.*) Hokkaido Prefectural Government took the lead by putting in force the ordinance "Hokkaido Ordinance on Prevention of Crossing by Growing of GMOs" (Hokkaido Government, 2005) in January 2006. Regulated by this ordinance, the growing of GMOs is now de facto under the moratorium in Hokkaido. Neither proponents nor opponents of GMO are satisfied with this ordinance. The ordinance regulates the test growing of GMOs by notification, and the commercial growing by approval. This means for proponents that the free growing of GMOs is prevented and for opponents that the commercial growing of GMOs could be approved. This ordinance is due to be revised in three years. This period will be a very important time period to promote the discussion between proponents and opponents of the growing of GMOs.
In order to execute an efficient risk communication, researchers' understanding of social factors as well as scientific literacy of the public should be improved. For this purpose, many opportunities of discussion involving various stakeholders – consumer groups, researchers, concerned citizens, farmers, and industry participants – would be necessary. We hope such an interactive risk communication would help proponents and opponents of GMO to overcome their differences and find a convergence to come to a shared understanding, which would contribute to reveal new possibilities in the future vision of regional agriculture.

References

Fiorino, D. J. (1989). Technical and Democratic Values. In *Risk Analysis*, 9 (3): 293-299.

Hokkaido Government. (2005). Hokkaido Ordinance on Prevention of Crossing by Growing of GMOs (Japanese). Available on http://www.pref.hokkaido.jp/nousei/ns-rtsak/shokuan/gm-jourei.html

MAFF. (2005)a. *Annual Report on Food, Agriculture and Rural Areas in Japan* (Japanese). Association of Agriculture and Forestry Statistics, Tokyo.

MAFF. (2005)b. Yearly Transition of Food Supply Per Capita. In *Food Balance Sheet of the Fiscal Year 2006* (Japanese). Available on http://www.kanbou.maff.go.jp/www/fbs/dat-fy16/H16besshi-1.pdf

MHLW. (The Ministry of Ministry of Health, Labour and Welfare of Japan). (2001). Labeling Genetically Modified Foods QandA (Japanese). Available on http://www.mhlw.go.jp/topics/0103/tp0329-2c.html

MHLW. (2006). List of the products whose safety assessments were completed by MHLW. Available on http://www.mhlw.go.jp/english/topics/food/pdf/sec01.pdf

NRC. (1989). *Improving Risk Communication*. National Academy Press, Washington, DC. 332p.

Norton, B. G. (2003). *Searching for Sustainability: Interdisciplinary Essays in the Philosophy of Conservation Biology*. Cambridge University Press, Cambridge. 554p.

Norton, B.G. (2005). *Sustainability: A Philosophy of Adaptive Ecosystem Management*. University of Chicago Press, Chicago. 607p.

Sassa, Y. (2006). Recent Trends of Local Governments on Growing GMOs (Japanese), *Iden*, 60(2), NTS, Tokyo. 25-29.

Yoshida, S. (2006). Risk Communication over Food Safety in Hokkaido (Japanese). *Iden*, 60(2). NTS, Tokyo. 60-64.

Uncertainty and precaution; challenges and implications for science and policy of DNA vaccines in aquaculture

Frøydis Gillund and Anne Ingeborg Myhr
Norwegian Institute of Gene Ecology, The Science Park, P. O. Box 6418, N-9294 Tromsø, Norway, annem@fagmed.uit.no

Abstract

There is a growing demand for fish and marine products for human and animal consumption. This demand has led to the rapid growth of aquaculture, which sometimes has been accompanied by an economic loss due to fish escape and diseases. Genetic engineering (GE) strategies, as edible vaccines and DNA vaccines may offer a technological solution for some of these problems. Hence, GE will undoubtedly be introduced into commercial aquaculture within a few years. However, the most prominent challenge to GE in a sustainable context is how to ensure environmental protection and at the same time achieve economic benefits. The current lack of scientific understanding with regard to benefits and risks with DNA vaccines are of both scientific and legal interest. Hence, how to handle lack of scientific information are core issues together with the problem of how to increase transparency and participation. In this paper we present the preliminary results from interviews of experts. Our intention was to elicit expert views on the present state of scientific understanding, and to identify and systematize uncertainty with regard to potential health and environmental effects by DNA vaccines. Our focus was on the scientific and regulatory aspects of risk assessment and management. We will give some suggestion in how information about uncertainty may help to direct further research, increase flexibility in decision-making and give guidance into how to take precautionary measures concerning introduction of DNA vaccines in aquaculture.

Background

A DNA vaccine consists of a bacterial plasmid with a strong viral promoter, the gene of interest, and a polyadenylation/ transcriptional termination sequence. The plasmid is grown in bacteria, purified, dissolved in a saline solution, and then administrated by direct intramuscular injection of naked DNA (ng and µg amounts) to activate protein expression *in vivo* and to ultimately induce an immune response and disease protection.

The advances in the field of DNA vaccines have the last years been profound. In 2003, the US Centre for Disease Control (CDC) and Prevention expedited delivery of an experimental veterinary DNA vaccine developed by the CDC and manufactured by Aldevron (Fargo, ND) (Bouchie, 2003). The target for vaccination was the wild Californian condor and the purpose was to protect this endangered species from becoming infected with the West Nile virus. In Canada, an IHNV DNA vaccine (Apex-IHN®) developed by Aqua Health, Ltd, (Canda), an affiliate of Novartis, was cleared for marketing by the Canadian Food Inspection Agency on July 15 (2005) (Novartis media release July 19, 2005). At present a number of experimental human medicinal DNA vaccines have entered phase 1 clinical trial. Hence, it may be expected that DNA vaccines will soon be of high relevance with regard to biosafety.

In our study we focuses on DNA vaccines to be distributed to fish. The development of DNA vaccines against infectious fish diseases has several attractive benefits: low cost, ease of production and improved quality control, heat stability, identical production processes for different vaccines, and the possibility of producing multivalent vaccines (Hew and Fletcher, 2001; Kwang, 2000). On the other hand, there is at present a limited scientific understanding of the fate of DNA vaccines after injection into the animal. To assure the quality of the knowledge input to policy making one need to deal explicitly with uncertainties in the scientific understanding. In this paper we aim to; i) illustrate how scientists can apply uncertainty analysis to identify and communicate to other scientists and to decision makers the scientific uncertainty involved, ii) that uncertainty analyses may help prioritise research efforts, and iii) illustrate how the distibution of DNA vaccines can be characterised by a number of different types of uncertainty.

Identification and systematization of uncertainty

Employment of model-based decision support, as for instance the Walker & Harremöes (W&H) framework (Walker *et al.*, 2003), may help to identify the types and levels of the uncertainty involved. The W&H framework has been developed by an international group of scientists with the purpose to provide a state of the art conceptual basis for the systematic evaluation of uncertainty in environmental decision-making. One of the main goals of the W&H framework is to stimulate better communication between the various actors in identification of areas for further research and in decision processes. In this framework, uncertainty is recognized at three dimensions;

1. Location (where the uncertainty manifests itself, (e.g. distinguish between context (ecological, technological, economic, social and political), expert judgement and considerations, models (model structure, model implementation, data, outputs etc.)).
2. Nature (degree of variability which can express whether uncertainty primarily stems from inherent system variability/complexity or from lack of knowledge and information).
3. Level (the severity of uncertainty that can be classified on a gradual scale from "knowing for certain" to "complete ignorance").

For instance, Krayer von Krauss *et al.* (2004) has demonstrated and tested the W&H framework with the purpose to identify scientists and other stakeholders judgement of uncertainty in risk assessment of GM crops. In these studies the focus was on potential adverse effects on agriculture and cultivation processes by release of herbicide resistant oilseed crops. Krayer von Krauss *et al.* interviewed seven experts in Canada and Denmark. To identify the experts view on location uncertainty the authors presented a diagram showing causal relationships and key parametres to the experts. With the purpose to identify the level and nature of uncertainty, the experts had to quantify the level and describe the nature of uncertainty on the key parametres in the diagram. By asking the experts to identify the nature of uncertainty it was possible to distinguish between uncertainty that may be reduced doing more research and ignorance that stems from systems variability or complexity.
Approaches that define and systematise the uncertainty involved, as the W&H framework may help to use scientific knowledge more efficiently, in directing further research and in guiding risk assessment and management processes.

Identification of uncertainty with the use of DNA vaccines

With the intention to identify experts scientific understanding of DNA vaccine distribution, we used the W&H integrated uncertainty analysis framework (Walker *et al.*, 2003). The uncertainty involved was illustrated in a system model and the model, as well as the explanation thereof, was based on a review performed by Myhr and Dalmo (2005). The system model served as the basis for inventorying the potential locations of uncertainty identified by interviewing the experts. The main area of uncertainty that was identified was related to potential:

- Unintended tissue distribution and prolonged persistence of plasmid DNA.
- Unintended environmental release of DNA vaccines.

Uptake, distribution and persistence of the DNA vaccine

Preliminary experiments have revealed that organs and tissues rich in leucocytes have accumulated intact DNA (plasmid DNA) for more than one month in salmon after intraperitoneal (i.p) and intramuscular (i.m) injections (Myhr and Dalmo, 2005). For instance, in sea bream, intact plasmid was found at the injection site two months after i.m injection (Verri *et al.*, 2003). Similar findings have been described for rainbow trout (Anderson *et al.*, 1996). Following on, it has been shown that not only muscle cells but cells in tissues very distal to the injection site (muscle) have expressed the transgene after plasmid injection (Romøren *et al.*, 2004). Moreover, it has been shown that glass catfish have been expressing a transgene as long as two years after injection (Dijkstra *et al.*, 2001). These reports illustrate that plasmid DNA can persist in fish for long time periods after the initially injection. Undoubtedly, there is an urgent need to analyse the longevity of DNA vaccines with respect to immunological parameters as well as to the risk of gene transfer.

Environmental release of DNA vaccines

DNA products could, be distributed unintentionally over vast areas and be effective across phylae because of the relative lack of physical and physiological barriers to such compounds. DNA is much more resistant to breakdown in the ecosystems, and after uptake in macro-organisms, than realised until recently. This is also the case for aquatic ecosystems (Heinemann and Roughan, 2000). There are, however, few published studies investigating the stability, horizontal transfer and uptake of DNA constructs in aquatic systems, including marine fish and mammals.

Systematization of uncertainty with the use of DNA vaccines

Often when scientists characterize scientific uncertainty they use statistical analysis (e.g., statistical confidence intervals, model output), while decision making is usually performed according to given standards for significance (e.g., Type I error probability levels of 0.05). In less controlled situations, scientific uncertainty is ascertained by other means, such as model prediction errors. However, in many cases the complexity and the novelty of the scientific problems being studied are such that quantitative analysis do not fully capture the uncertainty characterising these problems. Hence, the notion that uncertainty is only a statistical concept or represent insufficient data may leave out many important aspects of uncertainty (Wynne, 1992). For instance, uncertainty with regard to DNA vaccine distribution can be presented at different levels;

- Uncertainties that refer to situations where we do not know or cannot estimate the probability of hazard, but the hazards to consider are known. This may be due to the novelty of the activity, or to the variability or complexity involved. With regard to DNA vaccines this may be an allergic reaction to the vaccine, or that other unintended acute immune reactions arise.
- Ignorance that represents situations where the kind of hazard to measure is unknown, i.e. completely unexpected hazards may emerge. This has historically been experienced with for instance BSE, dioxins, and pesticides (EEA, 2001). With regard to DNA vaccine use may for instance unprecedented and unintended direct and indirect effects emerge. Direct effects may arise due to prolonged stability of the DNA vaccine that may facilitate that the DNA construct becomes integrated into the chromosome of the recipient animal or in microorganisms in the animal intestines. This may not have only a biological relevance but also policy relevance. For instance, the Norwegian Directorate for Nature Management has stated that a DNA-vaccinated animal is to be considered as GM as long as the added DNA

is present in the animals. This may have implications regarding the need for labelling and traceability. While indirect effects concerns for instance effects as contamination of wild gene pools or alterations in ecological relationships if the DNA vaccine may be released to the environment and taken up by other organisms.

Initiation of research in complex systems

By identification of uncertainty also research needs are revealed. In our interviews we identified that the areas of uncertainty and the reseach considered important was dependent on the background of the expert that was interviewed. Sarewitz (2004) has argued that scientific dissent in the case of higly complex and difficult to assess risk situations are due to different backgrounds /disiciplines, that may affect choice of hypothesis, methods and models, which give conflicting data and causes disagreements among scientists. Hence, the challenge is to manage the uncertainties that are characteristic of each field so that information of the highest possible quality can be obtained (Funtowicz and Ravetz, 1990). Reflecting on the role of scientific disunity in the interpretation of scientific uncertainty by DNA vaccination, the question arises how enhanced dialogue between competing disciplines can contribute to make explicit those values, interests, and implicit assumptions that represent the frame for each discipline's approach to scientific uncertainty. For instance, an enhanced dialogue can be facilitated by involvement of a wide base of scientific disciplines as well as independent scientific institutions in the gathering of scientific understanding. Involvement on a wide basis of scientific disciplines will (1) assist in the exploration of alternative problem framing and alternative indicators that can be used in risk assessment, (2) function as a source of knowledge, data and information –including information on uncertainties – that may be of relevance for risk assessment, and (3) assist in the evaluation and critical review of assumptions used, method, process and results. This will ensure diverse consideration where the different methods and models representing the different disciplines may be seen as compatible providers of information and models for studying the problem or the system. With more diversity in the risk assessment, more data will be generated and more responses will be available to understand complexity and changing conditions.

Conclusion

We have argued that there is a need to achieve wise management of uncertainties with regard to potential adverse effects by DNA vaccine use. This challenge has to be met by scientific conduct and approaches that aim to manage risk and uncertainty, taking into account the complexity of the ecological systems that the DNA vaccines and the vaccinated animals are to be released into. Broad risk assessments of DNA vaccines will include appreciation of uncertainty and complexity and entails application of the Precautionary Principle.

Acknowledgement

Support came from grant 157157/150 given by the Norwegian Research Council.

References

Anderson, E.D., Mourich D.V. and Leong, J.A.C. (1996). Gene expression in rainbow trout (*Oncorhynchus mykiss*) following intramuscular injection of DNA. *Molecular Marine Biology and Biotechnology*, 5:105-113.

Bouchie, A. (2003). DNA vaccine deployed for endangered condors. *Nature Biotechnology*, 21: 11.

Dijkstra, J.M., Okamoto, H., Ototake, M. and Nakanishi, T. (2001). Luciferase expression 2 years after DNA injection in glass catfish (*Kryptopterus bicirrhus*). *Fish and Shellfish Immunology,* 11; 199-202.

EEA: European Environment Agency (2002). *Late lessons from early warnings. The precautionary principle 1896-2000* [on line]. Available on http://reports.eea.eu.int/environmental_issue_report_2001_22/ [date of consultation 20/01/06].

Funtowicz, S.O. and Ravetz, J.R. (1990). *Uncertainty and Quality in Science for Policy*. Kluwer, Dordrecht, pp.7-16.

Heinemann, J.A. and Roughan, P.D. (2000). New hypotheses on the nature of horizontally mobile genes. *Annals New York Academy of Science,* 906: 169-186.

Hew, C.L. and Fletcher, G.L. (2002). The role of aquatic biotechnology in aquaculture. *Aquaculture,* 197; 191-204.

Krayer von Krauss M., Casman E. and Small M. (2004). Elicitation of expert judgments of uncertainty in the risk assessment of herbicide tolerant oilseed crops. *Journal of Risk Analysis,* 24; 1515-1527.

Kwang, J. (2000). Fishing for vaccines. *Nature Biotechnology* ,18: 1145-46.

Myhr, A.I. and Dalmo, R.A. (2005). Introduction of genetic engineering in aquaculture: ecological and ethical implications for science and governance. *Aquaculture,* 250: 542-554.

Novartis media release July 19, 2005 [on line]. Available on http://www.novartis.com/ [date of consultation 20/01/06].

Romøren, K., Thu, B.J. and Evensen, Ø. (2004). Expression of luciferase in selected organs following delivery of naked and formulated DNA to rainbow trout (*Oncorhynchus mykiss*) by different routes of administration. *Fish & Shellfish Immunology,* 16: 251-264.

Sarewitz, D. (2004). How science makes environmental controversies worse. *Environmental Science & Policy,* 7: 385-403.

Verri, T., Ingrosso, L., Chiloiro, R., Danieli, A., Zonno, V., Alifano, P., Romano, N., Scapigliati, G., S. Vilella, and Stornelli, C. (2003). Assessment of DNA vaccine potential for gilthead sea bream (*Sparus aurata*) by intramuscular injection of a reporter gene. *Fish & Shellfish Immunology,* 15: 283-295.

Walker, W.E., Harremoeës, P-. Rotmans, J., van der Sluijs, J.P., van Asselt, M.B.A., Janssen, P. and Kraye von Krauss, M.P. (2003). Defining uncertainty; a conceptual basis for uncertainty management in model based decision support. *Journal of Integrated Assessment,* 4: 5-17.

Wynne, B. (1992). Uncertainty and environmental learning: reconciving science and policy in the preventive paradigm. *Global Environmental Change,* 2:111-127.

Heuristics of precaution: Towards a substantial consideration of ethics in the regulation of agri-food biotechnology risks

Andoni Ibarra and Hannot Rodríguez
University of the Basque Country, Dpt. of Logic and Philosophy of Science, PO Box 1249, 20080 San Sebastian, Spain, andoni.ibarra@ehu.es

Abstract

The European regulatory framework of activities with uncertain risk for public health has been constituted since the past decade from a precautionary focus. The case of agri-food biotechnology, as a domain of strong public and scientific conflict regarding safety, is not an exception. In contexts of uncertainty, the precautionary principle (PP) prioritises the public health over the economic health. However, the institutional interpretation of PP considers it as a decision mechanism, which opts to manage uncertainty focusing on the safety side, maintaining the evidence available for decision-making outside the scope of the non-epistemic (i.e., ethical) considerations that shape the public discourse on genetic engineering safety. This dissociation between facts and values, scientific aspects and value assumptions, preserves the expert authority of the institutions: it doesn't do justice to the fact that the socio-economic values involved in the problem, besides conditioning risk acceptability, are also determinant for its own estimation. By the concept of "heuristics of precaution" we intend to show that PP, in its contexts of interpretation and application, works as a catalyser for the acknowledgement of new potential damage paths, converting the issue of the required evidence in a matter that should also be negotiable in terms of value.

Keywords: precautionary principle, uncertainty, heuristics of precaution, evidence, agri-food biotechnology

Introduction

This paper has three parts: the first describes the precautionary turn taken by the European regulatory framework regarding safety. The second part characterises the resort to the precautionary principle (PP, from now on) in relation to a socio-epistemic context that establishes the limits of its scope. Finally, in the last part, which is subdivided in two parts, we conclude that: (a) the very uncertainty is constituted according to a heterogeneity of factors (epistemic as well as non-epistemic); (b) the resort to PP in its interpretation and application contexts, and in the specific case of agri-food biotechnology, has an active part in this constitution, which we will call "heuristics of precaution". This means that the issues related to safety are opened to considerations of value, also in the most central aspects of those issues.

The "precautionary turn" of technological regulation

The European regulatory context has experienced, along the past and present decades, a turn towards precaution in relation to environmental and public health management. In the report *Communication from the Commission on the precautionary principle* (CEC, 2000) the European Commission elucidates PP's meaning and scope.
PP extends the preventive actions of risk avoidance to possible damages, which occurrence is uncertain, and on which there are reasonable indications that might exceed the chosen level of

protection (CEC, 2000: 8). Uncertainty is no more an excuse for not adopting measures tending to the avoidance of the possible damage, which realisation and severity are maintained open (CEC, 2000: 26).

PP goes beyond the traditional measures of administrative law and technology regulation, which tackle the risks just when the danger is evident or the damage is already produced (Bodansky, 1994).

At the conceptual level, a distinction operates between "prevention" and "precaution". Prevention prioritises the avoidance of a damaging fact over its reparation or mitigation. But when we act preventively we dispose of sufficient anticipated knowledge to know what will happen (at least what will probably happen) and opt for one or another course of action. Thus, risk analysis is a preventive tool as it acts on risks that have been probabilistically measured. Precaution, on the other hand, doesn't work on probabilities but on contexts of uncertainty. That is, "precaution" means here prevention with respect to a threat having an uncertain possibility (Hansson and Sandin, 2001: 28).

Precaution in context

This precautionary management of uncertainty must be performed, however, in accordance to some general principles of application[60], which relativise its scope by considering it in relation to a series of socio-epistemic moderating criteria (Jasanoff 2000). This means that PP is not a simple practical instance of the maximin rule of decision, which urges to act as if the worst thing that could happen will happen, choosing that alternative of action that has the best worst consequence. Thus, the criterion advocates to compare the minimum utilities of the different alternatives of action and to choose the one that has the minimum utility.

Elster (1983: 185-207) argues that to decide in accordance to the maximin rule under conditions of uncertainty is rational as far as the different alternatives of action have, all of them, the same best consequences. You would only have to choose the alternative with the less harmful consequence. However, it is not always the case that all the best consequences of the different options are equal. Besides, uncertainty does not signify that the different scenarios that can be foreseen have the same scientific consistency, what legitimates a distinction in relation to plausibility degrees (Godard, 2001).

PP shows, after all, a favourable bias to the maximisation of the minimum utilities (to avoid risks). This maximization intends to avoid the so-called Type II errors (false negatives) in scientific practice (Hansson, 1999), what provokes an increase of Type I errors (false positives).

Uncertainty in context

We have claimed that precautionary impulses are tempered by a spectrum of heterogeneous considerations that limits its practical scope. However, all these considerations are kept unaware to the scientific state of the facts, that is, to the "real" risk and uncertainty that, although susceptible to the heterogeneity of interpretations that might be made by some public claims institutionally acknowledged –which would advocate for precautionary measures–, are maintained safe from "strange" (i.e., ethical) interferences with respect to its determination (Levidow and Marris, 2001).

This institutional interpretation of uncertainty and, therefore, of PP's operativity and possibilities, were *de facto* overcame by the effective reading of the principle in the context of the conflicts

[60] The general principles of application are: proportionality, non-discrimination, consistency, examination of the benefits and costs of action or lack of action, and examination of scientific development (CEC, 2000: 18-21).

appeared in relation to technologies safety. Thus, we advocate that PP, in its diverse contexts of interpretation and application, is not only a mere expression of objective uncertainty but also a catalyser of potential damage paths previously unrecognised. We call this capacity of PP "heuristics of precaution".
It is not only the resort to PP subject to socio-economic considerations. Rather, it is the very uncertainty that is subject to an heterogeneous constitution in its context of application.

Malleable uncertainty

The fundamental thesis in (Wynne, 1992) is that uncertainty works according to tacit social commitments adopted according to a given corpus of knowledge. The institutional cultures tend to limit in this way the uncertain character of the problem to those uncertainties that can be controlled and solved. That is, the expert institutions recognize only the uncertainties they can manage. The uncertainty is presented here as a function of social and cultural factors conditioning its constitution. The level of uncertainty recognized would thus express a socio-cultural and political configuration, built on the available knowledge on risks (Hunt, 1994).
Douglas (2000) argues that scientific uncertainty constraints to constitute the "real" risk according to the relevance attributed to the non-epistemic consequences derived from the diagnostic error of a risk that is uncertain. These consequences take mainly the form of economic risks (overregulation risks for industry) and the form of public health and environmental risks. Therefore, the tendency of the preferences towards avoiding one or another type of risk, will condition the degree of severity of the risk finally established. The lack of scientific evidence is "filled in" with considerations that go beyond the purely epistemological context. Non-epistemic values totally condition the viability of scientific activity, further than what is acknowledged by those studies that limit this conditioning to factors which are external to science, as for example the choice of the problems to be considered, the methodological limits or the social consequences of research (Rescher, 1987).[61]
The attitude and the decisions related to the evidence are a function of the importance attributed to the possibility of erring at the moment to accept or reject the hypotheses concerning uncertain risks (Hansson, 1999: 918). Therefore, the scientific risk assessment cannot be fulfilled, if not made through the mediation of uncertain inferences based on the anticipatory appraisal of the potential errors of that assessment (Brunk *et al.*, 1991).
The risk evidence established acquires its status in its relationship to the dominant values and interests of the regulatory context (Winickoff *et al.*, 2005: 113). Scientific assessments are public and conditional, that is, relative to the contexts (Machamer and Douglas, 1999). Therefore, public claims are integrated in a stronger way into science and technology policies, because it is the very acceptable evidence what becomes a matter of public discussion (Levidow and Murphy, 2003: 62).
The evidence is not given. Thus, the reason for the existence of different regulatory attitudes (for example, between European countries and the United States of America) is not due to an understanding disparity of an equal factual base, or an absence of application of an hypothetical scientific method that might guarantee sound scientific results. Rather it is due to the existence of different regulatory cultures that have different risk frameworks (Jasanoff, 1995). That is, the scientific results and the attitudes towards them cannot be removed from the wider socio-institutional context (Winickoff *et al.*, 2005).

[61] On the other hand, in Douglas, 2000, the non-epistemic reasoning impregnates necessarily (to cover the "epistemological holes" that guide the search of a risk diagnosis) the three fundamental steps of scientific process: (i) the selection of a methodology; (ii) the characterization of the collected data; (iii) the interpretation of results.

The public and subjective contribution cannot be reduced, therefore, to the consideration of a public interpretation of a non-problematic expert knowledge, as a mean to clarify public preferences (that is, as expression of an ethical sensibility) regarding the established knowledge, but as a constitutive element of scientific activity.

Heuristics of precaution: the case of agri-food biotechnology

At the beginning of 1990, and in the context of an agricultural model of high productivity and economic competitiveness, the United States of America and Europe started the race towards the promotion of biotechnology applied to agri-food. This commitment conditioned a rather soft interpretation of regulatory demands to protect public health and the environment, which lead to an expert minimisation of the uncertain risks. However, the critical debate in relation to genetically engineered products and crops safety was increasing in both the public and scientific level. These critical voices demanded a more precautionary interpretation of the regulatory framework. This was reflected in an institutional wide-ranging consideration of problems and sources of uncertainty (Levidow, 2001).[62]

The wide recognition of the constituent character of uncertainty in European regulation of genetically modified organisms (GMO) owes very much to the public pressure towards the consideration of potential problems that -at the beginning- were not deemed scientifically relevant. The social conflict about the safety of genetically engineered technology was *de facto* an important element on the interpretation and application of the regulatory framework (Todt, 2002: 197-202). The scientific and systematic post-commercialisation monitoring of genetically engineered crops is the result of the public demand for risk assessments in "real-world" conditions. This fact represents the recognition of the irreducibility of uncertainty to laboratory ideal conditions. Thus, it is acknowledged that the only way to learn more about these systems behaviour (and, therefore, about their safety) is to take the lesson in the field, that is, at real scale (Todt, 2002: 101-3).

Thus, the progressive inclusion of PP in regulatory regimes represented an increase of the burden of the scientific evidence of safety, owing to a greater consideration of uncertainty through the acknowledgement of more complex causal paths of potential damage. That is, the "precautionary turn" brought together a criterion change with respect to the constitution of scientific evidence (Levidow, 2001: 865-8). The discussion about where to fix the limit of the sound research has here scientific-political tints (Silbergeld, 1991: 108-10).

Therefore, PP's functionality is not limited to manage a series of given uncertainties, since it promotes the recognition of a wider spectrum of uncertainties, and demands and stimulates at the same time new knowledge about more complex uncertainties. In this relies its heuristic character, which is reinforced in the context of public demand. Thus, the precautionary framework operates as a real catalyser, which promotes the renegotiation of already assumed uncertain instances.[63]

[62] The regulatory framework for genetically modified organisms (GMO) in the European Union stems from the so-called Directive 2001/18. This document that came into effect on October 17th, 2001, has replaced the Directive 90/220, in force since 1990. Basically, this new document has updated and reinforced the previous standards, demanding among other requirements: a more detailed assessment of environmental risk; an obligatory monitoring of results, after the commercialisation, of genetically modified products; the labelling and traceability in all the commercialisation phases of the products.

[63] Thus, PP is not an irrational risk management principle that denies science (e.g. Pieterman, 2001). Rather, it demands more science, since the full acknowledgement of uncertainty implies a scientific practice more sensitive to the possible error and more receptive towards a greater number of alternative hypotheses (Stirling, 2001).

References

Bodansky, D. (1994). The Precautionary Principle in US Environmental Law. In: *Interpreting the Precautionary Principle*. T. O'Riordan and J. Cameron (eds). Earthscan, London. 203-228.

Brunk, C., Haworth, L. and Lee, B. (1991). Is a Scientific Assessment of Risk Possible? Value Assumptions in the Canadian Alachlor Controversy. *Dialogue*, 30(3): 235-247.

Commission of the European Communities (CEC) (2000). *Communication from the Commission on the Precautionary Principle*. Brussels. 29 p. (http://europa.eu.int/comm/dgs/health_consumer/library/pub/pub07_en.pdf).

Douglas, H. (2000). Inductive Risk and Values in Science. *Philosophy of Science*, 67(4): 559-579.

Elster, J. (1983). *Explaining Technical Change*. CUP, Cambridge. 266 p.

Godard, O. (2001). Le principe de précaution entre debats et gestion des crises. *Cahier du Laboratoire d'Econométrie de l'Ecole Polytechnique*, n° 2001-010. 15 p . (http://ceco.polytechnique.fr/CAHIERS/pdf/2001-010.pdf).

Hansson, S.O. (1999). Adjusting Scientific Practices to the Precautionary Principle. *Human and Ecological Risk Assessment*, 5(5): 909-921.

Hansson, S.O. and Sandin, P. (2001). Persistence, Liability to Bioaccumulate, and the Precautionary Principle. Background Paper for the *New Strategy for the Risk Management of Chemicals (NewS)* Policy Forum (Falkenberg, 3-4 April 2001). 105 p. (http://www.infra.kth.se/~cr/NewS/NewS_SOH_backgr.pdf).

Hunt, J. (1994). The Social Construction of Precaution. In: *Interpreting the Precautionary Principle*. T. O'Riordan and J. Cameron (eds). Earthscan, London. 117-125.

Jasanoff, S. (1995). Product, Process, or Programme: Three Cultures and the Regulation of Biotechnology. In: *Resistance to New Technology*. M. Bauer (ed.). CUP, Cambridge. 311-331.

Jasanoff, S. (2000). Between Risk and Precaution – Reassessing the Future of GM Crops. *Journal of Risk Research*, 3(3): 277-282.

Levidow, L. (2001). Precautionary Uncertainty: Regulating GM Crops in Europe. *Social Studies of Science*, 31(6): 842-874.

Levidow, L. and Marris, C. (2001). Science and Governance in Europe: Lessons from the Case of Agricutural Biotechnology. *Science and Public Policy*, 28(5): 345-360.

Levidow, L. and Murphy, J. (2003). Reframing Regulatory Science: Trans-Atlantic Conflict over GM Crops. *Cahiers d'économie et sociologie rurales*, 68/69: 47-74.

Machamer, P. and Douglas, H. (1999). Cognitive and Social Values. *Science & Education*, 8(1): 45-54.

Pieterman, R. (2001). Culture in the Risk Society: An Essay on the Rise of a Precautionary Culture. *Zeitschrift für Rechtssoziologie*, 22(2): 145-168.

Rescher, N. (1987). Moral Limits of Scientific Research. In: N. Rescher, *Forbidden Knowledge*. Reidel, Dordrecht. 1-16.

Silbergeld, E.K. (1991). Risk Assessment and Risk Management: An Uneasy Divorce. In: *Acceptable Evidence: Science and Values in Risk Management*. D.G. Mayo and R.D. Hollander (eds). Oxford University Press, Oxford. 99-114.

Stirling, A. (2001). Science and the Rationality of Precaution. In: *The Role of Precaution in Chemical Policy*. E. Freytag, T. Jakl, G. Loibl and M. Wittmann (eds). Bundesministerium für Land- und Forstwirtschaft, Vienna. 88-105.

Todt, O. (2002). *Innovación y regulación; la influencia de los actores sociales en el cambio tecnológico. El caso de la ingeniería genética agrícola*. Universitat de València. Valencia (doctoral dissertation). 310 p.

Winickoff, D., Jasanoff, S., Busch, L., Grove-White, R. and Wynne, B. (2005). Adjudicating the GM Food Wars: Science, Risk, and Democracy in World Trade Law. *The Yale Journal of International Law*, 30(1): 81-123.

Wynne, B. (1992). Uncertainty and Environmental Learning: Reconceiving Science and Policy in the Preventive Paradigm. *Global Environmental Change*, 2(2): 111-127.

Negotiating signs of pleasure and pain: towards a democratic-deliberative model of animal welfare monitoring

Mara Miele and Adrian Bruce Evans
Cardiff School of City and Regional Planning, Cardiff University, Glamorgan Building, King Edward VII Avenue, Cardiff CF10 3WA, Wales, mielem@Cardiff.ac.uk

Abstract

In this paper we draw on interdisciplinary work conducted by the Welfare Quality project (a large EU project on farm animal welfare – www.welfarequality.net) to critically compare and contrast the different ways in which scientists and consumers 'frame' animal welfare. Furthermore, we examine how different ways of framing this issue embody different ethical concerns, different practical competencies and different techniques of information gathering and evaluation. Finally, given these differences, we address the issue of how one might best go about integrating consumer and scientific concerns.

Keywords: animal welfare, science-society dialogue, consumer concerns

The 'welfare quality' scientists' approach to welfare monitoring

The Welfare Quality project has brought together numerous animal scientists and veterinarians from a variety of different ideological backgrounds, as such any attempt to summarise or to distil a single 'Welfare Quality' approach to welfare monitoring runs the risk of unfairly glossing over important differences. Bearing this risk in mind we use this section to highlight three key characteristics/principles, which we believe are central to the Welfare Quality approach and we try to tease out the (often tacit) ethico-political understandings that underlie them. Furthermore, whilst we are supportive of many of the broad aims encompassed within these principles, in the spirit of positive co-operation we highlight some of the pitfalls and difficulties that scientists might face in achieving their goals.

Firstly, one of the key characteristics of the Welfare Quality approach to monitoring is that it is '*open*' to the views of different stakeholder groups. In other words, Welfare Quality scientists are keen to engage in '3rd order science', where there is a genuine upstream input from stakeholders. This strategy helps to foster one of the best features of scientific practice, namely an adherence to democratic, conversational forms of rationality (see Habermas 1981, Rorty 1989). However, this 'openness' is by no means a fait accompli, as a sustained effort is required in order to maintain a productive dialogue. Indeed, the only real way of assessing how effective this policy of openness has been is by looking back at the end of the project and asking: To what extent has there been a genuine two-way flow of information between science and society? To what extent have stakeholders concerns been incorporated into the actual monitoring scheme? To what extent have scientists' (and stakeholders') ideas been altered through the integration process? To what extent have scientists' (and stakeholders') methods of enquiry/competencies (see Stengers 1997) been altered through the integration process?

Secondly, Welfare Quality scientists are committed to developing a multi-criteria approach to animal welfare monitoring. This means that they reject the idea that it is possible to measure welfare via the 'holy grail' of a single indicator, such as longevity (Hurnik 1993) or corticosteroid levels (Barnett and Hemsworth 1990), but rather they believe that one might best assess welfare by combining a variety of different measures in a complex model (see also FAWC 1992, Fraser

1995). At present, Welfare Quality scientists are in the process of validating and assessing over 200 different measures of farm animal welfare. These measures have been grouped together into a list of ten broad areas of welfare concerns (see Tables 1 and 2, see also Keeling and Viessier

Table 1. Ten areas of welfare concern as formulated by welfare quality scientists.

1. Hunger, thirst or malnutrition
   This occurs when animals are denied a sufficient and appropriate diet or a sufficient and accessible water supply and can lead to dehydration, poor body condition and death.
2. Physical comfort and security
   Animals can become uncomfortable and have problems lying down, getting up and standing. This can occur when they are kept in inappropriately designed housing (e.g. insufficient space, poor ventilation, unsuitable flooring and bedding) or when they are transported in poorly designed or poorly ventilated vehicles.
3. Health: injuries
   Animals can suffer physical injuries, such as mutilations, broken bones, bruises or skin lesions, due to factors such as; uneven or slippery flooring, enclosures with sharp edges and environments that promote aggressive behaviours between animals.
4. Health: disease
   Animals can suffer a range of diseases (e.g. mastitis and metabolic disorders in cattle). Poor hygiene, irregular monitoring and insufficient treatment speeds can amplify these problems.
5. Pain (not related to injuries or disease)
   In addition to suffering pain from injuries and disease, animals can experience intense or prolonged pain due to inappropriate management, handling, slaughter, or surgical procedures (e.g. castration, dehorning) and as a result of intense aggressive encounters.
6. Normal/natural social behaviours
   Animals can be denied the opportunity to express natural, non-harmful, social behaviours, such as grooming each other and huddling for warmth. Separating females from their offspring and preventing sexual behaviour can bring about specific examples of this problem.
7. Normal/natural other behaviours
   Animals can be denied the possibility of expressing other intuitively desirable natural behaviours, such as exploration and play. The denial of these possibilities might lead to abnormal and/or harmful behaviours such as tongue rolling in cattle and feather pecking in chickens.
8. Human-animal relationship
   Poor relationships can be reflected in increased avoidance distances and fearful or aggressive animal behaviours. This can occur due to inappropriate handling techniques (e.g. slapping, kicking and the use of electric prods), or when farmers, animal transporters or slaughterhouse staff are either insufficiently skilled or possess unfavourable attitudes towards animals.
9. Negative emotions (apart from pain)
   Animals can experience emotions such as fear, distress, frustration or apathy, when they are kept in inappropriate physical or social environments (e.g. where there is over mixing, or not enough space to avoid an aggressive partner). These emotions can be reflected in behaviours such as panic, flight, social withdrawal and aggression and in certain vocalisations and behavioural disorders.
10. Positive emotions
    Poor management routines and a lack of environmental stimulation may prevent animals from expressing positive emotions. Positive emotions are difficult to assess but may be reflected in certain behaviours, such as play (especially in young animals) and by certain vocalizations.

Table 2. Parameters relating to the ten areas of welfare concern .

Specific parameters relating to each concern

In order to assess each of the ten broad concerns, scientists working on the project are in the process of identifying and measuring a series of welfare parameters. The table below provides a small illustrative selection of the parameters that researchers intend to use as a starting point for assessing the welfare of cattle. Over the course of the next five years researchers will develop and test a variety of different measures that relate to each of these parameters. Only measures that are deemed to be valid, reliable, repeatable and feasible to collect will be included in the final welfare assessment scheme.

Areas of Concern	Animal Based Parameters (Cattle)	Resource and Management Based Parameters (General)
Hunger, thirst or malnutrition	Body condition & dehydration Mortality	Provision of food and water on farm, during transport and prior to slaughter Management strategies
Physical comfort and security	Difficulties rising or lying Slipping and falling (on farm and during loading) Cleanliness of animal Panting after stress or effort	Housing design (e.g. space, flooring, bedding and litter) Air quality Duration of transport Method of slaughter
Health: injuries	Injuries on farm/at slaughter Fresh blood on floor Mortality and life expectancy	Handling strategies Log book of injured and culled animals Treatment procedures
Health: disease	Mortality and life expectancy Occurrence of disease Carcass damage	Log book of diseases, treatments and culls Identification and treatment
Pain	Lameness Routine mutilations (e.g. dehorning) Effectiveness of stunning Meat quality at slaughter	Presence of sharp edges Use of electric prod Stunning method
Normal/natural social behaviours	Frequency of allo-grooming (grooming each other) Occurrence of other natural social behaviours.	Grouping and regrouping of animals Physical contact with members of the same species
Normal/natural other behaviours	Abnormal behaviours (e.g. tongue-rolling) would receive a negative score	Presence of resources thought to be important
Human-animal relationship	Avoidance distance Fear Aggression	Attitudes and skills of farmers, drivers and slaughterhouse staff
Negative emotions	Fear (freezing, running away) Vocalization (on farm and at slaughter) Qualitative assessment	Stunning method
Positive emotions	Play (in young) Qualitative assessment	Environmental enrichment

2005 for a slightly altered list of 12 concerns). As one can see, these concerns are ideologically diverse as they incorporate measures from *'biological functioning'*, *'affective states'* and *'natural living'* approaches to welfare (see Fraser 2004), as well as indicators of welfare on-farm, during transport and at slaughter. Given the shear diversity of perspectives included in this monitoring scheme the ethical issues here do not revolve around the multi-criteria approach in itself, but rather the degree to which this diversity can be maintained within the final (reduced) monitoring scheme. In other words, the process through which certain measures are selected and de-selected from the final scheme, will present deep ethical and political challenges. As such, it is important that this process is transparent and that it utilises both quantitative statistical modelling and more qualitative expert (and possibly non-expert) debate/deliberation to arrive at the final model.
Thirdly, Welfare Quality scientists prioritise 'animal-based' measures of welfare. That is to say they prioritise indicators of the actual welfare status of the animal, such as lameness and the presence of disease, over resource or management-based indicators (such as available space, type of system etc.), which are deemed to only measure risk to welfare. These animal-based measures are much more difficult to develop than resource-based measures as they require a great deal of observational and experimental work, however they offer the possibilities of greater scientific validity and greater flexibility for farmers (who would be free to achieve animal-based targets in a variety of different ways rather than through the wholesale infrastructure changes that resource-based targets often demand). Whilst we believe that these animal-based measures represent a significant step forward in the monitoring of animal welfare, we feel that it is important to bring to light some of the ethical issues that this approach raises. First, by the very act of differentiating between animal-based and resource or environment-based measures, one runs the risks of separating off animals from their environments and of overlooking more process-based understandings that are attentive to the ways in which animals are inextricably linked or 'folded' (see Deleuze and Guattari 1988) with their environments (as in the *'natural living'* approaches to welfare). Second, one circumvents the (politically) difficult problem of making a priori ethical judgements about the nature of different types of farming system, as the welfare credentials of any given farming system are framed as an empirical issue that needs to be assessed by observing the animals within that system. Third, one focuses attention on the effects rather than the causes of good/bad welfare. Finally, the very fact that there are multiple means through which animal-based welfare targets can be achieved is itself subject to abuse (e.g. the overuse of anti-biotics as a measure to prevent disease in poor quality systems).

European consumers' understandings of farm animal welfare

One of the key aims for social scientists working on the Welfare Quality project has been to research European consumers' concerns about farm animal welfare. This has been achieved through the use of both qualitative and quantitative methods. Table 3 depicts a summary of some of the initial results from consumer focus groups that were carried out across seven study countries (for a more detailed account of the methods used during these focus groups see Miele and Evans 2005). In particular, Table 3 depicts consumers' spontaneous concerns regarding the welfare of farm animals. As one can see, consumers raise many important issues and there are strong overlaps between the concerns of consumers and those of scientists. Throughout the remainder of this section we attempt to delve more deeply into these issues by highlighting three more general themes/ethical orientations that we believe underlie and help to inform consumers' concerns.
Firstly, consumers from across Europe consistently held the belief that low intensity farming systems could provide better animal welfare than high intensity industrialised systems. Crucial to this positive perception of small-scale farming was the understanding of farmers as 'animal carers' (there are fewer farmers per animal in industrial production) and the greater

Table 3. European consumers' spontaneous concerns about farm animal welfare.

Spontaneous concerns	France	Italy	Holland	UK	Sweden	Norway	Hungary
Outdoor access, free range, extensive production, possibility to choose between indoors and outdoors		X	X	X	X	X	X
Space, natural space	X	X			X		
Natural feed, no artificial growth stimulants, lifespan, time for normal growth	X	X	X	X	X	X	X
Humane slaughter	X	X	X	X	X		
Transport (limited or avoided)	X	X	X	X	X	X	
Respect, care, physical comfort and security	X	X		X	X	X	
Good hygiene		X	X				X
Good quality of life	X	X		X		X	
Small scale production		X			X	X	
Breeding, genetic modification				X			
Products with someone 'accountable for' (farmer, vet.)		X				X	
No mutilations, no pain			X		X		
Natural light, fresh air		X	X				
Distractions (play)			X				
Animals as individuals (name)						X	
Natural reproduction		X					
No use of routine medicines				X			
Wild animals						X	
Company, love, happiness					X		X

opportunities for animals to be considered as 'individuals' (by being giving a name and by having a recognised family history), rather than as part of an undifferentiated mass. In addition, small-scale farming was often associated with memories of traditional, extensive farming, which consumers believed managed to achieve a good balance between the needs of food production and the quality of animals' lives. Furthermore, several consumers doubted whether many of the scientists' concerns (Table 1) could be successfully dealt with under conditions of high intensity production.

"I don't see how this [normal/natural social behaviour] could be assured in mass production farming." (Hungarian consumer)

"And their forms of behaviour. Because I'd say that means playing, flying, running around. How would that be possible in an industry?" (Hungarian consumer)

This presents a serious challenge to Welfare Quality scientists. On the one hand, if they are to take these concerns seriously, they must both question their own assertions that one cannot make a priori ethical judgements about the welfare credentials of different intensity production systems and they must ensure that scientific measures of 'natural living' conditions are included

in the final monitoring scheme. Alternatively, they must employ fair, non-biased arguments to convince consumers who have been acculturated into equating high intensity with poor welfare that this is not necessarily the case.
Secondly, one of the ethical dimensions that underlies consumers' welfare priorities relates to the iconic status that they grant 'wild' or 'natural' environments as the perfect habitats for ensuring good animal welfare. This in turn led to many consumers prioritising 'outdoor access' for animals, as this was considered to be the best compromise between the ideal state of life for animals (the wild) and the need to rear animals for human consumption (see also Edwards 2003). The idea of 'outdoor access' also stood as a proxy for a number of related issues such as animals being able to graze, groom, have a social life, have access to natural food and have exposure to natural light. This emphasis also goes some way to explaining the success of free-range and organic food products, which often make explicit appeals to 'the natural'.

"Animal welfare for farm animals is a nonsense, only for wild animals there is welfare" (Italian consumer)

"It's got to be an environment, I hate to use this word, but that's got some biodiversity which is more like a natural [environment]" (British consumer)

In contrast, Welfare Quality scientists are far more critical and reserved in their embrace of the 'natural', as they are perhaps more aware of the welfare risks (e.g. predation, disease) that 'natural' environments can pose. As such, the challenge for both social and natural scientists is to delve far more deeply into (and to continue to critically 'unpack') both consumer and scientific understandings of 'the natural' (see Macnaghten and Urry 1998), so that we might begin to establish just what specific dimensions are both valued *and* likely to be of real benefit for farm animals (these specific dimensions could range from 'fresh air' and a 'good view' to more 'natural' bedding and more 'natural' social groupings).
Thirdly, consumers tended to adopt a *holistic* approach to animal welfare, in other words they saw welfare as an overall state of well-being and they were less willing to break welfare down into (artificial) component parts. Furthermore, many rejected the idea that it was possible to rank welfare concerns, as they were all deemed to be equally important and intimately connected. This presents a serious challenge to Welfare Quality scientists, many of whom are faced with the task of prioritising, ranking and even 'trading-off' certain elements of welfare in order to produce a workable monitoring scheme.

Conclusions: Integrating insights from science and society

We would like to conclude by raising a number of important issues: First, although there are numerous similarities there are also significant differences in the ways in which consumers and scientists understand and 'frame' the issue of farm animal welfare. Consumers' concerns tend to be holistic, inspired by their experience of companion animals, more radical and less well informed on specific issues of animal biology and farming practices. In contrast, scientific concerns tend to be more quantitative, more-goal oriented and more realistic in their expectations of the types of improvements that can be achieved. Furthermore, whilst scientists believe that animal welfare can be broken down into a list of measurable parameters, for many consumers the 'quality of life' of animals is considered to form an undividable whole, as such consumers' primary method of evaluating welfare tends to be by considering to what extent a given farming system manages to re-create 'natural' living conditions.
Second, it is clear that we should encourage a flow of information from scientists to consumers. In particular, we should re-double our efforts to inform consumers about current farming

practices and current welfare science in an unbiased fashion. This will enable consumers to become more involved and be better able to contribute to welfare debates. Furthermore, it is vital that we ensure that food product information is *relevant* to consumer needs. In other words, we need to take into account the pluralism of consumers vis-à-vis their different ethical orientations to animals (e.g. consumers with a preference for organic products will require different welfare information than consumers with a preference for traditional rearing methods).
Finally, we must ensure that we achieve a genuine two-way dialogue between science and society by encouraging Welfare Quality scientists to continue to be 'open' to stakeholder influence at an early 'upstream stage' of their research. For it is only through a genuine two-way dialogue that we can arrive at a truly democratic monitoring scheme.

Acknowledgements

The present study is part of the Welfare Quality research project, which has been co-financed by the European Commission, within the 6th Framework Programme, contract No. FOOD-CT-2004-506508. The text represents the authors' views and does not necessarily represent a position of the Commission who will not be liable for the use made of such information.

References

Barnett, J. L. and Hemsworth, P.H. (1990). The validity of physiological and behavioural measures of animal welfare. *Applied Animal Behaviour Science*, 25: 177-187.

Deleuze, G. and Guattari, F. (1988). *A thousand plateaus. Capitalism and Schizophrenia*. The Athlone Press, London.

Edwards, S. (2003). *Animal welfare issues in animal production Proc. of the NJF's 22nd Congress "Nordic Agriculture in Perspective"* [on line]. Available on http://www.njf.dk/njf/reports/njfreports.htm [date of consultation: 24/04/06].

FAWC (Farm Animal Welfare Council) (1992). FAWC updates the five freedoms. *The Veterinary Record*, 17: 357.

Fraser, D. (1995). Science, values and animal welfare: exploring the 'inextricable connection. *Animal welfare 4*: 103-117.

Fraser, D. (2004). *Applying science to animal welfare standards*, in proceedings of global conference on animal welfare: an OIE initiative, Paris 23-25 February, 121-127.

Habermas, J. (1981). *The Theory of Communicative Action: Reason and the rationalization of society*. Beacon press, Boston.

Hurnik, J. F. (1993). Ethics and animal agriculture. *Journal of Agricultural and Environmental Ethics*, 6: 21-35.

Keeling, L. and Veissier, I. (2005). *Developing a monitoring system to assess welfare quality in cattle, pigs and chickens*, in Welfare Quality conference proceedings 17-18 November 2005, Brussels).

Macnaghten, P. and Urry, J. (1998). *Contested Natures*. Sage, London.

Miele, M and Evans, A.B. (2005). *European consumers' views about farm animal welfare*, in proceedings of the Welfare Quality conference, Brussels 17-18 November.

Rorty, R. (1989). *Contingency, Irony, Solidarity*. Cambridge University Press, Cambridge.

Stengers, I. (1997). *Power and Invention. Situating Science*. University of Minnesota Press, Minneapolis.

A critical appraisal of the UNESCO version of the Precautionary Principle

Bjørn Myskja[1] and Anne Ingeborg Myhr[2]
[1]*NTNU, Department of Philosophy, N-7491 Trondheim, Norway, bjorn.myskja@hf.ntnu.no*
[2]*Norwegian Institute of Gene Ecology, The Research park, N-9294 Tromsø, Norway, annem@fagmed.uit.no*

Abstract

A recent report by an UNESCO expert group on the Precautionary Principle suggests a new science-based approach for implementation of the principle. We intend to explore the advantages and disadvantages of this precautionary approach to the issue of DNA vaccines in aquaculture. Introduction of DNA vaccines hold promises for protection against diseases caused by pathogens. In aquaculture it is estimated a 10% loss due to problems of infections by various pathogens and another concern is the detrimental effects of the use of antibiotics. DNA vaccines may provide a technological solution to these problems. However, there are reasonable concerns that the benefits associated with distribution of DNA vaccines may also be accompanied with risks to environment and health. There is wide agreement that full certainty about negative and positive consequences of the use of genetic engineering in the environment is unattainable prior to fullscale employment of the technology. The complexity of natural and social systems involved implies that the information acquired in quantified risk assessments, however valuable, must be inadequate for evidence-based decisions. One controversial strategy for dealing with this kind of uncertainty concerning harm is the Precautionary Principle. We will argue that the UNESCO version of the Precautionary Principle represents a significant improvement compared to earlier attempts. We intend to highlight the advantages and disadvantages with the UNESCO version compared to other versions of the Precautionary Principle, using introduction of DNA vaccines as a case study. In addition, we discuss whether a combined scientific and ethical analysis that does not involve the concept of precaution will provide a sufficient basis for sound decision-making in areas of scientific uncertainty.

Keywords: the Precautionary Principle, ethics, scientific uncertainty, DNA vaccine, best scientific practice

Introduction

Edible and injectable DNA vaccines hold prospects for rapid immunization against a variety of diseases caused by preferentially intracellular pathogens (e.g. virus, some bacteria) that are difficult to eradicate with traditional vaccines and antibiotics. For instance, for a range of viral diseases there are at present no efficient vaccines based on either live, attenuated virus or vaccines containing recombinant viral antigens. DNA vaccines represent means to protect against diseases, and hence improve human and animal welfare, reduce antibiotic usage and spread of disease. Other potential usage of DNA vaccines involves treatment of diseases as cancer, autoimmunity and allergies. The development of DNA vaccines has several attractive benefits: low cost, ease of production and improved quality control, heat stability, identical production processes for different vaccines, and the possibility of producing multivalent vaccines (Kwang, 2000). On the other hand, there is at present a limited scientific understanding of the mechanisms underlying uptake, persistence and degradation of DNA vaccines after injection

into an animal. Main areas of uncertainty are with regard to the immunological impact, tissue distribution and persistence after injection and whether the DNA vaccine can be distributed to the environment (Myhr and Dalmo, 2005).
The present lack of scientific understanding of the health and environmental effects of distribution of DNA vaccines involves a challenge with regard to management and sustainability. In this paper we intend to explore the advantages and disadvantages of a precautionary approach to distribution of DNA vaccines in aquaculture. The Precautionary Principle has been accepted by many national governments as a basis for policy-making, and it has become important both in international environmental law and international treaties (CBD, 2000; EC, 2000, Freestone and Hey, 1996).
However, the application of the Precautionary Principle in risk assessment and management is at present a subject of heated scientific and public controversies. In the view of the critics, the use of the Precautionary Principle places additional regulatory burden on new technology utilisation, and thereby reduces returns from innovation, limits utilisation of technology worldwide and provides disincentives for research. On the other hand, advocates of the Precautionary Principle want to enhance safety procedures and to separate trade and environmental interests in decision-making. Furthermore, at present there are several definitions of the Precautionary Principle, which contributes to further difficulties for implementation in legislation and employment in actual cases.

The Precautionary Principle

The Precautionary Principle is a normative principle for making practical decisions under conditions of scientific uncertainty. Gardiner (2006) suggests that the following elements are essential to the principle:
- Threat of harm.
- Uncertainty of impact or causality.
- Precautionary response.

As we can see this raises a host of questions, with regard to how to define harm, how to assess and systematize uncertainty and when and how to initiate a precautionary response. In addition, the actual content of the Precautionary Principle and the practical implications of its implementation in policy issues are controversial. Several formulations of the Principle, ranging from ecocentric to anthropocentric, and from risk-adverse to risk-taking positions, have been put forward (see boxes 1 and 2). A weak version of the Precautionary Principle is often grounded in narrow utilitarian ethics, and its application involves risk/cost-benefit analyses. In this context, the Principle may be used as an option to manage risks when they have been identified through risk analysis. For instance, the Rio Declaration employs the weighing of costs and benefits (see Box 1), and similar wording has been reproduced in the preamble of the Convention on Biological Diversity and in article 3 of the Framework Convention on Climate Change.

Box 1. Weak version of the Precautionary Principle.

The Rio declaration
In order to protect the environment, the precautionary approach should be widely applied by States according to their capabilities. Where there are threats of serious or irreversible damage, lack of full scientific certainty shall not be used as a reason for postponing cost-effective measures to prevent environmental degradation (Agenda 21, 1992).

Box 2. Strong version of the Precautionary Principle.

The Wingspread statement
When an activity raises threats of harm to human health or the environment, precautionary measures should be taken even if some cause and effect relationships are not fully established scientifically (Raffensperger and Tickner, 1999).

Strong versions of the Precautionary Principle embrace inherent values of the environment and often, but not always, are founded in ecocentric views or duty-based concerns for non-human beings and ecosystems. A strong version is active in nature and obliges regulators to take action, for instance by implementation of risk management procedures. The Wingspread Statement is considered to represent a strong version of the Precautionary Principle (see Box 2).
Weak versions of the principle are criticized as vacuous, whereas strong versions are often considered too narrow, focusing on environmental *risk* without considering the costs of existing practices to the environment or taking into account relevant non-environmental concerns (Gardiner, 2006). For instance, employment of a weak version of the Precautionary Principle to DNA vaccines in aquaculture would focus on the proposed benefits while the potential risks to health and the environment would be downplayed due to the present lack of scientific understanding. While strong precautionary assessment of DNA vaccines would merely emphasize the risks of introducing the vaccines, neglecting the detrimental effects of the existing vaccination practices, as well as the potentially positive effect of more effective vaccines for food production. Full certainty about potential unintended health and environmental effects is unattainable prior to fullscale employment of the technology. The complexity of animal and environmental systems implies that the information achieved by doing laboratory research and in risk assessment is not adequate for evidence-based decisions. Hence a key issue becomes how to approach and acknowledge lack of scientific understanding.

Implementation of the Precautionary Principle; weak versions versus strong versions
The implementation of the Precautionary Principle requires that indications of adverse impacts are being documented in some way, and that risk-associated research is initiated. However, these indications must reach a certain threshold of scientific plausibility before a precautionary measure can be initiated. Article 15(1) of the Cartagena Protocol on Biosafety states that

> "Risk assessments undertaken pursuant to this Protocol shall be carried out in a scientifically sound manner ... Such risk assessments shall be based, at a minimum, on information provided in accordance with Article 8 and other available scientific evidence in order to identify and evaluate the possible adverse effects of living modified organisms on the conservation and sustainable use of biological diversity, taking also into account risks to human health."

The references to "available scientific evidence" and "scientifically sound manner" can be seen as a predetermined qualitative term, while for instance the EC communication on the Precautionary Principle (EC, 2000) have chosen to focus on the quality of the information. By demanding scientific evidence before employing the Precautionary Principle, the Biosafety Protocol requires documentation indicating that the GMO causes harm to health or the environment.
In contrast, a focus on the quality of information relates to what type of scientific understanding is known and what there is a lack of information about without the need to giving "degrees" or "levels" of proof. Hence, scientific uncertainty and indications of harm would be enough for

acceptance of employment of the Precautionary Principle. The first approach corresponds to a weak version of the principle, whereas the second indicates a strong version.
Thus, a demand that there be available scientific information to determine whether precautionary steps should be taken fails to evade the criticism of weak versions of the principle for being vacuous. On the other hand, the postponement of scientific assessments in the strong versions combined with the myopic focus on environmental risk, fails to balance the risks with benefits or, in other cases, risks of non-action.

Acknowledgement of uncertainty in the UNESCO version of the Precautionary Principle
The recent UNESCO report on the PP (UNESCO 2005) may indicate a promising middle ground between present strong and weak versions of the Precautionary Principle. By replacing scientific probability with plausibility, and by emphasizing evaluation of the consequences of both action and inaction in a participatory process, pitfalls of both weak and strong versions seems to be avoided:
When human activities may lead to morally unacceptable harm that is scientifically plausible but uncertain, actions shall be taken to avoid or diminish that harm.
Morally unacceptable harm refers to harm to humans or the environment that is
- threatening to human life or health, or
- serious and effectively irreversible, or
- inequitable to present or future generations, or
- imposed without adequate consideration of the human rights of those affected.

The judgment of plausibility should be grounded in scientific analysis. Analysis should be ongoing so that chosen actions are subject to review.
Uncertainty may apply to, but need not be limited to, causality or the bounds of the possible harm.
Actions are interventions that are undertaken before harm occurs that seek to avoid or diminish the harm. Actions should be chosen that are proportional to the seriousness of the potential harm, with consideration of their positive and negative consequences, and with an assessment of the moral implications of both action and inaction. The choice of action should be the result of a participatory process.
The definition is apparently open to several of the objections levelled at other versions of the principle. The problem of vacuousness seems to be reappearing in the expression "morally unacceptable harm". With regard DNA vaccines in aquaculture, this ultimately raises the following questions; for whom may DNA vaccines be unacceptable, and which moral theories or principles should be employed? The answer may be found in the final demand of a participatory process, including all affected parties. The report claims that "moral judgements are less subject to plurality and rest on a firmer basis than the ethical theories one adheres to" (p. 17) and expresses a belief in a yet unrevealed universal basis for ethics. Thus this alleged fact of moral agreement in particulars is basis for trusting the unanimity and correctness of the results of the participatory deliberation.
The report fails to reveal where they have drawn this assumption about agreement in particulars from, and the disagreement on how to deal with the more or less plausible harms from distribution of DNA vaccines hardly supports their case. On the contrary; most agree on the theoretical assumptions that irreversible damage is wrong, but disagree on the particulars, and even principles for handling them. There is no guarantee that staying with the cases will help us resolve the moral issues involved, and the reason for disagreement may not be moral, be it theoretical or practical. That is, given that we are able to construct meaningful distinctions between empirical facts and value judgments. The problem is the issue regarding interpretation of relevant data more than how to handle these data in practical reality.

Even if this definition of the Precautionary Principle escapes the charge of vacuousness, it stumbles on the assumption about agreement in moral judgments. We have no reason to expect such agreement because interpretations of the issues at stake differ, as is exemplified by the continuing debate between Europe and the USA, Canada and Argentine on the moral acceptability of the release of genetically modified organisms. Even if everybody agrees to the list of principles for delimiting relevant harm, we will disagree on how to balance different scientific scenarios. Furthermore, do the different scientific disciplines that are involved in the epistemic debate employ competing models or analogies for basic assumptions to frame the scope for further research. For instance with regard to genetically modified organisms, do the different scientific disciplines that are involved use competing analogies and models for basic assumptions to frame the scope for further research. Agricultural biotechnologists refers to the practice of conventional plant breeding, while ecologists refers to the experiences based introduction of exotic species to make up for lacking anticipatory knowledge. Since the principles and paradigms of the different scientific disciplines differ, they have no common ground to discuss means for gather new scientific understanding. When we have this situation, the easy conclusion is "better safe than sorry", and the criticism against strong versions of the principle becomes relevant for the UNESCO Precautionary Principle.
As the hope of agreement is little more than a hope, the consequence of the UNESCO approach is similar to strong versions. Admittedly, the principle emphasizes proportionality of action relevant to "the seriousness of the potential harm, with consideration of their positive and negative consequences, and with an assessment of the moral implications of both action and inaction." But given that we have no tool for weighing more or less plausible scenarios, scientific plausibility is sufficient, we are left with the seriousness of the worst case as the only guideline. This leads us to choose the devil we know, and pay undue weight to the suggested harm, disregarding the negative effects of present activity and the positive potentials of the proposed action.

May good decisions be made without employment of the Precautionary Principle?

An alternative to the Precautionary Principle is a return to a principle of best scientific practice. We do not need the Precautionary Principle to realize that the present climate changes leads to morally unacceptable consequences for future (or even present) inhabitants of the earth. We only need to pay heed to the words of the scientific community and combine it with common sense. The problem in the climate debate is not to agree on what to do, regardless of precaution, but to find ways to avoid versions of the problem of the commons. Likewise, in other matters, what we need is a wide, open and responsible debate based on best scientific knowledge. This calls for a privileged position for experts in the "participatory process", in accordance with Aristotle's theory of practical argumentation (Top. I, 1) where accepted opinions include those of the many, the majority, or the wise, and then the majority or the most respected of the wise. But, as biologists or other scientific expertise lack knowledge of the impact on society, also the expertise of ethicists and social scientists must be given a privileged position in the deliberation on how to deal with matters of uncertainty.

References

Aristotle (1960). *Topica*. Harvard: Harvard University Press.

CBD (2000). Cartagena Protocol on Biosafety to the Convention on Biological Diversity. Montreal: Secretariat of the Convention on Biological Diversity. Available on www.biodiv.org/doc/legal/cartagena-protocol-en-pdf (accessed April 2006).

EC (Commission of the European Communities) (2000) *Communication on the Precautionary Principle*. Available on europa.eu.int (accessed April 2006).

Foster, K. R., Vecchia, P. and Repacholi, M. H. (2000). Science and the Precautionary Principle, *Science,* 288: 979-981.

Freestone, D. and Hey, E. (1996) Origins and development of the Precautionary Principle, in Freestone, D. and E. Hey (ed.*), The Precautionary Principle and international law.* Kluwer Law International, Netherlands, p. 3-15.

Gardiner, S.M. (2006). A Core Precautionary Principle. *The Journal of Political Philosophy*, 14 (1): 33-60.

Kwang, J. (2000). Fishing for vaccines. *Nature Biotechnology*, 18: 1145-46.

Myhr, A.I. and Dalmo, R.A. (2005). Introduction of genetic engineering in aquaculture: ecological and ethical implications for science and governance. *Aquaculture,* 250: 542-554.

Raffensperger, C., Tickner, J. (ed.) (1999). *Protecting public health and the environment: Implementing the Precautionary Principle.* Island Press, Washington DC

UNESCO (2005) *The Precautionary Principle. World Commission on the Ethics of Scientific Knowledge and Technology*. Paris: UNESCO.

Biosecurity research and agroterrorism: Are there ethical issues at stake?

Matias Pasquali
Society in Science, Branco Weiss Fellow, AGROINNOVA, University of Torino, Current address: Dept. Plant Pathology, University of Minnesota, COAFES - 495 Borlaug Hall, 1991 Upper Buford Circle, St Paul, MN 55108-6030, USA, pasqu016@umn.edu; matias.pasquali@unito.it

Abstract

Food production and food protection are indissolubly interconnected. "In recent decades scientific research has created new and unexpected knowledge and technologies that offer unprecedented opportunities to improve human and animal health and environmental conditions. But some science and technology can be used for destructive purposes as well as for constructive purposes. Scientists have a special responsibility when it comes to problems of "dual use" and the misuse of science and technology" (from the InterAcademy Panel document, 2005). The paper describes recent efforts world wide to develop scientific networks working on the issue of agroterrorism and crop biosecurity. It then analyses some of the ethical issues arising from this kind of research, with special attention to dual use issue, freedom of publication and scientists responsibility. The paper argues that a new form of public science is rising. Thus the adoption of a pragmatic approach requiring a confrontation with society is suggested in order to achieve large agreements on possible restriction of freedom, on transparency and on the social role of academic research on security issues.

Keywords: security, dual use, code of conduct, scientist responsibility

Introduction

Food terrorism has been defined by the World health Organization (2003) as an "act or threat of deliberate contamination of food for human consumption with chemical, biological or radionuclear agents for the purpose of causing injury or death to civilian populations and/or disrupting social, economic, or political stability". A wider definition, including also deliberate infection of plants and animals used as food is actually the object of research efforts in the so called *agroterrorism* field. Thus, now, "food security" assumes a new concept, not only connected with food sufficiency but to potential contamination of food.
Historical reasons together with actual international tensions have increased attention towards the issue in the last 6/7 years (Gronwall, 2005). In recent years research efforts in US, and subsequently in Europe and in other areas of the world, have been addressing scientifically the risk of agroterrorism (Gullino and Pasquali, 2005).
This interest is connected to the crucial role that food production has in the economic and social sphere. Looking for example at the European agriculture and agro-food industry, it accounts for 15% of the European Union annual gross product, with production split nearly evenly between crops and livestock, making this sector economically important. Moreover the psychological and social role of food, the value of agricultural landscape as the last "guardian of nature", make it a potential target for terrorist attacks. Finally indirect and indiscriminate nature of an agroterror attack meshes perfectly with the perceived shift in terrorism goals, increasingly seeking the destruction of "enemy" societies (Foxwell, 2001), including therefore values connected with food and the environment.

Scientists dealing with the risk of crop biosecurity have to face some questions that never arose before within the agricultural research field.

The facts

High political interest in security issues has determined the increasing availability of funding in all the security related fields (see the VII European research framework at http://cordis.europa.eu.int/security for the European situation). The consequence is that academic scientists apply for grants in order to build research networks or to develop projects that are focused on the security issue. In the last century many nations (USA, URSS, Iraq, ...) invested in weaponization of plant pathogens as a mean for creating damages to potential enemies (Gronwall, 2005). But this kind of research, with specific offence purposes was carried out by military scientific groups. The novel attention to security is on the contrary changing the status of scientific initiative. Academic scientists are receiving funds for defensive research on novel detection strategies for pathogen, spread modelling, simulation of disasters, informatics infrastructures for rapid communication between experts, identification of dangerous species. Despite the number of scientists actually involved in this process is limited, the rising of funds in the sector is likely to increase the number of interested researchers.
What are the main issues related to scientific activity on security?

Dual use problem

Dual use research is defined as research that can be instrumental for a different purpose than the original one, usually connected to the potentiality to harm. So results of defensive research can be easily used for offensive purposes.
For example biological weapons can be produced entirely using materials, equipment and facilities that have other, completely legitimate uses and the technology used to provide a defence against biological weapons is largely the same as the technology that would be used to produce the weapons.
Looking specifically at the agricultural sciences, for example, the identification of DNA sequences connected to pathogenicity or survival factors of pathogens, the description of detection methods, the simulation of damages potentially caused by a pathogen can be easily manipulated/used for specifically perpetuate offence. Dual use research questions the responsibility of scientists and the level of caution that they have to assume when working on potential dual use outcomes.

The publication issue

The problem consist in the control of information that, if published, may harm society.
Mertonian ethos, first proposed in 1942, and in particular the Communalism rule that "requires that scientific knowledge should be public knowledge; that the results of research should be published; that there should be freedom of exchange of scientific information between scientists everywhere, and that scientist should be responsible to the scientific community for the trustworthiness of their published work" (Brown, 1986), gives a principle for dismissing the need of restriction in publication issues. This approach is often claimed by group of scientists discussing the issue.
Merton rules are anyway only hypothetical and idealistic, suggesting principles that should guide science development but have been proven far from applicable on the diverse existing typology of sciences. With more pragmatic consideration on possible outcomes, the majority of scientists would subscribe the statement by P. Campbell, the *Nature* editor: "We recognize that on occasions an editor may conclude that the potential harm of publication outweighs the potential societal benefits. Under such circumstances, the paper should be modified, or not be published".

Looking at the issue of freedom of publication and dissemination from a consequential point of view, historically, some reasons have been advocated for keeping low the level of secrecy. In 1970, the US Defense Science Board Task Force report on secrecy concluded that "more might be gained than lost if our nation were to adopt--unilaterally, if necessary--a policy of complete openness in all areas of information." A similar note was sounded in 1982 by a US National Academy of Sciences panel which reported that "security by accomplishment" offers greater promise than security through controls on information. So any ban that goes much beyond direct weapons research has rapidly diminishing benefit (Cohen, 2003).
Considerations on legitimacy in halting publication for security have been recently encountered in the agricultural field. A benefit/cost analysis has been recently adopted, when the PNAS has decided to publish an article regarding agroterrostic related research (Wein and Liu, 2005). Despite the objections of some government agencies such as the U.S. Department of Health and Human Services the editor in chief has opened the issue of the journal with these reasons:
"It is important to recognize that publishing terrorism-related analysis in the open scientific literature can make the nation safer in at least two different ways. First, science can make many important contributions to the design of our defences. Because science advances through the combination of knowledge in unexpected ways, the discoveries of each individual scientist must be made available to a wide variety of other scientists, who can then either build upon or criticize them... There is a second advantage to openness. Protecting ourselves optimally against terrorist acts will require that both national and state governments, as well as the public, be cognizant of the real dangers. If the types of calculations and analyses... are carried out only by government contractors in secrecy, not only are the many actors in the U.S. system who need to be alerted unlikely to be well informed, but also the federal government itself may become misled—either greatly overestimating or underestimating the seriousness of a particular danger relative to other concerns..." (Alberts, 2005)
It is suggested that information provided are not so dangerous and that the data are necessary for building important defence strategies by science itself.
Others criticize this approach showing that the increasing facility in building laboratories and in single man efforts in generating bio-terroristic weapon increase the need for secrecy.
What is the reaction in the agricultural sciences? The major plant pathology journal (Phytopathology, published by the Phytopathological American Society) has decided to adopt the solution chosen also by many other important scientific journals, that is to create a special panel of expert that will evaluate case by case the cost-benefit of a certain publication under the scrutiny of a security point of view (Policy guidelines of the APS Board, www.apsnet.org/members/ppb/PDFs/BiosecurityAPSPubBoardPolicy.pdf). Scientific community is proposing self-regulatory criteria for managing security issues.

Transfer of scientific samples

After the 1995 case of delivery by the American Type Culture Collection of a sample of *Yersinia pestis*, the agent of plague to a member of the White supremacist group Aryan nations, traditional openness of exchanging material has definitely been questioned. While human dangerous pathogens are now strictly monitored, more flexibility is still existing for plant pathogens. Novel requirements for obtaining strains in biological banks are easily falsifiable and at the same time generate discontent in the scientific community because generate more bureaucracy and slow down research. In many cases laboratory to laboratory material transfer bypasses strict regulations in order to speed up research procedure. Forcing regulation according to scientific tradition generates novel responsibilities for scientists. Again, as seen for the dual use problem and the publication issue, security is questioning the procedural methods of scientific enquiry.

What solutions?

In general security issues generates a tension in scientific community as testified by the discussion of the topic on scientific journals in these last years. Table 1 describes the main reasons for this tension.

The most favoured solution to the problems raised by research on security among scientists confirms that "when it comes to weighing up the potential risks and benefits of a piece of research, or deciding whether to publish a controversial result, scientists" (or editors) "must fill the gap by adopting their own principles for proper conduct" (Check, 2005).

Table 1. Reasons for the tensions in academic scientific practice when dealing with security issues.

1. Contemporary sciences are international "enterprises" where the national context is dismissed. On the contrary security policy, despite the will to internationalize defence strategies is still mainly, if not exclusively, a national issue.
2. Academic scientists are trained and educated still advocating the principle of a "pure" enquiry. Political will to generate practical results (patenting for example) has not completely changed the ideal nature of academic science as a purely free activity searching for knowledge. On the contrary the security concept require strict observance of needs and goals that are strongly politically influenced.
3. Traditional openness and self government attitude of academic science is intrinsically hostile to the concept of sensitive information and secrecy.
4. Academic science uses academic criteria for evaluation (publication mainly) but have to deal with potentially sensitive information (secrecy). While in a military scientific context secrecy and the military apparatus mitigate social responsibilities, in an academic context these responsibilities rely on single scientist and on its scientific community.

The code of conducts

Code of conducts are proposed as the self developed solution for managing research security issues by scientists.

According to promoters of ethical codes, international consensus of scientists is crucial. Moreover is necessary to ensure that all people and institutions involved in science are aware of their ethical obligations. Nonmaleficence, beneficence, respect for life, maintaining of trust, embedding ethics in science, establishing a high ethical tone in institutions, acknowledging individual and collective responsibilities, and the transmission of ethical values to colleagues are the principles advocated (Sommerville and Atlas, 2005).

Also some life-sciences organizations, such as the American Society for Microbiology and the Australian Society for Microbiology, have already adopted codes of ethics that refer to biological weapons and security issues.

Despite the will of self regulation and the need of advocating higher principles for guiding scientists conduct it should be pointed out the partial inefficacy of code of conducts:

- Social scientists and ethicists are sceptical on the overall effectiveness of professional and related corporate codes (Cash, 1987).
- little is known about their effectiveness in practice of stimulating certain behaviour
- few people consult really the code
- codes often leave much space for interpretation.

Learning from the GM case.
Security, and food security in particular, has clearly a strong impact on society. It touches economic, social and emotional sphere in a society.
The main solutions proposed by scientists include developing own guidance rules (ethical code) or creating a panel of scientific experts in biosecurity (for the editorial board of journals) together with appeal to increase the vigilance and the sense of responsibility in the scientific practice. One important aspect is missing from the agenda: the close influence and interrelatedness of science and society. The last fifteen years have shown that the main communication mistake on genetically modified organism was the self-closure of scientific community. Agricultural sciences are probably for importance of investments the last scientific sector involved in the security research, but they may lead, making treasure of the experience acquired with the GMO debate, a process of modification of the self-regulation attitude of scientific community. Scientist could take the initiative, as part of their projects on security, of establishing a preventive thorough confrontation with society at large, in order to build shared views on responsibilities connected to research on security. For example an argument that require a deep confrontation with society is the access to information with extremely dangerous potential. It will probably have to be restricted to professionally qualified groups. Certainly a new social contract will have to be formulated with those people whereby access is accompanied by active assurance of responsible use, including may be monitoring of relevant personal activities (Steinbruner and Okutani, 2004).
Public participation is the instrument to build trust in every scientific activity, more than ever if related to security. The strong opposition in the US by NGOs and local communities against the decision to build BSL-4 laboratories, required for working on dangerous diseases, is a signal for academic scientists involved in this kind of research to step by and start open a wide discussion on all the issues connected to security.
As outlined by Jasanoff (2005): "Democratic theory in the era of the knowledge society must actively take on board the involvement of citizens in the production, use, and interpretation of knowledge for public purposes". Food security definitively require this participatory effort. The modalities have to be defined but certainly scientists involved in this kind of research cannot avoid this challenge to their societal role.

Acknowledgements

This work was made possible by the Branco Weiss Fellowship support.

References

Alberts, B. (2005). Modeling attacks on the food supply. *PNAS*, 102: 9737-9738.

Brown, R.H.(1986) The Wisdom of Science. Cambridge, GB: Cambridge University Press. 194 p.

Cash, M. (1987). Codes of ethics, organizational behaviour and misbehaviour. *Research in Corporate Social Performance and Policy*, 9: 107-130.

Check, E. (2005). Biologists asked to breed a culture of responsibility in face of terrorism. *Nature*, 435: 860.

Cohen, P. (2003). Recipes for bioterror: censoring science. *New Scientist*, 9: 15.

Gronwall, K. G. (2005). A new role for sceintists in the biological weapon conventions. *Nature biotechnology* 23: 1213-1216.

Gullino, M.L. and Pasquali, M. (2005). Researches on agroterrorism in Europe in a global context. *APEC IAS Workshop*, 19-23 September 2005, Beijing.

Foxell, J.W. (2001). Current Trends in Agroterrorism (Antilivestock, Anticrop, and Antisoil Bioagricultural Terrorism) and Their Potential Impact on Food Security. *Studies in Conflict & Terrorism*, 24:107–129.

Iverson, M., Frankel, M. and Siage, S. (2003). Scientific societies and research integrity. *Sci Eng Ethics*, 9: 141-158.

Jasanoff, S. (2005). *Designs on Nature*. Princeton University Press, Princeton (USA) 344 p.

Somerville, M.A. and Atlas, R.M. (2005). Ethics: A Weapon to Counter Bioterrorism. *Science*, 307: 1881-1882.

Steinbruner, J. and Okutani, S. (2004). The protective oversight of biotechnology. *Biosecurity and Bioterrorism: Biodefense strategy, practice and Science*, 2: 273-280.

Wein, L.M. and Liu, Y. (2005). Analyzing a bioterror attack on the food supply: The case of botulinum toxin in milk. *PNAS*, 102: 9984-9989.

Part 5
Sustainability in food production

No-one at the helm: World trade, sustainability and farm animal welfare

Michael C. Appleby
World Society for the Protection of Animals, Compassion in World Farming, Eurogroup for Animal Welfare and Royal Society for Prevention of Cruelty to Animals, 89 Albert Embankment, London SE1 7TP, United Kingdom, michaelappleby@wspa.org.uk

Abstract

In World Trade Organisation (WTO) negotiations on rules for agricultural trade, issues such as sustainability and farm animal welfare are categorised as "non-trade concerns" and given scant consideration. This is partly because WTO rules mostly specify what is not allowed rather than what is allowed. Yet it is clear that there will be negative effects of trade on sustainability and welfare, and there is almost no other major organisation that can act to reduce those effects either. One important exception may be the World Organisation for Animal Health (OIE). No organisation is "in charge" of world trade, so there are few mechanisms to make improvements on issues that matter to people in fundamental, ethical and practical ways. However, there is nothing in WTO rules that explicitly prevents countries addressing these issues. We should therefore work with governments to promote sustainability and animal welfare, as concern for these continues to increase worldwide. One way of doing this for farm animal welfare would be to increase emphasis on the OIE welfare standards (so far established on transport and slaughter) and their adoption. It is also encouraging that the European Union continues to make positive statements about improving animal welfare, rather than reversing its protection of farm animals in the face of likely increases in trade.

Keywords: animal welfare, farm animals, sustainability, trade, World Trade Organization

Introduction

Agriculture is vital to every country in the world. It feeds people, it earns income and it has many other impacts. For these reasons, the 150 member countries of the World Trade Organisation (WTO) negotiate the Agreement on Agriculture (World Trade Organisation, 1995) to regulate agricultural trade. In 2001 a new round of talks was launched in Doha, Qatar, to renegotiate the Agreement on Agriculture with the intention of improving opportunities for development: this is called the Doha Development Agenda. Yet the WTO is often criticised – along with other Inter-Governmental Organisations – for damaging developing countries and causing many other problems. The range of issues affected by trade and potentially by the WTO is illustrated by the many themes of the EurSafe conference that are relevant (Table 1).
What the WTO is doing, what it is not doing and what it cannot do will be discussed with particular reference to sustainability and farm animal welfare.

Non-trade concerns

In the WTO, issues such as sustainability and farm animal welfare are categorised as "non-trade concerns." This name is arguably misleading, giving less emphasis to these issues than they merit: they could justifiably be called "trade concerns." Nevertheless, they are supposed to be

Table 1. Themes of the EurSafe conference affected by trade.

Sustainability in food production
Politics of consumption
Animal welfare and production
Social conflicts over resource management
Politics and power in the globalised food chain
Corporate responsibility and political governance in relation to food ethical issues
Cultural values in food production
Social distribution of responsibility for food ethical issues
Food, health and social inequality

considered in the talks. The Doha Ministerial Declaration of 2001, setting the Doha Development Agenda, said (paragraph 6):

> We recognize that under WTO rules no country should be prevented from taking measures for the protection of human, animal or plant life or health, or of the environment at the levels it considers appropriate, subject to the requirement that they are not applied in a manner which would constitute a means of arbitrary or unjustifiable discrimination between countries where the same conditions prevail, or a disguised restriction on international trade, and are otherwise in accordance with the provisions of the WTO Agreements.

And in relation to the Agreement on Agriculture (paragraph 13):

> We take note of the non-trade concerns reflected in the negotiating proposals submitted by Members and confirm that non-trade concerns will be taken into account in the negotiations as provided for in the Agreement on Agriculture.

However, there has been no indication that such consideration has been given to non-trade concerns as the talks move towards closure in summer 2006. There was no mention of non-trade concerns in the Declaration of the last Ministerial meeting, in Hong Kong in December 2005. This disregard is despite the fact that non-trade concerns are likely to have vital long-term effects both on trade and on other aims such as development. Conversely, it is clear that there will be negative effects of trade on non-trade concerns such as sustainability and farm animal welfare. For example, increased trade in animals and animal products both has direct impacts on the animals concerned and increases competition within and between countries and hence pressure to reduce production costs.

No-one is steering the ship

A major reason why non-trade concerns are given scant consideration must be that the talks are hugely complex, involving so many countries, even just considering what are seen as the main matters to be negotiated (market access, export support and domestic support, for both agricultural and other products). However, there is also a more fundamental reason. As with most rule-making processes, the WTO rules mostly specify what is not allowed rather than what is allowed. As such, it would actually be very difficult to build safeguards for sustainability or animal welfare into the WTO procedures.

In fact, there is almost no major organisation that has the power to reduce or prevent the negative effects of trade on sustainability and animal welfare. One important exception may be the World Organisation for Animal Health (OIE), which can place restrictions on livestock movements because of disease risks.

Despite common perceptions, neither the WTO nor any other organisation is "in charge" of world trade: no-one is at the helm, steering the ship. This means, disturbingly, that there are few mechanisms to make improvements on issues that matter to people in fundamental, ethical and practical ways.

Progress in countries and small groups of countries

However, there is nothing in the WTO procedures that explicitly prevents action being taken by countries to address these issues. On the contrary, the WTO's General Agreement on Tariffs and Trade says (Article XX) that:

Nothing in this agreement shall be construed to prevent the adoption or enforcement by any contracting party of measures:

a. necessary to protect public morals,
b. necessary to protect human, animal or plant health

and these principles were reiterated in the Doha Ministerial Declaration, quoted above.

As such, it is important that we should build on progress that has already been achieved, by working with governments to find ways to promote sustainability and animal welfare, in the context that concern for these continues to increase worldwide, albeit still unevenly. A number of animal protection organisations is working together along these lines, for example by producing a publication called "Animals and People First: Why good animal welfare is important for feeding people, for trade and for the future" (Appleby, 2005).

As just one example, the Kenya Veterinary Association (2006), with government support, organised an Open Learning Day for farmers and the general public in April 2006 on the theme of "Animal welfare for improved animal productivity and poverty alleviation."

Potential for progress worldwide

On a worldwide scale, one very positive development is the recognition by the World Organisation for Animal Health that animal health is affected by other aspects of animal welfare. It is therefore establishing the first global animal welfare standards – initially on transport and slaughter practices (OIE, 2005). It is imperative that ways should be found of increasing emphasis on these standards, and promoting their adoption among the 167 member countries.

It is also encouraging that the European Union continues to make positive statements about improving animal welfare, rather than reversing its protection of farm animals in the face of likely increases in trade. This includes promotion of animal welfare in bilateral trade agreements, most strongly in that between the European Union and Chile.

International trade is growing and will continue to have negative effects on sustainability and animal welfare. However, international communication is also growing, and bringing increasing pressure for animal and environmental protection in all countries. People concerned for animals and the environment hope, with some justification, that these positive effects are gathering pace.

References

Appleby, M.C. (2005). *Animals and People First: Why good animal welfare is important for feeding people, for trade and for the future.* World Society for the Protection of Animals, Eurogroup for Animal Welfare and RSPCA, London. 6p.

Kenya Veterinary Association. (2006). www.coastweek.com/2916-05.htm Accessed April 28, 2006.

OIE. (2005). www.oie.int/eng/press/en_050602.htm Accessed April 28, 2006.

World Trade Organisation. (1995). *Agreement on Agriculture.* World Trade Organisation, Geneva.

Integrative sustainability concept: Operationalisation for the food and agriculture sector

Rolf Meyer
Institute for Technology Assessment and Systems Analysis (ITAS), Research Centre Karlsruhe in the Helmholtz Association, P.O. Box 3640, D-76021 Karlsruhe, Germany, meyer@itas.fzk.de

Abstract

This paper outlines an integrative sustainability concept developed in the recent research project „Global sustainable development. Perspectives for Germany"of the Helmholtz Association of German Research Centres (HGF) under the direction of ITAS, deals with the operationalisation of this concept for the food and agriculture sector and compares this approach with other German sustainability studies. The basic idea of the integrative concept was to start from three constitutive elements of sustainable development (intra- and intergenerational justice, global perspective, anthropocentric view). In a next step, these constitutive elements were "translated" into three so-called general sustainability goals:

- securing human existence;
- maintaining society's productive potential;
- preserving society's options for development and action.

In a further step, the goals were specified by 15 substantial sustainability rules, building the core element of the concept. The substantial rules were used to describe the sustainability situation of Germany and of the German food and agriculture sector. The substantial rules were also the basis for indicators for the food and agriculture sector. These indicators map the different dimensions and were developed for the whole food chain, while other studies concentrate on agriculture and the ecological dimension. The future sustainability performance depends to a considerable extent on the development of fundamental societal framework conditions, which were analysed in three explorative scenarios (liberalisation and globalisation, modernisation, regionalisation and changes in values). The results of the scenario analysis show that with explorative scenarios the realisation problems of sustainability strategies can be better examined than with normative scenarios.

Keywords: sustainable development, agriculture, food, indicators, scenarios

Introduction

On the one hand "sustainable development" is today often used and broadly accepted in science, politics and nearly all societal groups as a guiding principle for societal development. On the other hand there are still quite different positions regarding concrete definitions and ways of implementing this principle.

This paper outlines essential results of the recent research project „Global sustainable development. Perspectives for Germany", which was carried out by several institutes of the Helmholtz Association of German Research Centres (HGF), under the direction of ITAS (see Kopfmüller *et al.* 2001; Coenen and Grunwald 2003). The main goals of this project were to make a substantial and innovative contribution to the scientific and public debate on the operationalisation of the sustainability concept in general and to contribute to its implementation in Germany.

The integrative concept

The so-called integrative concept of sustainable development was elaborated in the HGF project as a substantial result on its own and as a conceptual and analytical framework for further analyses. The core ideas of the Brundtland Report of the UN Commission on Environment and Development from 1987 (Hauff 1987) and the Rio documents from 1992 (i. e. Agenda 21, Rio Declaration) were used as fundamental background.

Constitutive elements of sustainable development

The basic idea of this integrative concept was – in contrast to other existing approaches – not to start from the separate "classical pillars" of societal development (economy, ecology, social and institutional aspects), but from three constitutive elements of sustainable development. The first element is the postulate of intra- and intergenerational justice. These two perspectives are strongly interrelated and have to be understood in equal terms. The present large inequalities and the increasing burden shift at the expense of future generations have to be seen as main causes for some of the pressing problems on the national and global level, as environmental degradation, poverty, debt or social and global instabilities. The second element is the global perspective due to the continuously intensified globalisation processes, the global character of many present problems, and the resulting requirements of improved and intensified responsible cooperation between rich and poor countries to solve these problems. Finally, the third element consists of the anthropocentric view, which means that human needs and their satisfaction are in the focus of the concept, considering the conservation of nature as a prerequisite for this.

General sustainability goals and substantial sustainability rules

In the next step, these constitutive elements were "translated" into three so-called general sustainability goals. The first goal is to secure human existence as the most fundamental goal and expression of the postulate of justice. The second goal is maintaining society's productive potential, including the material preconditions of societal development, in economic terms: natural, man-made and human capital. The third goal consists of preserving society's options for development and action, including immaterial aspects and needs. These goals are highly interdependent and underline the integrative character of the approach. In a further step, the goals were specified by substantial sustainability rules, understood as action-guiding principles, forming the core element of the concept (Table 1).

These sustainability rules describe the fundamental minimum requirements that present or future generations all over the world are legitimately entitled to see fulfilled. The total set of rules, each to be concretised by suitable indicators, constitutes a "reference line" which serves as basic orientation for future development and as a set of criteria for sustainability analyses regarding areas, societal activities or technologies. The rules are supposed to further specify the conditions which have to be met on the global level in order to realise the three general sustainability goals. Due to their characteristic as minimum requirements it is assumed, as a working hypothesis, that the rules are compatible with each other, that they have to be fulfilled simultaneously and that this is possible. This means that every rule applies only within the limits set by the others. Potential conflicts between rules or goals that might occur on contextual levels have to be looked at in each particular case and must be subject to weighting procedures. Their characteristic as minimum conditions also implies that other possibly desired goals - e.g. economic growth, wealth, luxury, etc. - are not independent goals, but only feasible to a degree that complies with the fulfilling of the rules.

The outlined conceptual framework differs clearly from other German sustainability studies, which are based on the economic, ecologic and social dimensions and partly prioritise the

Table 1. System of substantial sustainability rules.

General sustainability goals	*1. Securing human existence*	*2. Maintaining society's productive potential*	*3. Preserving society's options for development and action*
Sustainability rules	1.1 Protection of human health	2.1 Sustainable use of renewable resources	3.1 Equal access of all people to information, education and occupation
	1.2 Ensuring the satisfaction of basic needs (nutrition, housing, medical care etc.)	2.2 Sustainable use of non-renewable resources	3.2 Participation in societal decision-making processes
	1.3 Autonomous subsistence based on income from work	2.3 Sustainable use of the environment as a sink for waste and emissions	3.3 Conservation of cultural heritage and cultural diversity
	1.4 Just distribution of changes for using natural resources	2.4 Avoiding technical risks with potential catastrophic impacts	3.4 Conservation of the cultural function of nature
	1.5 Reduction of extreme income or wealth inequalities	2.5 Sustainable development of man-made, human and knowledge capital	3.5 Conservation of "social resources" (e.g. tolerance, solidarity or adequate conflict solution mechanisms)

Source: Coenen and Grunwald 2003, p. 68

ecological dimension (e.g. Bund and Misereor 1996; UBA 1997). The integration step should prevent, at least initially, controversial and inadequate debates on the appropriate relative weight between the development dimensions. Furthermore, the coherent and broad differentiation should help to avoid that the term "sustainable development" becomes or remains arbitrary.

Operationalisation for the food and agriculture sector

Based on the set of rules and respective indicators the present sustainability situation of Germany was analysed and evaluated. For this purpose a set of core indicators was selected to ensure analytical practicability, a better communication of results, and to meet the requirements of the specific characteristics of a country like Germany. The criteria used for the selection of indicators were the adequate reproduction of the core ideas of the rules, the possibility to set quantitative goals, the possibility to recognise the direction of increasing and decreasing sustainability, and finally the availability of suitable data. For the selected indicators quantitative goals were formulated, that were either adopted - in cases of already existing political decisions - or selected in view of actual debates or developed independently within the project.

Sustainability indicators for the food and agriculture sector

The first big sustainability studies for Germany worked – if at all – only with environment related indicators (e.g. Bund and Misereor 1996). In the national sustainability strategy of Germany, the two key indicators in the area of food and agriculture are the production area of organic farming and the surplus from the national nitrate balance, and therefore focus on agriculture.

Table 2. Sustainability indicators and goals for the food and agriculture sector in Germany.

Sustainability indicator	Development	Sustainability goal
Proportion of regular farms with an income equivalent to commercial wages	1996/97: 7% 2000/01: 28%	Further increase
EU substitutes for agricultural export	1995: 6.3 billion $ 1998: approx. 6 billion $	Complete reduction
Agricultural aid for developing countries	-	Increase
Primary energy consumption in the food and agriculture sector	1991: 2,735 PJ 2000: 2.,77 PJ	Further reduction
Greenhouse gas emissions of agriculture	N_2O emissions: 1990: 96,000 t/a; 1998: 84,000 t/a CH_4 emissions: 1990: 1,906,000 t/a; 1998: 1,560,000 t/a CO_2 emissions: 1991: 9,538,000 t/a; 1999: 9,048,334 t/a	40% reduction of the total greenhouse gas emissions from 1990 to 2020
Surplus of nitrate balance	1980: 128 kg N/ha/a 1999: 83 kg N/ha/a	Reduction to 50 kg N/ha/a
Pesticide risk indicator from SYNOPS model	(Data not yet available)	Reduction of risks associated with pesticides
Ammoniac emissions of agriculture	1990: 613,000 t/a 1999: 465,000 t/a	Critical load: 300,000 t NH_3/a
Proportion of persons with overweight of the total population (BMI over 25 kg/m^2)	Women: 1985: 49%; 1998: 52% Men: 1985: 65%; 1998: 69%	Trend change
Number of notifiable foodborne infections per year	1995: 192,400 1999: 197,340	Noticeable reduction
Average annual consumption of meat and sausage per person	1950: 26.2 kg; 1999: 63.5 kg; 2001: 60.7 kg	Reduction to a consumption of around 30 kg/a/person
Market share of the five leading retailers	1980: 26% ; 1999: 62%	Reduction of market dominating positions
Import of phosphate	P_2O_5 sales per year: 1993/1994: 415,000 t 2000/2001: 360,000 t	Reduction of imports, increase of phosphate recycling
Percentage of crop land exceeding the tolerable soil losses	46.6% of crop land in the medium and high categories of potential soil loss (> 10 t/ha/a)	Reduction of soil loss of 90% until 2010
Population of endangered livestock races	Promotion in agri-environmental measures 1997: 11,170 GVE; 1998: 10,200 GVE	Maintenance and increase of the endangered livestock populations

Source: Coenen and Grunwald 2003, p. 186

In the HGF project, an important step forward was taken by developing sustainability indicators for food and agriculture, based on the set of sustainability rules and on the analysis of the sustainability problems. These indicators cover key elements of the food chain and important elements from all dimensions. Table 2 shows the selected sustainability indicators for the German food and agriculture sector, their development in the past and the corresponding goals.
In the meantime, the Commission of the European Communities is developing – in a similar direction – a set of sustainable development indicators for monitoring the implementation of the EU Sustainable Development Strategy (EU Commission 2005), which includes indicators for different dimensions and elements of the food chain.

Explorative scenarios
The future sustainability performance and the resulting political requirements for action depend to a considerable extent on the development of fundamental societal and political framework conditions, which were analysed by three explorative scenarios (for Germany). In a first step, the scenarios were formulated and analysed for the national level, than in a second step they were broken down to the level of different "activity areas", thus also for the food and agriculture sector, and their effects studied on this level.
In the scenario "*liberalisation and globalisation*", further increasing globalisation processes and the dominance of market-based regulation mechanisms were assumed. For the food and agriculture sector, this means an increasing division of labour in food production. The analysis of the scenario consequences shows that the exiting sustainability deficits would increase.
In the scenario "*modernisation*" - interpretable as a business-as-usual development - a proactive governmental policy is assumed which intends to use the chances of globalisation and to minimise the risks. The internationalisation of the food sector will be slower and the structural changes in agriculture more moderate than in the first scenario. Some progress for environmental sustainability goals would be a consequence of the scenario "modernisation".
The scenario "*regionalisation and changes in values*" supposes changes of societal values towards environmental protection, public interest orientation, and increasing local economic structures as a reaction to problematic globalisation effects. A predominantly demand-side policy approach was assumed which focuses on problem solutions by governmental activities. For the food and agriculture sector, changes in diet patterns, willingness to pay for higher food quality and substitutes for environmental friendly agriculture were assumed. In consequence, a significant improvement for environmental goals would be achieved, but at same time a loss of export and working places in the food industry would result.
The results of the scenario analysis show that the sustainability goals are achieved to different degrees and with specific shortfalls. With the explorative scenarios, the realisation problems of sustainability strategies can be better examined than with normative scenarios.
Such normative scenarios were developed in important German sustainability studies (Bund and Misereor 1996; UBA 1997). They do not describe different paths towards more sustainability, but differ mostly in the degree of the achieved sustainability. This approach was adequate for their objective to show that sustainable development for Germany is possible. Other studies (e.g. EK Erdatmosphäre 1994, SRU 1996, Tappeser *et al.* 1999a, 1999b) proceed from a problem analysis and a vision for sustainability directly to possible options for action. The consequence of these approaches is that not so much can be learnt about restrictions and realisation problems of politics for sustainable development. After the need for a sustainability policy has been broadly recognised, more knowledge about the implementation – also under not so favourable circumstances – is needed.
With the HGF project, a starting point was made to examine the realisation problems of sustainability strategies. But further investigation is needed to understand better the influence of fundamental societal and political framework conditions on sustainability strategies for the

food and agriculture sector, to analyse conflicts between different sustainability goals, and to identify possible options for action which are robust under different and also unfavourable conditions.

References

BUND, Misereor (eds) (1996). Zukunftsfähiges Deutschland. Ein Beitrag zu einer global nachhaltigen Entwicklung. Birkhäuser Verlag, Basel. 453 p.

Coenen, R., Grunwald, A. (eds) (2003). Nachhaltigkeitsprobleme in Deutschland. Analyse und Lösungsstrategien. Edition sigma, Berlin. 544 p.

EK Erdatmosphäre - Enquete-Kommission "Schutz der Erdatmosphäre" (1994). Schutz der Grünen Erde. Klimaschutz durch umweltgerechte Landwirtschaft und Erhalt der Wälder. Dritter Bericht der Enquete-Kommission "Schutz der Erdatmosphäre" des 12. Deutschen Bundestages. Economica Verlag, Bonn. 702 p.

EU Commission (2005). Sustainable Development Indicators to monitor the implementation of the EU Sustainable Development Strategy. Commission Communication SEC(2005) 161 final of 09.02.2005

Hauff, V. (1987). Unsere gemeinsame Zukunft. Der Brundtland-Bericht der Weltkommission für Umwelt und Entwicklung. Eggenkamp Verlag, Greven. 421 p.

Kopfmüller, J., Brandl, V., Jörissen, J., Paetau, M., Banse, G., Coenen, R., Grunwald, A. (2001). Nachhaltige Entwicklung integrativ betrachtet. Konstitutive Elemente, Regeln, Indikatoren. Edition sigma, Berlin. 432 p.

SRU – Rat von Sachverständigen für Umweltfragen (1996). Sondergutachten "Konzepte einer dauerhaft-umweltgerechten Nutzung ländlicher Räume". Bundestags-Drucksache 13/4109

Tappeser, B., Baier, A., Dette, B., Ebinger, F., Jäger, M. (1999a). Globalisierung in der Speisekammer - Suche nach einer nachhaltigen Ernährung (Band 1). Öko-Institut, Freiburg. 163 p.

Tappeser, B., Baier, A., Dette, B., Ebinger, F., Jäger, M. (1999b). Globalisierung in der Speisekammer - Suche nach einer nachhaltigen Ernährung (Band 2). Öko-Institut, Freiburg. 91 p.

UBA - Umweltbundesamt (1997). Nachhaltiges Deutschland. Wege zu einer dauerhaft-umweltgerechten Entwicklung. Erich Schmidt Verlag, Berlin. 355 p.

Ethical and organic? Exploring the contradictions from a local community of small family farmers of southern Brazil

Raquel Moreno-Peñaranda
Energy and Resources Group, University of California, Berkeley, 310 Barrows Hall, Berkeley, CA 94720, USA, raquelmp@berkeley.edu

Abstract

The destructive trends of an increasingly uneven commodification of the world agro-food system are dramatically manifested in the global South –pitiable working conditions for contracting farmers, widespread pollution, child labor, or agrochemical poisoning are pervasive. Amid the spectacular demand for organic products worldwide, some small family farmers communities see in certified organic agriculture a promising way for linking small diversified producers to large prospective markets. However, the extent to which local mobilization process are able to transcend the institutional arena and effect a productive transformation in the community remains uncertain. Through a case study in Santa Rosa, a community of small family farmers in a tobacco growing region of southern Brazil, my research explores the ways in which "organic" is reconstructed at the local level. I argue that the characteristics of the organic momentum (global growth, a holistic nature-society vision, "natural" production techniques, international regulatory scope) constitutes an unprecedented conjuncture that is able to mobilize a diverse set of local actors in small family farmers communities. However, blending local claims of life quality improvement for the rural families with the certified organic market as "ethical and organic" can be a highly contradictory idiom from which to fight pervasive rural exclusion. How to reconcile exorbitant premiums paid by educated, wealthy urban consumers with claims of social justice for the excluded? How does the practice of input substitution jeopardize the central role of diversified management in small family farms?

Keywords: organic market, small family farmers, brazil, social mobilization, local community

Introduction

The destructive trends of an increasingly uneven commodification of the world agro-food system are dramatically manifested in the global South. Only three states in the south of Brazil (currently the second largest tobacco producer worldwide) account for almost 90% of the land devoted to tobacco production and for 93% of the tobacco produced in the country (Anuario Brasileiro do Fumo, 2003). Only in Santa Catarina State, where the community of Santa Rosa is located, 82% of the municipalities 54,000 producers are involved in integrated tobacco production and. Tobacco is produced in small production plots that take up an average of 20% of the total area of a small family farms.

Although tobacco represents a key source of income for small family farmers, pitiable working conditions for contracting farmers, widespread pollution, child labor, or agrochemical poisoning make tobacco contracts highly detrimental for the rural communities. Yet, is there any other option for the small diversified farmer but working under contract? An important body of knowledge points out the virtues of small family farming for food security, biodiversity preservation, environmental quality, and social justice. This multi-functionality of small farm agriculture has been used to challenge the conventional wisdom that small farms are backward and unproductive (Rosset, 1999). Despite they are in fact more productive, more efficient, and able to contribute more to local equitable economic development, the current trade liberalization in agriculture is seriously threatening their existence. Given the erosion of the regulatory capacity of the State in and increasingly global agro-food scenario, the consolidation of new forms of social mobilization is certainly the basis for effectively challenging its pernicious trends (Buttel, 1997). However, the extent to which local organizations are able to transcend the institutional arena and effect a productive transformation in the global agro-food chain remains uncertain.

The agro-food sector that is currently experiencing the fastest growing is organic food sales (IFOAM, 2000). Not only OECD countries but Brazil itself show sharp increases in the rate of organic products. Amid the spectacular demand for organic products worldwide, some small family farmers communities see in certified organic agriculture a promising way for linking small diversified producers to large prospective markets. Although figures of organic production in Brazil are still tentative, in Santa Catarina organic production involves 97 municipalities and 700 producers, most of them small family farmers (CEPA, 2002).

A case study in a community of small family farmers of southern Brazil

Santa Rosa, a community of small family farmers in the southern state of Santa Catarina was established in 1915 by an spontaneous group of German immigrants, during a period of government sponsored expansion of the frontier. Nowadays, the local economy is still mostly devoted to primary production –basically tobacco growing under multinational contracts, charcoal, and timber, together with self-subsistence farming. Throughout the last decade the community is been hosting an unusual experiment with ecological farming that has been broadly studied by Brazilian scholars. An association comprised of a few influential urban professionals who emigrated from Santa Rosa decades ago, together and some relatives and friends still living in the community, was established in 1996 with the aim to "promote the quality of life of the local producers through ecological farming." Only two years later, a third of the population of Santa Rosa was involved in organic production, and an ambitious project of local development comprising more that 50 diversified small scale agro-industries was being launched in the region coordinated by the ecological farming association. Together with local private capital, several local, state, and federal programs, international development agencies, non-profit organizations, and prestigious academic institutions, have supported numerous development projects in collaboration with Santa Rosa's association for ecological farming.

Ethical and organic: Reconciling nature and profit

In Santa Rosa, the labels of the organic products commercialized through the association read: "Ethical and Organic". A more detailed look to the jar of honey, the pot of jelly, or the piece of cheese, gives us a little bit more information about that "ethical" and "organic" product: an organic certification stamp from an international certification agency, and a note concerning a group of associated farmers who welcome you in their small agro-industries. The organic stamp leaves no doubt as to the kind of agricultural inputs used to produce the fruits, the sugar, or the milk. However, how should the other word –ethical– be interpreted?

Ecological economics offers us some clues to interpret Santa Rosa's labels. Based on the work of Georgescu-Roegen (1971) on the economic process and the entropy law, any (economic) process in which energy and matter are transformed is by definition a thermodynamic process. Therefore, the production of any good implies the creation of some amount of entropy, or disorder, in the system, making the given process intrinsically irreversible –at least in part. Considering the earth as our system, what we use our natural resources for inevitably compromises alternative uses of them. In a finite world, or a world with scarce natural resources, the only way to ensure

a truly sustainable economy (across places and throughout generations) is to minimize entropy by producing efficiently and to maximize per capita use by distributing access equitably.
Following Martínez-Alier's (1994) ecological critique of modern or industrial agriculture, the 'higher' productivity of this type of production hides huge ecological costs that are not measured by the market, and therefore ignored in traditional economic analysis -which considers them as externalities. Thus the actual costs of production and inputs of this type of agriculture are actually miscalculated since they do not include the externalities and the damage to the production conditions. One can argue that it is precisely in the global South that some of the most pernicious 'externalities' (pollution, awful working conditions) endure. However, social mobilization against the commodification of nature is not a deterministic process, therefore every community, at any given point, processes those grievances in a quite unique way. The reminder of the section offers some insights as to the ways in which long term claims of marginalization and exclusion of small family farmers were reconstructed by Santa Rosa's community as "ethical and organic" through their inclusion in the broader scenario of organic agriculture

Organic dimensions: Striking global prospects for restricted local options

It was during a business trip to Europe and the USA in the beginning of the 1990s that Hans, owner of a mid-size supermarket in a middle class neighborhood in Santa Catarina State's capital, Florianopolis, first saw the flourishing organic market of California and the artisan products with the *agriculture biologique* label produced in some bucolic French rural landscapes. Coming himself from a small family farm in Santa Rosa, a community with which he kept strong personal ties, he immediately saw an extraordinary business opportunity to link the small family farms of his community to the incipient organic market of Brazil. As he puts it, *"we could be pioneers at a national, even international level"*.
Hans supermarket, as many other similar stores in Southern Brazil cities, had been commercializing products directly from small family farms for decades. Cheese, sausage, beans, potatoes, bread, and even wine produced in small family farms of the region have traditionally been part of the products urban consumers would have access to through the broad denomination of *produtos coloniais* or products from the small family farms of European immigrants. The commercialization of *produtos coloniais* has never been significant in Santa Rosa (lack of know-how, infrastructure, networks, etc), were tobacco and other forms of cash production predominate. However, by reconstructing those traditional products (produced in a quasi-organic way if we examine them through current international organic standards) as organic products the commercial possibilities expand geometrically. Specially back in the mid 1990s, when organics was more than an incipient market in Brazil.

A holistic vision: Reconciling traditional agriculture and cash production

The extent to which organic principles translate into truly sustainable practices has been widely discussed -specially within the context of the USA context (Guthman, 2004). Despite the restrictions that institutionalization forms might impose to certified organic agriculture, the social process that initiated the organic movement is reflected on the organic principles worldwide by incorporating both biological and social notions of management. According to the Ministry of Agriculture of Brazil, an organic system of agro-industrial production is
" each system in which technologies that optimize the use of natural and socio-economic resources are used, respecting the cultural integrity and pursuing self-sustainability in time and space, maximizing social benefits, minimizing the dependency of non-renewable energies, eliminating the use of agrochemicals and other artificial toxic supplies, genetically modified organisms/transgenics, and ionizing radiation in every phase of the productive process, storage and consumption, enhancing human health, and assuring the transparency of all the production and transformation processes"

One can imagine then how small family farmers of the global South can perceive some space for themselves within organic agriculture. Notions of cultural integrity, self-sustainability, social justice, and human health are clearly opening space for the potential incorporation of rural communities in which traditional forms of agriculture still subsists (FAO, 2002). Small diversified producers from Santa Rosa were able to identify themselves with the ecological optimization principle by appealing to the "traditional" way of doing agriculture still persistent in their community. But even more importantly, it is the participation of farmers in cash production activities such us tobacco what allows them to articulate claims of socio-economic optimization. As one farmer puts it *"growing tobacco is a terrible job, it is all about chemicals, pesticides. Nobody likes it. You can ask anyone, nobody is going to say that they like it, because it is just for the money that you grow it"*. Therefore, organic appears as the ideal platform from which to redeem two conflicting local realities: the good ecological performance and poor economic returns of traditional agriculture, and the negative ecological tradeoffs but good economic return of cash production.

Organic production: Managing nature with nature in a just way

Organic farming is often referred to as a way of producing by managing nature with nature. According to international standards (IFOAM, 2000), natural broadly refers to anything that does not contain synthetic chemicals, genetic modifications, or radiation. According to this definition, an autochthonous species of beetle is as natural as a complex and expensive biological pesticide. Besides, standards do not exclude the (massive) use of "unnatural" oil to fuel the machines on the field, or to transport products across continents. One can argue that, in fact, organic farming is more about what cannot be used that about how to manage what it can be used.

The initial link between small family farming and ecological agriculture is found in the whole-systems approach of agroecology. As the discipline that provides the basic ecological principles for how to study, design and manage agroecosystems that are both productive and natural resource conserving, economically viable, as well as socially just and culturally sensitive (Altieri, 1987). In an agroecological strategy, management efforts are directed toward highlighting the conservation and enhancement of local agricultural resources by enhancing farmer participation, use of traditional knowledge, and adaptation of farm enterprises to fit local needs and socioeconomic and biophysical conditions. Throughout Latin America, and specially in the southern states of Brazil in which small diversified are abundant, agroecological management strongly resonates to the structure and functioning of small family farms.

Agroecology somehow resonates with the traditional way of doing agriculture, as many interviewed farmers acknowledge: *"you should not use chemical fertilizers, only animal manure you could use. Look, this system is not very difficult. It's almost the same as what we used to do"*. Although the organic way of producing is much less fitted to the singularities of Santa Rosa's small diversified family farms, the framework is broad enough to include them. In addition, the experience farmers have working with cash crops was vital to construct the vision that joining organic farming would not be a very complicated endeavor. As a young farmers poses it: *"No, it's no so difficult, it's no so complicated. It is just that before we used fertilizer for the corn, and now we don't use it. The system hasn't changed that much"*. By combining both agroecological and cash production experiences, Santa Rosa's community reconstructs organic as easy, something the community should be ready for.

Organic contradictions

In sum, the link between "organic" and "ethical" is based on the local perception that the small family farm possesses a unique combination of traits and values that ultimately make it the rightful heir of sustainable agriculture. Through the interactions with the global momentum for organic agriculture, and shaped by he local ways of doing agriculture, the community creates its own personal translation of social justice and sustainable farming. Organic proves to be an excellent opportunity to resolve historical grievances of rural exclusion and marginalization in a way that is both sustainable and just. Using Latour's (1993) analysis of the relationship between nature and culture, organic gives the small farmers an unprecedented opportunity: to leave behind both the backwardness of pre-modern traditional agriculture and the unjust tradeoffs of modern agriculture, by joining the organic in a reflexive or post-modern way.

However, blending local claims of life quality improvement for the rural families with the certified organic market as "ethical and organic" proves a highly contradictory idiom from which to fight pervasive rural exclusion. How to reconcile exorbitant premiums paid by educated, wealthy urban consumers with claims of social justice for the excluded? How does the practice of input substitution jeopardize the central role of diversified management in small family farms? In sum, the post-modern transitioning of pre-modern small family producers of the global South towards more sustainable forms of agriculture seems to pose more complex issues than that of conventional modern farmers of the North.

References

Altieri, M.A. (1987). *Agroecology. The Science of Sustainable Agriculture.* Boulder: Westview.

Anuário Brasileiro do Fumo (2003). *Anuário Brasileiro do Fumo n. 7.* Gazeta Grupo de Comunicações, Santa Cruz do Sul.

Buttel, F.H. (1997). *Some Observations on Agro-Food Change and the Future of Agricultural Sustainability Movements.* In: David Goodman and Michael J. Watts Eds. *Globalising Food: Agrarian Questions and Global Restructuring.* Routledge, London.

CEPA (2002). *Agricultura Orgânica em Santa Catarina.* Instituto de Planejamento e Economia Agrícola de Santa Catarina, Florianópolis.

FAO (2002). *Organic Agriculture, Environment and Food Security.* Food and Agriculture Organization of the United Nations, Rome.

Georgescu-Roegen, N. (1971). *The Entropy Law and the Economic Process.* Harvard University Press, Cambridge.

Guthman, J. (2004). *Agrarian Dreams: The Paradox of Organic Farming in California.* University of California Press, Berkeley.

IFOAM (2000) *IFOAM Basic Standards.* International Federation of Organic Agriculture Movements. Tholey-Theley, Germany.

Latour, B. (1993) *We Have Never Been Modern.* Harvester Wheatsheaf, New York.

Martínez-Alier, J. (1994) *De la Economía Ecológica al Ecologismo Popular.* Icaria, Barcelona.

Rosset, P.M. (1999). The Multiple Functions and Benefits of Small Farm Agriculture in the Context of Global Trade Negotiations. In: Global Exchange Policy Brief No 4. Global Exchange, Oakland.

The new scientific totalitarianism: An overview on the discourse over genetically modified crops

Asterios Tsioumanis and G. Perentzis
School of Natural Resources and Enterprises Management, University of Ioannina, G. Seferi 2, 30100, Greece, steriost@auth.gr

Believe nothing, no matter where you read it,
or who said it, no matter if I have said it,
unless it agrees with your own reason
and your own common sense.

Buddha (563BC-483BC)

Abstract

Scepticism is the philosophical current which over-emphasises doubt and the relativity of human knowledge, while dogmatism underestimates the relativity of knowledge and lays claim to knowledge of absolute truths. Addressing the term dogmatism as an appeal to the authority of one interpretative scheme over the other, the popular argument that science should be the ultimate judge on the GM food debate is discussed. The pro-biotech sector underlines that the campaign against transgenic crops is not based on scientific evidence calling for an immediate halt of dogmatic opposition. Departing from analyzing problematic relations that render scientific facts open to critique, such as the continuous commercialization of science, the paper focuses on the essence of scientific knowledge in its crystal-clear form and argues that the provisional character of all scientific knowledge should not be underestimated. Proclaiming that technological advances are inevitable has been proven in the past to be both false and arrogant, while calls to the infallibility of science are typical of a scientific elitism that denies what 'all serious studies of scientific and technological change have shown, namely, that technological changes of any significance involve intense social interaction, competition, conflict, and negotiation in which the eventual outcomes are highly contingent'. Analyzing similar cases of modern, state-of-the-art technologies provides useful insight while a philosophic glance on the immoralist character in the modus operandi of scientific knowledge adds to the argumentation. Granting scientific research and knowledge the role of the final judge to any dispute on matters of determining policies, is a kind of dogma and therefore it should be submitted to further critique. The willingness to disregard past human experience, leaving no space for doubt on the very existence of scientific infallibility, and the effort to focus solely on the validity of scientific data can be described as a kind of totalitarianism which skillfully dismisses all the questions that is unable to answer.

Keywords: GM debate, sound science, decision-making

Introduction

Scepticism is the philosophical current which over-emphasises doubt and the relativity of human knowledge, while dogmatism underestimates the relativity of knowledge and lays claim to knowledge of absolute truths. Scepticism, embarking from the notions of paradox and doubt in ancient Greek philosophy has developed a complex relationship with rationality, subjectivity and knowledge. Dogmatism may be defined as the arrogant assertion of opinions as truths. When cornered with unavoidable facts it is likely that a dogmatist would count on the philosophy of relativism to support his/her claims. Addressing the term dogmatism as an appeal to the authority of one interpretative scheme over the other, the popular argument that science should be the ultimate judge on the GM food debate is discussed.

Brief notes on the GM debate

The genetically modified varieties that exist today are related to a great extent to the resistance of the crops. Therefore, defenders of the introduction of genetically modified crops claim that the use of modern biotechnology will lead to an increase of the productivity in the field of agriculture followed by an overall economic growth. The increase in agricultural productivity will not be the result of the expansion of the cultivations to new fields, but the consequence of the high resistance of the crops and thus, of their increased yields. These innovations could be very important for the developing countries and in particular for these countries that need a boost in agricultural productivity.

Controversy still surrounds the introduction of genetically modified crops. From the growing body of literature a series of potential risks may be identified, which has led to scepticism, especially in Europe. Briefly, adding to health concerns that have been expressed mainly through allergens and antibiotic resistance, environmental concerns have been focusing on threats to biological diversity, genetic pollution, effects on non-target species and the probability of the formation of resistant insects or weeds as a result of physical selection. Socioeconomic concerns include the lack of benefit sharing agreements, cross-country appropriation of genetic material, the intellectual property rights system and the granting of patents on biological material, the oligopoly in the market structure of the new technology. Finally there are ethical concerns related to the "unnaturalness" of the modern technology, the definition of all non-human organisms "telos" as serving human interests interrelated with religious beliefs.

In the course of the debate the calls for "sound science" or "good science" to become the centre in the decision-making process grew. Dane Scott (2004) draws from the rhetoric of Trewaras and Prakash and statements of President Bush, in order to point out that *the assertion that sound science must decide the GM debate is a standard strategy of pro-GM industrialists, scientists and politicians*. It is true that scepticism on the introduction of GM crops has been accused as based on irrational, unscientific fears. Following this line of thinking, the negative public attitudes towards biotechnology, often encountered in literature, are the result of little or biased flow of information. The phenomenon is not unique referring to genetic engineering and its applications. As all informed on environmental policy know, it is quite hard to come across any policy proclamation that does not make a specific reference on the role of "sound science".

The trans-temporal character of scientific knowledge

While "sound science" can unquestionably provide objective facts and data as well as estimates on potential benefits and risks, it is a great conceptual leap to accept that hard certainties that science may produce could replace uncertainties generated from other domains of everyday life. Calls for the predominance of science on decision-making seem to underestimate a few distinct elements of the essence of scientific knowledge.

Scientific information always includes some degree of uncertainty. One cannot ignore that throughout the history of modern science, there have been several examples that lead to the conclusion that the prospect of a single scientific truth is at least deniable: starting from the beginning of the 20th century with the introduction of quantum mechanics and the quarrel with the classical Newtonian theory as it was reformed by the general theory of relativity.

Scientific knowledge is not only uncertain, but also dynamic in the sense that via time, mainstream theories that were considered true for centuries were proved to be wrong. Sabbato (1951) approaches this dynamic character as inherent mortality of human knowledge pointing out that emerging knowledge often presupposes the deconstruction of previous scientific "truths". Even in shorter time spans, the modern scientific community has been divided over scientific data as for instance the discussion on the sources, which contribute to the depletion of the ozone layer. The role of anthropogenic CFC emissions to the depletion of the ozone layer has been challenged even by prominent scientists. The same stands true for the global warming which has been the focus of debate in the scientific community for at least 15 years.
During technological revolutions of ambiguous advantages and associated risks like nuclear energy, pesticides or the SDI, risk assessing has been utilized. It is not seldom in risk assessing that the estimated result is heavily influenced by the method chosen. In addition, the Popperian *no society can predict, scientifically, its own future states of knowledge* (Popper, 1971), should be used as an additional risk assessment factor.
The proponents of modern scientific elitism suggest that science, as an autonomous apparatus, will provide one absolute answer and this shall be accepted since it constitutes an objective outcome. Even disregarding for an instance, the provisional character of scientific knowledge and accepting the above procedure as attainable, the question on alternatives would still be valid. Considering technological progress as a one-dimension process is a typical argument that has been recurring in modern history. However, this fails to see *what all serious studies of scientific and technological change have shown, namely, that technological changes of any significance involve intense social interaction, competition, conflict, and negotiation in which the eventual outcomes are highly contingent* (Winner, 2003).

Beyond Sabbato's immoralist character of scientific knowledge

Having in mind the disputes in the scientific community and the results that have been overturned in the course of the modern era, one should be very careful to point out that, based on scientific data alone, something is desirable for the society as a whole. At the GM context, Scott (2004) underlines that *science provides the facts and probabilities about benefits and risks of GMOs. Nonetheless, science cannot decide what "benefits" should be pursued and what "risks" ought to be taken.*
Technological advances provide a picture of what is technically attainable, not socially desirable. Sabbato (1951) argues on the immoralist character in the modus operandi of scientific knowledge pointing out that *science on its own guarantees nothing as when it comes to the realization of a major achievement, its moral concerns are absent.* In addition, no major technological advancement is safe from misuse, deviating from its original objectives.
Even the argument that scientific knowledge and technological know-how provide a whole picture of what is technically attainable may be submitted to further critique. Sabbato rejects the political neutrality of scientific knowledge as *in the development of science, humans act under a complicated set of ideas, emotions and prejudices that characterize their very nature* (Sabbato, 1951). Winner advocates that *within the making and application of new technologies, there are always competing interests, contesting positions on basic principles, and numerous branching points in which people choose among several options, giving form to the instrumentalities finally realized, discarding others that may have seemed attractive.* Lewontin (1991) underlines that the choice of a specific research hypothesis among thousands, the ideas that science utilizes to confront the problems, even the formation of the problems themselves are significantly influenced by predispositions derived from the society as a whole.
While the direction of scientific research is an enormous and controversial issue per se, research directly related to the applications of modern biotechnology is attached to even more problematic

relationships. Opportunities steaming out of recent development in genetic engineering together with the liberalisation of agricultural input and output trade, lead private companies to huge investments in the sector. In the expectation of future profits, it is fair to draw the conclusion that relevant research and development will be strongly influenced by commercial interest. However, this process creates new obstacles in approaching the direction of future scientific research as there is no guarantee that the social and private optimum will be identified. Thus, even in terms of technical feasibility, science can decide what is attainable, given a specific set of research questions, which are chosen under a complex system of interactions that, by no means, may be considered to function without bias.

Is the term "scientific totalitarianism" a bit harsh?

Even if the essence of science leaves space for skepticism regarding its role in societal decisions, in the GM debate, similar line of arguments were subjected to fierce critique. Although the utilization of the term scientific totalitarianism may prove controversial, the augmenting pressure to accept the existence of a single, catholic scientific "truth" provides a great incentive to do so. The controversy mainly derives from the definition of the term. Totalitarianism is a twentieth century notion, introduced by the Italian philosopher Giovanni Gentile and has been used to describe regimes in which the state regulates nearly every aspect of public and private behaviour. Prior to the emergence of totalitarianism, the totalitarian impulse, defined as the ambition of the state to extend its authority to realms where it has no authority, may be identified.
The term totalitarianism is addressed in a Marcusian context in order to describe *not only a terroristic political coordination of society, but also a non-terroristic economic-technical coordination which operates through the manipulation of needs by vested interests* (Marcuse, 1964). Totalitarianism may thus be mirrored in production and distribution systems and even be compatible with a certain degree of pluralism. Drawing a semantic parallel, concerning scientific totalitarianism, the impulse would be the ambition of science to extend its authority. Through the coordination in an economic sense, *the emergence of an effective opposition is precluded* (Marcuse, 1964). The extent to which such a coordination is or will be apparent, will finally decide whether the term scientific totalitarianism may be used in the future.

Conclusive remarks

Bearing in mind the complexity of the issue and the variety of parameters involved in such a process, it makes sense to deny the argument that scientific data alone can provide the answer to the future demands and needs of humanity. Thus, case specifically, in the GM debate, there can be no arbitrary other than genuine social dialogue on the objectives and the means to meet them. The notion that scientific facts can constitute the ultimate judge in deciding the debate is over-simplistic and dismisses questions that are unable to answer.
All scientific knowledge is provisional. Everything that science "knows", even the most mundane facts and long-established theories, is subject to reexamination as new information comes in (The Scientific American, 2002). Granting scientific research and knowledge the role of the final judge to any dispute on matters of determining policies, is a kind of dogma and therefore it should be submitted to further critique. The willingness to disregard past human experience, leaving no space for doubt on the very existence of scientific infallibility, and the effort to focus solely on the validity of scientific data can be described as a kind of totalitarianism In a rather more complex process, there is need for a more widespread form of control of society in the directions of scientific research since the latter has been transformed into an area of political and economic struggle rather than the benign pursuit of the truth.

In addition to the effort to separate "sound science" from illogical fears, Martina McGloughlin's (2004) response to a previous article by Altieri and Rosset (2003) attempts to draw a line between politics and scientific knowledge; *their arguments are primarily directed against Western-type capitalism and associated institutions (e.g. intellectual property rights, the WTO). Biotechnology is used as a Trojan Horse...The developing and developed world will need and use biotechnology in many ways during this century. Those with political battles to fight may want to use other, more appropriate fora to fight them.*

The uncomfortable reality however is that even genetically modified crops shall lead to productivity gains that exceed the most optimistic estimations, unless our civic societies can come up quickly with an economic system that allocates resources more equitably and more efficiently than the present one, there is no guarantee that 50 years from now challenges will be any different. *Neither in the line of thinking, nor in the field of reality, may the technological system of a given society be separated from what society actually is* (Kastoriadis, 1977).

References

Altieri, M.A and Rosset P. (2003), *Ten reasons why biotechnology will not ensure food security, protect the environment, and reduce poverty in the developing World,* AgBioForum, vol.2, number 3 and 4, Article 3

Kastoriadis K.(1977).Thoughts on development and rationality, in *Mythe de developpement*, C. Mendes (ed.), Seuil Publications, Paris

Lewontin, R. C. (1991), Biology as Ideology: The Dogma of DNA, Stoddart Publishing Co. Ltd.

Marcuse, H. (1964). *One-deminesional Man*, Boston: Beacon: 1-18.

McGloughlin, M. (2004). *Ten reasons why biotechnology will be important to the developing world,* AgBioForum, vol.2, number 3 and 4, Article 4.

Popper, K. (1971), *The Open Society and its enemies*, Princeton Univerisity Press.

Sabbato, E. (1951), *Hombres y Engranajes. Reflexiones sobre el dinero, la razón y el derrumbe de nuestro tiempo.* Buenos Aires, Sur, 1951. Edición definitiva: Barcelona, Seix Barral, 1991.

Scott, D.(2004). Sound science, practical reasoning and the genetically modified organisms debate, in *Science, Ethics and Society*, J. De Tavernier and S. Aerts (eds.), Centre for Agricultural Bio- and Environmental Ethics: 173-176.

The Scientific American (2002), Editorial, December 2002.

Winner L. (2003). Are humans obsolete?, *The Hedgehog review*, vol. 3, no.3

Part 6
The tool-box for assessing ethical issues

Development of an Animal Disease Intervention Matrix (ADIM)

Stefan Aerts, Johan Evers and Dirk Lips
Centre for Science, Technology and Ethics, Kasteelpark Arenberg 30, 3001 Leuven, Belgium, Stefan.Aerts@biw.kuleuven.be

Abstract

Very different animal disease control scenario may be identified as the best, depending on the starting position and the considerations taken aboard. If not all concerns are allowed, comparison between different possible scenarios is impossible. We will argue that this is the most important reason for the polarised debate. It is difficult to weight typically ethical considerations versus "hard" numbers such as (predicted) economic losses. Nonetheless, the former are also important. We describe and discuss a decision support system and ethical elicitation tool, the Animal Disease Intervention Matrix (ADIM), aimed at identifying the ethically best animal disease intervention scenario. The ADIM compares different control methods and scenarios by a layered system of objectives and indicators. A set of indicators is used to evaluate (and score) to what extend a method or scenario meets with all objectives of animal disease control. These objective-scores are then clustered to yield an overall scenario-score (the ADIM-score). It is reasonable to question the role of this system in the decision making process: decision system or decision support system. We defend an intermediate position in which any deviation from the ADIM should be clearly and elaborately accounted for. This system can (and will) be used in ad hoc situations, but it will also be used as an evaluation and scenario building tool that may give incentives to change current legislation. A simulation indicates that a full vaccination policy for H5N1 Avian Influenza has considerable ethical advantages over stamping out or intermediate scenarios.

Keywords: animal disease, ethical matrix, animal production, ethical tool

Introduction

Since the early 90's the European Union installed a non-vaccination policy for Foot and Mouth Disease (FMD), Classical Swine Fever (CSF) and Avian Influenza (AI) (EU Directives 90/423/EEC, 91/685/EEC and 92/40/EEC). This decision opened the door for an increasing number of disease outbreaks in the EU during the last decade. Some of these outbreaks have had serious implications for the social and economic life in the countries concerned (e.g. FMD in the UK, AI in The Netherlands). With every new outbreak the public outcry grows. It seems that the culling of (an ever growing number of) animals to protect the country's animal disease status is no longer acceptable to the general public. Furthermore, the economic consequences of such disease outbreaks might begin to outweigh the benefits of the non-vaccination policy as the cost of the 2001 FMD outbreak was estimated € 280 million in the Netherlands and € 4 billion in the UK (Anonymous, 2001). The recent H5N1 threat already had economic (and other) consequences before any European commercial poultry was infected. Even in this case, where there could be real effects on public health, European citizens react quite heavily when seeing the eradication methods used in South-East Asia.
This clearly indicates that, although poultry production is inherently a commercial activity, not all actions are acceptable to the public. On an ethical account, from a moderate anthropocentric approach (De Tavernier *et al.*, 2005), it is clear that – for a number of reasons, which we will discuss later – economic nor public health concerns can justify all actions taken to eradicate

animal diseases. Although this is clear in a general discussion, it is often difficult to identify the (ethically) best option in a specific case.

General approach

Animal disease control or eradication is a complex issue with serious ethical implications (in a wide range of fields). There is also no "objective", mathematical approach to the issue, i.e. any decision taken will involve normative, ethical assumptions. This will evidently be a difficult point for the people that have to make these decisions, as these are often veterinarians and other 'disease specialists', who are all but trained in matters of ethical decision making. Often only indirect or limited information is available at the time where decisions have to be made, e.g. there can be a lack of knowledge about the disease agent (e.g. with H5N1 AI), it is usually unknown when (and where) an outbreak will take place and possible effects of an outbreak can be difficult to estimate.

There has been some interesting work dedicated to alleviating the difficulties in decision making processes through a matrix approach (Bartussek, 2001; Bracke, 2001, Mepham and Tomkins, 2003). Although these systems vary enormously in shape and aim, there are some general similarities that are of interest to the current discussion. Most importantly, all are layered systems that use some sort of goal-indicator structure (Vemuri, 1978; Østergaard and Hansen, 1995) that enables the segmentation (and simplification) of the problem. In contrast to Mepham and Tomkins' system, Bartussek and Bracke attempt to come to one final conclusion and therefore need to address the weighting issue. We will use this type of approach to construct what we will call an Animal Disease Eradication Matrix (ADIM).

Using indicators, we need to clearly set out the different characteristics of such an indicator. Without going deeply into this discussion, we state that a good indicator should be a compromise between scientific correctness and practical usefulness. It should at least be clearly connected to the topic, have a scientific basis, be quantifiable and give information about possible solutions or corrections (see also Østergaard and Hansen, 1995). At first these requirements seem rather basic, but it will quickly become apparent that in the matters we discuss here, it is even difficult to find indicators that meet with these basic requirements. To be applicable in practice, data should also be available, the indicator should be easy and cheap to use etc.

The ADIM

The ADIM – as it is today – is aimed at identifying the ethically best *eradication* scenario. This does not imply that other measures (e.g. prevention) are less important. We have developed this first in order to be prepared for any – expected or unexpected – disease outbreak.

In this paper, we define a disease control *method* as a single (chain of) event(s) used upon one animal, a disease control *scenario* (or strategy) as a combination of one or more methods used to eradicate an animal disease (e.g. at national level).

As we have outlined above, the ADIM has a layered design in which indicators are used to evaluate different goals or 'objectives' whose scores then yield the general ADIM score. This is summarized in Figure 1.

First, we have segmented the general problem into smaller parts: the different objectives that a good disease control scenario should achieve. Through intensive discussions with all stakeholders and disease control experts, we have identified 15 objectives:

1. Protecting the health of control personnel and farmers.
2. Protecting public health.
3. Protecting animal health.
4. Ensuring animal welfare.

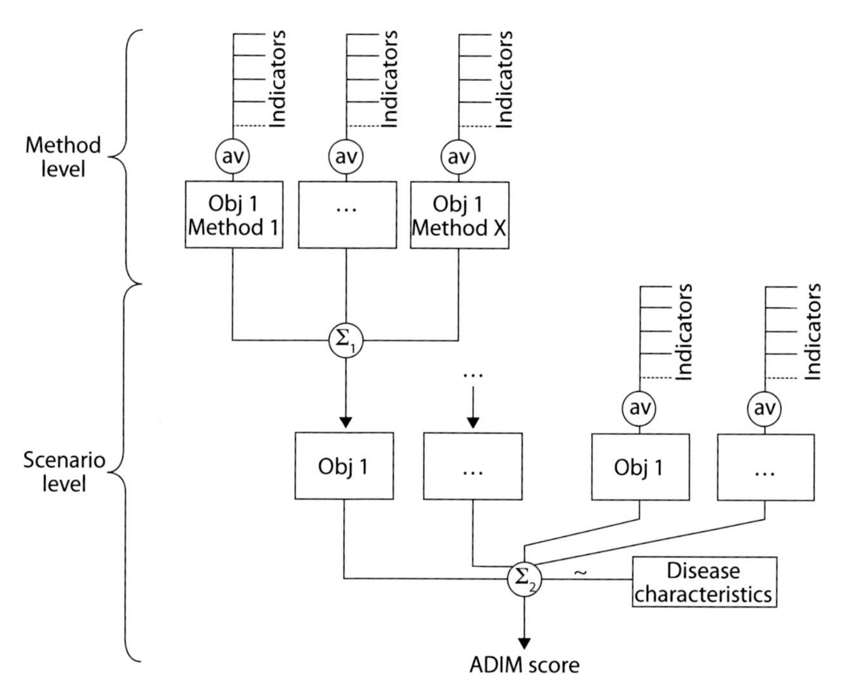

Figure 1. General overview of the ADIM. The whiskers represent the different indicators per objective, which are averaged ('av') to yield the objective-score for a method (or scenario). If necessary (objective 1-8) these scores at method level are used to calculate an objective-score at scenario-level ('Σ_1'). The objective-scores at scenario level are then weighted and summed ('Σ_2') to yield the overall ADIM score. Weighting depends on the disease characteristics.

5. Respecting the human-animal bond.
6. Limiting environmental damage.
7. Limiting the psychological impact on the farmer.
8. Limiting the psychological impact on the control personnel.
9. Respecting food.
10. Limiting disturbance of social life.
11. Limiting economic losses in agriculture.
12. Limiting economic losses in non-agricultural sectors.
13. Ensuring practicality.
14. Ensuring food security.
15. Protecting valuable animals.

It is clear that not all these objectives can be evaluated at the same level. Some of these objectives depend on individual disease control methods (objectives 1 to 8), others will have to be evaluated by taking the general scenario into account (objectives 9 to 15). This has its computational repercussions when trying to reach a general ADIM score, as scores for objectives that are evaluated at method-level can only be used in a general ADIM score after a correction is made for the amount of animals subjected to that method; i.e. a second weighting procedure within the ADIM. In Figure 1 this is represented by 'Σ_1'.

As we have mentioned above, the different ethically relevant criteria (i.e. the objectives) are not all of equal weight. Some will be more important in general (e.g. protecting animal health), some are only important with some diseases (e.g. protecting public health). We consider the characteristics of the disease, assessed through its own set of indicators, as the most important factor influencing the weight of a certain objective within the evaluation. All this is represented by 'Σ_2' in Figure 1. Conceptually this is equivalent to the possibility of (partial) compensation between different objectives.

Secondly, all objectives were assigned a set of indicators. For objectives that are scored at method level, these indicators are applied with all methods. We will not overburden this paper with a list of all indicators, but suffice with stating that at present all objectives have between three and nine indicators. New indicators can easily be integrated into the system, provided that they add new information. If there is too much overlap between indicators of an objective, this will only dilute the score. This does not preclude an indicator being used to score more than one objective as it can be informative for different objectives.

It is important to note that a clear distinction between an assessment of the disease and the disease control scenario is necessary. If one wants to distinguish between good and bad disease control scenarios, information about the disease should be kept apart. Indeed, the disease characteristics do not change the characteristics of the control scenario, but they do determine *why* some scenarios are better than others (i.e. they determine the relative weight of the objectives).

We have applied this system to the case of H5N1 AI. Figure 2 is a graphical representation of the ADIM-evaluation of 4 AI control scenarios. S1 is a 100% national preventive vaccination

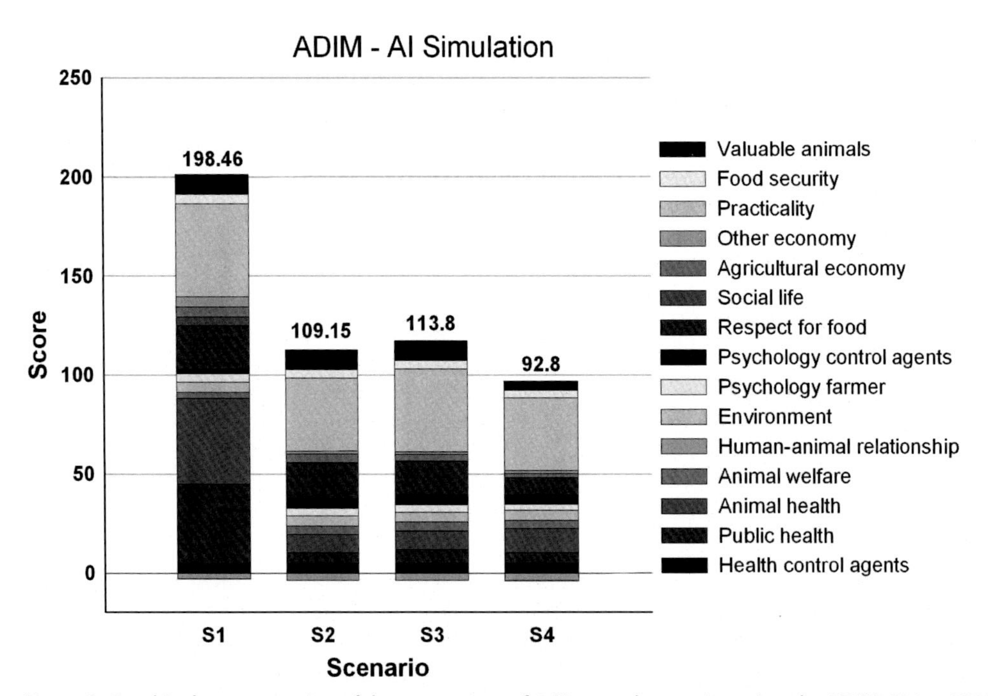

Figure 2. Graphical representation of the assessment of 4 AI control scenarios using the ADIM. S1 is a 100% national preventive vaccination scenario, S4 the 'classical' stamping out scenario, S2 and S3 or intermediate scenarios. Scores below zero have to be subtracted to reach the final scenario score (indicated above the bars). The different layers represent partial scores (per objective).

scenario, S2 is a scenario with emergency vaccination of non-production animals in a 10 km zone around an infected farm and a stamping out of all susceptible animals in a 3 km zone (3 day national standstill, 30 days in 20 km zone), S3 is identical to S2 except for the national emergency vaccination of non-production and free-range animals and S4 is the 'classical' scenario without vaccination (stamping out of utility animals in a 3 km zone, of hobby animals in 1 km zone, 10 day national standstill, at least 30 days in 20 km zone). Note that all scores below zero have to be subtracted to reach the final scenario score. The different layers represent partial scores (per objective). We conclude that a general vaccination policy has significant ethical advantages over the classical or intermediate scenarios.

Discussion

The development of the ADIM does not relieve all ethical and practical concerns that are connected to animal disease control. We believe that there are two groups of concerns to be addressed: (1) concerns about the position of the ADIM in society and (2) conceptual choices and restraints.

Most importantly, one should discuss whether such a system can (in practice and in principle) be used to take decisions about animal disease control. Should this remain a decision support system or should this be a real decision system – given some further development? The latter option appears to lead to some type of 'computerized' society in which decisions are made and laid upon us without any real human intervention. The fundamental question is who is to take responsibility, who is to take the final decision? Intuitively it seems that there needs to be someone who takes the final responsibility, but on the other hand, if we believe this system to be relevant and correct, why would we want to leave open the option to choose an ethically worse option? We feel that we need to defend an intermediate position in which the final decision is taken by a government official who clearly and elaborately accounts for any deviation from the best ADIM option.

This reasoning can be expanded towards the connection between ADIM and legislation. It is possible to construct and evaluate a scenario that is legally prohibited but ranks high amongst different options (see Figure 2). In a democracy, one can not ignore the primacy of legislation and therefore we can not insist on implementing such an scenario, but this will surely be a strong argument in favour of adapting existing legislation.

The last of 'external' concerns connected to the ADIM is the possibility of manipulation. There is a fine line between leaving room for adapting the ADIM to improve the system and opening the door for manipulative changes. We feel that a procedure involving as many stakeholders as possible (as we have used) is a relatively elegant way to avoid such manipulation. Additionally it may be advisable to only 'update' the system every three or five years.

With respect to the question whether all conceptual choices in the ADIM will lead to scientific certainty, it is clear that this is not the case, and this need not be. As we have already stated earlier, in animal disease control, there are no 'objective' approaches (as much depends upon fundamental positions about the position of humans and animals) and therefore no system can be expected to deliver a scientifically certain evaluation (which is unattainable anyway). We have developed this system precisely to be able to circumvent this clash of positions that clouds the real discussion. By stepping back from scientific certainty and allowing a certain amount of simplification, decisions can be made in an issue as complex as animal disease control.

Conclusion

Animal disease control is an enormously complex issue that raises a plethora of social and ethical questions, to which there are often different answers depending on some fundamental

practical and ethical choices. This makes decision making difficult for those responsible in practice, resulting in ethically doubtful approaches and great public outcry.
The ADIM is an ethical matrix system developed to be able to evaluate different scenarios for animal disease control that recognises these different positions but gives an overall score for every scenario. This system brings governments a tool to take more ethically just decisions about animal disease control and allows for scenario-building exercises in order to improve current (and future) control scenarios.
From the H5N1 AI example we conclude that a general vaccination policy has significant ethical advantages over stamping out scenarios. The development of better vaccins will even improve this score and should thus be stimulated. The current non-vaccination policy is clearly ethically indefensible.

Acknowledgement

The development of the ADIM was made possible through the support of the Belgian Federal Agency for the Safety of the Food Chain and the cooperation of its Consultative Committee.

References

Anonymous. (2001). *Final report of the International Conference on Control and Prevention of Foot and Mouth Disease, 12-13 December 2001*. EU Presidency, Brussels. 21 p.

Bartussek, H. (2001). An historical account of the development of the Animal Needs Index ANI-35L as part of the attempt to promote and regulate farm animal welfare in Austria: an example of the interaction between animal welfare science and society. *Acta Agriculturae Scandinavica, Section A, Animal Science.* Suppl. 30: 34-41.

Bracke, M.B.M. (2001). *Modelling of animal welfare. The development of a decision support system to assess the welfare status of pregnant sows*. Wageningen University, Wageningen.

De Tavernier, J., Lips, D. and Aerts, S. (2005). *Dier en welzijn*. LannooCampus, Leuven. 202 p.

Mepham, B. and Tomkins, S. (2003). *The Ethical Matrix* [online]. Available on: www.ethicalmatrix.net [Date of consultation: 05/05/2005].

Østergaard, V. and Hansen, J.P. (1995). Indicators - A method to describe sustainability of livestock farming systems. In: *Agricultural Sciences for Biodiversity and Sustainability in Development Countries - Proceedings of a workshop*. Dolberg, F. and Petersen, P.H. (Eds.). National Institute of Animal Science, Tjele.

Vemuri, V. (1978). *Modeling of complex systems - An introduction*. New York, Academic Press. 446 p.

The process of developing and implementing criteria for sustainable agriculture: A cross-case analysis

Menno Binnekamp, Paul Ingenbleek and Silvia Goddijn
Wageningen University, Marketing and Consumer Behaviour Group, Hollandseweg 1, 6706 KN Wageningen, The Netherlands, Menno.Binnekamp@wur.nl

Abstract

Codes of conduct organizations, like Organic farming or EurepGap, rapidly increase their impact on agribusiness. Yet, our understanding of these new institutional arrangements is limited. In particular, the question how criteria are developed and implemented by these codes of conduct organizations (CCOs) is of interest in order to understand how these organizations can contribute to animal welfare. A multiple case study was conducted, analyzing and comparing four cases that incorporate animal welfare issues: Freedom Food, Organic, EurepGAP and 'Scharrel'-eggs. The theory-building case studies eventuate in propositions on the criteria setting process in CCOs. In effect, setting norms comes down to weighing animal welfare (and possibly other ethical values) against market opportunities. The broader the sustainability goals of the CCO are and the more international the scope is, the more subcultures are present, making the process complex. When a CCO is based on positive values (in stead of negative values like opposing against the bio-industry) and is in hands of private ownership (and hence criteria are not embodied in the law) it seems that a process of continuous upgrading is more likely. When the organization, besides setting norms and controlling them, uses 'artifacts' like symbols or brands, that strengthen the sense amongst farmers that they belong to a specific group, these farmers will have a more positive and active attitude towards the CCO. On the basis of these findings, other code of conducts organizations may avoid pitfalls and maximize potential levels of animal welfare.

Keywords: animal welfare, certification, sustainability, implementation

Introduction

In agribusiness increasingly code of conduct organizations (CCO) are emerging, like for example Organic farming, EurepGap, or Utz Kapeh. These are organizations where various groups or stakeholders with different backgrounds, e.g. producers, societal organizations, retailers or governments, jointly define specific rules for producing a product in a social responsible way. Ingenbleek and Meulenberg (2005, p. 3) define a CCO for sustainable agriculture as 'a non-governmental organization that develops one or more formal statements of rules of conduct regarding environmental and/or social domains of sustainable agriculture that producers voluntarily agree to implement.' The need for a CCO is felt when rules or legislation are absent, or fall short (according to those who establish the CCO), or are difficult to control. CCOs develop sets of requirements that agricultural producers should implement in order to increase sustainability. These requirements are targeted at agricultural producers as they actually put the sustainability objectives of CCOs into practice (for example, by adopting more environmentally friendly production methods).

Though CCOs rapidly increase their impact on agribusiness, our understanding of these new institutional arrangements is limited. In particular, the question how criteria are developed and implemented by these organizations has not been subject to prior research. This process is of interest as it can improve understanding how these organizations can contribute to issues like ethics and animal welfare.

Approach

In order to study the development and implementation process, we use an organizational culture approach. Although they are forms of governance that emerge from an intricate network of stakeholders (Kooiman 1993), CCOs are in fact an organization (Ingenbleek and Meulenberg 2005). The final goal of a CCO for sustainable agriculture is to improve sustainable development on one or more domains of sustainability. To this respect, they aim to change behavior – conduct– of primary producers. Organizational culture literature makes clear where behaviors stem from: from the culture of the organization, or "how we do things around here". CCOs are often established by different stakeholders that have different missions and whose organizations are marked by different cultures based on different values. Think for example of a collaboration between private companies, animal activists, environmental pressure groups, and governments. Based on a joined interest, they may together participate in a CCO. The background of each stakeholder and their individual interest in a CCO determine the framework for setting the criteria in a CCO. The joined interest in a CCO may, either or not, be based on shared values. Shared values will facilitate the process of developing requirements, whereas different values may hinder it: different values are likely to result in different norms, and norms are the basis of conduct.

Similarly, shared values may result in a joined language, and symbols that are supposed to have a strong influence on behavior, facilitating the implementation process of the code of conduct among farmers. In fact, the set of requirements as it is formulated and presented to farmers, can be seen as an artifact of an organizational culture. It may for example reflect a joined belief of doing the right things for the world, or reflect a hierarchical culture in which a powerful market party tells others how to behave.

Because CCOs are based on a network of stakeholders and CCOs may have different or even conflicting subcultures, the processes of developing and implementing criteria may be different processes across CCOs. They may encounter different problems and have different capabilities to deal with these problems. As a consequence, the CCO may have in the end a relatively larger or smaller impact on sustainability. This is why it is important to increase our insights in these processes.

Case study methods

In this study we will focus specifically on requirements regarding animal welfare, for two reasons: (1) a single issue within sustainability simplifies a cross-case comparison, and (2) animal welfare currently receives much attention in Europe. This enables us to trace back the processes by which the criteria were developed and implemented.

To our knowledge, our study is the first to focus on the process of developing and implementing criteria on sustainability issues by CCOs. CCOs may encounter different problems and have different capabilities to deal with these problems. As a consequence, in the end the CCO may have a relatively large or small impact on sustainability. Our purpose: to reveal how CCOs develop and implement criteria on sustainability issues, and why they do it this way. These are typical *why* and *how* questions that require a depth of analysis that can be achieved through case study research (Eisenhardt, 1989; Yin, 1994). The theory-building case study will result propositions that theorize the criteria setting process within CCOs.

Case Selection

Initially, CCOs were selected based on a broad desk research including scientific and public sources. Subsequently, we consulted three consultants with expertise in different fields of agriculture to make the list more complete. This resulted in an initial list of 64 organizations. Next, CCOs were excluded based on several requirements the CCOs had to meet: In order to understand impact on animal welfare issues, at least a (legal) minimum on this issue had to be incorporated in the CCO. Moreover, for practical considerations, CCOs had to be present in the agrifood business in North-Western Europe. Finally, cases had to be different from each other in order to allow for an insightful cross case analysis. Hence, the selected CCOs differ from each other on two dimensions: they have either a broad or small pallet of sustainability criteria, and set high or low criteria on animal welfare in particular (see Figure I). Four cases were selected for an in-depth and cross-case analysis, each representing a quadrant of Figure I.

The final selection of cases include: Freedom Food, Organic products, EurepGAP and 'Scharrel'-eggs (barn eggs).

Freedom Food: The Royal Society for the Prevention of Cruelty to Animals (RSPCA, UK) desired to develop standards specifically for animal welfare and founded Freedom Food (FF) in 1994. The goal of FF is to achieve a higher level of animal welfare for as many husbandry animals as possible.

Organic: The increased use of chemicals (artificial manure, pesticides) in agriculture gave cause for some farmers and consumers to develop an 'own' agriculture where quality of life was the central issue. Organic cattle farming means in practice: a limited number of animals per hectare, free range and organic on-farm feeding.

EurepGAP: EurepGAP is a cooperation between European retailers, started in 1997 that aims to develop widely accepted standards for retail food suppliers. The criteria are broad: they cover both food safety as sustainability issues like animal welfare, environment and labour conditions.

'Scharrel'-eggs: In the 1970s a group of stakeholders brainstormed about a so-called 'welfare-egg', where chickens are able to walk around freely (inside a barn). This Dutch so-called 'scharrel'-egg set the standards for the European barn egg scheme.

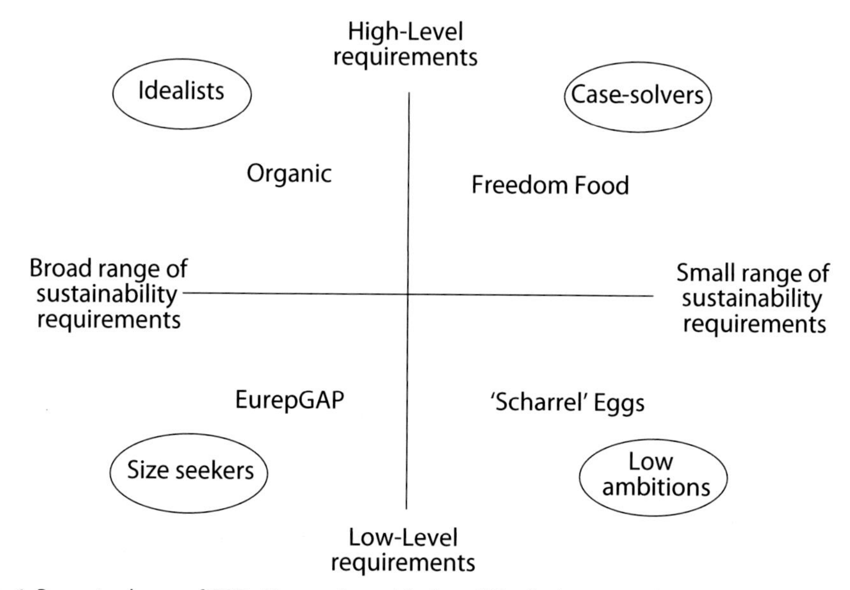

Figure 1. Perceptual map of CCOs (Source: Ingenbleek and Meulenberg, 2006).

Data collection and interpretation

Data were collected by an intensive desk research of articles, research reports and public sources like websites. In addition, interviews were conducted with several experts and persons directly involved in the (criteria setting process of the) CCO. In total, 16 interviews were held, based on the questions in the framework as depicted in Figure 2. Subsequently, four extensive case descriptions were made by the researchers and analyzed independently (Eisenhardt, 1989). In the following cross case comparison, 5 theoretical propositions about the criteria setting process in CCOs were formulated and discussed. For this shortened paper, we will limit to the description of the cases above and focus on the cross-case analysis and the discussion.

Organizational culture	Research questions
Shared Values	Which stakeholders are involved in the CCO? What are their driving values to participate? How different are the values of different stakeholders
Norms	How are values transferred to norms in the process of developing criteria? Which norms have become dominant in the CCO? Are norms (potentially) conflicting, how do stakeholders deal with conflicting norms? How strict are norms? What happens when they are surpassed
Artifacts	Have stakeholders developed a joint language or terminology? What does the written statements of conduct (the actual code of conduct) reflect? Are all artifactsconsistent with each orther, or do they reflect subcultures? Is behavior of primary producers guided by anything else than the code, like a newsletter, etc.
Conduct	Do primary producers feel that they are part of a group? Are there different types of behavior amomg primary producers? Are rules of conduct violated, why? Is behavior proactive or reactive regarding the code of conduct?

Figure 2. Conceptual framework for case analysis.

Case Comparison

1. The criteria setting process comes down to weighing ethical values against market opportunities

In all cases the choice emerged between setting high, more idealistic norms, or jump into opportunities that emerged in the market during the process of setting criteria. In the cases where ethical values play a more important role, criteria are set at a relative high level, consciously taking into account that this may hinder swift market development (Organic; Freedom Food). For example, the idealistic nature of Organic has a direct effect on its costs resulting in an average market share of 2-3%. Freedom Food strives for improved welfare for as many animals as possible and therefore did not grant exclusivity to Tesco. Contrastingly, when market opportunities are regarded as important, the criteria are set at a relative modest level. When the battery egg emerged in The Netherlands, efforts were made to develop a 'welfare egg'. However, these were surpassed by a more practical system (using existing old barns), that ensured market growth possibilities: the 'scharrel'-egg. EurepGAP entails mainly minimum legal requirements to ensure most producers can live up to these criteria.

2. Pro-values ensure continuous improvements; anti-values do not

When a CCO is based on positive or pro-values (continuous improvement of animal welfare) it becomes property of an organisation/discussion platform. When a CCO is highly based on negative or anti-values (e.g. against the bio-industry) the discussion falls silent as soon as the alternative has been realised. The latter is found in the 'scharrel'-eggs case. This was a direct reaction to the battery eggs. There was no public debate about the norms: the pragmatic criteria of 'scharrel' were a good alternative for all stakeholders and the criteria are not subject to changes.

The other CCOs are based on pro-values. The organic movement was formed by engaged producers that shared equal values. Standards are incorporated in both public as private laws, which are subject to continuous improvements. EurepGAP anticipates on globalisation in retailing, their standards are to be revised every two years. Freedom Food is erected to improve animal welfare specifically, and tries to incorporate the latest scientific insights.

3. Private ownership ensures continuous sharpening of the criteria; incorporating in regulation does not

When the criteria are incorporated integrally in regulation ('scharrel'-eggs) this will lead to dissolution of the discussion platform. Directly applying new insights becomes hard, as in fact laws need to be adjusted. In CCOs that are governed by private organizations (EurepGAP, Freedom Food), the discussion continues and the criteria can be sharpened over time without lengthy formal procedures.

EurepGAP intends to sharpen criteria every 2 years. Freedom Food is also a private organization, where criteria are evaluated every 6 months. On the other hand, 'scharrel' was incorporated in law soon after the set-up. There is no driving force behind the scheme that would lead to a continuous increasing of the criteria. In the case of organic, a compromise was found by using both public as private laws to remain some flexibility.

4. Artifacts and a bottom-up approach ensure higher sense of belonging to a group, pro-activity and less opportunistic behaviour

CCOs can be sustained by labels, certifications and other artefacts, and when in addition the primary producers are actively involved in the criteria setting process, farmers feel as if they are part of a certain group or movement. Farmers are more proactive in setting and implementing criteria and less opportunistic behaviour is observed. Organic agriculture was started by engaged farmers: the organic sector shows active participating groups that both represent farmers as steer their behaviour. When Freedom Food started to let the producer join the visits to the retailer, the monitoring of the RSPCA was accepted better than before. EurepGAP shows a growing conscious to involve the sector and the farmers in the process.

In the case of 'scharrel' it is questionable whether there is high involvement or group feeling. It can not be said there is high involvement, the sense of belonging to a group or subcultures. At times, opportunistic behaviour can be observed.

5. A broader pallet of sustainability goals and/or an international composition of the CCO lead to more subcultures and increase the complexity of criteria setting

When stakeholders attempt to include multiple sustainability goals, like animal welfare, environment and labour conditions, the setting of the criteria becomes more complex due

to a lack of a clear focus and contradicting goals. Moreover, if the CCO works across borders, complexity increases due to cultural and geographical differences.
Organic attempts to include various issues, like for example animal welfare (cows outdoor in the meadows) against environment (emission of minerals) and at the same time strive for the ideal goals. This led to a difficult attempt to standardize rules, permitting for national differences. EurepGAP also works with a multitude of sustainable issues, also across borders where there is a contrast between North-West European retailers and Southern or Eastern states. Contrasting, Freedom Food and 'Scharrel' were erected within a country's scope (UK resp. Netherlands). Moreover, both CCO's focus on animal welfare solely (and setting basic minima for other issues), which simplifies the process of setting criteria.

Limitations and implications

The case study was generalized theoretically, yet in order to generalize the conclusions empirically outside the scope of animal welfare, more evidence is needed. In sustainability areas like for example Fair Trade, environment or fishing, additional case studies can be executed in order to generalize our findings into other domains.
However, based on the findings up to here, we can state that new initiatives on sustainability certification should:

1. trade off market opportunities against their ethical values in a way that fits the mission of the organization (and not let the short-run market "dictate" the norms)
2. base the mission not only on a "common enemy" (like the bio-industry), but on an ideal that continues when the enemy is conquered
3. not make formalisation of criteria in laws a goal, because it inhibits the CCO from continuous learning and continuous improvements of sustainability
4. make farmers a party in the negotiation process. Not only because they implement criteria, but also because it increases commitment and thus decreases free-riding and opportunistic behaviour
5. be prepared for a long and complex criteria formulation process when organizations have a broad mission in the sustainability domain

Taking these implications into account, CCOs in development can learn from them in order to maximize their impact on animal welfare or other sustainability issues.

Acknowledgements

The present study is based on research that has been carried out in the context of programme 434 "Space for natural behaviour and transparency" of the Dutch Ministry of Agriculture, Nature and Food Quality, and the Welfare Quality research project, which has been co-financed by the European Commission within the 6th Framework Programme, Contract No. FOOD-CT-2004-506508. The text represents the authors' views and does not necessarily represent a position of the Commission who will not be liable for the use made of such information.

References

Eisenhardt, K.M. (1989). Building theories from case study research. Academy of anagement Review, 10, 803-813.

Kooiman, J. (ed.) (1993). Modern governance; New government-society interactions. London: Sage.

Ingenbleek, P.T.M. and Meulenberg, M.T.G. (forthcoming), "The Battle between "Good" and "Better". A Strategic Perspective on Codes of Conduct for Sustainable Agriculture. Agribusiness (forthcoming)

Yin, R.K. (1994). Case study research: design and methods. Thousand Oaks, CA: Sage Publications.

'Animal Ethics Dilemma': A computer supported learning tool

Trine Dich, Tina Hansen, Anne Algers, Alison Hanlon, Hillar Loor and Peter Sandøe
The Royal Veterinary and Agricultural University, Institute of Food and Resource Economics, Rolighedsvej 26, DK-1958 Frederiksberg C, Denmark, td@kvl.dk

Abstract

In many Veterinary Schools there is a growing focus on animal ethics and it has become a part of their curricula. But how can veterinary students learn to relate to ethics? How can we motivate the students? How can we clarify the ethical issues relating to various forms of animal use? How can a student understand his/her own ethical standpoints – and those of others? These and many other questions are addressed by means of 'Animal Ethics Dilemma', a computer supported learning tool developed primarily for veterinary students and professionals who work with animals. It is not a stand alone tool for teaching animal ethics but is a supplement to other ways of teaching. The programme is constructed as a computerized role game with five cases that the students can play or explore through several different ethical perspectives. Film- and theatre narration techniques are used to enhance student motivation. The students are also allowed to create their own cases. This paper will firstly give a brief presentation of the programme. Secondly the underlying idea behind the programme and ethics learning will be presented. Finally perspectives and further development of the programme will be discussed. This paper may usefully be read together with the abstract of the poster "Presentation of the computer programme 'Animal Ethics Dilemma'".

Keywords: animal ethics, learning, computer game, dramaturgy, provocation engine.

Introduction

The computer supported teaching programme, 'Animal Ethics Dilemma' (Hanlon *et al.*, 2006) is a completely new tool, launched in June 2006, which brings together animal ethics, philosophy, pedagogy, narration, and dramaturgy. The result is a learning tool to improve the student's ability to understand and relate to ethical issues that arise in animal use. The programme is freely available on the internet at www.aedilemma.net.
The programme has been developed in a joint project between the institutions to which the authors are attached: The Royal Veterinary and Agricultural University, Denmark, Swedish University of Agricultural Sciences, University College Dublin, Ireland, Imcode Partners AB, Sweden. The research group "Flexibility in Learning/Creative Environments", Malmö University, Sweden was involved in designing and evaluating an early version of the programme.
The programme is developed in three languages: English, Swedish, and Danish in order to help the students in the respective countries to achieve an in depth understanding of the demanding topic by using the language they are most familiar with.

Presentation of the programme

The programme aims to make the student understand how ethical choices may be seen as reflections of an underlying ethical theory. The programme operates with the following five theories: utilitarianism, contractarianism, relational ethics, animal rights, and respect for nature. These theories are not only presented as abstract entities. The content of these theories is in the programme unfolded in relation to more or less real life choices.

The student experiences a role play centred round five cases: "Blind Hens", "Dog Euthanasia", "ANDi the GM monkey", "Slaughter", and "Rehabilitation of Seals". The cases are fictitious, but based on real dilemmas. Each case presents different situations with a number of alternative choices. The choice made by the student leads to a new dilemma where he/she has to make a new choice – provoking and challenging the student's ideas.
When the student first enters the game she is presented with a pre-test, to determine her 'personal profile'. The pre-test comprises 12 questions with a choice of answers. The answers are taken to reflect different ethical theories. Once the student has selected her/his choice of answers to the 12 questions, a "Personal profile" is generated in the programme. Then the student can begin to play the game. The profile is used to navigate the game by using an opposing view to that of the profile, as the opening question for the case. After finishing a case, the profile will be updated, and the changes visually shown. The new profile will be the starting point for any new case chosen. Accordingly, the programme can be viewed as a provocation engine that adapts to and challenges the views of the student.
Each case has four levels. Within each of the first three levels, there is a statement followed by four or five responses. The statements and responses have been written from different ethical perspectives, creating a labyrinth of ethical dilemmas. The fourth level serves to round up the case by bringing the story to an end based upon the previous responses.
The cases also include visual information such as photos, diagrams and video clips in order to support the text information and utilise the possibilities of the media.

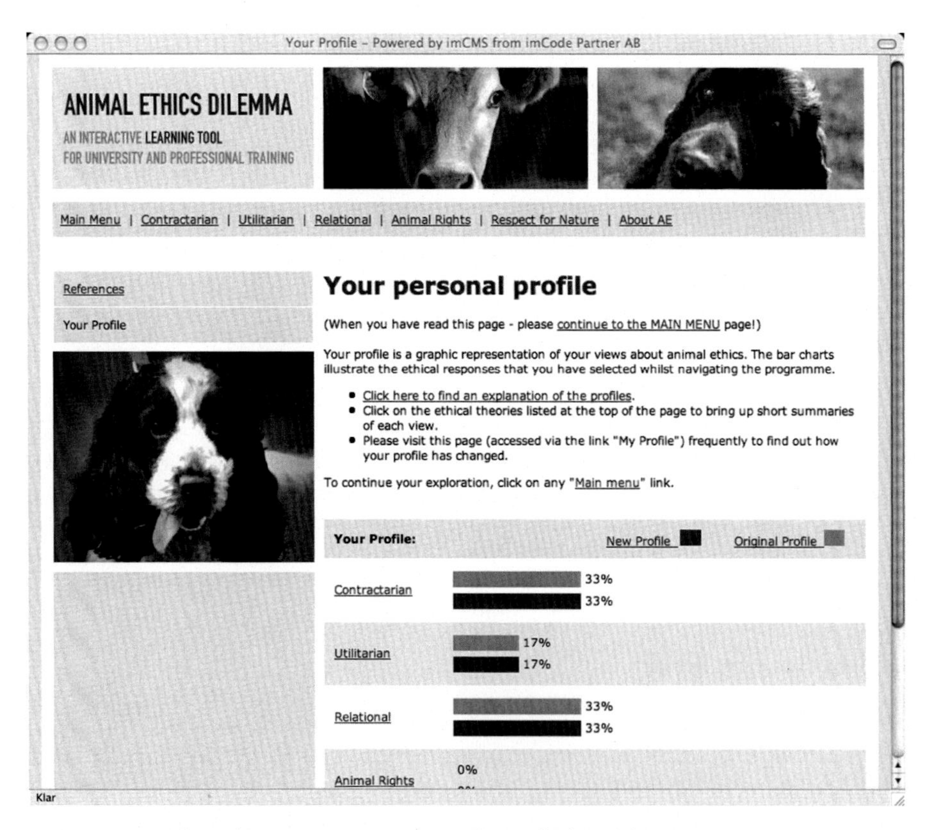

Figure 1. 'Main menu' from the computer programme "Animal Ethics Dilemma".

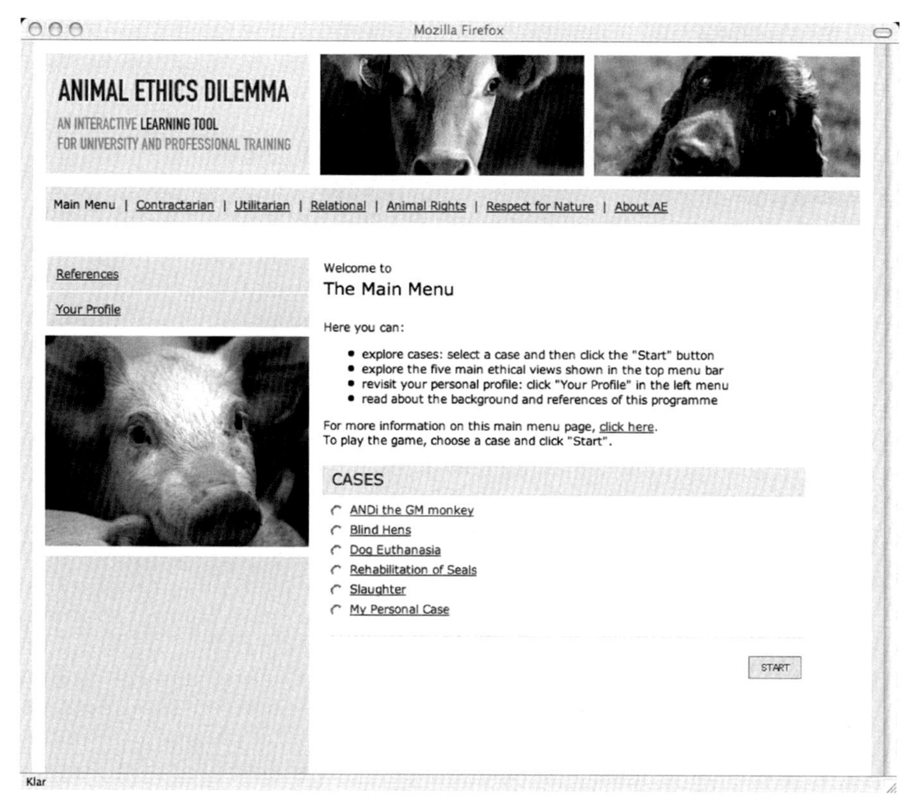

Figure 2. 'Personal profil' from the computer programme "Animal Ethics Dilemma".

The student can select a variety of ways to explore 'Animal Ethics Dilemma'. For example, she can navigate the programme using the "Assist me" option, which explains the basis of the assignment of ethical theories to the various answers by providing pop-up boxes, the "Why?" response. Background text and a glossary of terminology, "References", are available for the student to explore via links in the text. The student will also be encouraged to make 'blind tests', to attempt to navigate the game using one ethical perspective, e.g. choose the utilitarian argument in a sequence of situations and arguments.
Selected access to a programme template is also available, "Create your own case", enabling students or lecturers to create their own cases.

Idea behind the programme and ethics learning

The programme should be considered as a tool to support the learning process, not as a stand alone tool for teaching animal ethics. There will still be a need for teacher-student interaction whether it will be in cyberspace or in the classroom.
Integrating film- and theatre narration into the educational material is a new approach that deliberately uses basic dramaturgical notions such as 'conflict', 'turning point', 'identification', and 'driving force' to enhance student motivation (Pearce, 1997):
'Conflict' is a basic term in dramaturgy that is used for every situation where the characters involved meet resistance, a challenge, or a threat to their views. A conflict introduces a need to overcome this obstacle, to make a new choice. And in the choice there is always the question:

What is at stake? What would be the best choice? The more that is at stake, the stronger the conflict – and also the greater possibilities to engage the viewer. From a pedagogical point of view, the dramaturgical term 'conflict' can be used to introduce situations where it is necessary to solve a problem, and therefore the programme supports the notion of problem based learning (PBL). Learning takes place while solving conflicts and reflecting upon different solutions. In the programme, a conflict is brought up every time the student has to make a choice based upon a challenging viewpoint.

The dilemmas - for example between breeding blind hens which do not feather peck and cannibalise each other or having seeing hens with higher mortality rates – make the student feel the pressures of conflicting concerns. The students will not only experience that their own opinion is one of many possible standpoints, but may also be able to appreciate why others may think differently. Thus, an important learning outcome will hopefully be that the student develops a better understanding of his or her own ethical standpoint and a better understanding for the views that other persons or stakeholders may have.

'Turning point' means that the narration takes another direction than is expected, in which new facts are brought into the story to cast new light upon our beliefs. From a pedagogical point of view, this can be used to deliberately introduce surprising elements. These elements make learning much more interesting – and force the student to reflect. Also the students will see that one can become wiser regarding what is ethically right and wrong. One may start with a certain point of view and to one's own surprise find oneself situated with a more or less different view. This may also make one more tolerant to other points of view. In the programme, stories will take a 'turn' based on the student's answers in the former situation.

'Identification' is a question of how the audience relates to the characters. The stronger the audience's individual identification, the stronger the audience's experience will be. From a pedagogical point of view this can be used to increase motivation, one of the most important aspects of a learning situation. In the programme, identification is found in role plays where the student is given a certain role in different situations in each case.

The term 'driving force' is used in several ways. It is a term for the factors that force the story to continue and the audience to desire more and find out what the next step in the narration will be.

Driving force is also used to describe the sum of a specific character's reasons for acting in a certain way. If John sees his 2 year old daughter run towards the street with a lot of traffic dangerously close to her, he will at that moment have a very strong driving force to stop talking to his dying mother and instead run after the child: This even if the mother dies and he endangers his own life in the traffic.

Strong conflicts, an interesting and relevant story, internal dilemmas and many other factors create a strong driving force in the story. This in turn creates curiosity and motivation in the audience. In an interactive context a strong driving force creates curiosity to continue exploring the programme.

From a pedagogical point of view this means more motivated students and better learning.

All this could be incorporated in conventional teaching material (e.g. through role plays) but the computer media opens up for new dimensions. For instance, the student has the possibility of doing her/his own exploration of the material as many times and at any time as desired. And the interactive opportunity to "Create your own case" is based on the assumption that you learn more about a subject if you have to teach it. If the student chooses to create her/his own case, she/he will have to study in depth to be able to create the situations, questions, and arguments that relate to every ethical view. This would also be a challenging assignment for a student group to develop a case together. Finally it will make it possible for the teacher to create a case that is adapted or relevant to a specific issue, which has been brought up in society or among the students.

Perspectives

As mentioned earlier, the programme is meant only to be a supplementary and supporting teaching tool. Till now the group of authors has only had experiences in using beta versions of the programme in their teaching, but based on the preliminary feedback it can be concluded the students recognise the learning benefits of the tool.
Part of the programme has been tested by veterinary students twice. In 2003 a qualitative evaluation was carried out. The aim was to investigate if the provocative dramaturgy was comprehensible and acceptable as a tool in a learning situation. The conclusion was that the provocative dramaturgy was seen as an exciting way to learn about animal ethics and it was a good tool to motivate the students to reflect on ethical questions related to their profession. In 2005 a quantitative evaluation was carried out. The aim was to test the graphical interface in the programme and the professional benefit related to animal ethics. The survey showed that students find the programme easy to use; the cases are interesting and the programme contributes to the understanding of different ethical viewpoints.
One immediate point, however, is to discuss how to incorporate the programme in a learning process. When will it be appropriate and meaningful to the students? Are there any assumptions that have to be fulfilled before the tool can be used? And how is it possible to evaluate the effects of the programme on student learning ?
A discussion on the first two points will require further experience from using the programme. Regarding the third point it is of course important that the tool is used in connection with courses or teaching programmes that share the underlying pluralistic view on ethics. With respect to the last point, it is intended that an evaluation of student learning could be based on the student's ability to create his/her own case. This will give an indication of the student's understanding of the ethical theories and whether she/he is able to apply them to a given situation.
Further development of the programme will take place over the coming years. In the programme is included an electronic evaluation form, and the responses from both students and teachers will be used to make any changes and additions to the programme. New cases could also be added based upon or inspired from teachers' and students' own cases if they are made available to the authors via the evaluation form.

References

Bonime, A. and Pohlman, K.C. (1998). *Writing for new media*. John Wiley and Sons, Inc. p.153-170.

Dich, T., Hansen, T., Christiansen, S.B., Kaltoft, P. and Sandøe, P. (2005). Teaching ethics to agricultural and veterinary students: experiences from Denmark. In: *Animal Bioethics. Principles and Teaching Methods.* Marie, M., Edwards, S., Gandini, G., Reiss, M. and Borell, E. von (eds.). Wageningen Academic Publichers, The Netherlands. p.245-258.

Hanlon, A.J. (2005). An introduction to Problem-Based Learning and its application to an animal bioethics curriculum. In: *Animal Bioethics. Principles and Teaching Methods.* Marie, M., Edwards, S., Gandini, G., Reiss, M. and Borell, E. von (eds.). Wageningen Academic Publichers, The Netherlands. p.283-295.

Hanlon, A.J., Dich, T., Hansen, T., Loor, H., Sandøe, P., Algers, A. (2006). *Animal Ethics Dilemma – an Interactive Learning Tool for University and Professional Training* [on line]. Available on www.aedilemma.net [24/04/2006].

Pearce, C. (1997). *The Interactive Book*. Macmillan Technical P. Indianapolis, Indiana. p.119-123,329-345,420-425,490-494.

Rawles, K. (2000). Why do vets need to know about ethics? In: *Veterinary Ethics. An Introduction.* Legood, G. (ed.). Continuum, London & N.Y. p.3-16.

Wimberley, D. and Samsel, J. (1995). *Interactive Writers Handbook.* The Carronade Group, Los Angeles. p.173–189,191-217.

Giving ethical advice in a pluralist context: What is the problem?

Ellen-Marie Forsberg
The National Committees for Research Ethics, P.O. Box 522 Sentrum, 0105 Oslo, Norway

Abstract

The ethical matrix approach was developed by Prof Ben Mepham and his colleagues at the University of Nottingham in the early 1990s. Since then the approach has received increasing attention and has been used by several researchers in different projects related to assessing ethical impacts of different food production technologies and other policy options of societal concern. At the last EurSafe conference Prof Mepham reviewed the most important applications of and objections to the method. In this paper I would like to present results from my doctoral project on justification of ethical advice, in particular justification related to the ethical matrix approach. The project involves assessment of coherentist and pragmatist approaches to moral justification, and their appropriateness in this context. I will present a more ambitious version of the matrix approach than what Mepham advocates, involving not only assessment of ethical concerns, but reaching judgements on ethical issues in a participatory process. A central problem is how to prioritise between values within the framework of value pluralism. The particular version of the ethical matrix approach to be presented builds on applications of the method performed at the Norwegian Research Ethics Committees, but will include recommendations for revisions of the method in light of metaethical concerns about justification.

Keywords: ethical matrix, pluralism, justification, coherentism, pragmatism

Introduction

These are pluralist times; we live in states which acknowledge the rights of individuals to pursue different goals and life projects (within some boundaries). Correspondingly, when doing public ethical assessments there is an assumption that a pluralism of values must be respected. Isaiah Berlin argued in 1959 that all substantial ethical theories and worldviews are controversial. So we need to make value choices without reference to any dogmas; neither of a religious or theoretical sort. But how can values choices be justified if there are no given standard with which to assess them? This is a fundamental question of *value pluralism*.

A solution to this question has been to develop so-called ethical tools. These are to be decision making aids that help justify value choices without recourse to substantive theories. These tools might have different functions; some simply to map values systematically, while others to indicate a justified solution.

The ethical matrix method is a well-known assessment tool in agricultural and food ethics. This tool stands in an intuitionist tradition from, in particular, David Ross, and via Beauchamp and Childress' principle based ethics. Both Ross and Beauchamp and Childress advocate a fundamental pluralism of moral principles and in the matrix this is expressed through three[64] principles structuring the ethical concerns. This pluralism of principles might be called *principled pluralism*. Like all intuitionist approaches the ethical matrix is sometimes characterised as a checklist of ethical concerns. This is due to the fact that it does not provide an overarching ethical system which determines how the principles and their specifications are to be weighed up

[64] Although sometimes four.

against each other. Ross argued that this kind of system was incompatible with the irreducibility pluralism of intuitionism.[65]

Value pluralism and principled pluralism are related in different ways. One might characterise value pluralism as a position in political philosophy (often associated with liberalism), while principled pluralism is a position in meta-ethics (on the nature of our duties). When using the ethical matrix method in public ethics assessments both kinds of pluralism will influence how it can be used in a justified manner. I will here argue that the ethical matrix method can be used to assist making conclusions on issues when used in a participatory process, and I will show how this is compatible with commitments both to value pluralism and principled pluralism.

Principled pluralism and value pluralism

Ross talks about five (originally seven) clusters of principles which exert certain demands on us. These are self-evident and we know them by intuition, hence the name intuitionism. Although some of these are stringent and usually have more weight than others (cf. McNaughton (2002) for argumentation about this) there are no general rules for how the principles are to be balanced. To determine what Ross calls our 'actual duty' (in contrast to the seven principles which ascribe *prima facie* duties) we use perception. This must not be understood as simply 'seeing' the correct answer, but as a form of reflection, as coming 'in the long run, after consideration, to think one duty more pressing than the other' (2002, p. 31). The matrix method has inherited the intuitionist view that there are more than one fundamental principle and that there are no general priority rules between these principles.

The background of the matrix method is Beauchamp and Childress' principle based ethics and these authors are also intuitionist. Beauchamp and Childress are more explicit than Ross on how conflict between moral principles can be solved in intuitionism. They suggest that we can either specify the principles so that the apparent conflict dissolves or so that the balancing between them is much more limited. The matrix method has inherited both these methods from Beauchamp and Childress; the content of the matrix is specifications of the three general principles (related to each affected party's situation). Balancing can then be done between these specifications.

Beauchamp and Childress have made some practical rules to avoid purely subjective balancing, but otherwise there are no principles that decide how the specifications are to be weighted against each other. This would be incompatible with the intuitionist roots. It would also be incompatible with the fact of value pluralism.

Value pluralism is pluralism from the point of view of the good, rather than from the point of view of the right. There are two versions of value pluralism: one metaphysical and one ethical, although the two are sometimes combined. The metaphysical value pluralism is a doctrine about the nature of The Good, namely that there is no one source of value (as Plato claims), but in fact several. The ethical version consists in the claim that there are a variety of life projects, all worthy of equal respect. These two doctrines are interwoven in the sense that if there really is only one Good and only one life project can be inferred from this Good, then there is no need to respect any other life projects as inherently valuable and worthy of equal respect and protection. However, there is no necessity in taking a stand on the metaphysics of value to be committed to protecting the pluralism of life projects. This might be defended simply on the *possibility* that there should exist several inherent goods. However, the metaphysical claim does have some consequences for the acceptability of ethical theories. If there is no single source of value, normative theories than build upon this, for instance utilitarianism, cannot be accepted

[65] However, modern intuitionist like Audi (19..) argue that it is possible. I will stick with Ross, since this is the roots of the matrix method.

as methods to solve moral conflict. So, value pluralism disallow normative decision principles that build upon the conception of a single good, and it also commits to protecting the manifold of life projects.

Value pluralism is often associated with liberalism. Political choices are being made in liberal societies all the time, so it is clear that value pluralism does not require abstinence from value based action.[66] The question is then whether or not liberalism itself is a normative position built upon commitment to freedom as the ultimate value. There is an extensive debate about this which I will not rehearse here[67]. But if one succeeds to show that there exists a framework for normative questions, which is not itself built upon a notion of the good, this should not be a problem for value pluralists. In fact, some try to show that this is so. [Refs.] I try in my thesis to show that a Peircean deliberative approach might be such a neutral framework; and that deliberation in participatory processes is a way to perform a balancing of the different principles that provides us with substantial reasons for claiming justified conclusions to cases.

My main reason for claiming this as a solution is that this is an entirely general epistemological position on how we can reach truth, in science, in everyday life, and in ethics. Although it has a normative aspect in the sense that if you want to reach true opinion, then this is how you should proceed, it seems to have no *moral* commitments. It then seems also to be compatible with a pluralism about the right. In fact, Habermas takes a similar stance. He defines the domain of the right as the true area of moral questions, and identifies this with the question of justice. However, his main argument for his discourse ethics is not from justice, but similarly from the procedure to reach truth. So this seems defensible.

However, is really the ethical matrix a framework of principles of action? In the Norwegian version of the method the original matrix is called a value matrix, and the content is defined in statements about desired goods. The reason for this lies mainly in the fact that this terminology is more user-friendly. And it need not be a theoretical problem. Also Ross admits that there is a prima facie duty to realise intrinsic goods (and we can assume the corresponding negative duty not to harm these goods). Although the procedure to determine the values in the matrix is not equal to Ross' procedure to determine what the intrinsic goods are, we can still assume a certain correspondence. The difference would be that Ross claims that there are duties that do not correspond to intrinsic goods, while this is not included in the matrix. So we assume that the duties are implicit in the matrix.

The epistemological solution – how does it fare?

Let us briefly consider one epistemological solution, Peircean pragmatism (in Misak's version).[68] Misak advocates a theory about truth that also holds for truth of moral enquiry. Let us imagine that an ethics committee deliberates about a situation where two prima facie principles conflict. The committee ends up with defending that one of these should be prioritised at the expense of the other. Misak has resources to show how the committee can claim that this choice is the right on. They can claim this because they have critically assessed the belief and found that this is the choice that best has withstood objections scrutiny, and whose assumptions and consequences can be accepted. This involves listening to the opinions of all affected parties. Minimum criteria for good inquiry are taking experience and argument seriously. An experientially-guided belief is

[66] Although value pluralism is rightly associated with liberalism, it need not be understood in a liberalist context. I will return to a different framework for value pluralism.

[67] See for instance Larmore, Nagel, etc.

[68] Weak coherentism (like Beauchamp and Childress propose, building on Rawls' and Daniels' account of reflective equilibrium) is a different option. I show in my thesis how this solution does not work for intuitionist methods.

a belief that forces itself upon a subject; it is a conviction. So 'those engaged in moral deliberation who denigrate or ignore the experiences of those who are of a certain skin colour, gender, class, or religion are also adopting a method unlikely to reach the truth' (2000, p. 82). Therefore it follows that asserting something (implicitly to be true) automatically commits one to open up for deliberation including all affected parties. A statement that 'survives' the rational argumentation with affected parties is both true and justified, in addition to legitimate in the eyes of the participants and perhaps also in the eyes of the authorities and the public. Single individuals cannot determine what the true belief is; true beliefs are what the community of inquirers would converge upon. A hallucination might seem true to us, but would not be accepted as a true experience of the world, by the community. But convergence is not a strict requirement: 'the expectation of convergence is a normative presupposition of inquiry, differing in strength as our needs for convergence differ' (p. 99). In fact, she claims that we should often generate disagreement to make sure that all different views are raised and discussed. Needless to say, this resonates with the practice we have had in Norway with using the ethical matrix in a participatory process.

Is this an acceptable method for solving moral conflict in the context of principled pluralism and value pluralism? Principled pluralism involves respecting a plurality of principles. The extent of participation here seems to make this the case. Note that justification is only temporary; the assessment of the belief continues and if someone points out that some fundamental principle has not been considered, then the belief is not justified until the new argument has been scrutinized. But when this has been said; we would hesitate to call something justified, and more so for calling it true, if we were not reasonably sure we had considered most aspects of importance. In any case, pluralism seems to be central in this approach [Ref from Misak]. In addition to this, there is no decision rule prioritising between the different principles, except the commandment to critically enquire about the belief together, and this commandment is justified simply be reference to this is what we do when we want to reach the truth.

What about value pluralism? It seems that the inclusion of people with different life projects would be a strength, and not a problem, in this approach. The approach seems completely neutral. But isn't there a bias towards democratic values? Could we end up, for instance, recommending less civil rights to the advantage of a strong leader? Misak would answer to this that the process could indeed have this result, but that it is very unlikely because the participants would probably tend to defend their rights. But given an extraordinary situation (for instance wartime) there is nothing in principle against this result.

Could we justify results from this kind of process to a religious fanatic? Would he not say that truth was giving to him through divine revelation, and that the feeble capacities of mortals would be nothing but irrelevant? Misak's answer here would be to point to how the concept of truth is used in 'men's conversation' (Peirce, ...). Although some religious fanatics may have a deviating understanding of the concept, there will be widespread agreement that truth is intimately connected to enquiry, experience and evidence. Religious revelation may give insight in divine matters on a personal level, but this is in our practices called faith. If the fanatic participates in the discourse like other participants, raising his views and arguing for them, then he is out for truth and he is welcome in the process.

There does therefore not seem to be anything in this approach that contradicts principled pluralism or value pluralism. Alasdair MacIntyre argues in his book Whose Justice? Which Rationality? That rationality is always historical and social, and in that sense not neutral or universally valid. Misak agrees with this and does not argue that the rationality concept we find connected to the concept of truth in our culture is something outside history. However, she argues that if we claim something to be true in our societies, this is how we go about justifying it.

Conclusion

It seems the epistemological solution is compatible with principled pluralism and value pluralism, at least when it is combined with a pragmatic naturalism. Therefore it seems reasonable to design a variety of the ethical matrix process that uses the resources that lie in Peircean approach to be able to use the matrix methodology to come to substantial conclusion on a case to case basis.

References

Beauchamp, T and Childress, J. 2001 (1979) The Principles of Biomedical Ethics. Oxford University Press

Larmore, C. 1996. The Morals of Modernity. Cambridge University Press.

Lukes, S. 1991. Moral Conflict and Politics. Clarendon Press, Oxford

Misak, C. 2000. Truth, Politics, Morality. Pragmatism and deliberation. Routledge.

Nagel, T. 1987. Moral Conflict and Political Legitimacy. Philosophy & Public Affairs. Vol 16, pp. 215-240

Values behind biodiversity: A study of environmental controversies related to salmon farming

Arne Sveinson Haugen
Centre for technology, innovation and culture, University of Oslo, Box 1108 Blindern, 0317 Oslo, Norway, a.s.haugen@tik.uio.no

Abstract

It is discussed possible reasons behind or values behind biodiversity that might explain why it is important to conserve biodiversity. In this connection, some questions related to how wild and farmed salmon might be valued is used to illustrate how traditional inter-human ethical positions might stand in relation to nature-environmental positions when handling environmental controversies.

Keywords: salmon farming, environmental ethics, values, biodiversity

Introduction

Biodiversity conservation is often seen as a goal in itself and a main reason for why ecosystems and species should be protected (Lemons and Brown 1995). It sounds however like a tautology to say that ecosystems, species and animals, which all actually are segments that contribute to constitute biodiversity, should be protected because biodiversity conservation is important. If this should make any sense, there must be some other reasons behind or values behind biodiversity that make it reasonable on separate grounds to say that it is important to conserve biodiversity. This is why this paper focuses on the values behind biodiversity.

The paper is based on a search for values behind biodiversity done in my PhD study of environmental controversies related to the salmon farming industry in Norway. The main research questions asked are how humans may value ecological entities, and what kinds of moral status values these entities might have. The term ecological entities refers to different kinds of single living organisms and over-individual entities such as species, communities and ecosystems. The value considerations of this study is based on traditional and contemporary philosophical and ethical theories with focus on how Aristotle, Aquinas, Descartes, Kant (Clarke and Linzey, 1990; Armstrong and Botzler, 1993), Singer, Regan, Goodpaster, Taylor, Leopold, Callicott (Zimmerman, 1993), Westra (1994, 1998) and Norton (2003) would ascribe moral status value to ecological entities.

In an effort to illustrate how the research might be used in analyses of environmental controversies, following three questions are formulated:

> On what ethical grounds might it be fair that the negative impacts of agriculture on original or primary terrestrial ecosystems, which has resulted in the formation of culture landscapes, is accepted, while at the same time the farming salmon and its impacts on aquatic ecosystems is not?

> On what ethical grounds might it be fair that a government committee, like the Norwegian Wild Salmon Committee, is appointed in an effort to secure the wild salmon's future existence, while at the same time the minnow[69] is perceived as a threat that should be combated?
>
> On what ethical grounds might it be fair to be concerned with the animal welfare of domesticated animals, such as farmed salmon, while at the same time enjoying salmon game fishing is accepted?

The answers to these questions will probably vary from person to person, and between groups and societies, dependent on their world-views and value grounds (Norton 2003). Since it will be a too big order within the frames of present paper to handle all the possible varieties of perceptions, the scope is limited to a discussion of what ethical grounds the answers might be based.

In short, an answer which will be argued for, is that traditional inter-human ethics of the anthropocentric kinds are the only ethical ground on which the issues addressed in the three questions might be perceived as fair. This means ethical positions where nonhuman ecological entities not are perceived as ends in themselves.

Values and ethical positions

A basic part of the answer argued for is the classical sorting of ecological entities into those that are valued non-instrumental as ends in themselves and those that are valued instrumental as means to other ends (Shrader-Frechette and McCoy, 1995).

Universal inner worth, particular inner worth, or instrumental value

Specifications of what might be non-instrumental and what might be instrumental ways of valuing ecological entities is based on Rolston's (1988) inventory list of how nature is valuable to humans. This list contains a number of value types which in a short version can be termed economic, recreational, aesthetic, scientific, evolutionary, symbolic and religious. These value types are further in my PhD study grouped into three value domains termed life, biological and cultural. The life domain contains the life value of Rolston's list. Belonging to this domain are ecological entities that are valued non-instrumentally as ends in themselves. Such ecological entities are said to have a universal inner worth. The biological domain contains ecological entities that according to Rolston's list represent life-support values, such as water, food and shelter. These are thus pure instrumental valuations of ecological entities as means to humans' survival, growth and reproduction. The rest of the elements of the inventory list are grouped under the cultural domain. Except for the economic value type, which is categorised as instrumental, it is more unclear how the other value types should be categorised.

On one side they might be perceived not to be purely instrumental valuation of entities as means to other ends. On the other side, it is difficult to see that they represent valuations of entities as ends in themselves of the universal inner worth value category. The kind of valuations applied here might be compared with the way artworks are valued. For some people an artwork might be perceived to have an inner worth, while others will see no value in it, unless may be for purely economic instrumental reasons. This kind of valuations within the cultural domain, which depend on how single humans or groups of humans subjectively are appreciating, valuing or respecting ecological entities, will be described to represent a particular inner worth. It is

[69] Minnow is a small carp fish living in fresh water lakes and watercourses, which was endemic to some parts of Eastern Norway, Troms and Finnmark. The minnow has the last hundred years been spread to most parts of Norway and represents now a strong competitor to many local fresh water fish stocks.

called particular because it represents valuations of ecological entities that are more subjectively personal or group related than the universal and more objective valuation of life as an end in itself. The particular inner worth value category, which by some might be perceived to represent a non-instrumental value, will strictly speaking be perceived as instrumental.

Traditional inter-human ethics, natural-environmental ethics

The value grounds are eventually synthesised into two ethical positions. On one side are positions termed traditional inter-human ethics, which are about ethics between humans only. On the other side are positions termed nature-environmental ethics, which are about ethics involving nonhuman ecological entities.

Universal inner worth is decisive for the dividing of ethical positions into the traditional inter-human ethics and the nature-environmental ethics. An ecological entity with a universal inner worth is ranked to have a higher value status than those with a particular inner worth or a pure instrumental value. Ecological entities with a universal inner worth might however in some situations also be perceived to have particular inner worth or pure instrumental value. Since the highest ranked status should be respected most, this might equivalently with the Kantian ethics be expressed as a categorical imperative saying that humans should act so as to use ecological entities with a universal inner worth always at the same time as an end, never merely as entities with a particular inner worth or a pure instrumental value. The relation between particular inner worth and pure particular value is less clear. In situations where ecological entities which not have a universal inner worth, are valued purely instrumentally as necessary and basic life-support sources, then this might have priority before any kind of particular inner worth. If on the other hand such ecological entities are valued purely instrumentally for non-basic and luxury purposes, then any kind of particular inner worth might be given priority.

Discussion

Culture landscapes versus salmon farming impacts

If nature-environmental ethics is applied in relation to the first question about culture landscapes versus salmon farming, then in principle any human impact or influence on ecological entities and natural processes would be ethically wrong. This means that maintaining culture landscapes, which are results of human impacts, should be just as ethically questionable as impacts on aquatic ecosystems caused by salmon farming. If the main concern really should be the universal inner worth of ecological entities, it sound like a contradiction if the efforts to protect culture landscapes are done by fighting against the natural invasion, succession and ecological development that may occur when traditional agriculture activities come to an end.

If traditional inter-human ethics where ecological entities have no universal inner worth is applied, however, then it can be argued that culture landscapes might represent a better alternative for fulfilment of humans' needs and preferences. This is because cultural landscapes in addition to sustainable food production contribute with cultural values of the particular inner worth kind such as aesthetic sceneries, interesting and beautiful biodiversity, rustic and village style buildings, recreational and adventures activities, traditional events, and local and national identities.

The salmon farming in this context is more associated with area use and pollution related incidents such as escaped farmed salmon, genetic interactions with wild salmon, salmon lice, fish diseases and effluents resulting in contaminated water, eutrofication and seabed degradation. Further can be added the threat that salmon farming might represents to the wild fish stocks world wide primarily due to the use of fish raw materials in fish feed production, which at the end might create food security and life-support problems both locally and globally.

If salmon farming, however, could make it feasible to believe that it when it comes to sustainability and risk elements, could develop to be comparable with traditional domesticated terrestrial meat production, then they should get a relatively stronger ethical record. But still it probably will not receive the same value status as culture landscapes. Salmon farming most probably have to realise that it in the near future, and most probably neither, will be perceived as a positive contributor to such as beautiful, fascinating and adventurous biodiversity.
The salmon farming case might however be strong indirectly in relation to particular inner worth of the cultural domain if their activities would result in long-term income bases that similar to or better than alternative activities should create possibilities for more people to live in rural coastal areas and to enjoy the ecosystems and biodiversity available there.

Wild salmon versus minnow

If the wild salmon and minnow on the basis of real nature-environmental ethical positions should be regarded, on the basis of being fish species, possibly to have the ability to feel pain, they both should be ascribed an equal universal inner worth. Then any actions that should interfere with the integrity of any one of them would be ethically wrong. This means that any action or interference that would have a negative impact on the minnow species would be just as bad as any negative impact on the salmon species. This would make it difficult to defend any measures taken to reduce the distribution of or to combat the minnow species. Further, if the minnow species should be endangered, then the motivation for establishing a government approved committee to secure its future existence should be just as good as for the wild salmon.
If traditional inter-human ethics is applied, then things such as the excitement of game fishing, practicing of wild life in general, fascination by the wild salmon's biology, enjoyment of its culinary palatability, and appreciation of income created by wild salmon related activities might favour the wild salmon. An ethically acceptable reason for why minnow should be combated would then be that it represents a threat to the viability of other fish stocks such as fresh water trout, which like wild salmon is highly valued by many people.

Salmon game fishing versus animal welfare

Then comes the third questions, which is about animal welfare versus game fishing. If it on the basis of nature-environmental ethics should be so that farmed salmon should be treated well, their living conditions should be optimal, harm should be avoided when alive and slaughtering processes should be as painless as possible, then this most probably should apply also for wild salmon.
A reflection in this connection is to imagine how fishing traditions had been developed if salmon or fish in general was screaming and yelling when forced out of the water and landed by the fishing gear. Question marks would then certainly be put on most fishing techniques, like in salmon game fishing where angling hooks are used, and the salmon often has to fight for a long period before it is landed. If this then only is done for the pleasure of the game fisher, without any kind of food supply objectives, and even, as for catch and release fishing, letting the salmon out again in the water, then the case seems to be relatively weak in relation to nature-environmental ethical considerations.
Again if traditional inter-human ethics is applied, then the cultural values of the particular inner worth kinds would apply again.

Conclusion

As a conclusion it is found reasonable to say that ethical considerations which might appear to be based on nature-environmental ethical positions, often are of the traditional inter-human ethical kind with a focus on the cultural values perceived to represent a particular inner worth.

This might thus be a reason for why environmental interests might approve and support culture landscapes, while salmon farming is criticised, or that wild salmon is worth a government appointed committee to secure its future existence, while the minnow is combated, or animal welfare of farmed salmon is an issue while many at the same time can enjoy salmon game fishing.
A question then is if there in any practical situations will exist a nature-environmental ethics, or if it at the end boils down to care for the environmental conditions that are necessary to secure and satisfy the lives and well-being of humans. These might be such as fresh air, sufficient water supply, access to basic food sources, comfortable housing, good road systems and infrastructure, nice parks, relaxed coffee bars and other goods and services that humans need and prefer for their well-being and flourishing.

References

Armstrong, S.J. and Botzler, R.G. (1993). *Environmental Ethics: Divergence and Convergence*. McGraw-Hill, New York. 570 p.

Clarke, P.A.B. and Linzey, A. (1990). *Political Theory and Animal Rights*, Pluto Press, London. 186 p.

Lemons, J. and Brown, D.A. (1995). *Sustainable Development: Science, Ethics, and Public Policy*, Kluwer Academic Publishers, Dordrecht. 281 p.

Norton, B.G. (2003). *Searchin for Sustainability: Interdisiplinary Essays in the Philosophy of Conservation Biology*, Cambridge University Press, Cambridge. 554 p.

Rolston, H.III (1988). *Environmental Ethics: Duties to and Values in the Natural World*, Temple University Press, Philadelphia. 373 p.

Shrader-Frechette, K.S.and McCoy, E.D. (1995). *Method in ecology: Strategies for consevation*, Cambridge University Press, New York. 328 p.

Westra, L. (1994). *An Environmental Proposal for Ethics: The Principle of Integrity*. Rowman & Littlefield Publishers, Maryland. 235 p.

Westra, L. (1998). *Living in Integrity: A Global Ethics to Restore a Fragmented Earth*. Rowman & Littlefield Publishers, Maryland. 269 p.

Zimmerman, M.E. (1993). *Environmental Philosophy: From Animal Rights to Radical Ecology*. Prentice-Hall, New Jersey. 453 p.

Part 7
Cultural values in food production and consumption

Factors influencing halal meat consumption: An application of the theory of planned behaviour

Karijn Bonne[1], Iris Vermeir[1], Florence Bergeaud-Blackler[2] and Wim Verbeke[3]
[1]Hogeschool Gent, Department of Business Studies and Public Administration, Voskenslaan 270, B-9000 Gent, Belgium; Karijn.Bonne@hogent.be; Iris.Vermeir@hogent.be
[2]Université de Méditerrané, Laboratoire d'Anthropologie, UMR 6578, Bd Jean Moulin 27, 13385 Marseille cedex 05, France ; Fbb@aof.org
[3]Ghent University, Department of Agricultural Economics, Coupure links 653, B-9000 Gent, Belgium; Wim.Verbeke@UGent.be

Abstract

This study aims at investigating the determinants of halal meat consumption within a Muslim migration population in France using the Theory of Planned Behaviour as a conceptual framework. Also the role of self-identity as a Muslim and dietary acculturation in the host culture is explored. Therefore, a survey with 576 Muslims in France was performed. Data are analysed by means of independent samples' t-tests, correlations and stepwise multiple regression. In general, a positive personal attitude towards the consumption of halal meat, the influence of peers and the perceived control over consuming halal meat predict the intended consumption of halal meat among Muslims. After adding two characteristics that are associated with religious food decisions within a migration context (i.e. self-identity and dietary acculturation), analysis showed that low acculturated Muslims rely completely on their personal attitude towards halal meat, whereas high dietary acculturated Muslims rely on attitude and perceived control when intending to consume halal meat. Muslims with a low Muslim self-identity intend to eat halal meat not only because they find it very important, but also as a personal moral obligation. Muslims with a high Muslim self-identity are more influenced by peers and the degree of control they believe to have over eating halal meat. We can conclude that the more individuals are dietary acculturated and the less they identify with Islam, the more the Theory of Planned Behaviour becomes valid for Muslim consumers in France, or also, the more eating halal meat should be regarded as a reasoned action.

Keywords: theory of planned behaviour, france, halal, meat, religion

Introduction

Earlier research has given ample evidence that religion, as one of the environmental factors, can influence food consumption decisions (e.g. Shatenstein and Ghadirian, 1997; Asp, 1999). Most religions forbid certain foods, except for Christianity that has no food taboos (Sack, 2001: 218). Although religion may impose strict dietary laws, the amount of people following them may vary considerably. For instance, it is estimated that 90% of Buddhist and Hindus (Dindyal, 2003), 75% of Muslims versus only 16% of Jews in the US strictly follow their religious dietary laws (Hussaini, 1993). Factors explaining differences in adherence to religious dietary prescription pertain among others to social structures, e.g. origin, immigration, and generation differences (Limage, 2000; Saint-Blancat, 2004; Ababou, 2005; Bergeaud-Blackler and Bonne, 2006). The consumption of products of animal origin, and more specifically meat and meat products is the most strictly regulated or prohibited in cases where religious considerations prevail

(Shatenstein and Ghadirian, 1997). Therefore, this research is designed to explore consumers' meat consumption decision-making within a population of Muslim migrates in France.
In addition to the five pillars in Islam, Muslims have to follow a set of dietary laws intended to advance their well being. These laws determine which foods are "lawful" or permitted for Muslims. Unlawful or haram is the consumption of alcohol, pork, blood, dead meat and meat which has not been slaughtered according to Islamic rulings. All other foods are permitted or halal.
Despite the potential importance of religion in consumer behaviour, empirical research is scarce. Previous studies have predominantly studied the relationship between religious variables and attitudes/behaviours from sociological and psychological points of view (Delener, 1994). Furthermore, no study ever reported an investigation of the religious dimension and its correlation with food decision making. Therefore, we measured the influence of the classical components of the Theory of Planned Behaviour on intention to consume meat within an ethnic minority population of Muslims originating especially from North-Africa and currently living in France. We argue that meat consumption decisions within a religious context could differ significantly from purchase situations where religion does not play a key role. By extending the TPB model with self-identity and dietary acculturation, the influence of the cultural and more specific religious context is investigated.

Conceptual framework

According to the theory of planned behaviour (TPB; Ajzen, 1985; 1991), behavioural intention is determined by attitude, subjective norm and perceived behavioural control. Attitude is the psychological tendency that is expressed by evaluating a particular entity with some degree of favour or disfavour (Eagly and Chaiken, 1995). Subjective norm assesses on the one hand the social pressure on individuals to perform or not to perform certain behaviour and on the other hand, the personal feelings of moral obligation or responsibility to perform or refuse to perform certain behaviour. Perceived behavioural control is described as perceptions of the extent to which the behaviour is considered to be controllable. It assesses the degree to which people perceive that they actually have control over enacting the behaviour of interest (Liou and Contento, 2001). Control factors such as perceived availability may facilitate or inhibit the performance of behaviour (Conner and Armitage, 1998; Verbeke and Lopez, 2005). In addition, Conner and Armitage (1998) suggest incorporating habit measuring the degree of automaticy of ones behaviour.
Self-identity and dietary acculturation are added to investigate whether consumers who are high versus low dietary acculturated or with a high versus low Muslim self-identity rely on other individual characteristics to make their meat choice decisions.
Self-identity can be interpreted as a label that people use to describe themselves. It reflects the extent to which an actor sees him- or herself as fulfilling the criteria for any societal role, for example "someone who is concerned with green issues" (Sparks and Shepherd, 1992). The expectation is that the lower the self-identification with Islam, the stronger will be the role of individual factors like personal attitude and perceived behavioural control.
Dietary acculturation refers to the process that occurs when members of a minority group adopt the eating pattern or food choices of the host country (Negy and Woods, 1992). We hypothesise that the predictiveness for behavioural intention improves with the degree of dietary acculturation in the host culture. The resulting conceptual framework is presented in Figure 1.

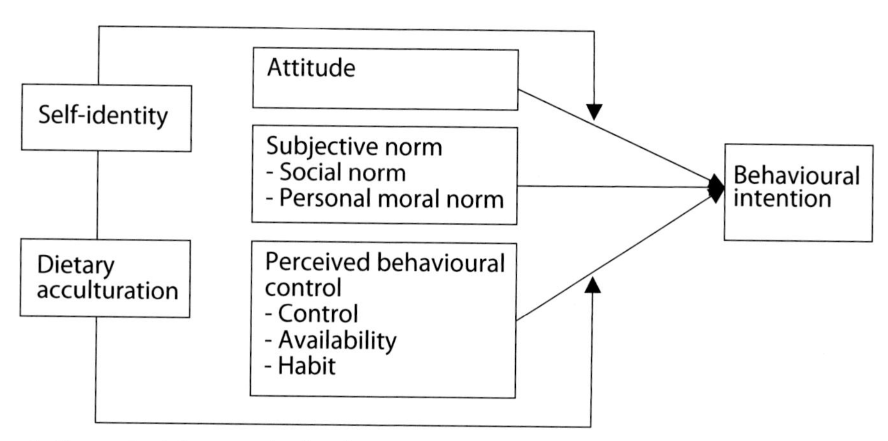

Figure 1. Conceptual framework: the theory of planned behaviour with application to halal meat consumption.

Research method

Survey data were collected through questionnaires in France during a yearly meeting of Muslims (22ième rencontre annuel des musulmans de France) in Paris organised by the UOIF (Union of French Islamic organisations) between the 25 and the 28 of March 2005. For interviewing participants, respondents were selected based on convenience or the judgments of the researchers and poll-takers (convenience sample), and were then invited to complete the self-administered questionnaire taking 20-40 minutes. In total, more than 600 surveys were completed of which 576 were valid for analysis. The characteristics of the respondents show that slightly more men (53.1%) than women (46.9%) completed the survey. With respect to age, our sample consisted mainly of younger respondents (73.2% <35 years) (M= 31.1 years; SD=10.1) and more single (54.4%) completed the survey compared to married or living together respondents (40.3%). Respondents originated mainly from Morocco (37.7%) and Algeria (32.6%), the two main migration populations in France. First generation, those born abroad, compose of 47% of the sample and second (those born in France or who came at or before the age of 6) or third generation account for 52%.

Results and discussion

The Cronbach's alpha for social norms and availability was respectively .91 and .80 and only one factor was extracted for both concepts individually. Behavioural intention, attitude, personal moral norm, control and dietary acculturation were measured using 1 item.
Correlation analyses are presented in Table 1. They show a significant positive relation between intention and attitude, social norms and control. Attitudes are positively correlated with social norms and control. A positive significant correlation is, furthermore, found between social norm and personal moral norm on one hand and habit on the other hand. In addition, availability is positively correlated with the perception of control.
With respect to dietary acculturation, we found that Muslims tend to retain their dietary behaviour (acculturation score: M= 1.77) although the sample has an average stay in France of 20.5 years (SD= 11.54) and 53% of the sample is born in France. This finding corroborates with Liou and Contento (2001) who have found a low average acculturation among Chinese Americans and

Table 1. Mean scores, Standard deviations and correlations.

	M	SD	1	2	3	4	5	6	7	8
1. Intention	3.61	1.49	-							
2. Attitude	3.76	.659	.209**	-						
3. Social norms	3.62	1.49	.097*	.095*	-					
4. Personal moral norm	3.19	1.41	.053	.068	.036	-				
5. Control	2.93	1.26	.120**	.212**	.039	-.018	-			
6. Availability	2.27	1.09	.009	.042	.047	-.049	.175**	-		
7. Habit	1.88	1.78	.005	-.059	.097*	.282**	-.057	.034	-	
8. Self-Identity	3.83	.505	.017	.259**	.063	-.003	.101*	.015	.008	-
9. Dietary acculturation	1.77	1.04	-.099*	-.218**	-.146**	.034	-.108*	.011	.041	-.053

* < .05; ** < .01

by Park *et al.* (2003) who reported that food habits may change most slowly when individuals migrate to other cultures.
Self-identity is positively correlated with the importance attached to halal meat on one hand and the perception of control on the other hand.
Finally, some interesting correlations are found between the TPB concepts and time of residence or the generation the individual belongs to. Previous research suggests that consuming halal meat is habitual behaviour for second generation Muslims (Bonne and Verbeke, 2006; Bergeaud-Blackler and Bonne, 2006). In this study, a significant correlation was found between generation and habit ($r=.129$, $p<.01$) and supported by an independent samples' t-test resulting in the second and third generation Muslims to rely more on habit when buying halal meat then first generation ($t(576)=4,13$, $p<.001$). Muslims born in France are also more convinced that eating halal meat is a personal choice ($t(576)=5.38$, $p<.001$). Time of residence and acculturation are only slightly correlated ($r=.094$, $p<0.05$) confirming Verbeke and Lopez (2005).

Determinants of behavioural intention

In the original TPB model (Table 2), attitude towards halal meat; social norm and perceived control are significant predictors of intention, while personal moral norm, availability and habit are not significant. In contrast with Verbeke and Lopez (2005) who found that lack of ethnic ingredients is a barrier for Hispanics in Belgium to retain their eating habits, availability of halal meat is no barrier for Muslims to consume meat. Vermeir and Verbeke (2006) argue that a consumer who is highly involved with a product, as is the case for halal meat consumers, is less interested in the availability of a product as they are willing to invest more effort to obtain their product. In addition, habit seems to have no influence on intention to eat halal meat. Probably, consumers consider halal meat consumption as a reasoned process which can explain the nonsignificance of our automatic processing item (i.e. habit). A highly important product is usually attained through a reasoned buying process (Vermeir and Verbeke, 2005).
Next, we analysed the possible differential predictive value of the TBP components depending on the level of individual's dietary acculturation and self-identity (Table 2). After classifying respondents in different dietary acculturation and self-identity categories using median split (low, high), results showed that only attitude towards halal meat predict intentions for low dietary acculturated consumers ($R^2=.027$), while high acculturated consumers' intentions are predicted by control in addition to attitude ($R^2=.113$). Muslims retaining their dietary habits are solely influenced by the importance attached to halal meat contrary to the more acculturated

Table 2. Multiple stepwise regression predicting intention to consume halal meat, total sample (n=576) and for different levels of dietary acculturation and self-identity.

			Dietary acculturation				Self-Identity			
			Low		High		Low		High	
	R^2 .062		R^2 .027		R^2 .113		R^2 .236		R^2 .045	
	α	P value	β	P value	β	P value	β	P value	β	P value
Attitude	.182	<.05	.108	<.05	.209	<.01	.330	<.01	.151	<.05
Subjective norm										
• *Social norm*	.080	<.1	Ns		Ns		Ns		.083	<.1
• *Personal moral norm*	Ns		Ns		Ns	<.05	.295	<.05	Ns	
Perceived behavioural control										
• *Control*	.091	<.05	Ns		.152	<.05	Ns		.088	<.1
• *Availability*	Ns		Ns		Ns		Ns		Ns	
• *Habit*	Ns		Ns		Ns		Ns		Ns	

Muslims who's intention to consume halal meat are also determined by the control they think to have over the effective consumption of halal meat and the importance of halal meat. Subjective norm did not influence behavioural intentions for both high and low acculturated consumers thereby contradicting the findings of Liou and Contento (2001).
Both consumers who consider themselves to be more ($R^2 = .045$) or less ($R^2 = .236$) Muslim are guided by their importance attached to halal meat when intending to consume halal meat. Furthermore, consumers who consider themselves less as being a Muslim, believe that their consumption pattern is a matter of personal choice, while consumer with a higher Muslim identity consider more the opinion of other important persons and institutions (in addition to their feeling of control). More (vs. less) 'religious' consumers are more sensitive to the norms and rules prescribed by their religion, while less (vs. more) 'religious' consumers make more 'egocentric' (i.e. considering one's own opinion instead of other one's opinion) consumption decisions.
We can conclude that the more individuals are acculturated and the less they identify with Islam, the more the Theory of Planned Behaviour becomes valid for Muslim consumers in France.

Conclusion

The purpose of this study was twofold. First, the Theory of Planned Behaviour was used to investigate halal food consumption in France. In general, a positive attitude towards the importance of halal meat, the influence of peers and the perceived control over consuming halal meat predict the intended consumption of halal meat among Muslims. Possible low availability of halal meat does not seem to be a barrier for consuming halal meat; neither is intention strongly determined by habit.
The consumption of halal meat for Muslims is very different from the consumption of 'regular' meat for non-Muslims. The religious associations attached to halal meat probably make this decision more important for the Muslim consumer, which could lead to a different consumption process. Therefore, and secondly, we added two characteristics that are associated with religious

food decisions within a migration context (i.e. food acculturation and self-identity) that could help us understand the concept of religious food decisions. When considering the degree of dietary acculturation, low acculturated Muslims rely completely on their positive attitude towards halal meat whereas high dietary acculturated Muslims rely on attitude and perceived control when intending to consume halal meat. Finally, Muslims with a low Muslim self-identity intent to eat halal meat not only because they find it very important but also as a personal choice whereas Muslims with a high Muslim self-identity are rather influenced by peers, their attitude towards halal meat and the degree of control they think to have over eating halal meat. Future research could investigate other individual characteristics, such as trust and moral obligation, influencing religious food decisions.

References

Ababou, M. (2005). The impact of age, generation and sex variables on religious beliefs and practices in Morocco. *Social Compass,* 52 (1): 31-44.

Ajzen, I. (1985). From intention to action: a theory of planned behaviour. In Kuhl, J. and Beckman, J. (eds.), *Action control: from cognition to behaviour*. Springer, New-York. 11-39.

Ajzen, I. (1991). The theory of planned behaviour. *Organizational Behaviour and Human Decision Processes,* 50: 179 – 211.

Asp, E. H. (1999). Factors influencing food decisions made by individual consumers. *Food Policy,* 24: 287-294.

Bergeaud-Blackler, F. and Bonne, K. (2006). D'une consommation occasionnelle à un régime halal: quelles conséquences sur la santé". *Migrations Santé* (forthcoming).

Bonne, K. and Verbeke, W. (2006). Muslim consumer's attitude towards meat consumption in Belgium: insights from a means-end chain approach. *Anthropology of food* (forthcoming).

Conner, M.T. and Armitage, C.J. (1998). Extending the theory of planned behaviour: a review and avenues for further research. *Journal of Applied Social Psychology,* 28 (15): 1429-1464.

Delener, N. (1994). Religious contrasts in consumer decision behaviour patterns: their dimensions and marketing implications. *European Journal of Marketing,* 28 (5): pp. 36-53.

Dindyal, S. (2003). How personal factors, including culture and ethnicity, affect the choices and selection of food we make. *Internet Journal of Third World Medicine.* 1 (2): 27-33.

Eagly, A.H., and Chaiken, S. (1995). Attitude strength, attitude structure and resistance to change. In Petty, R.E. and Krosnick, J.A. (eds.), *Attitude strength: antecedents and consequences,* Lawrence Erlbaum Associates, Mahwah, pp. 413-432.

Hussaini, M.M. (1993). Halal Haram lists. Why they do not work, available on: http://www.soundvision.com/info/halalhealthy/halal.list.asp (23/08/04) .

Limage, L. J. (2000). Education and Muslim identity: the case of France. *Comparative Education,* 36 (1): 73-94.

Liou, D. and Contento, I.R. (2001). Usefulness of psychosocial theory variables in explaining fat-related dietary behaviour in Chinese Americans: association with degree of acculturation, *Journal of Nutrition Education,* 33 (6): 322-331.

Negy, C. and Woods, D.J. (1992). The importance of acculturation in understanding research with Hispanic Americans. *Hispanic Journal of Behavioural Science,* 14: 224-247.

Park, S., Paik, H.Y. and Ok, S.W. (2003). Mother's acculturation and eating behaviours of Korean families in California. *Journal of Nutrition Education and Behaviour,* 35 (3): 142-147.

Sack, D. (2001), *Whitebread Protestants: Food and religion in American culture,* First Palgrave,
New York. 272 p.

Saint-Blancat, C. (2004). La transmission de l'islam auprès des nouvelles générations de la diaspora, *Social Compass,* 51 (2): 235-247.

Shatenstein, B. and Ghadirian, P. (1997). Influences on diet, health behaviours and their outcome in select ethno cultural and religious groups. *Nutrition*, 14 (2): 223-230.

Sparks, P. and Sheperd, R. (1992). Self-Identity and the theory of planned behaviour: assessing the role of identification with green consumerism. *Social Psychology Quarterly*, 55 (4): 388-399.

Verbeke, W. and Lopez, G. P. (2005). Ethnic food attitudes and behaviour among Belgians and Hispanics living in Belgium. *British Food Journal*, 107 (11): 823-840.

Vermeir, I., and Verbeke, W. (2005), "Sustainable food consumption, involvement, certainty and values: an application of the theory of planned behaviour", *Working paper in the Working paper series of the Faculty of Economics and Business Administration*, Ghent, Belgium

Vermeir, I., and Verbeke W. (2006), "Impact of values, involvement and perceptions on consumer attitudes and intentions towards sustainable consumption", *Journal of Agricultural and Environmental Ethics*, Vol. 19 No. 2, pp.169-194.

Ethics and transformation of Polish food chain

Marek Gaworski
Department of Production Management and Engineering, Warsaw Agricultural University, Nowoursynowska 166, 02-787 Warsaw, Poland, gaworski@alpha.sggw.waw.pl

Abstract

Transition of the Polish economy towards market system has initiated many discussions on criteria taken into account to assess the changes observed in all stages of food chain. One of the mentioned criteria is quality. There are many factors which contribute to the high quality of Polish agricultural products, i.e. clean environment, a considerable number of farms involved in agricultural production based on traditional methods with a relatively low usage of agro chemicals as well as animal production characterized by low intensity. The agricultural and environmental conditions in Poland, as well as biological variety are favourable for organic farming development. But such perfect picture can be disturbed by needs to achieve higher and higher economic efficiency in food chain. As a result there is tendency to develop more industrial methods of agricultural production and processing with development and extension of convenience and functional foods as an example showing follows the global trends. Such situation creates necessity to propagate suitable knowledge and information about effects and consequences resulting from the observed tendencies to enlighten consumers on the proceeded changes and show ethical approach to consumer interests. As an example we have in Poland the Try Fine Food Programme aimed to inform consumers of high quality food products. Concluding, the food chain should be assisted by chain of ethical activities. Moreover, it is proposition to create system of ethical assessment of all stages in food chain after the example of the HACCP (Hazard Analysis and Critical Control Point).

Keywords: ethics, food, HACCP, Poland, transition

Introduction

For sustainable development of the food production system not only material but also immaterial aspects and consequences coming from food economy transformation should be taken into account. The food production system considered globally and as an example covering particular countries can be field of many discussions to look for some solutions, which generate higher and higher consumption safety. The protection of the consumer belongs to the ethics of consumption and deals with the good life of the consumer as mentioned by Coff (2003).

Some data about Poland, Polish agriculture and food sector

Poland is a country located in the central part of Europe, with the total area of 312,700 km^2 and the population of more than 38.2 million. Since May 01, 2004, Poland has been a member of the European Union. In the enlarged European Union (25 countries), Poland covers 7.9% of the total area, and its inhabitants constitute 8.5% of the EU population. Such data indicate high potential of Poland considered from view-point of consumption needs. Poland comes in the 8th position in Europe, and the 30th in the world, as far as the number of people is concerned. Moreover, Poland is characterised by significant potential of agriculture. The area of arable land constitutes about 52% of the total area of the country, i.e. Poland takes the leading position in the EU, just

after France and Spain, with the land area similar to Germany. The soil-climatic conditions are unfavourable, as compared to most of the European countries.
Polish agricultural is characterised by a large diversification of farms including scale of production. The sizes of farm cover areas from one hectare to several thousand hectares.
Moreover, it is necessary to indicate, that rural areas are inhabited by about 38.5% of the total population. We have large number of people working in agriculture and relatively low usage of industrial means of production. Polish villages are characterised by a spread settlement area and as a result high costs of infrastructure and reduced possibilities to take up off-farming activities. The share of people professionally active in agriculture is over 18% in Poland, against only 5% in the European Union.
Poland is the country with great biological diversity and large diversification of regional natural conditions. A considerable group of farmers, especially owners of small area farms is involved in agricultural activities, where traditional production methods are preferred, mainly to ensure food supplies for their own families and needs. In most of the Polish farms animal production is mostly of low intensity, which is not harmful to the surrounded environment. As an example, the average size of dairy cow herds is little more than 3 animals in Poland, i.e. about 10-times less than average value in Europe.

Ethical questions and principles in food sector

Including specific features of Polish agriculture and food sector there is possible to give some questions concerning relationships between food production, consumers and ethics.
The most significant principles in the undertaken activities should be consumer protection and food safety assurance. When consumer and its food are protected it is possible to tell that ethical aspects are included in food chain. But it is necessary to take into account specific conditions find in particular regions and countries. Such specific conditions can be met in Poland. There is high potential of agriculture, high number of inhabitants and as a result high responsibility on the field of protection and assessment of ethical activities.
The important factor considered from view-point of consumer protection is possibility to introduce rules of organic farming. There are many factors which contribute to the high quality of Polish agricultural products, i.e. clean environment, a considerable number of farms involved in agricultural production based on traditional methods with a relatively low usage of agro chemicals as well as animal production characterized by low intensity. High diversity of agriculture and less intensity decide about more friendly contact between farmers and livestock, as suggested by Gaworski *et al.* (2000). The agricultural and environmental conditions in Poland, as well as biological variety are favourable for organic farming development.
And what about food ? Small-size farms have smaller incomes and as a result they buy and use less fertiliser and plant protection chemicals so there are possibilities to produce higher quality plant material as a raw material for food processing.
On the other hand lower incomes in small-size farms, especially with animal production decide about limited possibilities to equip them e.g. dairy farms with such technical implements as cooling machines, which guarantee keeping the proper quality of milk. Reduced possibilities to keep rules of cooling chain and high quality of milk can be considered as source of risk for consumers and lack of ethical approach to food safety assurance.
Such perfect picture can be disturbed by needs to achieve higher and higher economic efficiency in food chain. As a result there is tendency to develop more industrial methods of agricultural production and processing with development and extension of convenience and functional foods as an example showing follows the global trends.
Such situation creates necessity to propagate suitable knowledge and information about effects and consequences resulting from the observed tendencies to enlighten consumers on the

proceeded changes and show ethical approach to consumer interests. As an example we have in Poland the Try Fine Food Programme aimed to inform consumers of high quality food products. It is only the producers who have the right to apply for the Try Fine Food quality mark and the application is free of charge.

Ethical questions and participants of food production process

It is necessary to indicate that more modern technical equipment can help the farmers and those employed in food processing sector to increase productivity, but progressive automation of agricultural processes is accompanied by such important changes as:

- decrease in number of workplaces and as a result increase in unemployment, depopulation of villages and weakening of human interrelationships;
- decrease in physical load of farmers, but unfortunately increase in their psychical load.

When decrease in physical load of man is observed with higher and higher automation there are needs for better qualifications of workers and higher responsibility for the more expensive and high effective technical systems. Thus, an increase in psychical load decides on the lower psychical comfort of workers and possible health problems, stress, etc.

Replacement of labour with modern technical infrastructure creates ethical problems and conflicts in the field of agrifood system, resulting from lost of work places and other ones.

To solve the considered problems and ensure sustainable development of society and surrounded environment it is suggested that according to Aristotle's doctrine the food production should aim at a "golden mean" (Gaworski and Nowacki 2004). Moreover, including the interests of current and future generations it seems to be important to create interdisciplinary groups of specialists to intensify economic and social studies, to indicate ways of most effective and sustainable development of society, surrounded environment and improvement of world food safety, what is actively and objectively undertaken by the EurSafe members, as mentioned by Jensen and Sandøe (2001).

Ethical questions and possibilities to supervise food production process

There are many discussions how to develop and explain ethical problems in agriculture and food production. It is possible to indicate various approaches to ethical aspects and connect them with technological and technical view-points, social (consumer) requirements, environmental needs and other ones.

It seems to be important in order to determine the key targets concerning development and assessment of food chain in connexion with ethics. When we consider interests of consumers and environment, it is necessary first of all to take into account aspects of food safety assurance. According to the EU Directive on Food Hygiene (93/43/EEC) (Article 3, no. 2), food processors should establish a system for food safety assurance based on the principles of HACCP.

The HACCP (Hazard Analysis and Critical Control Points) is the tool of quality assurance system and should be a part of any quality system in each activity undertaken on the field of food production. With the HACCP system, as a tool to assess hazards it is possible to identify specific hazards and preventative measures for their control to ensure the food safety. It is important, that HACCP system can be applied in whole food chain from the level of raw material production to final consumer. The emphasis is put on prevention of identifiable hazard rather than on removal of contaminated products. Some plans concerning HACCP procedure should be based on risk assessment principles and activities, which need clear definitions.

In typical procedure of the HACCP analysis there is possible to give many detailed questions, like:

- are microbiological standards fulfilled on all steps of food production chain?
- are suitable hygiene procedures included in the undertaken activities, which lead to preparing food products? etc.

At the same time there is possible put the following, similar to the given above question:

- are ethical standards fulfilled on all steps of food production chain?

But what are the ethical standards in such considerations? In order to find the answer it is necessary to analyse the food production chain and determine steps, where is possible to connect consequences resulting from disregard of some quality standards with their ethical assessment. But how to measure such consequences? It can be measured by level of hazard. After hazard analysis, critical ethical points (CEPs) can be identified. CEPs are activities, where ethical analysis can be exercised to achieve a quantifiable reduction in a hazard or its stabilization, which can lead to acceptable safe food product and satisfaction and safety of consumer.

References

Coff C. (2003). *Consumers ford ethics: tracing the production story*. Proceedings of the 4th Congress of the European Society for Agricultural and Food Ethics "EurSafe 2003", Toulouse, France, p. 54-56.

Gaworski, M. Hutnik, E. and Nowacki, T. (2000). *A study of the transformation of the agri-system and of the creation of an ethical farming model.* Proceedings of the 2nd Congress of the European Society for Agricultural and Food Ethics "EurSafe 2000", Copenhagen, Denmark, p. 115-120.

Gaworski, M. and Nowacki, T. (2004). *Man and engineering: source of ethical conflicts ?* Proceedings of the 5th Congress of the European Society for Agricultural and Food Ethics "EurSafe 2004", Leuven, Belgium, p. 341.

Jensen K.K. and Sandøe P. (2001). *Food safety and ethics: the interplay between science and values.* Proceedings of the 3rd Congress of the European Society for Agricultural and Food Ethics "EurSafe 2001", Florence, Italy, p. 15-20.

Innocent food and dirty money: The moral economy of food production[70]

Stig S. Gezelius
Norwegian Agricultural Economics Research Institute (NILF), P.O. Box 8024 Dep., NO-0030 Oslo, Norway, stig.gezelius@nilf.no

Abstract

A study of North Atlantic fishers revealed that the morality of compliance in commercial fisheries differed greatly from that which could be observed in subsistence fisheries. While violations of government regulations were generally accepted in food fisheries, they were met with moral unease or informal sanctions in commercial fisheries. These findings are discussed in relation to the moral meanings of money and food, and subsequently accounted for in the light of a theory of two moral spheres of economic activity. These spheres constitute distinct frameworks for moral interpretation, and represent highly-different conditions for interference by the state.

Keywords: food, money, morality, compliance

Introduction

A comparative study of compliance in fishing communities in Norway and Newfoundland revealed a striking moral distinction between commercial and household fishing (Gezelius, 2004). The fisher families of these communities had traditionally fished for subsistence as well as for the market. Their fishing activities had become strictly regulated by the government, and there were significant incentives to violate regulations. The communities were intimate and transparent, and exercised efficient social control over their members. The ever-present surveillance of the community represented, in most cases, a much more efficient control than the occasional visits of the state's fish inspectors. Consequently, the fishermen's actions in terms of compliance largely corresponded to the moral norms of their communities.
One of the most striking findings in this study was that the moral vigilance of the fishing community proved much greater in relation to commercial fishing than in relation to subsistence harvesting. This paper attempts to answer two questions: why is commercial harvesting more morally-sensitive than harvesting for subsistence, and what does this entail for the state's attempt to regulate food production?

The findings in the Newfoundland case

The cod fisheries of eastern Newfoundland were closed in 1992, following many years of overfishing. The stocks have not recovered since, and the moratoria were still in effect when I carried out my fieldwork in 1998. The cod moratorium also included food fisheries, and Canadian authorities enforced this ban strictly. Poaching implied a genuine risk of being prosecuted and fined. It was almost unanimously agreed among fishermen that the stock was vulnerable and that extreme caution, and thus strict management, was necessary.

[70] This paper is an abstract of data and ideas outlined in an article by the author, entitled 'Food, Money, and Morals: Compliance among Natural Resource Harvesters', Human Ecology 32(5), 2004.

The Newfoundland community was located in an area with great density of cod, which represented a constant incentive for poaching. However, poaching for the purpose of sale was strongly condemned among the fishers. Unregulated commercial fishing was seen as incompatible with resource conservation, people who poached for money were seen as free riders on a collective sacrifice to rebuild the stock. A government income support programme ensured that the fishers maintained their material standard of living, and poaching for the purpose of money was perceived as being motivated by greed. The fishermen claimed that their continued compliance depended on the viable household economy ensured by this income support.

Those who poached for money were also met with a significant amount of fear. There were rumours of threats and fear of physical acts of vengeance following a blow of the whistle. The villagers avoided social contact with commercial poachers due to a concern for their own personal reputations. Commercial poacher thus emerged as criminals excluded from the larger collectivity of fishers.

However, the moral requirement for compliance did not include food fisheries. In contrast to commercial fishing, people believed that there should have been a free fishery for food, as this could not pose a threat to the stock. Food was associated with small-scale fishing, while money was associated with harvesting on a large scale. In contrast to commercial poachers, people who poaching for food were not met with informal sanctions but rather with sympathy and understanding. Consequently, household poaching could take place fairly openly, and poached cod could for instance be exchanged as gifts among villagers. While the villagers requested more enforcement in terms of commercial poaching, they reacted with indignation and fury when household poachers were arrested and fined.

The findings in the Norwegian case

The Norwegian community was largely dependent on local saithe fisheries, and a series of seasonal closures created problems for many vessels in this community when I carried out the fieldwork in 1997. Most vessels had managed to secure a reasonably good season when the fishery closed, and some vessels also had the opportunity to fish for pelagic species. The boats thus ceased fishing saithe and went into pelagic fisheries or stayed by the wharf. It was also clear that they were morally expected to do so. Breaking the law in situations that could not be classified as *force majeure* was clearly not accepted. Fishermen who were perceived as breaking rules "on a large scale" in order to maximise personal profit became objects of back-biting and social degradation. However, a few vessels were particularly severely affected as they had had poor seasons and lacked acceptable alternative fishing possibilities. A couple of these boats continued fishing for groundfish. The general understanding in the community was that these boats concealed illegal catches of saithe through falsification of sales notes.

The most striking aspect of the data was that these fishermen did not become objects of significant moral blame among their law-abiding colleagues. Their behaviour generated moral unease, doubt and some conflict in the community, and this moral ambiguity left the collectivity incapable of generating efficient sanctions. This moral ambiguity was rooted in the conflict between two moral norms: a perceived moral obligation to obey the law, and a perceived right to make a reasonable living from fishing. The moral "grey zone" which emerged in the conflict between these two moral norms, created moral elbowroom for some degree of illegal adaptation by vessels experiencing financial hardships.

The moral vigilance connected with commercial fishing, could not be found in food fisheries. Drift net fishing for salmon has been illegal in Norway since 1989, and affected several fishermen in this community. Illegal salmon fisheries for the purpose of sale are not accepted among the fishermen, but salmon poaching for the purpose of food is quite another matter. In contrast to violations in commercial fisheries, illegal food fishery is not experienced as controversial, and is not met with discussions or gossip. Several fishermen occasionally poach salmon for food, and these activities are not connected with extensive secrecy.

The moral meanings of food and money

The morality of compliance in the two communities can be viewed as a line stretching from food to excessive monetary income, as outlined in Figure 1. Violations of fisheries law gradually become more problematic as they approach the domain of large-scale commerce. In the following, we will address three factors that might explain this: the relationship between production scale and conservation of common pool resources; the morality of exchange; and historical ideological bases.

Production scale and scarce common pool resources

The bodily need for food can be regarded as naturally limited. When the resource harvester is satisfied and safe in terms of food supply, it is not rational to continue increasing the scale of subsistence-oriented resource harvesting. A subsistence economy may thus be experienced as putting a natural limit to the scale of resource harvesting activities, and thus as compatible with the conservation of scarce common pool resources. The Newfoundland villagers thus did not see illegal food fishery as a threat to the stock.

By contrast, the exchange possibilities accompanying money enable the actor to pursue an almost unlimited range of goals. Consequently, money sets no natural limit as to when further extraction of common pool resources ceases to be rational. Commercial fishing, in contrast to subsistence harvesting, may thus be associated with the logic of the herdsmen in Garret Hardin's commons (Hardin, 1968). It is consequently subjected to moral norms requiring that regulations concerning the extraction of common pool resources be complied with.

The morality of exchange

Exchange of specific goods, such as food, requires complementary needs. It is therefore most suited in settings where goods can be transformed into long-term obligation or love and respect (see e.g. Faris, 1972). Such exchange consequently requires trust and durable social ties. Subsistence production is hence generally associated with exchange among close relations. It may therefore be associated with strict social control and a wide range of moral responsibilities.

Money allows exchanges to be settled immediately, and thus only requires a temporary relationship and a modest level of trust. Consequently, money allows exchange with strangers, and thus the looser networks of the market. This may be experienced as reducing the level of social control and range of moral responsibilities involved.

The nature of money-based exchange has turned money into a symbol of strangeness. For example, using money as a medium of exchange among friends is easily experienced as offending. Strangeness as such is associated with moral danger (Hogg and Abrams, 1988; Tajfel, 1978). Money thus becomes a symbol of the moral peril associated with stranger relationships. By contrast, exchange of food symbolises intimacy and thus builds emotional ties. The cultural

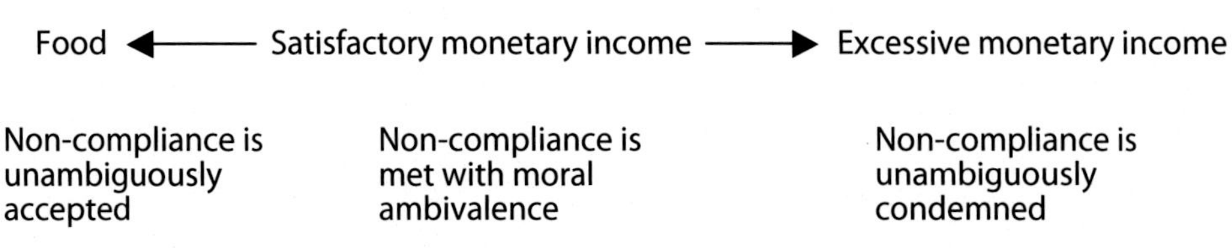

Figure 1. The morality of compliance.

barriers against using money as a means of exchange among close relations entails that economic activities restricted to such relations are usually informal, and detached from the state.

Ideological foundations

The fact that commerce puts no natural limit to the rational scale of resource extraction, means that money may be perceived as a potential medium through which greed can unfold as economic action. Greed is not only associated with potential harm to the common good, but has also been viewed historically as a reprehensible motive in its own right. The idea that desire for material wealth corrupts the human soul pervades western moral and religious thinking. It can be found in the works of Plato, which inspired Augustin and thus also influenced the Christian Church (Doyle, 1999; Plato, 1985, Book VIII, pp. 549-551). The Bible quite consistently holds that greed is a sin and worshiping of false gods. Moderation is seen as a moral virtue (Eph. 5:5; Col. 3:5; Isa. 56:11; Luke 12:15; Prov. 1:19; Prov. 15:27; Rotter, 1979).

The idea that money and trade are associated with greed is equally as old. Aristotle saw money as connected with unlimited desire, while idealising subsistence production. This view influenced the Church through Thomas Aquinas, and can also be found in several famous passages in the Bible (Aristotle, 1981; Bloch and Parry, 1989; Parry, 1989; I Timothy 6:8-10; Luke 16: 13). "Mammon" is commonly used as a synonym for money, while it originally refers to the devil of covetousness or the false god of greed (Harris, 1989). It is well known that certain Christian movements developed widespread acceptance of trade and profit- making in the early days of capitalism, and no one would dispute that trade is a morally legitimate enterprise today (Tawney, 1936; Weber, 1930). However, the legacy of moral thinking described above has arguably left ideological traces, in the form a certain moral vigilance associated with money, in modern secularised societies (see also Bloch and Parry, 1989).

Moral spheres of economic activity

The latent moral impurity of money only becomes manifest in relation to acts that are morally ambiguous or immoral at the outset. If someone breaks a rule, he becomes an object of particularly strong abhorrence if his motive is money. By contrast, if he breaks the same rule for the purpose of food, he might very well be excused. It can be argued that this is the dynamic of morality we have seen in relation to poaching in our two communities.

We have thereby suggested that the morality of compliance observed among Newfoundland and Norwegian fishers form part of a much more widespread moral pattern. In the following, I will attempt to account for the empirical findings by suggesting the existence of two moral spheres of economic activity in economies based on harvesting of natural resources for both household and the market.

One of these spheres is perceived as morally safe ground, where activities are met with a moral "green light". We may thus refer to it as the "green sphere" of economic activity. This sphere is linked to subsistence harvesting, small-scale operations and the provision of necessities. The other sphere is perceived as morally dangerous. Activities in this sphere are not seen as immanently immoral, but they entail a danger of immorality which requires moral caution. This sphere will thus be referred to as the "yellow sphere" of economic activity. The yellow sphere is connected with commerce, large-scale operations and search for maximum profit. The two spheres constitute a system for the moral classification of acts. This system is outlined in Table 1.

In the Norwegian community, we saw that regulations were sometimes violated in commercial fisheries without triggering informal sanctions, although they were met with significant uneasiness. One the one hand, these violations were committed in order to ensure a monetary outcome, which is a hallmark of the yellow sphere. On the other hand, they only aimed to

Table 1. The moral spheres of economic activity.

Green sphere	Yellow sphere
Production for subsistence. Food.	Production for the market. Money.
Exchange among close relations.	Exchange with strangers.
Naturally focused on basic needs. Temperance.	Allows focus on infinite wants. Greed. Potential for spiritual corruption.
Small-scale production.	Large-scale production.
Moderate use of common pool resources.	Potential depletion of common pool resources.
Morally safe.	Morally dangerous.
State regulation perceived as inappropriate. Non-compliance.	State regulation perceived as appropriate. Compliance.

ensure a moderate and necessary outcome, which is a hallmark of the green sphere. Moreover, these violations were committed by small inshore vessels. The Norwegian inshore fishing fleet is accompanied by a large offshore fleet, and commercial inshore fisheries thus emerge as being of a comparatively small scale, and consequently as only a moderate case of yellow sphere activity. The unsanctioned violations observed in the Norwegian community thus contained elements of both spheres. Therefore, these illegal incidents triggered moral doubt and uneasiness rather than unambiguous condemnation.
Therefore, classification according to the two moral spheres may not be made in a strictly dichotomous manner. It can often emerge as placement along a continuum where activities are perceived as more or less yellow or green according to the number and type of criteria they fulfil and the extent to which they fulfil them. We may thus hypothesise that the collectivity's moral vigilance increases as the economic activity gets more yellow. We can also hypothesise that state interference is less likely to be accepted among food producers the greener their activities are perceived to be.

References

(1965). *Holy Bible.* King James (Authorized Version). The British and Foreign Bible Society, Wellington.

Aristotle. (1981). *The Politics.* Penguin Books, London. 319 p.

Bloch, M., and Parry, J. (1989). Introduction: Money and the morality of exchange. In: *Money and the Morality of Exchange.* Bloch, M., and Parry, J. (eds.). Cambridge University Press, Cambridge, pp. 1-32.

Doyle, K. O. (1999). The Social Meanings of Money and Property: In Search of a Talisman. SAGE Publications, Thousand Oaks.

Faris, J. C. (1972). Cat Harbour: A Newfoundland fishing settlement. ISER, St. John's. 184 p.

Gezelius, Stig S. (2004). Food, Money and Morals - Compliance among Natural Resource Harvesters. *Human Ecology,* 32(5): 615-634.

Hardin, G. (1968). The Tragedy of the Commons. *Science* :1243-1248.

Harris, O. (1989). The earth and the state. In: *Money and the Morality of Exchange.* Bloch, M., and Parry, J. (eds.). Cambridge University Press, Cambridge. pp. 232-268.

Hogg, M. A., and Abrams, D. (1988). Social Identifications: A Social Psychology of Intergroup Relations and Group Processes. Routledge, London. 268 p.

Parry, J. (1989). On the moral perils of exchange. In: *Money and the Morality of Exchange.* Bloch, M., and Parry, J. (eds.). Cambridge University Press, Cambridge. pp. 64-93.

Plato (1985). *The Republic,* W. W. Norton & Company, New York. 317 p.

Rotter, H. (1979). Moderation. In: *Consise Dictionary of Christian Ethics.* Stoeckle, B. (ed.). Burns & Oates, London, pp. 184-185.

Tajfel, H. (1978). Social Categorization, Social Identity and Social Comparison. In Tajfel, H. (ed.), *Differentiation between Social Groups*, Academic Press, London, pp. 61-76.

Tawney, R. H. (1936). *Religion and the Rise of Capitalism*, John Murray, London.

Weber, M. (1930). The Protestant Ethic and the Spirit of Capitalism, Unwin Hyman, London.

The cultural taste of *Gamalost fra Vik*: Reflections about the meanings of a local cheese

Atle Wehn Hegnes, Virginie Amilien and Hanne Torjusen
SIFO, National Institute for Consumer Research, Sandakerveien 24 C, Bygg B, Box 4682 Nydalen, 0405 Oslo, Norway, atle.hegnes@sifo.no

Abstract

This paper focus on *Gamalost fra Vik* - the old Norwegian cheese - as a natural, cultural, industrial, economic and political food product in Norway. After a contextual presentation of the case study describing the cheese in a field of tensions, we outline the way dimensions and paradoxes are embedded in the comprehensions related to this product. The paper is based on qualitative interviews with consumers, producers and marketing actors in 2005, as well as a public discourse analysis around *Gamalost fra Vik*. Our main research questions are: How do consumers experience *Gamalost fra Vik*? What are possible ethical implications of discrepancies between conceptions and presentations of the cheese by different actors? By exploring the distance between the actual product consumers purchase and eat, and the associations arising from eating it, our findings indicate that *Gamalost fra Vik* is understood in diverse ways by different consumers. Different consumers also tend to use different strategies in understanding and appropriating Gamalost. Whereas participants in the group with "enthusiasts" use active strategies to achieve their preferred taste and experience, such as physical modifications of the product and "mental techniques", "ordinary" consumers have other expectations and typically apply a defensive strategy of not buying the product because of its reputation. After exploring and analyzing the meaning of *Gamalost fra Vik*, we end by questioning and problematising the ethical implications of discrepancies between presentations and conceptions.

Keywords: *Gamalost fra Vik*, consumer conception, marketing, self-seduction, ethics

Introduction

The intention of this paper is to observe the conceptual dimensions that a traditional Norwegian cheese is built on and then to explore how these dimensions are perceived by the consumers. To better understand our cheese product, we conducted interviews with different market actors (such as producers, milk industry and marketing professionals) and consumers. Discussions with these groups of actors emphasize the ethical implications of the different dimensions of the product, and enable us to explore whether there are any conspicuous mismatches between consumers` ideas of what they are eating and the product they actually are consuming.
Gamalost (old cheese) is a traditional product primarily known from the western part of Norway. Initially the cheese was made at mountain farms, but whereas the Norwegian dairy system was rationalised the last century, Gamalost production left the hand made mountain tradition to a more industrial, competitive and centralized fabrication. The cheese is brownish, has a sharp flavour and is today ripened by the mould specie *Mucor mucedo*. Traditionally the cheese used to contain a variety of moulds, often including some of the *Penicillium* type, which gave the cheeses a blue or greenish colour in the centre (Fosså 2000). Despite its name, Gamalost is ready for consumption after a couple of weeks. Nowadays the annual production is about 120 tonnes, compared to 500 tonnes in the 1970's. 96% of the Gamalost is distributed through grocery stores (Brendehaug and Groven 2004). Trying to differentiate the product and increase the demand

and interest for Gamalost, the Vik dairy and its owner, Tine (the Norwegian Dairy Cooperative), together with the municipality of Vik, applied for PGI[71] in 2005.
Gamalost is surrounded by myths. Several myths were established before Gamalost became an industrialized product, but in the aftermath these myths are now also transferred to the modern product. Some examples: If Gamalost was ripened in the bed straw of a (dairy made) virgin this would give a divine taste and also improve sexual function for males eating this cheese (Moberg 1997). Although the tradition of Gamalost being made at the mountain farms by dairymaids is almost gone, the myth which attributes the Gamalost to improve sexual function has been transferred to the industrially made product. Gamalost is also believed to have medical and healing properties, as well as adverse effects: *you could loose your hair if you ate it.*
Although the sale has decreased during the last years *Gamalost fra Vik* has for several reasons received increased attention from scholars. Among the factors contributing to an extensive attention to a relatively marginal product are the mobilisation for the future existence of the local dairy in Vik through "The Gamalost Project"; new agricultural policy with increased focus on local food in Norway and by Tine; and application for PGI labelling of the Gamalost . Although there has been a rising focus on Gamalost in terms of research and documentation, most of those documents are of ethnological or historical character: they focus on the history (Fosså 2000, Notaker 2000) the myths (Moberg 1997), they describe the traditional way of production (Sæbø 2003), they document the industrial turn of production (Bergfjord 1997), and the present status and potential as well as challenges for increasing the sales (Brendehaug and Groven 2004). Although several of these contributions acknowledge the marginal position of Gamalost, few attempts have been made to provide knowledge about how consumers perceive this product.
This paper is based mainly on qualitative interviews with a semi-structured interview guide, face to face interviews with market actors and focus groups with consumers carried out in 2005, as well as a public discourse analysis and historical sources. Four focus groups (each including 6-10 persons) were carried out. Two groups of respondents were recruited from Vik, the village to which the Gamalost application for geographical indication is linked, and two from the capital - Oslo. Two kinds of focus groups were carried out at each location: One with people who had a special interest in Gamalost (enthusiasts) and one with people who, as far as we know, had no special knowledge about or special interest in Gamalost (ordinary consumers). We chose people with different, age, sex, family situation etc. although the group with enthusiasts had a higher than average age. In all, nearly 30 persons participated in the focus groups. Every focus group started with tasting Gamalost, and later other cheeses, in order to direct the focus towards taste and their own impressions about a concrete product.

Consumers conceptions of *Gamalost fra Vik*

It is a paradox that both the history, representing the progress towards a standardised product and stable quality, as well as the myths connected to the product constitutes the reputation or experience of Gamalost simultaneously. How do the consumers we spoke with apprehend, and consider, this duality?
The term Gamalost is understood in many different ways and gives raise to different connotations. A duality became evident during our talks with our informants. Whereas some immediately associate Gamalost from Vik to the dairy in Vik, others refer to a more imaginary Gamalost they "remember" from before. These differences appear both at a collective and individual level: the same person could regard Gamalost as a modern and traditional product at the

[71] In the case of the PGI (Protected Geographical Indication) the geographical link must occur in at least one of the stages of production, processing or preparation. Furthermore, the product can benefit from a good reputation. (http://europa.eu.int/comm/agriculture/foodqual/quali1_en.htm)

same time. To bypass this ambivalence, participants in the "enthusiast" groups use different strategies actively to achieve their preferred taste and experience. Those strategies can involve physical modifications of the product to achieve the perfect taste and experience, while others follow a "personal" myth, linked to nostalgia and individual apprehension of the product. The physical modification is often done by a nostalgic consumer who wishes to re-create a taste by transforming the cheese with time or ingredients, like herbs or alcohol. The mental adaptation consists of "bringing" the cheese out in an ideal situation, up in the mountain or on a fishing trip. These differences in strategies may be ascribed to the level of knowledge consumers have about the product. Enthusiasts often have thorough familiarity with the traditional Gamalost or *Gamalost fra Vik*. Trying to achieve the original taste they have to physically moderate the cheese according to how they imagine or remember the traditional cheese tasted like. More ordinary consumers do not have the same reference to the traditional product. Neither do they have a specific strategy for transforming the product. Their knowledge is built on a collective myth, based on rumours and old stories which emphasize the bad smell and the magical aspect of Gamalost.
An informant[72] – who claimed that *"a day without Gamalost is a bad day"* - describes with great enthusiasm the "old" Gamalost – which he learned to like at a young age during the summers in the mountains of the West coast. After having traced down private producers in the past, he has resigned to the fact that it is now hard to come across any other cheese than the one produced by the dairy in Vik – the "TINE cheese". He is not very fond of this cheese as compared to the traditional ones:

> *"This is not Gammelost. It is a hygienic industrial product"* (...) *"this is a hygienic product made not to scare the bacteria controllers and such. So it is in fact very "anti cheese", really."*

But he uses an offensive strategy: Rather than being deceived he uses the industrialized product actively to achieve what to him represents the perfect Gamalost. He treats it with much affection at home (letting it age, adding wine or coffee etc.), and eats it while dreaming of the "original" Gamalost from the mountain farms:

> *"I buy the one from Vik in Sogn, but I keep it and then, with a little butter you might in a way get rid of the worst "Vik in Sogn" taste and then you can dream about Gamalost "*

There is no doubt that he is totally aware of the difference between the product he eats and the cheese he dreams about while eating it. There is no deception, but rather a kind of conscious visualising in order to increase his pleasure from the cheese. The duality of the product is almost an added value, because he can play with and appropriate the food he bought.
The ordinary consumers, who do not know the product from before, have other expectations. Several informants had never tasted Gamalost before our focus group. They told first about the bad smell and the terrible taste, but when they first ate it together with us they said that *"it was actually good...not so bad"*. The defensive strategy, consisting of not buying the product because of its reputation, changed when they got more knowledge through taste and discussion. Nevertheless, they do not feel any special duality around the Gamalost.

Consumption between self-seduction and rationality

Different aspects of myths and modernity seem to play important roles in influencing consumers associations concerning Gamalost. Whereas myths represent nostalgia and the search for an

[72] Focus group in Oslo, "enthusiasts"

idealized old time, modern/technical traits are represented through industrial production. While myths are important for satisfying the need for product authenticity, the modern traits are important for satisfying consumers' demands for stable product quality and safety. Consumer's conception of Gamalost lies in the point of intersection between myths and modern traits. Although the consumers tend to know that this is an industrial product, consuming Gamalost often includes a disconnection from modern traits and appears to be an exit from modernity helping to recapture the idealized atmosphere of old times as they are represented in myths. In the case of enthusiasts, their kind of consumption can be described as a crossover action between dream and rationality, where rational is understood in the sense that they are fully aware that the product they are consuming is an industrialized product. By the same time they use the product to re-represent something that disappeared, through a kind of self-seduction. The consumers are thus not only seduced by marketers, but also by themselves. The product seduces the consumer, but the consumer wants to be seduced and plays an active role in the seduction (Belk, Ger and Askegaard 2003). As our study uncovers a prevailing ambiguity related to *Gamalost fra Vik*, what can be said to be the ethical implications of this condition?

Ethical implications of discrepancies between presentations and conceptions of the cheese

Several ethical dimensions and implications can be related to the discrepancies between presentations and conceptions of Gamalost. In light of the above-mentioned paradoxes we will raise some ethical questions regarding the tension between marketers' communication, myths and different consumer's interpretation and comprehension of Gamalost.

Market actors involved in Gamalost can mainly be seen as supporting the discrepancies between presentations and conceptions of the cheese.[73] While they are producing Gamalost in the industrial way, they are also mostly underlining the traditional and mythical aspect of this product.

As there today is only one place of production of Gamalost: is it then possible to satisfy both kinds of consumers at the same time with the same product? In this sense marketing of Gamalost can be said to be stuck with a paradox. If they continue to focus on the mythical traits, they might mislead potential consumers, and themselves, to believe that the myth is the real history of Gamalost. Turning the problem around: If they stop focusing on the myth and point at the fact that Gamalost is an industrialized product, they might at the same time undermine the importance of Gamalost as a product preserving the past in a positive manner. This could impair the 'enthusiast' consumer's possibilities to reach their desired experience.

In addition to the marketers' role, we underlined the consumer's role with self seduction, but it may also be worth questioning the implications of researchers' intervention in this field: How may producers and other actors react to such a highlighting? Does the consumer want to be reminded that *Gamalost fra Vik* is a highly industrialized product? Does the researcher interaction complicate their way of practices or will they experience such information as helpful to better understand their own consumption? Or perhaps consumers and marketers never access this kind of information or just find it unimportant and not relevant for their market strategies or consumption?

[73] One exception is worth mentioning. People working in the production of Gamalost in Vik emphasised more strongly the modern aspects than the myths.

Conclusion

The tension and fusion of modern and traditional traits become evident when we study *Gamalost fra Vik*, both in the way the product is promoted and the way that it is understood by the consumers. Whereas marketing and myths of Gamalost emphasise and connote old days and a product with long traditions, present day production methods are highly modern and standardised. The consumers we interviewed vary in their response to whether Gamalost is really traditional and authentic or an industrialised modern product. To bypass this ambivalence, consumers of the "enthusiast" group use different strategies to achieve their preferred taste and experience. The ordinary consumers have other expectations and hold a defensive strategy, often consisting of not buying the product because of its reputation. Contrary to the participants in the group with "enthusiasts" they do not feel any special duality around this product. As a matter of fact, level of knowledge and information appear as major tools in consumers´ conceptions and representations of the product.
Ethical questions can nevertheless be related to the discrepancies between presentations and conceptions of Gamalost, especially because marketing campaigns are oriented towards the "new consumer", who does not have a special knowledge from before. However, a campaign based on standardisation and hygiene, would obviously neither be a proper reference, nor an exciting image for the Gamalost, in spite of its nutritional characteristics. Is it actually possible to satisfy both kinds of consumers at the same time with the same product? Focusing on the mythical traits would mislead potential consumers to believe that the myth is the real history of Gamalost, while avoiding references to the myth would mislead about the historical and identity value of the cheese. The ethical dilemma is then reinforced by the researchers' interventions in this field, which obviously have some implications...

References

Belk, R. W., Ger, G. and Askegaard, S. (2003). The Fire of Desire: A Multisited Inquiry into Consumer Passion. *Journal of Consumer Research.* 30:326-351.

Bergfjord, K. (1997). *Til jordbrugets ophjælp – ei reise i soga til meieriet i Vik gjennom 100 år.* Sogn og Fjordane Meieri, Vik i Sogn.

Brendehaug, E. and Groven, K. (2004). Gamalost – Old Norwegian Cheese. In: *Mountain food products in Europe: Case studies describing 18 initiatives within the framework of the European project.* EUROMONTANA. 57-66.

Fosså, O. (2000). How Old is Old Cheese? Gamalost in Coffin-shaped Boxes and Eccentric Jars. In: *Milk beyond the dairy. Proceedings of the Oxford Symposium on Food and Cookery 1999.* Walker H. (ed). Prospect Books, Totnes. 144-160.

Moberg, J. (1997), *Gamalosten.* Det Norske Samlaget, Oslo. 91 p.

Notaker, H. (2000). *Ganens Makt.* Aschehoug, Oslo. 333 p.

Sæbø, A. I. (2003). *Gamalost: stølsost og meieriost.* Vik lokalhistoriske arkiv, Vik I Sogn. 27 p.

Slow food lessons for a fast food nation

Christopher D. Merrett, William Maakestad, Heather McIlvaine-Newsad, and Patrick McLaughlin[74]
The Program for the Study of Ethics, 414 Stipes Hall, Western Illinois University, Macomb, IL 61455-3035, USA, wj-maakestad@wiu.edu

Abstract

During the latter half of the twentieth century, American agriculture underwent a technological revolution which transformed the American food system. While these changes provided benefits such as lower-cost food, it also generated concerns among many people who worried that the unquestioning embrace of technology would hurt rural communities and the environment, while raising questions about food quality and food safety. This paper examines how direct marketing strategies such as community supported agriculture (CSA) may offer a more sustainable alternative to large scale production agriculture, and provides one example of such a model.

Introduction

Western Illinois, the region in which we live, is dominated by industrial agriculture, primarily corn and soybeans. Despite this agricultural heritage, most local consumers eat food that travels thousands of miles. Eric Schlosser (2002) critiques this "productivist" agriculture in *Fast Food Nation*. He describes how the drive to sell cheap, processed food compromises food safety, rural communities, local flavors, and the connection between farmers and communities. Opposing industrial agriculture, we advocate small-scale "civic agriculture", to borrow Lyson's (2000) term, which develops relationships between local producers and consumers. The rise of "civic agriculture" coincides with the rise of the post-productivist "slow food" movement. According to the Slow Food Movement (2004), emphasizing the agricultural "productivity [and] the 'Fast Life' has changed our way of being and threatens our environment and our landscapes." As an antidote to the perils of Fast Food and the 'Fast Life', "Slow Food is working not only to protect the historic, artistic and environmental heritage of places of gastronomic pleasure (cafés, inns, bistros), but also to safeguard the food and agricultural heritage (crop biodiversity, artisan techniques, sustainable agriculture, rural development, food traditions)."
In the spirit of civic agriculture and the Slow Food Movement, we report on one research and one outreach activity designed to educate consumers about local produce. First, we report on a survey of 746 CSA farmers to identify their marketing strategies. Results show an array of strategies including word of mouth, advertising in local media, approaching local restaurants, newsletters, and displays at farmers markets. Then we report on a strategy to promote local produce in western Illinois. Each year, we host a "Taste of the Tri-States" (TOTTs) at a local up-scale restaurant. The menu consists of appetizers, entrée, desserts, wine and beer produced entirely within the tri-state region of Illinois, Iowa and Missouri. At this dinner, we also have locally-produced artwork and music played by local musicians. We market this event as "Food for the Body and Soul," promoting the natural talents and resources of the region. Proceeds go to the local public radio station, which advertises itself as "public radio for the tri states."

[74] Authors are respectively, Professor and Director of the Illinois Institute for Rural Affairs, Professor of Management, Associate Professor of Anthropology, and Peace Corps Fellow, all at Western Illinois University. Portions of this paper are based on a previously published technical report by McLaughlin and Merrett (2002).

Community Supported Agriculture

CSAs originated in Japan over thirty years ago (Broydo 1997). The first American CSA started in Massachusetts in 1985, growing to over 1,000 by 2002, with 150,000 consumers (Broydo 1997; Iowa Café 2001) (Fig. 1).

CSAs represent a partnership between community members and local farms. The community member pays a "subscription" fee to the CSA farmer for a season's supply of produce. The farmer in turn provides the subscriber with fresh, local produce every week during the growing season. This agreement provides the community with fresh, locally grown food. It provides the farmer with a guaranteed market and capital to buy seeds, fertilizer and equipment (Cone and Myhre 2000). The agreement allows risks and benefits to be shared between the consumers and the farmer (Hinrichs 2000).

CSA farmers struggle to develop and retain their subscriber base, with attrition rates often exceeding 50% each year (Cone and Myhre 2000). In response, we conducted a national survey of 746 CSAs between November and December 2002. We received 373, giving a 50% response rate. This paper reports on overall trends. It then explores how CSAs create and sustain a viable subscriber base. An examination of the CSA start dates shows that this is a new phenomenon that has grown rapidly over the past decade (Fig. 2). The results show that the average CSA operator is 44.6 years old. This is ten years younger than a traditional farmer, which is 54 years of age (USDA 1997). The gender breakdown also reveals a difference between CSAs and traditional farm operations. Women operated 52% of the CSAs. Only 8.6% of traditional farms are run be women (USDA 1997).

We asked farmers to rate the factors prompting them to start a CSA on a scale of 1 to 5, where 1 = not important and 5 = very important (Table 2). While profitability was identified, the key factor was the provision of good food to the local community. Other factors include the promotion of sustainable agriculture and participating in a local food system which reconnects the producer and consumer.

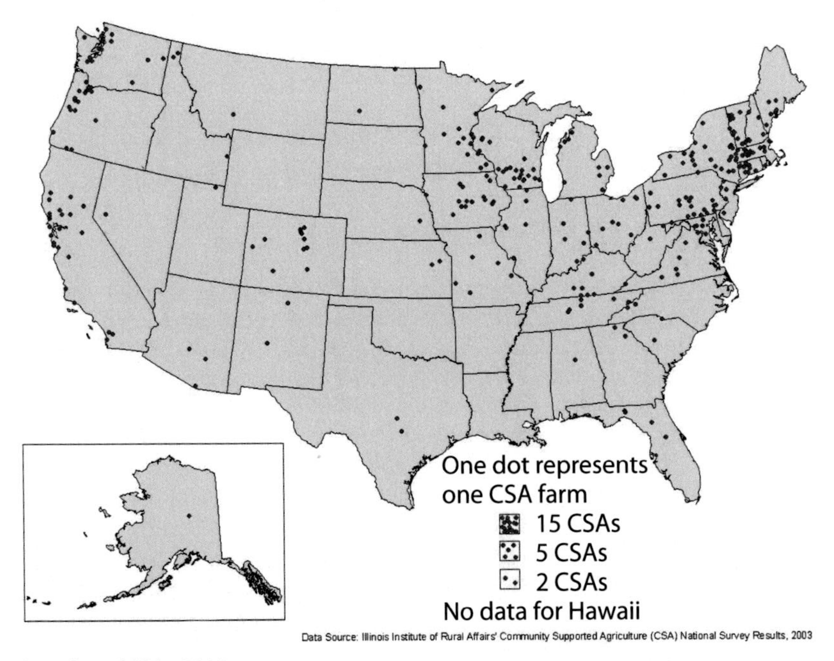

Figure 1. Location of CSAs, 2002.

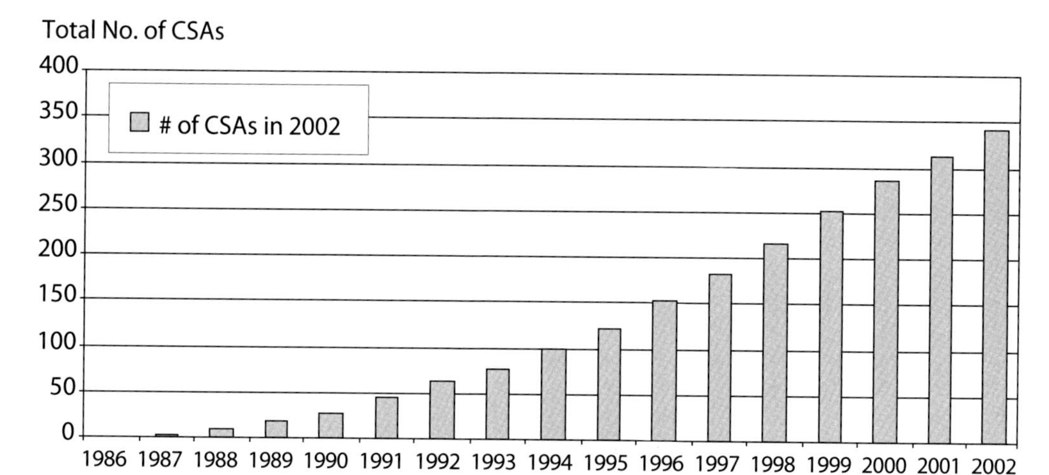

Figure 2. Cumulative frequency for CSA start dates in the survey.

Table 1. Summary statistics for CSAs and CSA operators.

Average age of a CSA operator	44.6 years
Gender of CSA operators	48% male / 52% female
Average start date	1996.5
Average acreage in first year	3.6
Average acreage in current year	8.6
Median number of subscribers in first season	20.0
Median number of subscribers in current season	50.0
Average subscription price in first season ($)	318.75
Average subscription price in current season ($)	399.72
Percent of total family income from CSA	25.1%

Table 2. Ranking of factors prompting farmers to start a CSA.

Motivating factor	Mean
Providing wholesome food to community	4.79**
Promoting sustainable agriculture	4.65**
Land stewardship	4.53**
Lifestyle	4.32**
Generating a sense of community	4.25**
Profit	3.40**
Keeping farmland within the family	3.11**

A chi-square test was used to determine if the mean responses were significantly non-random.
** Results are significantly non-random at $p < .01$.

Previous research shows that operating a CSA is difficult work. We asked CSA farmers to rank challenges on a scale of 1 to 5 where 1 = not important and 5 = very important (Table 3). Results show that three interrelated factors are the most common obstacles. Retaining the customer base is important because if the base shrinks, then profits will decline. In this survey, respondents reported median retention rates of 75.0%.

A related question asked operators why some people did not renew their subscriptions. Respondents were asked to rank reasons on a scale of 1 to 5 where 1 = least common and 5 = most common. The most commonly identified problem is that subscribers don't have time to get their CSA delivery and prepare dinner using fresh ingredients (Table 4). Of course, this is the whole message of the slow food movement which is to take the time to prepare meals with fresh ingredients.

To investigate how CSAs recruit new business, we asked respondents to identify the specific ways that they market their farm. The results show that most CSAs engage in a range of activities to sell their produce (Table 5). The fact that the most frequent strategy is simply relying on "word of mouth" suggests that CSAs are not involved in sophisticated marketing strategies. Besides

Table 3. Ranking obstacles to CSA success.

Obstacle	Mean
Retaining subscribers	3.32**
Low profits	3.31**
Marketing	3.02*
Variety of produce	2.94
Competition with grocery stores	2.43**
Persuading subscribers to work on farm	2.12**
Competition with other CSAs	1.85**

A chi-square test was used to determine if the mean responses were significantly non-random.
** Results are significantly non-random at $p < .01$.
* Results are significantly non-random at $p < .05$.

Table 4. Ranking the reasons why people do not renew their CSA subscriptions.

Reason for not renewing subscription	Mean
Subscriber does not have time	3.52**
Subscriber moved away	3.37**
Inconvenience	3.14**
Too far for pick up or delivery	2.63**
Unfamiliar food varieties	2.62**
Not enough product choice	1.92**
Does not feel price is a good value	1.83**
Subscriber thought there was not enough food	1.65**

A chi-square test was used to determine if the mean responses were significantly non-random.
** Results are significantly non-random at $p < .01$.
* Results are significantly non-random at $p < .05$.

Table 5. Marketing strategies used by CSA operators.

Marketing strategy	Percent of CSAs using strategy
Word of mouth	88.8
Internet webpage	44.4
Farmers' market	38.2
Participating in community events (e.g. county fair)	35.6
Open house—Inviting community to visit	35.3
Newspaper advertisements	27.5
Roadside stands	9.1
Radio	7.2

this strategy there is no other marketing plan used by a majority of CSAs. This finding, coupled with the high turnover rates suggests that CSA operators might find it easier to find subscribers by employing a better marketing plan. The lack of well-developed marketing strategy is likely due to the fact that just 53 percent of CSA respondents reported having a formal business plan. An underdeveloped marketing strategy would be symptomatic of a CSA without an overall business plan.

A promising finding is that the second most frequent strategy reported is to market over the internet. While there are still fewer than half of the CSAs with an internet presence, this situation is likely to change. Recent stories in USA Today and on CNN noted that many small farms, especially those with an agritourism focus find that a web presence is essential for success (Groppe 2004).

Taste of the Tri States

The fact that many CSAs have minimal or no formal marketing strategy raises questions. How can a small business enterprise survive with little or no attention to marketing? This is a particularly pressing issue because research suggests that CSA farm location may directly affect its viability, age, size and number of subscribers (Lyson and Welsh 1993). CSA farms located in more rural settings are more likely to have a harder time selling the CSA concept to the local community (Hendrickson and Ostrom n.d.). Because of these added challenges to CSAs in rural regions, we undertook a project to promote the CSA concept in western Illinois.

The project we developed is called the Taste of the Tri States (TOTTS): Food for the Body and Soul. The subtitle of the event reflects a "slow food movement" ethos because we are trying to market the resources of our region including locally-grown food, local music and locally produced art. The prix fixé meal, including appetizers and salads, entrées, desserts and beverages are provisioned (to the greatest extent possible) from producers in the tri-state region of Illinois, Missouri and Iowa.

We are in our fourth year of organizing the event which takes place at a local upscale restaurant called Magnolia's. The chef and owner Lisa Ward works with all of the suppliers to ensure that ingredients are available for the planned menu. Last year, the menu included vegetarian and non-vegetarian entrées, salads and fresh bread. Fresh vegetables came from two local CSAs. The vegetarian entrée included polenta and stir-fried vegetables. The dessert included an apple tart with fresh cream and apples from a local orchard. The non-vegetarian entrée was a "reef and beef" dinner which included fresh water prawn produced in Illinois and Angus beef steak produced by a local livestock cooperative. Beer and wine came from breweries and wineries in Illinois and Missouri. We also offered apple cider as a non-alcoholic alternative.

While our primary target is to help market locally grown foods, we are also interested in promoting artisans in related economic sectors. Though Macomb is a small college town, there is an active arts community. There is a nascent arts cooperative which is attempting to buy and refurbish space in the downtown area for studio and exhibit space. This effort is also part of plan to revitalize the downtown area which has lost significant retail business due to strip development at the outskirts of town. We provide space for members of the arts community to display their work, including paintings and photography. In addition, they are provided with an opportunity to discuss their progress and to possibly solicit funds to further redevelop the regional arts center initiative planned for downtown Macomb.

In addition to locally produced food and art, we also invite a number of local musicians and groups to perform during the evening. In the past, we have had performers playing jazz and classical music. The proceeds from the evening go to our local National Public Radio station, WIUM, which advertises itself as "Public Radio for the Tri States" because while it broadcasts from Western Illinois University, its coverage area includes parts of Iowa and Missouri.

By sponsoring this event, we are able to educate people about locally grown food, the work of local artists, and the range of music played by local musicians while providing money for our local NPR station. The NPR station provides news coverage of the event and hence, provides further marketing for local producers. At a broader level, this event also brings together a range of people who might not ordinarily have a chance to meet. Farmers, artists, musicians, and fans of public radio spend time celebrating the wealth of resources in the region.

Conclusion

The CSA presents the possibility that consumers and farmers can move away from the negative effects of and concerns about the current industrialized farm system, and towards a more community based, environmentally sound, sustainable practice of agriculture. The CSA idea allows for smaller, decentralized farms to move towards an agriculture system based on principles of land stewardship and local economic development. In addition, CSAs provide an avenue for increased social development through community participation and interaction. That is why the CSA provides such as good foundation for the TOTTs event. Though the movement is not one that will replace the conventional agro-industrial system, it is one that most certainly can compliment it by providing farmers with an opportunity to shift their efforts to a more locally driven food system.

Though CSAs have been very successful, there are challenges that need to be addressed. These challenges include a high member turnover and an increased work load for the farmer. These challenges can be overcome by an increase in member involvement and education. The two remedies seem to go hand-in-hand. Increased community involvement comes with education. It is and will continue to be the challenge of the both the farmer and CSA farm members to develop innovative ideas to encourage this participation. For the goal is a "community" supported farm, where all members are active in supporting a food system, which is sustainable from an economical, social and environmental perspective.

References

Broydo, Leora. 1997. Buying the farm. *Mother Jones*. (March/April). http://www.motherjones.com/mother_jones/MA97/homeplanet.html. May 31, 2004.

Cone, Cynthia and Andrea Myhre. 2000. Community-supported agriculture: A sustainable alternative to industrial agriculture? *Human Organization*. 59: 187-197.

Groppe, Maureen. 2004. Farmers strike pay dirt with Web. USA Today. May 18. http://www.usatoday.com/news/nation/2004-05-18-webfarms_x.htm. Accessed: May 31, 2004.

Hendrickson, John and March Ostrom. n.d. Managing a CSA farm. Center for Integrated Agricultural Systems. Research Brief #40. http://www.wisc.edu/cias/pubs/briefs/040.html. Accessed: May 31, 2004.

Hinrichs, Clair. 2000. Embeddedness and local food systems. *Journal of Rural Studies*. 16: 295-303.

Iowa Café. 2001. *Building a Better Future for Rural Iowa. Workshop Manual on Community Agriculture and Food Enterprises*. Ames, IA: Iowa State University.

Lyson, Thomas and Rick Welsh. 1993. The production function, crop diversity, and the debate between conventional and sustainable agriculture. *Rural Sociology*. 58: 424-439.

Lyson, Thomas. 2000. Moving toward CIVIC agriculture. *Choices*. 15 (3): 42-45.

McLaughlin, Patrick and Christopher D. Merrett. 2002. Community supported agriculture: Connecting farmers and communities for rural development. *Rural Research Report*. 13 (10): 1-8.

Slow Food Movement. 2004. *All about Slow Food*. http://www.slowfood.com/eng/sf_cose/sf_cose.lasso. Accessed: May 31, 2004.

USDA. 1997. *Census of Agriculture*. Historical Highlights. http://www.nass.usda.gov/census/census97/.pdf. Accessed: May 31, 2004.

Olive oil and the cultural politics of 'Denominaciones de Origen' in Andalusia

Nefissa Naguib
Department of Social Anthropology, University of Bergen, Nefissa.naguib@sosatanr.uib.no

Abstract

This paper will explore Spain's '*Denominaciones de Origen*', a protective legislation dedicated to the authenticity of traditional agricultural products and their geographical place of origin. The commercialisation of one of these 'productos Andaluçes', olive oil, provides the ethnographic framework through which the paper explores *Denominaciones de Origen*'s cultural politics as they are played out in the smart superstores in the South of Spain. The paper associates itself with the politics of 'origin' within current debates regarding the European Union's concern with regional culinary diversities.

Keywords: Andalusia, olive oil, denomination of origin, culinary traditions, 'productos Andaluçes'

Introduction

During the 1990s an increasing number of supermarkets in Andalusia have been devoting more shelf place to foods labelled 'productos artesanos', 'productos de la tierra', 'de la abuela' or 'patrimonios gastronòmicos locales'; many are quality sealed with 'productos Andaluces' and '*Denominaciones de Origen*' (DO) authentification stamp. Through quality inspection by the Instituto Nacional de Denominaciones de Origin (INDO) and their DO stamp, products are afforded protective legislation dedicated to the authenticity of traditional agricultural products and their geographical place of origin. One of these 'productos Andaluçes', olive oil, provides the ethnographic framework through which I will explore cultural politics as it is talked about among local people and displayed in superstores in the South of Spain.
Spain has 4.7 million acres of olive trees under cultivation and is the largest producer and exporter of olive oil in the world. Since 1917 -18 she produced, on average, over one-half of the world's supply of olive oil; today she produces about one-third of the world's olive oil. Andalusia accounts for 75% of Spain's olive cultivation; almost 3 million acres of olives and ten of the largest oil mills are located in the region, accounting for 2/3 of the country' olive oil production. In the early 1970s Spain passed Europe's first law permitting extra virgin olive oil to have a DO label, like its wine. According to the Ministry of Agriculture, Fisheries and Food, a denomination is a geographic space, and one key element within that is the variety of olives grown. Another is the local ecosystem or micro-climate; furthermore, the tradition of making fine olive oil is paramount (http://www.mapa.es/). At the time of researching this paper the INDO certified the following olive growing geographical areas in Andalusia: Sierra del Segura and Sierra Magina in the province of Jaen; and Baen and Priego de Córdoba, both in the province of Cordoba. With the DO authentification stamp the Andalusian olive oil, also referred to as 'oro líquido'[75] (Granados 2000), is not only distinguished as 'productos Andaluçes'; it also reflects on the region's recent prosperity, an image that challenges the idea of Andalusia as poor, coarse and simple. Filled with the nostalgia of heritage, the cultural politics behind DO are also concerned

[75] Gold liquid

with commercialization and the marketability of products which are culturally traceable, thus distinguishing regional oils as legislated quality products. The first part of my paper is grounded in what olive oil means for local people and how meanings have changed in relation to its recent appropriation as a reference of 'productos Andaluçes'. I concentrate on this example because it provides an ethnographic window into the promotional politics of '*Denominaciones de Origen*'outlined in the second part of the paper.

Buying olive oil

Back in the 1980s when the South of Spain was experiencing an influx of Scandinavians buying second homes, the olive oil sales personnel in the supermarkets were ladies in flamenco dresses offering tourists Melba toast dipped in olive oil. Since the 1990s olive oils with DO labels are taking more space, and the sales personnel have exchanged their folkloric dress for the more clinical look of white overcoats. While in the 1980s the majority of interested customers in the superstores were non-Spanish, today it's more usual to find Spanish tourists and locals enthusiastically discussing the 'aroma', 'maturity', 'softness', 'spiciness' or 'colour' of a specific olive oil with sales personnel from producers with DO labelled bottles. The customers degust the oil either by drinking it from small cups or soaking a crust of country bread in it. This change was taking place at a time when the region's level of autonomy and standard of living were dramatically higher than they have been for many centuries. This was also a time of growing interest in Mediterranean cuisine; information posters and pamphlets illustrate the unique benefits of Mediterranean diets, and ongoing campaigns feature oil as the primary ingredient in a healthy life. It is not unusual to see posters or pamphlets in the supermarkets texts resembling this one: "Dribbling oil on bread with a bit of tomato wasn't gourmet -- it was sustenance. Before olive oil became the oil choice of dieticians and doctors, the Spanish diet revolved around it. Spain, over the years, held on to its Mediterranean and Spanish traditions, and is now the first olive oil producing country in the world!"[76]

This change in choosing 'productos Andaluçes' certified with a DO stamp is increasingly discussed by those in my neighbourhood, which lies between Malaga and Marbella. They used to get their home rations from family or acquaintances' olive presses, following several restrictions concerning private oil pressing more people buy their oil from the superstores. When neighbours talked about their choice of olive oil, they frequently referred to that "Denominaciones de Origin was not important before, but now we want to make sure its from here. So we look at the quality label."

Olive oil is one of the bases of the Andalusian cuisine, which is known for its 'fritos' and celebrated for its chilled soup 'gazpacho'. In fact, it all starts with breakfast, when toasted bread is drizzled in virgin oil to eat with morning coffee. Given its dietary importance, it is perhaps not surprising that it is also adopted as a cure for a number of health ailments, from constipation to bad skin to bad heart. But more important than such factors in the coalescence of the regional heritage of prolonged poverty and oppression and Andalusia's unfulfilled potential is, above all, the perception among several of my neighbours that Andalusia has been exploited and neglected. For many the cultural and socioeconomic predicaments are closely linked; as one of them put it "Now the English want olive oil on their bread". This is seen as a symbol of the development of new, revitalised forms of regional culture which are distinct from the rest of Europe; reclaim the region's unique culinary heritage and its beneficial role in improving Europe's health while at the same time adapting to the realities of post modern Andalusia.

[76] Freely translated from a poster in a supermarket in a Malaga neighbourhood.

'Origin'

"Throughout Europe there is an enormous range of great foods. However, when a product acquires a reputation extending beyond national borders it can find itself in competition with products which pass themselves off as the genuine article and take the same name. This unfair competition not only discourages producers but also misleads consumers. That is why, in 1992, the European Union created systems known as PDO (Protected Designation of Origin), PGI (Protected Geographical Indication) and TSG (Traditional Speciality Guaranteed) to promote and protect food products" (http://europa.eu.int/comm/agriculture/foodqual/quali1_en.htm). During the 1990s the EU policy was to increase knowledge of Europe's diverse cultural heritage and the distinctiveness of European peoples (EC culture 2000 program). The idea of a common European culture based on expressions of local and regional cultural diversity has become a unifying symbol (Delanty 1998). Member states were given the task of encouraging the spread of culture in society (Cederman 2001:17), and food heritage has become and essential part of this task.

The traceability of agricultural products is among the most dynamic sectors in the EU's promotion of food safety and Europe's diverse culinary traditions. Demands to protect local regional culinary traditions combined with food security issues encouraged growing awareness among locals to look out for certified food. Scholars have noted how local products, for example the French idea of 'terroir', present rich symbolic variations through their site, representation, smell and taste, including their ability to be elaborated and absorbed in local cuisine (Barjolle 1998). As the case of my neighbourhood illustrated, an interest in 'productos Andaluçes'is a topic of fascination to the general public, as evidenced by the discussions at the stands in the supermarkets and by the huge numbers of published cookbooks dedicated to regional cuisine with titles that include 'productos de la tierra', 'la cocina del abuela' or 'patrimonios gastronòmicos locales'.

Anthropologists have long called attention to the fact that "eating is never purely biological activity" but, rather, one of many arenas in which we invest "a basic activity with social meaning(s)" that are "symbolic... communicated symbolically" and that "also have histories" (Mintz 1996). Regional memories of a 'rougher' Anadulsia are embedded in the above local responses to certified olive oil. '*Denominaciones de Origen'* is also about the continuity between past and present. We might agree that heritage provides historical depth and a permanent pattern in a perpetually changing world. Yet, beyond that continuity, beyond that legacy which is passed on, heritage is part of the present, and at the same time holds promises for the future (Naguib 2006); *Denominaciones de Origen* products in the supermarkets indicate clearly that heritage is no longer considered solely as a link between past and present, but also as a reservoir of what is natural and authentic because it concerns a limited area where natural conditions and history provide distinctiveness. Furthermore, *Denominaciones de Origen* conveys not only the reputation of Andalusia's culture, history, and skill. In seeking to protect this cultural heritage, *Denominaciones de Origen*, has effectively become a 'clearing house' for distinctive culinary concern, initially in Andalucía, but latterly the rest of Spain. The main means by which knowledge about olive oil as 'productos Andaluçes'is disseminated is in displays along the corridors in the superstores and the daily stands, representing specific geographic areas, which offer the degustation of regionally controlled oils.

Concluding reflections

This paper drew on olive oil as "a total social fact" (Mauss 1967), as an indicator of peoples' economic, symbolic, social and health concerns. As mentioned above, olive oil is one the most essential ingredients in the Andalusian diet and a key to understanding local responses. I began by tracing the recent changes of olive oil's 'social life'(Appadurai 1987), as it has moved from an *'alimentario'* purchased directly from family or acquaintances' mills to a *Denominaciones de Origen* label promoted as 'productos Andaluçes' and obtaining more shelf place in local supermarkets. As a direct result of more information available in the shops and media, olive oil has also acquired new meaning as a symbol of Andalusian heritage, consumed by tourists and locals alike and contributing to the healthy Mediterranean way of life. *Denominaciones de Origen* has succeeded in creating the cultural space for the performance of a new kind of consumer politics.

Labelling the traceability of olive oil can be interpreted within the realms of advertising, commerce and cultural critique. It is as much concerned with the comodification of rural nostalgia as with the actual protection of regional and local material cultures of productions and the memories to which they are attached; all this within the context of generating certified food. Equally, we also know that the implementation of EU guidelines for consumer health and food safety as well as agricultural and environmental policy effects a standardisation of legal frameworks in food production that are perceived as a threat in the several distinctive local and regional cultures. Yet, traceability of local and regional cuisines promotes competitive assets both economically and in terms of the assertion of cultural identities. The DO label promotes the idea of cultural diversity in line with the EU by urging local consumers to buy heritage-marketed foods which are health beneficial and 'safe'.

Acknowledgements

Bjorn-Erik Leerberg told me to write about olive oil, thank you. In Spain my teacher Gema Toro Rubia helped during my fieldwork and is still patiently trying to teach me Spanish. Christine Amadou provided discussions on Europe's culinary complexities. I am as always grateful to my dear friend Nancy Frank for reading and commenting what I write, thank you also for sharing your passion for foodways with me.

References

Appadurai, A., ed. (1986). *The Social Life of Things*. Cambridge, Cambridge University Press

Barjolle, D (1998). *Le Lien au terroir, des travaux de recherché*, online http://www.origin-food.org/pdf/wp1/wp1-ch.pdf

Cederman, L-E (2001). *Constructing Europe's Identity*. Boulder, Lynne Rienner Publishers

Counihan, C. and Van Esterik P. (1997). *Food and Culture*. London, Routledge

Council Regulation (EC) No510/2006 of 20 March 2006 on the protection of geographical indications and designations of origin for agricultural products and foodstuffs.

Delanty, G (1998). Redefining political culture in Europe today. In *Political Symbols, Symbolic Politics; European Identities in Transformation*. U Hedetoft (ed), pp. 24 – 43. Hants: Ashgate Publishing Company

Diner, R. H. (2002). *Hungering for America*. London, Harvard University Press.

Douglas, M. (1982). *In the Active Voice*. London: Routledge

Goody, J. (1982). *Cooking, Cuisine and Class*. Cambridge, Cambridge University Press

Granados, J.Á. (2000). *Enciclopedia del Aceite de Oliva*. Barcelona, Planeta.

Jacobsen, E (2004). The Rhetoric of food. In *The Politics of Food*. M.Lien (ed). pp.59 - 78. Oxford, Berg Publishers

Lien, M. (2004) The Politics of Food: Introduction. In *The Politics of Food*. M.Lien (ed). pp.1 – 17. Oxford, Berg Publishers

Mauss, M.(1967). *The Gift*. New York, Norton.

Ministry of Agriculture, Fisheries and Food, *http://www.mapa.es/, http://www.mapa.es/es/alimentacion/pags/Denominacion/consulta.asp*

(date of consultation: 03.04.2006)

Mintz, S. (1996) *Tasting Food, Tasting Freedom*. Boston, Beacon Press

Naguib, N. (2006) The Fragile Tale of Egyptian Jewish Cuisine: Food Memoirs of Claudia Roden and Colette Rossant. *Food and Foodways*, Vol 14, Issue 1. pp 35-53

Sutton D, (2001). *Remembrance of Repasts*. Oxford, Berg.

Terrio, S, (2000). *Crafting the Culture and History of French Chocolate*. Berkeley, University of California Press.

Lookin' for some down-home cookin'? A case study of household pork production in the Czech Republic

Zbynek Ulcak
Masaryk University, Faculty of Social Studies, Department of Environmental Studies, Jostova 10, CZ-602 00 Brno, Czech Republic, ulcak@fss.muni.cz

Abstract

Traditional agricultural practices are often seen as friendly to the landscape and natural environment, and the lifestyle associated with farming often serves as romantic example of rural idyll (Berry 1997, 2002, Henderson, Van En 1999, Charles 2003). Is this true, or is this view just another instance of the romanticisation of the countryside and rural landscapes, as described by David Pepper (1996)? This paper describes the roots of the tradition of small-scale home pork production in the Czech Republic, and analyses the meaning of the tradition as it is practised today. Special attention is given to the presentation of the tradition in the media and visual arts, namely in the work of Josef Lada, a Czech painter from the first half on 20th century who in his work combined an idyllic view of Czech village life with a slight accent of parody. Today the tradition exists often simply in the form of a "pig roast" – a social event involving the slaughter, preparation and cooking of a pig purchased from large commercial breeders. This results in an interesting contradiction, because "homemade food" generally has an ideation value of high quality reaching the formal standards for organic farming, and is very often associated with a high standard of animal welfare. The paper discusses the often-contradictory reality of the tradition as it is practised today.

Keywords: pig, animal slaughter, organic agriculture, ecological luxury, intrinsic values

Introduction

Proponents of the contemporary concept of ecological or organic agriculture often tend to admire traditional farming practices of the past as being nature-, landscape-, and animal-friendly. The lifestyle associated with these practices often serves as a romantic example of rural idyll (Berry 1997, 2002, Henderson, Van En 1999, Charles 2003, Stibral, 2005). So such images reflect reality, or are these views just another instance of the romanticisation of the countryside and rural lifestyle as described by David Pepper (1996)? Some authors (Frewer *et al.*, 2005) suggest that consumers are generally more sensitive to extrinsic quality factors associated with the methods of production (e.g. type of animal housing). This seems to be an attitude which is less demanding on consumer awareness of intrinsic values, resulting from the knowledge of basic animal needs and animal rights, as expressed in the "Five Freedoms" concept and adequate responses to them (Spedding, 2000). This may be an important factor influencing consumer trust in food. These questions are raised by studying the roots and present state of household pork production in the Czech Republic, and the cultural images associated with the very process of pig husbandry and slaughtering.

Self-supply or luxury?

Dvoraková-Janů (1999) and Frankova, Dvorakova-Janu (2003) suggest various reasons for food self-supply in modern households. For some households self-supply is still motivated by economic reasons, but more often other aspects are mentioned – leisure activity, local tradition in association with social control, community life (hobby gardeners and breeder associations). But from the point of time- and cost-effectiveness this activity remains mainly irrational and inefficient – people usually invest more time and money into the home food production compared to the market price of the product of similar quality (often including organic products). Therefore, home food production can be considered an "ecological luxury" as proposed by Librova (2003). This luxury is characteristic for its small ecological footprint, use of soft technologies, reflected intrinsic value of nature and social dimension in the form of altruism, in comparison to the "predatory luxury" which often may seem to respect nature, but in fact tends to appropriate everything rare for its own use (e.g. "eco-tourism", building new houses in the countryside).

Roots and present state

In the Czech lands, keeping a pig and slaughtering it at home is an old tradition described as early as the 17th century by the Czech pedagogue and theologian Comenius in his Latin encyclopedia Janua linguarium reserata. Javurek (2006) summarises some of the rituals mentioned by Comenius, which may be understood as paying respect to an animal – like the hog being ritually "sentenced" to death and slaughtered only after his "will" was read.
Although these rituals vanished a long time ago, the tradition of keeping or killing a pig has continued until the present time. This did not change even during the period of communist rule and the dominant collective style of farming (1948-1989), when individual ownership of swine was allowed, but in some periods regulated to a single pig. This can be illustrated by the decree No. 48/1963 (Sbirka, 1963) of the "Central Direction for Agricultural Products Purchase." Whilst this state regulatory body ceased to exist in 1967, the decree is still a valid part of the Czech legal system, though it is not enforced anymore. It regulated the formal conditions under which slaughtering was allowed by different categories of farmers and public institutions, but it also permitted the slaughter of one hog per year even for non-owners of agricultural lands – an illustration of how deeply the tradition was respected. It must be mentioned that this had been happening in Czechoslovakia, which, compared to some other nations with centrally regulated economies, was not of a mainly rural character and where there was no serious shortage of public supply of animal products. Rearing a pig was nevertheless a welcomed contribution to a household's economy and a good way of using household by-products as fodder for the animals.
At present around 4 million pigs from large-scale farms are killed annually in commercial slaughterhouses in the Czech Republic. To this number should be added an estimated 450 thousand house-slaughtered pigs (Czech, 2006).

The pig as instrumental animal

The popularity of the pig roast itself is increased by its presentation in the media and some Czech cultural icons – including painters of the 19th and early 20th century such as Mikolas Ales, Josef Manes, Vlastimil Rada and Josef Lada. These artists often illustrated children's primary textbooks together with the rhymes for different seasons, where winter is often accompanied by the theme of the pig roast. Particular attention should be paid to the works of Josef Lada (1887-1957), who developed an unique style sometimes misunderstood as a naïve copying of reality, but which in fact created a new reality, admired by the modern artists of his time (Picasso, Filla, Tichy, Capek). As stressed by Pecinkova (1998:156): "*While Czech painters develop the surrealistic lessons and discover unconscious dramas of human existence, Lada happily draws flowers, birdies, and kills piggies in his museum. But total opposites often have common grounds. Lada's works are in their ways related to the Jungian world of archetypes, but speak a different language than was the artistic one of the international avant-garde."* In his works Lada often returns to the world of his village childhood with its traditional events, village feasts and balls, and children's games. It is in the works of these artists that we can see the roots of the pig being perceived as "instrumental animal" (Swabe *et al.*, 2005) by the majority of people who in modern society are totally separated from any form of agricultural activity. Instrumental animals are more strongly transformed into objects than others – thanks to their size or function, people are more sensitive to their deaths (for example, a cow versus a chicken). In Czech society nevertheless it seems that in the case of the pig, the perception is the opposite – it is the death and the processing of its products that make this animal favourable. As presented in detail by Idel (2000), this was not the case in the 18th and 19th centuries in the central European region, when the reputation of the pig was very bad.

Animal welfare

The above-mentioned high number of household-slaughtered pigs nevertheless does not mean that all these animals are home-fed for most of their lives. In fact, buying a hog bred to maturity by a commercial breeder has become a common practice. It is therefore mainly the slaughter as a family or community event that is really of importance. Organising a whole-day "pig-roast" event, including the killing, is often listed in countryside hotel and restaurant menus and is also offered as business party.

Household and conventional

In this point an interesting contradiction can be seen. Household-produced pork nowadays has the ideation value of good quality meat from an animal that lived happily in a pigsty with straw and was fed with home-grown potatoes and cereals. The fact that the husbandry conditions in large commercial breeding facilities are widely criticised does not seem to matter when a pig is bought for a home slaughter. Ostensibly obvious extrinsic values, mentioned above, are thus easily forgotten.

Household and organic

But there are other aspects how the ideas of "home-produced wholesome food" are misunderstood. Such products generally have an ideation value of high quality reaching the formal standards for organic farming and they are often associated with a high level of animal welfare (see Council, 2006). But in the case of household pig production in the Czech Republic this is not the case. If the pigs are home fed, their living conditions are probably even less suitable for providing the appropriate responses to the above-mentioned Five freedoms of animal welfare than in large scale commercial pig factories. This applies namely to skilled, knowledgeable and conscientious stockmanship, considerate handling and transport, and in the most striking example, the appropriate environmental design of husbandry systems.

Conclusions

It is argued that in the case of household pork production and slaughtering in the Czech Republic the "home-made" character of not only meat products, but a whole social event, is more of a "label" (see McEachern, Schröder, 2002) which provides the consumer with a false idea of rural authenticity and wholesomeness of the food. This label has its roots in cultural traditions influenced by artistic and media presentation of the pig as an instrumental animal and its slaughter and processing as a fashionable event. Despite the practice's possible characterisation as an ecological luxury, the actual method of household pig production is more often an example of a predatory luxury which considers neither the extrinsic nor the intrinsic values of the animal.

Acknowledgements

The author wishes to acknowledge the technical assistance of Pavel Klvac and Radim Lokoc, and the editing assistance of Benjamin Vail.

References

Berry, W. (1997). *The Agricultural Crisis: A Crisis of Culture*. The Myrin Institute, New York. 41 p.

Berry, W. (2002). *The Art of the Common Place: The Agrarian Essays*. Counterpoint, Washington. 330 p.

Charles, The Prince of Wales. (2003). In The Name of Progress. *Resurgence*, 220: 26-27.

Council Regulation (EEC) N. 2092/91 of 24 June 1991 on organic production of agricultural products and indications referring thereto on agricultural products and foodstuffs. (1991). [on line]. Available on http.//europa.eu.int/eur-lex/en/consleg/pdf/1991/en_1991R2092_do_001.pdf [date of consultation: 15/04/06]

Czech Statistical Office (2006). *Monthly livestock slaughter.* [on line]. Available on http.//www.czso.cz/eng/edicniplan.nsf/s/2005-2 [date of consultation: 15/04/06]

Dvorakova-Janu, V. (1999). *Lide a jidlo (People and Food).* ISV Nakladatelstvi, Prague. 182 p.

Frankova, S., Dvorakova-Janu, V. (2003). *Psychologie vyzivy a socialni aspekty jidla. (Psychology of Nutrition and Social Aspects of Food.* Karolinum, Prague. 256 p.

Frewer, L.J., Kale, A., Van De Kroon, S.M.A., De Lauwere, C. (2005). Consumer attitudes towards the development of animal-friendly husbandry systems. *Journal of Agricultural and Environmental Ethics,* 18: 345-367

Henderson, E., Van En, R. (1999). *Sharing the Harvest: A guide to community supported agriculture.* Chelsea Green Publishing Company, White River Junction. 254 p.

Idel, A. (2000). Animal welfare aspects of the utilization of domestic animals in the German speaking area during the 18th and 19th century. In: *IFOAM 2000 – The World Grows Organic. Proceedings 13th International IFOAM Scientific Conference.* Alföldi, T., Lockeretz, W., Niggli, U. (eds). Vdf Hochschulverlag, Zurich. 320-323

Javurek, V. (2006). *Venkovske hody (Village Feast* [on line]. Available on http.//www.muzeum-teplice.cz/historie/em0404.htm [date of consultation: 15/04/06]

Librova, H. (2003). *Vlazni a vahavi: Kapitoly o ekologickem luxusu (The Half-hearted and the Hesitant: Chapters on Ecological Luxury).* DOPLNEK, Brno. 313 p.

McEachern, M.G., Schröder, M. J.A. (2002). The role of livestock production ethics in consumer values towards meat. *Journal of Agricultural and Environmental Ethics,* 15:221-337

Pecinkova, P. (1998). *Josef Lada.* Gallery, Prague. 203 p.

Pepper, D. (1996). *Modern Environmentalism: An Introduction*. Routledge, London. 376 p.

Sbirka zákonů č. 43/1963. (Law Digest No. 43/1963). (1963). 28: 200-203

Spedding, C. (2000). *Animal Welfare.* Earthscan, London.188 p.

Stibral. K. (2005) *Proc je priroda krasna? Esteticke vnimani prirody v novoveku (Why is nature beautiful? Aesthetic perception of nature in modern period).* Dokoran, Prague, 202 p.

Swabe, J.M., Rutgerts, B.L.J.E., Noordhuizen, E.N. (2005). Cultural attitudes towards killing animals. In: *The Human-animal relationship. Forever and a day.* De Jonge, F., van den Bos, R. (eds). Royal van Gorcum BV, Assen. 123-139

How animal friendly should our agricultural practice become? A moral anti-realist argumentation for veganism

Tatjana Visak
Ethics Institute, Heidelberglaan 2, 3584 CS Utrecht, Netherlands, t.visak@ethics.uu.nl

Abstract

There is a discussion in Europe about the distribution of responsibility in striving for a morally more acceptable agriculture and food sector. Only after being clearer about our aims, however, can we discuss the best distribution of responsibility. First, we need to clarify what kind of agricultural practice we would like to have. How animal friendly and sustainable, for instance, should our agricultural practice become? That is the central issue of this paper. This normative issue will be discussed from a particular meta-ethical perspective, namely that of moral anti-realism. I don't think that normative theories should be used for showing us moral truth, because there is no such thing as moral truth. There are no mind-independent moral facts out there in the world. Nor are there mind-dependent moral facts, whatever that last category should mean. (Mostly it refers to the idea that rationality would lead us to "moral truth".) The meta-ethical background idea of moral anti-realism will lead me to slightly adapt the way I think and talk about ethics and to avoid moral realist language. In several ways, deontological language seems to be linked to the false belief in moral realism. I will argue in favor of consequentialist language and discuss the normative issue of agriculture and food ethics accordingly. I will argue in favor of veganism, based upon the argument that veganism has the best consequences for animals, nature/ the environment and people.

Keywords: veganism, anti-realism, deontology, animal ethics, consequentialism

Introduction

Before distributing responsibilities in the agriculture and food sector, we have to be clearer about our aim. What kind of agricultural practice do we want to have? How animal friendly and sustainable should our agricultural practice become? Ethical analyses can be very useful in dealing with this question, but one thing which they can not do is lead us to moral truth. In this paper I will present a moral anti-realist argumentation in favour of veganism.

Moral anti-realism

My moral anti-realism is based on the observation that no satisfying answer can be given to the question `What makes moral claims true?´ Everyone who makes moral claims and believes that they are true must at least implicitly believe that something makes these claims true. Moral principles connect moral properties with physical properties such as people, actions or institutions. What establishes the connection between evaluative properties and value-neutral properties? If no satisfying answer to this question can be given, we have to conclude that moral realism is false and that there is no fact of the matter concerning what's right and what's wrong. Kantians hold that there are no moral facts 'out there', but that they are revealed by the structure of argumentation of rational actors. They think that moral claims can be true, but not really true. If you are a realist, then I ask "What makes your foundational moral principles true?" What, in the case of Rawls, makes it true that what is just is what we would choose from behind a veil of ignorance? Is this analytic, conceptually entailed by what me mean by "just"? Is this an empirical truth? Is this a synthetic a priori metaphysical truth? If so, how did we come to know this? By intuition? How do we know that our intuitions are reliable? If you are an anti-realist then I ask what value there is in asserting (as if it were true, or really true) that what is just is what we would choose from behind a veil of ignorance? What good does this do us if someone can just say "Well, according to me and my intuitive sense of justice, that's NOT what justice is!" Or if someone can just say that what he happens to want is what we would choose from behind a veil of ignorance? This kind of moral theory doesn't do anything to help us resolve real moral disagreements.

Projectivism

There is a lot of empirical evidence, for instance from the fields of anthropology, moral psychology and recently from brain imagining techniques, which confirms that at least a big part of our moral choices are based on sentiments. Note that this does not rule out rationalism, understood as the idea that 'real' moral judgments *should* be based on reasons. It only tells us that many moral judgments actually happen to be based on something else. According to Greene, the moral properties we think to find in the world are really projections of our minds.
We are unaware of our tendency to project, because in the course of our evolution, we have developed to do some things quickly and effortlessly. For instance, we know immediately whether an adult face is female or male, without being aware of the complex and subtle differentiations that lead us to that conclusion. To us, a face just *looks* male or female. At the same time, we find it hard to complete much easier tasks such as remembering the capitals of the states. That is partly because, as Greene puts it *"the forces of natural selection that have shaped our brains have lavished more R&D on some of our abilities and less on others."* Our moral intuitions are not intended to tell us how the world is in and of itself, but to help us to survive as individuals in a social context. The advantage of being moral in the sense of taking the interests of others into account is bigger for us than its costs (Axelrod, 1984, Wright, 2000). Cognitive processes that are easy are likely to result in a perceptual phenomenology.

Moral intuitions: Blessing or burden?

If the evidence is right, at least a great deal of our moral judgments are based on evolutionary developed and culturally shaped intuitions. Why is this problematic? First of all, this is a naturalistic explanation of why being a moral realist is the default option. Furthermore, the evidence of the sentimental basis of moral judgment confronts a moral realist with the task of arguing either that these are no 'real' moral judgments or that people ought to have the moral intuitions we happen to have. All I said so far is no knock-out argumentation against moral realism. However, a better understanding of the origin of our moral intuitions might help us to take a step back and look critically at our moral intuitions instead of simply taking them to be the messengers of moral truth.
The good thing about our moral intuitions is that they have helped us to survive and in our daily lives they still lead us in a quick and effortless way to decisions which very often are good. For instance, people are endowed with an intuitive repugnance against direct personal violation and in most of the situations this is a good thing. The problem is that we lack an intuitive repugnance against violence in indirect situations, because indirect violence has only recently become possible due to technological developments and our intuitions could not catch up with those developments. This might lead to judgments which are arbitrary from a moral point of view. Empirical evidence points to several categories of situations in which people are prone to make judgments based on affective responses which are arbitrary and cannot be rationally

explained or supported. Besides the bias of up-close versus impersonal violations, people are, for instance, prone to condemn harmless violations which they find disgusting and they are less likely to take the needs of the 'distant needy' into account. As Greene puts it: "*We are using Stone Age moral psychology in a Nuclear Age and the costs begin to rise as the world becomes smaller.*"

Consequentialism versus deontology

From a moral anti-realist perspective, deontological moral language is not attractive. References to moral rights, moral duties, moral status and so on usually fit together with a kind of moral realism, including constructivism. Furthermore, brain imaging techniques show that deontological reactions to moral dilemma's usually come very quickly and are associated with the emotional centers of the brain. It is suggested that these reactions can be understood as affect-based 'alarm bells' which simply tell us 'do this' or 'don't do this'. The problem is, as I explained above, that in some occasions those 'alarm bells' lead to decisions which, on reflection, are wrong. Consequentialist judgments, to the contrary, are based on cognitive activity. They reflect an inclusive weighing process and are therefore less arbitrary and more justifiable as a practical guideline.
Greene advises to refrain from moral realist language. He claims that "*assertions about rights, wrongs, duties, and obligations (over)state a position, but they do nothing to defend it. They tell you what to do, but they don't tell you why, and they certainly don't give someone who is inclined otherwise a reason to change her mind.*" It makes sense in a debate to point to relevant empirical information which the other party might have overlooked. It can also be useful to make someone see the issue from another perspective. This can enrich the view of the other. An ethicist can propose and discuss values and aims and evaluate whether some behaviour is counterproductive to those values and aims. Ethicists can play an important role in pointing to arbitrariness. This might seem a minor role, but don't forget that the major advances in moral development (such as an abolishment of slavery and woman's rights) are due to an eradication of arbitrariness in certain areas. With Greene, I prefer consequentialism as a practical guideline, because it does not suffer from the arbitrariness of deontological reasoning.

Practical guideline for solving moral problems

On the basis of the above mentioned considerations I propose not to refer to moral truth when discussing how we should live or, for the present purpose, what kind of agricultural practice we should strive for. 'Moral' as I propose to use it, does not refer to moral truth. However, even without believing in moral truth, people generally will be concerned with taking the interests of others into account. Even non-human primates do this, although they probably know nothing about meta-ethics. People who call themselves moral anti-realists are shown to distinguish moral from conventional transgressions and judge moral transgressions to be more serious, authority-independent and based on welfare considerations (Nichols, 2004). Even very young children distinguish moral transgressions on these grounds (Nucci, 2001). That is because of the crucial role that sentiment plays in moral judgment and in the social liability of moral norms (Nichols, 2004). 'Moral', as I propose to use it, refers to serving (or refraining from undermining) the interests of others (Greene, 2002).
The idea is roughly that no-one likes to be in pain, so we should try not to inflict pain. We fare well if our interests are being taken into account, so we should try to take each others interests into account. This needs to be specified, but the basic idea is intuitively plausible and does not seem to be far-fetched. We do not need to refer to concepts such as 'inherent worth' or 'rights' which, in turn, would need a sound normative basis and would be far-fetched as a mere practical guideline, without any aspiration for moral truth. Instead of quarrelling whether animals do

have moral rights, I would just say: "Look, you have different possibilities to feed yourself. If you can chose healthy and tasty food which is produced without harming others, why should you chose food which seriously harms others?" How can you possibly defend to adopt a diet which seriously harms others? Empirical considerations about the consequences of different options are crucial for my evaluation of possible agricultural practices.

Arguments in favour of veganism

One third of the available agricultural land in the world is used for livestock and half of the grain is fed to livestock. Within 50 years, the scarcity of agricultural land will become a problem when growing numbers of people will have to be fed. It is already impossible to feed the world on an average Western diet. There is simply not enough land available. The production of food from animals is highly inefficient and amounts to wasting food. If the plant food would be consumed directly by humans, much more humans could be adequately fed. In the production of milk, which is the most efficient conversion of plant food, up to 45% of the energy of the food is wasted. About seven times the amount of soy is needed if it is fed to animals instead of directly to humans.

A vegetarian diet requires twice as much land than a vegan diet. A typical Western diet requires even five times the amount of land. Europe imports 70% of the protein of the feed, which leads people in poor countries to grow cashcrops instead of food for themselves.

Livestock production leads to global warming, widespread pollution, deforestation, land degradation and the extinction of species. Ammonia emissions from manure storage and application lead to localized acid rain and ailing forests. Livestock production uses more energy and water than the production of vegan food.

Out of concern with all the people who need to eat something, I think we should not waste the available agricultural land and food. The consequences of adopting a vegan agriculture and food system would be preferable, because there would be plenty of healthy food available to feed the world and a lot of water, energy and land could be saved for other uses. A vegan agriculture is much less devastating for nature and the environment and a healthy environment is better for people and animals.

In addition, livestock production leads to the unnecessary suffering and slaughter of billions of animals. It should be clear that taking the interests of others into account includes the interests of the animals. Our interests in taste, convenience and economics are no proper justifications for the consumption of animal products, because the satisfaction of those interests does not outweigh the enormous harm that is caused to animals and humans. This holds especially if we are aware of the alternatives that are available for providing in human interests and the alternatives that are likely to be developed once livestock production loses terrain.

If a strict vegetarian diet could not support people's health, this would, of course be an argument against veganism. However, well-planned vegan diets can be perfectly healthy, as is confirmed by the American Dietetic Association (2003) and by other nutritionists (Dwyer and Loew, 1994, Reed Mangels and Havala, 1994).

Conclusion

If we want to take the interests of others into account, a vegan agriculture and food system is preferable. A vegan agriculture and diet is more sustainable and animal friendly and it is the best option for feeding the world. My argumentation is not based on references to moral truth, nor on moral realist concepts such as 'inherent worth', 'moral status' or 'moral rights', but on an evaluation of the real-life consequences of our actions. Much more could be said about the relevant empirical data, and I would be pleased if those crucial aspects of our daily choices would get more attention.

References

American Dietetic Association (2003). Position Paper Vegetarian Diets. Available on http://www.vegetarierbund.de [4/4/06]

Axelrod, R. (1984). *The Evolution of Cooperation*. New York, Basic Books.

Davis, S. L. (2003). The Least Harm Principle may require that Humans consume a Diet containing Large Herbivores, not a Vegan Diet. *Journal of Agricultural and Environmental Ethics*, 16: 387-394.

Dwyer, J., Loew, F. M. (1994) Nutritional Risks of Vegan Diets to Women and Children: Are they preventable? *Journal of Agricultural and Environmental Ethics*, 7 (1) 87-109.

Greene, J. D. (2002) *The terrible, horrible, no good, very bad truth about morality and what to do about it*, diss. Princeton University. 19-20.

Greene, J.D. (forthcoming) The secret joke of Kant's soul. *The Psychology and Biology of Morality*. Walter Sinnott-Armstrong, ed.

Nucci, L. (2001) *Education in the moral domain*. Cambridge. Cambridge University Press.

Nichols, S. (2004) *Sentimental Rules. On the natural foundations of moral judgment*. New York. Oxford University Press.

Paxton George, K. (1990) So Animal a Human ..., or the Moral Relevance of Being an Omnivore. *Journal of Agricultural and Environmental Ethics*, 3 (2): 172-186.

Wright, R. (1994). *Nonzero: The Logic of Human Destiny*. New York. Pantheon.

Reed Mangels, A., Havala, S., Vegan Diets for Women, Infants and Children. *Journal of Agricultural and Environmental Ethics*, 7(1) 111-122.

Part 8
Emerging technologies in food production

Genetics and tailor-made diets: Some ethical issues

Angus Dawson
Centre for Professional Ethics, Keele University, Staffordshire, ST5 5BG, United Kingdom, a.j.dawson@keele.ac.uk

Abstract

This paper has two aims. The first aim is to seek to clear up a significant confusion in some recent discussions about the role of genetics in tailor-made diets. This confusion relates to the difference between screening and testing for genetic traits related to dietary factors. Screening involves large-scale interventions in an asymptomatic population. Testing involves a focus on particular individuals chosen because of family history, medical symptoms, particular lifestyle etc. This paper explores this distinction and illustrates its importance in the area of genetic/ food interactions. It is argued that this distinction makes a significant difference to the ethical arguments about such interventions. Screening is far less likely to occur, at least partly because of the ethical problems, so discussion should focus largely on testing. The second aim is to outline and critically review the different ethical arguments that might be advanced about the use of genetic testing related to tailor-made diets. Once again it is suggested that we need to take care to specify what we wish to discuss. For example, both the particular type of test and the nature of the genetic trait we are thinking about will make a difference to the ethical acceptability of the testing. It is concluded that we need to be more precise in what we mean when we construct our ethical arguments about tailor-made diets.

Keywords: genetics; tailor-made diets; screening and testing; public health; ethical arguments

Introduction

The possession of some genes may mean that individuals will be harmed (or at increased risk of harm) if they consume certain foods. In such cases it may be possible to prevent (or reduce the risk of) such harm through adaptations to an individual's diet. Tests for such genes are not yet available, but there has been some discussion in the food ethics literature of this issue. The best discussion so far is that of Meijboom *et al.* (2003), but these issues are also mentioned in a recent report from the Food Ethics Council (2005). Most of the comment is quite negative about the prospects for such tests.

In this paper I have three tasks. The first is clarificatory, in that I argue that it is important to be clear about whether the intervention we are talking about is a form of screening or testing. Discussion is often confused on this issue, but it is important as it makes a difference to the arguments. I believe that we are in fact talking about testing in most cases. If this is true, objections based on screening programmes should not be used. Second, I outline the arguments that might be used in support of such testing. These are often left out or not fairly outlined in the literature. Thirdly, I will outline the arguments against, and subject them to critical analysis.

Screening and testing

One difficulty in discussing this issue is that we do not yet have any actual cases to subject to analysis. The danger is that this makes it too easy to dismiss all cases by focusing on unlikely scenarios. One place to start is to think about whether there is really anything new in this case. There are in fact a number of possible analogies or models that could be used for discussion. For example, the routine neonatal phenylketonuria (PKU) screening available in most developed world countries is a case of general screening, where a positive result has dietary implications for the affected individual as a means of preventing harm (in this case devastating psychological impairments). However, PKU screening does not look like a direct parallel case, as the most likely type of intervention foreseen for food-related genes is not going to be detecting *disease*, but, rather, various *risk factors* for disease. Perhaps the testing of blood cholesterol followed by diet modification provides a closer parallel to the most likely genes-diet intervention. This offers us a way into clarification of the issues as we can ask whether the proposed intervention is likely to be a form of screening or testing?

I want to suggest these two interventions are different in some important respects and that it certainly makes a difference to the ethical issues. How are they different? Screening is an intervention aimed at improving a population's health, through a reduction in disease across that population. Screening aims to identify those at risk for a specific condition in an asymptomatic population. This provides the grounds for a traditional ethical argument to suggest screening is ethically problematic, as the calculation of harms and benefits will differ, because the intervention is focused on *preventing* the development of conditions rather than *curing* them (Verweij 2000). For example, a screening programme may cause anxiety to a large number of people for little benefit. For this reason, many hold that screening programmes will only be ethical if they meet a number of specified conditions (Wilson & Jungner, 1968; Shickle, 1999). By contrast, testing is focused on individuals chosen because of their symptoms or some other characteristic (e.g. a risk factor such as a particular life-style or family history of the condition), or the individual themselves may request the test. Well-know example of screening programmes include those for cervical and breast cancer. Examples of genetic testing will include such things as Huntingdon's and BRCA1 and 2. In the latter cases screening is considered uneconomic and perhaps unethical so the focus is upon testing those held to be at increased risk.

How would such issues fit in the food context? Screening might be justifiable in some cases, but it seems unlikely to be a major issue. It is far more likely that the focus will be on individuals because they have some reason for concern such as a family history or symptoms of a particular disorder. The closer the relevant intervention is to screening the more the general public health arguments relating to screening will come into play. However, we need to take care not to focus on talk of screening inappropriately, as the arguments about screening will not apply in all food cases. It is far more likely that tailor-made diets will be recommended on the basis of genetic testing of individuals rather than a screening programme focused on a whole population.

In conclusion, there are likely to be a range of possibilities for such testing based on a set of variables including whether someone already has the disease; whether they are in an asymptomatic early stage; or whether they have a risk factor for a disease. Other variables relate to the genetics – are we talking of a single gene that brings about the disease (with 100% affected) or a multi-gene combination that pre-disposes the carriers to increased risk (by, say, 24% or 8%)? What percentage of the population is affected with the relevant gene(s)? How serious is the condition brought about by gene-diet interaction (is it life-threatening or just mildly annoying)? How easy is the diet to follow? These factors in their different combinations will influence decisions about whether to focus on screening or testing. In some cases – where a condition is serious enough – a PKU-type universal screening programme might be justified. In other cases it won't. The most important conclusion is that this is not a simple issue, so it makes little sense to be in favour or against such diet-related testing as a matter of principle.

Arguments for

As we have already seen, there will be a range of conditions and interventions according to the nature of the particular gene-diet interaction we are considering. However, it seems hard to deny that there may be some scenarios where it will be beneficial for an individual to have information about their gene(s) where available, although this will of course have to be weighed against other considerations. I will sketch two positive arguments in favour of gene-testing in relation to diet modification here.

First, an argument in favour of using such tests might appeal to the knowledge to be gained about the individual's genetic status. It can be argued that such information will enhance an individual's autonomy and provide the basis for informed decisions about their diet. The supporter of such an argument need not be committed to autonomy being the most important value, to be prioritised above all others. All this argument requires is a commitment to individual autonomy being *an* important value, something to be given serious consideration in any deliberations about public policy. This argument emphasises the importance of individuals deciding for themselves whether or not to run any risk.

Second, a cost-benefits argument can be constructed appealing to the idea of reducing the risk of harm to the individual. Once someone knows the relevant information about the risks that their genetic make-up might provide, they can choose to reduce or avoid the harm that would otherwise occur. Again, like the autonomy argument, this need not be an absolutist argument. The idea would be that the potential benefits would have to be weighed against any harms. In some cases, possible harms arising from the intervention might outweigh any benefits, but we cannot decide *a priori* how such judgments will turn out. We should reach a balanced and considered judgment in the light of the relevant empirical facts and normative considerations.

Arguments against

There are many possible arguments against the use of such genetic tests. I will sketch a number below, and suggest some quick responses. Of course, there is much more to be said here.

First, it might be argued that there are no good tests available. This may be true at this particular point, but it is purely a contingent fact. Tests are bound to appear soon. Whilst there are real issues here about the accuracy of any test (e.g. false positives and false negatives), we have no reason to assume that the production of accurate tests is impossible.

Second, there will be poor compliance after testing. However, is this a strong objection? Surely the same could be said of the public's response to many drugs or lifestyle advice. Should we really not do PKU screening because some individuals fail to follow the suggested diet? Should we really not try and discourage smoking through the provision of information about the risks because not all will quit?

Third, people just won't understand because the information given will be too complex. This is likely to be true as the information about genetics and diets will probably be quite technical. However, we do not require medical patients to understand all about their condition and the relevant treatment. They only need to understand in broad terms. Do we really need to require more in relation to such testing? If so, why? In addition, we have good reason to be sceptical about the possibility of gaining an informed consent in general (Dawson, 2005) as well as in relation to screening (Nijsingh, 2006).

Fourth, it might be argued that there is something special about food, perhaps because what we eat and how we eat is part of our identity, and such diet-modification will erode this. The main problem with this argument is its vagueness. Do you become a different 'person' by becoming a vegetarian? Is our identity really so bound up with what we eat? Even if it is, it seems unlikely that anyone will make conforming to the new proposed diet compulsory. It will be open to people to choose not to follow the recommended advice just as they are free to ignore the present advice about eating more fruit and vegetables.

Fifth, perhaps the worry is really about the way that such tests may be marketed. This may be a general worry about the exploitation of health issues in a free-market capitalist economy, but this is a topic beyond the scope of this paper. Of more particular concern are the potential problems with direct-to-customer marketing. This may be a legitimate concern, particularly given the possibility of purchasing such tests over the internet. However, this is hardly a problem that arises only for such genetic tests. Most people are likely to access such tests through a health care practitioner, so careful regulation might be an answer.

Sixth, there may be a danger than when priority is given to genetic testing in relation to diet-related issues that other public health activities such as health promotion advice will be reduced or eliminated. But why think this? Why should such testing not be complementary to existing services rather than a replacement? Indeed, such testing might actually have a positive impact upon the take-up of such advice, as it may provide a focus for life-style change, and make it less abstract and more directly relevant. There might be a particular worry that in publicly funded health systems such testing might be seen as a 'quick fix' and given priority over other measures. However, this seems unlikely unless the relevant gene(s) are common enough for screening or testing to be cost-effective. If this is so, perhaps it should indeed be given priority.

Seventh, there is the reverse of the harms and benefits argument given in support of such testing above. It might be argued that harms and benefits calculation do not work out in favour of testing. For example, it might be the case that because the test does not identifying a disease but only a predisposition then it is not worth doing. However, if I can counteract that predisposition it might well be worth it (at least to me). Why should I wait until I am diseased before doing something? Is it not better to (at least give me the choice to) prevent a problem developing if I can? So the issue comes back to elements such as the seriousness of the condition, the prevalence in the population, the degree of risk reduction, and the nature of the action to be taken in response to the intervention etc. It is easy to imagine situations where the harm/benefit calculation might well be positive. Indeed, it might turn out that the arguments in favour of screening for diet-related genes may be stronger than for screening for many genetic disorders (as part of routine antenatal and neonatal care). If we can identify a particular gene that predisposes the bearer to colorectal cancer, is there really anything wrong in screening when the best thing to do to combat the risk (once known about) is to modify the diet?

Conclusions

In conclusion, I have argued that we need to be clear about what we are talking about (whether such intervention are cases of screening and/or testing) as this makes a difference to the arguments. There is a real danger of exaggeration because of speculation about what such tests will be like. We should not just dismiss such new developments through the application of general anti-technology arguments: such technology can bring benefits as well as harms. Even where such tests do not yet exist we can model the characteristics of different possible gene and diet interactions. The only fair conclusion is that it is makes no sense to be *against* such screening and/or testing as such. The core to any assessment has to be careful deliberation over the harms and benefits. Where we lack the relevant empirical evidence we should suspend judgment and not be tempted to hold *a priori* that such testing is unethical.

References

Dawson A. (2005). 'Informed consent: bioethical ideal and empirical reality'. In: *Bioethics and Social Reality*. Hayry, M., Takala, T., & Herissone-Kelly, P. (eds.). Rodopi, Amsterdam/New York.

Food Ethics Council. (2005). *Getting Personal: Shifting Responsibilities for Dietary Health*. FEC, Brighton.

Shickle, D. (1999). 'The Wilson and Jungner principles of screening and genetic testing'. In: *The Ethics of Genetic Screening*. Chadwick, R., Shickle, D., Ten Have, H. & Wiesing, U. (eds.). Kluwer, Dordrecht/Boston. 1-34.

Meijboom, F.L.B., Verweij, M.F., & Brom, F.W.A. (2002). 'You eat what you are: moral dimensions of diets tailored to one's genes'. *Journal of Agricultural and Environmental Ethics*; 16: 557-568.

Nijsingh, N. (2006). 'Informed consent and the expansion of newborn screening'. In: *Ethics, Prevention, and Public Health*. Dawson, A. & Verweij, M. (eds.). Oxford University Press, Oxford/New York.

Verweij, M.F. (2000). *Preventive Medicine Between Obligation and Aspiration*. Kluwer, Dordrecht/Boston.

Wilson, J.M.G., & Jungner, G. (1968). *Principles and Practice of Screening for Disease*. World Health Organization, Geneva.

Nanofood: Lessons to be learnt from the debate on GM crops?

Mette Ebbesen
Centre for Bioethics & Interdisciplinary Nanoscience Center (iNANO), University of Aarhus, Taasingegade 3, Building 1443, DK-8000 Aarhus C, Denmark, meb@teo.au.dk

Abstract

Basic research into nanoscience has shown tremendous potential for the use of nanotechnology to improve food safety and nutritional composition, so-called nanofood. Nanotechnology may for instance provide solutions to nanoscale biosensors for pathogen detection and to delivery systems for bioactive ingredients in foodstuffs through improved knowledge of food materials and their uptake at the nanoscale. However, researchers and society in general need to be aware of the risk that nanofood may suffer the same destiny as Genetically Modified (GM) crops, which have been boycotted by consumers in many parts of the world. This paper outlines the lessons to be learnt from the public debate on GM crops. Public acceptance of nanotechnology is assumed to depend on confidence, which is created through information, education, openness and debate. However, empirical studies indicate that public attitudes towards biotechnology are shaped not only by information, education, openness and debate but also by risk perception and by moral and democratic considerations. This paper shows that from these empirical studies we can learn that public information on nanofood should address political, sociological and ethical aspects to meet the public requirements. The humanities and the social sciences do research into several of these aspects, for instance, they reflect on the objectives we wish to realise by introducing new technology and the values at stake. These reflections aim not to build trust and acceptance in the public, but to critically assess new technology so that the public can make informed judgement.

Keywords: nanofood, GM crops, ethics, public, information

Introduction

Two of the keystones in modern food technology are safety of groceries and healthy nutrition. Both fields are considered high priority areas in the food industry and at the governing authorities. A contamination of foodstuff may have indescribable consequences, not only for the consumers, but also economically for the implicated companies. Unhealthy nutrition is one of the most essential factors for development of several chronic diseases and of reduced life quality. Importantly, basic research into nanoscience has shown tremendous potential for the use of nanotechnology to improve food safety and nutritional composition, so-called nanofood.
On one hand, the use of nanotechnology within food and nutrition research has remarkable potential. However, on the other hand, reports on strategies for nanotechnology research point out the importance of avoiding that nanotechnology suffers the same destiny as GM crops which have been boycotted by the consumers (NSF 2001: 63; Royal Netherlands Academy of Art and Sciences 2004: 27). According to these reports nanotechnology can avoid the crisis of GM crops by informing the public about scientific and technological developments; and moreover by including the public in discussions of the pros and cons of nanotechnology.
This paper explores which factors shape public attitudes towards new technology and it explores which lessons the field of nanofood can learn from the debate on GM crops. Essentially, the paper focuses on the role of the humanities and social sciences regarding the implementation of nanotechnology.

The use of nanotechnology within food and nutrition research

Nanotechnology refers to a cluster of technologies directed at making, studying and manipulating structures at the nanometre scale. The prefix 'nano' comes from the Greek word nanos meaning dwarf, i.e. nano refers to something small. Nano designates 10^{-9}, which means that one nanometre (nm) is one thousand millionth of a metre. Research on the nanoscale is not new in the sense that researchers have studied atoms and molecules for well over a century. However, it is not this 'old-fashioned' nanotechnology that creates so much interest. What is new about nanotechnology is the fact that researchers are now capable of handling and characterizing nanostructures by means of advanced microscopes and thus are gaining the power to alter physical structure at the atomic level. A variety of disciplines contribute to nanotechnology, such as molecular biology, biology, medicine, food and nutrition research, chemistry, physics, electronics, engineering and materials science. These disciplines co-operate, share knowledge and develop a culture beyond traditional disciplinary boundaries to explore and fulfil some of the visions and goals of nanotechnology. For instance, in the field of food and nutrition research various disciplines contribute with their specific expertise to develop nanoscale biosensors for pathogen detection and to develop nanostructured surfaces that can reduce bacteria growth on food processing equipment (Figure 1).

Examples of the use of nanotechnology within food and nutrition research:

- Nanorobots with biosensors for detection of fungal spores, insects, bacteria or viruses
- Development of nanostructured surfaces that can reduce bacteria growth on food processing equipment
- Development of thinner, stronger and cheaper packaging material
- Research into the effect of micro and macronutrients in foodstuffs on the human health
- Development of tasty food products with a more desirable nutrient profile
- Development of delivery systems for bioactive ingredients in foodstuffs through improved knowledge of food materials at the nanoscale

---

Figure 1. Examples of the use of nanotechnology within food and nutrition research (Nanofood Consortium 2005).

The humanities and social sciences as a means of gaining public acceptance

It appears from European and American reports that particular efforts are devoted to integrating the humanities and the social sciences into the interdisciplinary approach to nanotechnology. The overall objective is to gain the general public's acceptance of nanotechnology in order not to provoke a consumer boycott, as it happened with GM crops and foods. It is stated implicitly that this accept depends on the public's confidence in the technology and that the confidence is created on the basis of information, education, openness and debate. Thus, in a European report it says: "Without a serious communication effort, nanotechnology innovations could face an unjust negative public reception. An effective two-way dialogue is indispensable, whereby the general public's views are taken into account and may be seen to influence decisions concerning R&D[77] policy. The public trust and acceptance of nanotechnology will be crucial for its long-term

---

[77] Research and Development (R&D).

development and allow us to profit from its potential benefits. It is evident that the scientific community will have to improve its communication skills" (EU Commission 2004a: 19).
An American report states that the integration of researchers within the humanities and social sciences can establish a dialogue between nanotechnologists and the public. According to the report, this dialogue will assist in maximising the social benefits of the technology and in minimising the risk of debilitating public controversies: "The inclusion of social scientists and humanistic scholars, such as philosophers of ethics, in the social process of setting visions for nanotechnology is an important step for the NNI[78]. As scientists or dedicated scholars in their own right, they can respect the professional integrity of nanoscientists and nanotechnologists, while contributing a fresh perspective. Given appropriate support, they could inform themselves deeply enough about a particular nanotechnology to have a well-grounded evaluation. At the same time, they are professionally trained representatives of the public interest and capable of functioning as communicators between nanotechnologists and the public or government officials. Their input may help maximize the societal benefits of the technology while reducing the possibility of debilitating public controversies" (NSF 2001: 15).
In the American report, it is mentioned that informing the public is not enough; the public have to be educated to perceive the advantages of nanotechnology (NSF 2001: 100-101).
Thus, it is assumed that informing and educating the public will create trust and consequently an acceptance of nanotechnology. In that way, according to the American report, research into the societal implications of nanotechnology will boost the success of nanotechnology, and hence it will be possible to take advantage of the benefits of nanotechnology sooner, more effectively and with greater confidence (NSF 2001: 2). Hence, it is not assumed that information about nanotechnology may lead to scepticism. The public must perceive and be convinced of the benefits of the introduction of nanotechnology and no importance is attached on the public's informed judgment. However, a few EU reports assume the citizen's right to informed judgement. But in these reports it is also stressed that educating people in science and technology must be prioritised in order to obtain this informed judgement (European Commission 2004b: 7-18).
Nonetheless, researchers point out that information and education are not the only factors influencing the public attitudes towards new technology. Returning to the Europeans' sceptical attitude towards GM foods, there is disagreement whether the scepticism is exclusively due to lack of information and education. If we first look at the results of the so-called Eurobarometer survey on the European's attitudes towards GM crops and foods, it shows an increasing scepticism from 1996-1999 about GM crops (a rise from 20% to 32%) and about GM foods (a rise from 39% to 52%), respectively. In contrast, the figures were relatively stable from 1999 to 2002 (European Commission 2003). However, regarding the application of biotechnology in medical science, the Europeans' attitudes were positive in 2002: E.g. only 9% were opposed to genetic testing and 17% to cloning of human cells (European Commission 2003). Hence, the general public's attitudes varied according to the specific biotechnological application. Applications within the plant and food area were assessed considerably more negatively than applications in the field of medicine. The ethicist *Bryn Williams-Jones* from the University of Cambridge writes: "Indeed, there tends to be widespread positive public regard for technologies that appear to have a clear benefit and minimal or at least well understood risks (*e.g.* biotechnologies that improve health care, such as genetic diagnostics or bio-pharmaceuticals). But when the benefits are dubious and the risks are potentially very serious and not well understood, as in the case of GM foods, then the public as consumer of new technologies may be very wary. The lesson for a nascent field such as nanotechnology – in which there are as yet few applications, but which is receiving billions of dollars of public monies – is that there must be broad and genuine public engagement in determining the scope and possible futures for this field" (William-Jones 2004).

[78] National Nanotechnology Initiative (NNI).

The Eurobarometer surveys, which are based on responses from approx. 1000 individuals in each EU country, depict how different perceptions of biotechnology are distributed among the population on EU level and within the individual countries. However, these quantitative investigations are not sufficient to explain *why* the general public responds the way it does. As mentioned above, the reports on research into nanotechnology blame the general public's lack of knowledge of new technology for the boycott of GM food products (Royal Netherlands Academy of Arts and Sciences 2004: 27). Taking the studies on the Europeans' knowledge of GM crops and foods into consideration, it is fair to point out the lack of knowledge, for instance 64% of the European population believed that GM tomatoes contain genes as opposed to ordinary tomatoes (European Commission 2003). However, science sociologist *Claire Marris* emphasises that studies have shown that a greater insight into GM organisms does not necessarily lead to a more positive attitude; on the contrary, it makes the public more sceptical and polarised (Marris 2001; Sjoeberg 2004). Marris dismisses it as a myth that persons who are against GM foods are irrational, and that they would accept GM foods if they knew more about biotechnology (Marris 2001). In the debate on GM crops and foods there has been much focus on the public's confidence in the experts. The argument goes that without confidence in the experts the public will misunderstand risks and uncertainties. The public will then be persuaded by the opposing organisations using eye-catching headlines. Consequently, risk communications by trusted experts has long been offered as the solution to public scepticism (Gaskell *et al.* 2003). However, a Swedish study shows that confidence in experts only plays a small role in connection with the public perception of risk. Topics like 'intervention in nature' and moral considerations generally mean a lot more (Sjoeberg 2004). Researchers claim that the European population's perception of risk in connection with GM foods is much broader than the technical-scientific perception communicated by experts. In the public mind, risk also involves moral considerations (is it right doing this?), democratic considerations (who is funding and controlling biotechnology?) and uncertainties (will there be any yet unknown adverse consequences?) (Gaskell *et al.* 2003). This is also the conclusion of a Danish qualitative investigation made in year 2000 based on focus group interviews. The interviewees' assessment of risk included considerations on the possible violation of the order of nature, violation of the eigenvalue of nature and of God's creation. The respondents also mentioned power relations, democratic rights and the possible application of biotechnology to prevent poverty in developing countries (Lassen and Jamison 2006; Tveit *et al.* 2003: 9-14). The referred studies indicate that viewed from a traditional (technical-scientific) risk assessment perspective, the use of new technology may be unproblematic. However, the application of the new technology may yet still be rejected by the public on social, economic, ethical and political grounds.

The studies referred to above indicate that social, economical, ethical and political dimensions of implementation of new technology are important to the public. A lesson to be learnt from the introduction of GM foods regarding the implementation of nanotechnology may hence be that information to the public on nanotechnology should encompass more aspects than specific technical-scientific facts. It must also deal with political, sociological and ethical aspects of nanotechnology.

The critical function of the humanities and social sciences

As described above, it appears from reports on nanotechnology that the role of the humanities and social sciences is to maximise the societal advantages of nanotechnology, boost nanotechnology and reduce the possibility of debilitating public controversies. This entails e.g. that ethics is reduced to a tool or a means to an instrumental end, which can be expressed as a reduction of ethics to a PR agent for the laboratory. I object that this is a narrow apprehension of the role of the humanities and the social sciences to focus on creating trust in and acceptance of nanotechnology in the general public. The humanities and social sciences have a critical function. I believe that the function of philosophy and ethics regarding the implementation of any kind of new technology is to ask the fundamental questions such as: What impact will this new technology have on humanity? What is a good life? And will this new technology impact the realisation of a good life? The aim of posing these questions is not to build trust and acceptance in the public, but to make a critical assessment of new technology so that the public may make an informed judgement. This critical assessment does not have to be a negative one. Ethics is not only a demarcator saying thus far and no further. Instead, ethics may be viewed as a co-player firstly discussing the needs and goals of the public and society, and secondly serving as a framework to guide society towards these goals. As for nanotechnology, it should be contemplated, which goals we wish to obtain by means of technology. Is it the goals stated in the reports on nanotechnology research strategies? Or is it totally different goals? To mention a specific example, some reports state that the aim of research into nanotechnology is to improve human quality of life (European Commission 2004a: 1). But what does it mean to improve human quality of life? An American report claims that the answer lies in the improvement of human capabilities and performance, which includes improving work efficiency and learning and enhancing individual sensory and cognitive capabilities while at the same time respecting fundamental values (NSF/DOC 2002: ix-x). Ethics could contribute to a reflection on whether improving human quality of life really equals improving its capabilities and performance, and in the first place whether it is possible to improve human subjects without compromising fundamental ethical values.

Closing remarks

The above may be an idealisation of the role of ethics regarding the implementation of nanotechnology. Maybe in reality, nanotechnologists are most likely to include an ethical dimension in their research projects, since it is required in their applications for research funding. However, ethics should take advantage of the fact that there is a market for ethics in connection with the current research into nanotechnology; ethics should take advantage of the fact that it can be voiced and heard. The ethicist must then retain his or her critical sense and provide the society with impartial information on nanotechnology. I believe that nanotechnologists have a duty to take part in ethical discussions within both the professional and the political context. Nanotechnologists are not only practitioners of a profession dealing with ethically relevant matters. They are also citizens with a special expertise creating a specific responsibility. Within both the professional and political context, ethical reflection demands an interdisciplinary co-operation between nanotechnologists and ethicists. To create the optimum basis for this kind of co-operation, it is necessary to establish interdisciplinary research environments integrating the humanities and the social sciences, where for instance ethicists and nanotechnologists are in daily dialogue facilitating ethical reflection as an integral part of the research process of nanotechnology.

References

European Commission (2003). *Europeans and Biotechnology in 2002*. Eurobarometer 58.0.

European Commission (2004a). *Towards a European Strategy for Nanotechnology*.

European Commission (2004b). *Science and Society Action Plan*.

Gaskell, G., Allum, N., Bauer, M., Jackson, J., Howard, S. and Lindsey, N. (2003). Climate Change for Biotechnology? UK Public Opinion 1991-2002. *AgBioForum*, 6 (1&2): 55-67.

Lassen, J. and Jamison, A. (2006). Genetic Technologies Meet the Public. *Science, Technology, & Human Values*, 31 (1): 8-28.

Marris, C. (2001). Public views on GMOs: Deconstructing the myths. Stakeholders in the GMO debate often describe public opinion as irrational. But do they really understand the public? *EMBO Reports*, 2(7):545-8.

NanoFOOD Consortium (2005). Conference on nanofood, University of Aarhus, Denmark, 7 June 2005.

NSF (2001). *Societal Implications of Nanoscience and Nanotechnology*. Report of a workshop run by the National Science Foundation, 28-29 September 2000. Roco, M. C. and Bainbridge, W. S. (eds.). Kluwer Academic Publishers, Boston.

NSF/DOC (2002). *Converging Technologies for Improving Human Performance*, Nanotechnology, Biotechnology, Information Technology and Cognitive Science. NSF/DOC-sponsored report. Roco, M. C. and Bainbridge, W. S. (eds.).

Royal Netherlands Academy of Arts and Sciences (2004). *How big can small actual be?* Study Group on the Consequences of Nanotechnology.

Sjoeberg, L. (2004). Principles of risk perception applied to gene technology. *EMBO Reports*, 5: S47-s51.

Tveit, G., Madsen, K. H. and Sandoe, P. (eds.) (2003). *Biotechnology and the public*. Danish Centre for Bioethics and Risk Assessment.

Williams-Jones, B. (2004). A Spoonful of Trust Helps the Nanotech Go Down. *Health Law Review*, 12 (3):10-13.

Farm animal cloning: The role of the concept of animal integrity in debating and regulating the technology

Mickey Gjerris and Peter Sandøe
Danish Centre for Bioethics and Risk Assessment, Royal Veterinary and Agricultural, University, Rolighedsvej 25, DK-1958 Frederiksberg C, Denmark, mgj@kvl.dk

Abstract

Farm animal cloning is moving out of the labs and into the marketplace. The area to benefit most from this will be biomedical research, where the technology can be used to produce disease models and bioreactors; but there are also plans to utilize the technology in the food production system, mainly through the cloning of animals with outstanding commercial value for breeding purposes. Only a few countries now have legislation covering farm animal cloning (Denmark, Norway), but several other bodies with regulatory authority are considering measures to control the technology (the EU, US/FDA). As momentum towards regulation builds, we need to ensure that *all* the ethical concerns raised by cloning are properly considered. The standard framework concentrates on risks to human health and the environment and concerns about animal welfare, but to our mind, some ethical concerns are not captured by this. These concerns are evident both in work on applied ethics and in surveys of public attitudes to biotechnology. They can be gathered under the heading *integrity*. This concept, we argue, has rather different roles to play in the debate and the legislative process. In the debate, the open-endedness of the concept allows it to collect several issues (e.g. naturalness, dominion). In the legal context, where the boundary between acceptable and unacceptable practices must be more precise, a more carefully defined concept of integrity is required. The two notions of integrity will be discussed using the Danish legislation as an example.

Keywords: animals, cloning, ethics, regulation, integrity

Introduction

Ever since Dolly was presented back in 1997 (Wilmut *et al.* 1997) cloning has been controversial. The tendency so far has been to assume that cloning can be regulated indirectly through existing regulation on animal welfare, consumer protection and such like. Now, however, more and more countries and political bodies (the US and EU) are contemplating direct regulation. Recently Denmark and Norway passed laws on animal cloning. Mindful of lessons learned from the GM debate in the 1990s, both scientists, industry and governmental bodies are aware that it will be necessary to regulate cloning in a way that reflects the ethical concerns of the public. These include familiar concerns about risks to human health and the environment (which are to some extent met in scientific risk analysis and management) and concerns about animal welfare. But other concerns about animal cloning are less readily dealt with in conventional risk and animal welfare management. These relate to the *integrity* of animals, the unnaturalness of the cloning process and the ever-growing dominion of man over nature.
In this article we will explore how the concept of integrity can play a productive role in ethical discussions of animal cloning leading to regulation. We shall suggest that the concept of integrity will play rather different roles in the debate, on the one hand, and the legislative process, on the other. We shall also try to place cloning technology in a broad context. We shall use the recently passed Danish legislation to show how the concept of integrity can be utilized.

Moves towards regulation

Until now animal cloning has mostly been neglected from a regulatory/legislative point of view. In almost all countries and regions where research is taking place, the technology is regulated indirectly by regulatory apparatus on animal welfare and consumer protection (Lassen *et al.* 2005). Legislation on animal cloning per se has been passed only in Denmark (The Danish Ministry of Justice 2005) and Norway (The Norwegian Ministry of Justice and the Police 2004). There are several reasons for this. One is that the technology is still rather new. Another is that, since the technology presumably produces identical copies of animals, it has been assumed that the products from cloned animals and their offspring would be substantially equivalent with products from non-cloned animals and therefore - following the framework of the GM legislation - there would be no need for further regulation. A third factor is that any animal welfare problems would appear to be covered by existing rules and recommendations on animal research. However, both nationally and internationally, evaluations of the current regulatory mechanisms are now being initiated. The sudden interest in animal cloning from the legislative bodies is the result of a growing belief among people working within the political systems in the EU and the US that although the success-rates remain low, and the costs therefore remain high (Vajta and Gjerris 2006), cloning is almost ready to leave the labs and enter the market for commercial use in biomedicine and – to some degree – food production. The untidy jumble of legislation currently applying to cloning is considered one of the main barriers for this move. It is believed that clear and uniform legislation will remove this barrier (Galli *et al.* 2004). With current success rates it is, however, hard to see cloned animals playing more than a peripheral role in the food industry. Equally, biomedical applications of cloning (e.g. in bioreactors) have proved much harder to develop than was initially expected (Vajta and Gjerris 2006). The current interest in legislation may therefore be as much a matter of paving the way for investment in the development of cloning technology.

At the same time there is a growing recognition by scientists, industry and governmental bodies that the 'leap' (as it will be seen) from lab to commercialization will reignite ethical debate about animal cloning. From a public perspective it is one thing that a small number of animals are produced for research purposes in confined laboratories, but quite another when it is announced that cloned animals will be commercialized in huge numbers. This reaction is natural and understandable. The closer something gets to us, the more strongly we react. And since public reaction was not properly anticipated when GM plants were introduced commercially, with serious consequences, it seems prudent to anticipate it more carefully this time around.

Why consider integrity?

When new technologies are about to reach the marketplace it is customary to evaluate them from a societal perspective, and perhaps even to promote them as subjects of public discussion. Concerns that are typically discussed within this framework relate to animal welfare (where that may be affected) and risks to human health and the environment. But with a number of biotechnologies, including GM crops and cloned animals, other kinds of concern can also be heard in the discussion. These are concerns about the unnaturalness of the technologies, the violation of the integrity of living beings, and the increased dominion of man over nature. Such concerns also appear in most research on public perceptions of biotechnology (Einsiedel 2005, Lassen *et al.* 2006).To simplify matters, we can gather these worries under the heading of *integrity*.

From a democratic perspective it seems obvious that it would be wrong to exclude widely held concerns about the integrity of animals from public discussion. Whether or not these concerns will actually have any influence on subsequent regulation is another question: that depends on their validity. Looking at matters from a more pragmatic view, it is worth recalling that the attempt to dismiss integrity as irrational and unimportant in the debate on GM crops taught us that public acceptance of a technology can depend on things other than its safeness. And from a societal point of view it seems prudent to seek debate that engages the concerns people actually have. Again, this much should be clear from the GM crops case: to create socially robust regulation of biotechnology it is a good idea to include actual public concerns in any discussions.

The Danish legislation

In both the Danish and Norwegian animal cloning regulation (Gamborg 2005) attempts are made to meet concerns beyond those that typically appear in a risk analysis and animal welfare framework. The Danish legislation is especially noteworthy, since the concept of integrity played a significant role in the preparatory work leading up to the law. Thus the committee which undertook this work concluded (The Danish Ministry of Science, Technology and Development 2003):

The creation and use of cloned and genetically modified animals gives rise to ethical considerations... there is first of all the animals' welfare to consider. Besides, there may be risks to the environment if genetically modified animals are released; and there may be hazards to human health if using the techniques leads to the development and spreading to humans of new disease carrying viruses. The committee attaches great importance to these concerns while remaining aware of other ethical considerations that are less clearly related to the traditional scientific way of thinking. These considerations are comprised by the committee under the concept of integrity...The committee stresses that considerations on integrity should be included in the ethical evaluation but notes at the same time that the concept of integrity causes controversy and cultural conflict.

The committee further states that these considerations are to be weighed against possible benefits of the technology (to humans and animals) in accordance with *a principle of proportionality*. This principle demands that any use of cloning or genetically modification of animals should be justifiable in terms of the coveted objectives (The Danish Ministry of Science, Technology and Development 2003). Thus the Danish regulation prohibits the use of these technologies *unless* it can be argued that the objectives are of substantial value to society (The Danish Ministry of Justice 2005).

The importance of the concept of integrity in the Danish legislation becomes clear when one considers that it genuinely adds to the burden of evidence. Even if animal cloning is shown not to impair animal welfare and not to be a risk to human health or the environment, it still has to be argued that the objectives of the cloning are of substantial value and outweigh the breach of animal integrity that the technology requires. In other words, the substantial value criterion forces anybody wishing to use the technology to make a considerably stronger case. There can, of course, be argument about what has substantial value, but it cannot be argued that it is unnecessary to consider the matter. So the Danish legislation limits animal biotechnology in ways the traditional risk-based framework does not, by not only requiring absence of negative effects but also the presence of substantial benefits.

Integrity's two roles

As is recognized in the passage from the Danish report above, the concept of integrity "...[is] less clearly related to the traditional scientific way of thinking... and...causes controversy...".

This goes a long way to explaining why the concept, although frequently analysed in the ethical literature on animal biotechnology (see Bovenkerk *et al.* 2002, Rollin 2002, Gavrell Ortiz 2004 and Gjerris 2006) often seems to be superficially treated in official reports and policy work (see NAS 2002, Seamark 2003, Claxton *et al.* 2004 and Blaszak 2004). There are, however, as we have shown, good reasons both to in a genuine way to include the concept in the debate and to contemplate its possible integration into regulatory mechanisms.

Integrity collects ethical concerns ignored in scientifically based risk analysis. These include a sense that a technology is unnatural, or crosses ethical boundaries of technical intervention in animals, or involves excessive human control over natural processes and non-human life – a control that is ethically problematic, because it tends to reduce nature to some kind of raw material for human consumption (Gjerris 2006). The open-endedness of the concept of integrity accommodates the complexity of the concerns that are, or should be, up for discussion; and that is exactly the point. If we recognise this complexity the ethical debate about cloning technology will be more successful in reflecting the concerns it raises among us all (Lassen 2005, Einsiedel 2005). There is, obviously, no democratic duty to agree with any and all concerns simply because somebody nurtures them, but we do have a duty to provide ample opportunity for all concerns to be heard and discussed.

The journey from debate to regulation can be seen as a move from an open-ended interpretation of concepts to an agreed understanding of those same concepts – such agreement must arise if the various concerns are to be translated into a legal framework. There is no doubt that the ethical concerns falling under the heading of integrity are seen as legitimate by a significant proportion of the public in the Western world. And as shown in our short presentation of the Danish legislation, these concerns can be integrated in a meaningful way into legislation. There the notion of integrity serves as a reminder, within the legislative framework itself, that a technology as cloning should only be used if its acknowledged objectives warrant interference with the integrity that animals undeniably appear to possess.

To advocates of the concept of integrity, this might seem too little; to advocates of a strictly science-based approach to regulation, it might seem too much. Nevertheless the approach sketched here will help to ensure that only applications of cloning technology that are in broad terms socially acceptable will be introduced. This might leave pro-cloners wishing for wider applications of the technology, but in the end it could prevent wholesale public rejection of cloning, precisely because the regulation of the technology is as socially robust as it could be.

References

Blaszak K (2004): *The governance of welfare of animals involved in biotechnology in Victoria – current regulatory framework, relevant committees and emerging issues.* Bureau of Animal Welfare, Department Primary Industries. Australia

Bovenkerk B, Brom FWA and Babs JVDB (2002): Brave New Birds: The Use of 'Animal Integrity' in Animal Ethics. *The Hastings Center Report* 32

Claxton J, Sachez E and Matthiessen-Guyader L (2004): Ethical, legal and social aspects of farm animal cloning in the 6th framework programme for research. *Cloning and Stem Cells* 6: 178-181

Einsiedel EF (2005): Public perceptions of transgenic animals. *Rev. sci. tech. Off. int. Epiz.* 24 (1): 149-157

Galli C, Duchi R, Lagutina I and Lazzari G (2004): A European perspective on animal cloning and government regulation. *IEEE Engineering in Medicine and Biology Magazine* 23(2): 52-54

Gamborg C (2005): *Challenges in regulating farm animal cloning: An assessment of regulatory approaches and legal framework.* Cloning in Public, www.sl.kvl.dk/cloninginpublic

Gavrell Ortiz SE (2004); Beyond Welfare: Animal Integrity, Animal Dignity, and Genetic Engineering. *Ethics & the Environment.* 9 (1): 94-120

Lassen J (2005): *Public perceptions of farm animal cloning in Europe*. Cloning in Public, www.sl.kvl.dk/cloninginpublic

Lassen J, Gjerris M and Sandøe P (2006): After Dolly – Ethical limits to the use of biotechnology on animals. *Theriogenology* 65: 992–1004

National Research Council of the National Academies (NAS)(2002): *Animal Biotechnology. Science-Based Concerns*. Academy of Sciences, USA

Rollin, BE (2004): Ethics and Species Integrity. *The American Journal of Bioethics*, 3 (3): 15-17

Seamark RF (2003): *Review of the current status of the extent and use of cloning in animal production in Australia and New Zealand*. SA Consulting

The Danish Ministry of Justice (2005): *Lov om kloning og genmodificering af dyr m.v. af 14. Juni 2005* [Animal cloning and genetic modification Act of 14 June 2005]

The Danish Ministry of Science, Technology and Development (2003): *Genmodificerede og klonede dyr* [Genetically modified and cloned animals]

The Norwegian Ministry of Justice and the Police (2004): *Act No 22 of May 7 2004 amending Act No 38 of April 2 1993*

Vajta G and Gjerris M (2006): Science and technology of farm animal cloning: State of the art. *Animal Reproduction Science*. 92, 210-230

Wilmut I, Schnieke AE, McWhir J, Kind AJ and Campbell KH (1997): Viable offspring derived from fetal és adult mammalian cells. *Nature*, 385: 810-813

The transatlantic conflict over GM food: Cultural background

Tatiana Kochetkova
Utrecht Universiteit voor Humanistiek, Van Asch van Wijckskade 28, 3500 AT Utrecht, The Netherlands, tk@uvh.nl

Abstract

During the last two decades, an unexpected conflict over GM food arose, with particularly strong transatlantic consequences. Both the defenders and the opponents of GM food are found on both sides of the Atlantic, however, the opponents, affiliated with NGO's and ecological parties, have more success in Europe, while the defenders, bound to GM multinationals, are stronger on the American side. This paper explores the cultural grounds of this conflict. It examines the academic, legal and political argumentation on GM food to uncover the underlying metaphors. It will claim the current debate to be a rationalization of the top of an iceberg, the underwater part of which includes visions on food, nature and human relation to Earth. The talk includes the analysis of the metaphors on GM food, like 'Frankenstein foods' or 'golden rice', and food as a 'program' or 'mechanism'. The conflict is seen as existing between, on the one hand, the Enlightenment ideology, equating living entities to mechanical ones, i.e., objects of engineering, and, on the other hand, the organic approach, rooted in traditional world views, which attributes importance not only to the final result but to the actual processing of organisms into food as well. The debate involves the metaphors of 'pure' versus 'impure' food, 'natural' versus 'artificial', with overtones traceable to alternative visions of morality. A contrast is evident between Europeans, inclined to accept the safety of traditional foodstuffs like raw milk cheese, and Americans, more attracted to experiments with new food processing technologies.

Keywords: GM food, WTO dispute, risk analysis, cultural attitudes, rationality

Introduction

The rapid commercial use of genetic manipulation has turned it into a factor in legal and political transatlantic conflicts. On one side of this conflict are the major exporters of GM crops: Canada, Argentina and the US, blaming Europeans for their unwillingness their products, while on the other side is the European Community, with its wide and culturally rooted public opposition to GM crops. Even though there is an entire spectrum of opinions on GM crops on both sides of the Atlantic, the majority of opinions in Europe are on the precautionary side, while the majority of Americans appear supportive of GM.

At first sight, this controversy looks like a scientific disagreement between two different interpretations of risk from GM crops. This is, however, only part of the story, and one has to agree with Lawrence Busch that 'the current debates between the European Union and the United States over the Precautionary Principle and Risk Analysis (or what is sometimes referred to as the familiarity principle), respectively, are far less about what constitutes sound science than they are about what interests are at stake and what values are taken to be paramount in particular historical situations' (2002: 18). The other dimensions of this controversy include economic interests at stake and political pressure of GM multinationals, interpretation of international legal treatises on food export, such as the Biosafety protocol, anti-GM movement, the recent WTO dispute on GM crops, divergent cultural stereotypes, and, finally, the difference in the underlying ideologies and world views.

We shall try to find here which elements in this controversy are significant, i.e., which factors are deeper causes and which are secondary manifestations. Are the real causes conflicting economic interests, competition for global political dominance, or divergent ideologies? With this goal in view, let us consider the major dimensions of this debate:

- A disagreement in the scientific assessment of GM risks.
- Conflicts of economic and political interests, expressed in the WTO debate on GM crops.
- Difference in cultural orientations, values and world views.

Scientific assessment of risks

The European Community and the US interpret the risks of GM crops in two different ways. To the Europeans (and also to the Japanese), the important yardstick is the *precautionary principle*, i.e. the idea of scientific uncertainty concerning the possible side effects of GM on human health and environment. To the Americans, the crucial dimension is the *familiarity* or *substantial equivalence principle*, i.e. the idea that GM food should be judged according to the same criteria as conventional food, and that current scientific methods are sufficient to ascertain their safety.

The precautionary principle applied to gene manipulation is based on the fact that the final position of an inserted gene in DNA is currently unpredictable. Moreover, it is not clear how it will interact with the surrounding genes, which – given the complexity in assessing the role of genes and gene combinations, and also of the so-called 'junk DNA', in determining the phenotype – makes their effects likewise unpredictable. An example: the successful insertion of a gene for blue colour in roses did not result in a blue flower because the necessary chemical environment for this gene's expression was missing.

In contrast to this, the familiarity or substantial equivalence principle claims that current methods, as used when judging conventional food, are sufficient to determine the level of danger in every case. This is, in principle, a somewhat risky position: certain parts in the debate claim that there were adverse health effects in several specific cases, and also that the level of monitoring of US GM crop plantations was unsatisfactory. There is also the risk that relevant information on possible adverse effects may be withheld as CBI (confidential business information).[79]

These two principles can be seen as different interpretations of risk, which correlate on both sides with the interests and values involved, as will be seen below.

The legal and political disagreement of GM crops

The most obvious manifestation of this conflict is the WTO dispute on GM crops. It started in May 2003 with a formal complaint from Canada, Argentina and United States against the European Union's 'illegal moratorium' on GM crops trade. The formal accusation against EU was that it had not applied its own legislation and had taken too long to decide, thus hindering the import of GM crops, so that the economic interests of US farmers and multinationals has suffered.

The difference in interpretation is an argument in the WTO debate: the accusing side assumes that interpretations of GM risks with the familiarity principle should be accepted everywhere as a matter definitively decided by scientists rather than a disputable question. Such a position seems to undermine the usual right of every country or group to theirs own view on such topics.

In practice substantial equivalence resulted in the absence of mandatory labelling of GM crops on the American continent, and also that American public in practice cannot differentiate between GM and non-GM crops and consumes both without knowing which is which. As for the EU, its precautionary principle in practice involves more strict safety controls, mandatory

[79] Amicus Curiae Submission. 1 June 2004.

producer transparency and the defence of consumer's choice. The EU also defends the legality of a plurality of approaches to GM crops: from *laissez-faire* to complete bans, including the intermediary option of regulatory and pre-market assessment for each country, the EU choice. The EU has active support in the WTO dispute from the global NGO network as well as from some leading world scientists (cf. the *amicus curiae* submissions to the debate, in support of the legitimacy of the EU position in this case, and of its scientific and environmental soundness). Despite this worldwide public support, current developments in the debate suggest that the WTO is leaning against the EU. Although the discussion is not yet closed, its most probable outcome can be deduced from the recent interim report from 31 March 2006, in which the WTO blames the EU for violating the Sanitary and Phytosanitary Agreement, which led to the so-called moratorium on GM products and consequent undue delay. The WTO has refused to apply in this case the Biosafety Protocol, which the EU has signed, because the US did not sign it. In my view, this is a good example of preferring formal law to actual justice. The course of the GM dispute may be seen as questioning the impartialness of the WTO itself.

The divergence in the interpretation of facts springs from the divergence in values and priorities. The interests behind this dispute seem to be the following: the US (and Argentina and Canada) act on behalf of their own economic interests and express the expectations of GM business. The EU represents the interests and values of the European public, and also reflects, in a moderated way, the approach of ecological NGOs worldwide. The major motives of the European Community regulatory framework are the protection of health and the environment, as well as ensuring their citizens' right to choose. Thus, the debate seems to reflect a conflict between economic interests and the value of health and environment.

The producer-consumer conflict is easy to spot. The stance of consumers was expressed by a Dutch scientist at the Genomics Momentum 2004 conference: 'if the benefits are zero, any risk is too much.' Many European consumers, and in this regard they are close to the Japanese, consider genmodification technology as well as chemical technology for foods as 'market and trade-oriented but not as consumer- or environment oriented, since it serves for cost efficiency and for long-distance transportation', by Yoshiki Otsuka (2004). The priorities of consumers in Europe emphasise health and environmental safety of food. This concern is explained by the previous controversies in food industry, when modern technologies on agriculture resulted in various toxic contaminations, like pesticides, carcinogenic chemicals for preservatives and sweeteners and post-harvest chemicals. This experience has lead the public to see technological innovation in food production as a potential danger, and resulted in associating safe food with naturalness, purity from chemical or other technological contamination – with organic and domestic agriculture.

The cultural background of the debate. The metaphor of 'Frankenstein foods'

This orientation of the part of the European public can be also traced with media analysis. According to the content analysis of the French, American and Japanese press by Kyoko Sato (2004), the same basic values appear. Both the French *Le Monde* and the Japanese *Asahi Shimbun* more frequently use the frame of scientific uncertainty and unknown risk in relation to GM crops than the *New York Times*. In the case of articles on Starlink corn, a GM crop designed for animal feed and found in human food, the American newspaper has mostly focused on the economic consequences of this scandal, the Japanese one looked at the human health aspect and only sometimes mentioned the economic costs, and the French used the market frame even less frequently, focusing at the health aspects of the event. Further, both the *Asahi Shimbun* and *Le Monde* have sometimes framed GM crops as a 'threat to culture or traditional practices' (Kyoko Sato), while *New York Times* have never done it. As follows from this study,

- The general acceptance of GM food in the US might be explained by 'the prevalence of economic concern and lack of concern for risk'.
- It is possible that French rejection of GM food is driven largely by concern for risk.
- Japanese opposition to GM food is likely to be accounted for by the prevalence of safety concerns and the risk and uncertainty frame.

One might try to trace the link between these cultural orientations and the underlying difference in worldviews. Thus, as I will argue, the familiarity principle with respect to GM crops, which is associated with focus on the economic dimension (the view of industry), is also connected with classical rationalist metaphysics (appeared in 17-18th centuries) that opposes the rational subject to a passive material object. In our case this is how plants are viewed: a predictable object of genetic manipulation. This is also the view of genetic reductionism that sees plants as mechanisms, 'programmed' by their DNA. From this appears the metaphor of 'genetic engineering' that may produce the impression of pre-calculated genetic changes with known outcomes. Despite the full range of all positions everywhere, it seems that the defining in American position on GM crops is still the reductionist view, enabling the treatment of GM crops from familiarity principle.

An alternative vision, on which the precautionary principle is based, is more in line with the idea of science being aware of the limits of its knowledge. This alternative approach is akin to ecological vision, Gaia approach, and anti-reductionism. It sees DNA as one of the factors in determining the phenotype, along with protein structure. Hence, plants appear as self-organising systems, with probably infinite complexity, of whose genmanipulation we have only limited knowledge. This worldview find in each living being a manifestation of life as such, which final nature is unknown, while we still can study its various reflections, as genetic, protein, phenotypic, or even mental.

While genetic reductionism and rationalist metaphysics correspond to the assumptions of classical physics, also focused on the stability of the world structure, this anti-reductionist vision is more consistent with more recent ideas about physics. Nalimov (1991), for instance, claims that current physics suggests plenty of alternative models of the cosmos without giving us the possibility of determining in which one we actually live: a contemporary rediscovery of Nicolas de Cusa's '*docta ignorantia*' or wise ignorance. And if the Japanese appear closer to the French and to Europeans in general in their assessment of GM crops, which might be attributed to their strong cultural and spiritual tradition, which makes them more aware of the element of complexity and unpredictability, the unknown in the universe.

In the context of the awareness about the limits of scientific knowledge, one can better interpret the metaphor of 'Frankenstein foods' sometimes applied to GM crops. Mary Shelley's figure of Frankenstein – a scientist who created a monster from parts of dead people and managed to give him life, but then was not able to control his creation. In the metaphor of 'Frankenstein foods' one can read the fear that a scientific creation can start having a life of its own, qualities of its own, which were neither predicted nor desirable. One of the meanings of this metaphor might be the arrogance of Frankenstein, who thought he knew enough to control life, while he was in fact playing with uncertain factors. The problem might be not in the game itself, but in the lack of awareness that this is a game with the unknown.

To conclude, we suppose that behind the difference in interpretations of risk, behind the economic and political conflict of interests, even deeper divergence in world visions might lie: one party sees the world basically as a predictable object of scientific knowledge and technological manipulation, and the other stresses the limits of today's knowledge and sees the world as encompassed by the unknown.

References

Batten D. (2002). Frankenstein Foods? *Creation* 24 (4): 10-13.

Bruce Donald (2002). Finding a Balance Over Precaution. *Journal of Agricultural and Environmental Ethics*, Volume 15/1: 7-16.

Busch L. (2002). The Homiletics of Risk. *Journal of Agricultural and Environmental Ethics*, 15/1: 17-29.

Douglas, M. (1966). *Purity and Danger: An Analysis of the Concept of Pollution and Taboo*. Boston: Routledge.

Capra, F. (2003). *The Hidden Connections*. Flamingo: New York.

Klee K. (1999). Frankenstein Foods? *Newsweek*, September 13.

Kyoko S. What Are Genetically Modified Foods About?: Comparative Analysis of How Newspapers Frame GM Foods in the United States, France and Japan. Department of Sociology, Princeton University, published on the web.

Lamont, M. (1989). The Power-Culture Link in a Comparative Perspective. *Comparative Social Research* 11: 131-150.

Lamont, M. and L. Thévenot (2000). Introduction: Toward a Renewed Comparative Cultural Sociology. In: Michéle L. and L. Thévenot (eds). Rethinking Comparative Cultural Sociology: Polities and Repertoires of Evaluation in France and the United States. Cambridge: Cambridge University Press and Paris: Presses de la Maison des Sciences de l'Homme.

Lean G. (2005) Frankenstein Foods: GM Industry Puts Human Gene into Rice. *Independent Digital (UK)*.

Nalimov, V. V. (1991). Requirements to the Transformation of the Image of Science. *The Messanger of Moscow State University* 5: 7, Moscow. 21–34.

Otsuka Yoshiki (2004). Issues on food safety from a perspective of information ecology, Musashi Institute of Technology, web-publication.

Sheingate, A. (2001). *The Rise of the Agricultural Welfare State: Institutions and Interest Group Power in the United States, France, and Japan*. Princeton University Press.

Thévenot, L., M. Moody and C. Lafaye. (2000). Forms of Valuing Nature: Arguments and Modes of Justification in French and American Environmental Disputes. In: *Rethinking Comparative Cultural Sociology: Polities and Repertoires of Evaluation in France and the United States*. M. Lamont and L. Thévenot (eds). Cambridge University Press and Presses de la Maison des Sciences de l'Homme.

Nutrigenomics: A bridge too far, for now?

Albert E.J. McGill
James Martin Institute for Science and Civilization, Said Business School, Oxford University, Park End Street, Oxford OX1 1HP, United Kingdom, albert.mcgill@sbs.ox.ac.uk

Abstract

Nutrigenomics and nutrigenetics have been defined and explained and their relationship to functional foods and nutraceuticals outlined. The development of genotype testing and related dietary services have been explored from the viewpoints of the food industry, the consumer and the potential impact on public health initiatives. Evaluation of the present state of nutrigenomic informed dietary advice has been made by clinical and nutritional experts and, in their opinion, the science is not yet well enough developed to make significant contributions to dietary guidelines. Skepticism among ethicists of the present value of nutrigenomics in public health initiatives is recorded.

Keywords: nutrigenomics, nutrigenetics, diet, food, health

Introduction

Nutritional genomics is the study of how food and genes interact, with food being a complex mixture of thousands of different compounds and the human body being run by around 40,000 genes. Associations between diet and chronic disease have long been recognizes through epidemiological studies. New genomic technologies are now allowing more to be discovered about the basis of these associations through studying the functional interactions of food with the genome at the molecular, cellular and systemic levels, (nutrigenomics) and the ways in which individuals respond differently to different diets according to their individual genetic make-up (nutrigenetics). In his overview of the subject, Gibney (2005) regrets the new terminology, pointing out that "molecular nutrition" instead of nutrigenomics would have been more in keeping with the established disciplines such as "molecular biology" or "molecular medicine". Unfortunately, popularity or notoriety may derive faster from a more dramatic name in the eyes of both public and practitioners (compare "home economics" with "ekotrophology"). The two new areas are of significant importance potentially to the health-care system and to the food industry, generating the hint of a controversy. This is also very important to those seeking to understand the biology underlying normal homeostasis and disease.

The human genome project has provided the background information and new tools that enable researchers to take a more global view of diet-gene interactions, but many of the technologies are new, still being refined and a rethink of standard approaches is being required (Astley, 2006). The establishment of the European Nutrigenomics Organization (NuGO) involving 22 partner institutions from 10 EU member states with funding of 17.3 million Euros over six years, will spread excellence in the field of nutritional genomics. This will complement the work of biomedical and pharmacological research communities in genome based development of curative therapies.

At the same time services are being advertised on the internet for genotyping as a basis for the recommendation and delivery of personalized dietary advice and even a service for the manufacture and delivery of meals, similar to those sold as part of weight control regimes.

While Gibney's overview of the field is interesting, the most authoritative evaluation has been given in a report by the Public Health Genetics Unit of Cambridge University (PHGU) (2005), which has cast doubt on the value of any individualized dietary advice based on present genomic information.

Are current advertised services and diets based on nutrigenomic information fanciful or grossly misleading? Is the food industry rushing into this new market too early, in search of quick profits, as happened with dietary supplement?(SNE, 2004). What are the ethical implications of general exchange of individuals' genetic information? Will this information and its related dietary advice be used to judge compliance with the more individual responsibility for diet-related health outlined in progressive Department of Health white papers (DOH) (2004; 2005)?

Methodology

The Pros and Cons of applying current nutrigenomic information for dietary advice seem to derive from the Food Industry and medical geneticists, with many scientists and a large part of the general public found somewhere between the two extremes.

The Food Industry, its spin-off companies and its new entrepreneurs have a responsibility to their shareholders and financial backers to make a profit on their operations. New raw materials, processes, products and services provide fertile ground for these developments. One of the earlier developments in this field was Functional foods, closely followed by Nutraceuticals. Karla (2003) defines (and redefines) both concepts as well as clarifying their differences from dietary supplements. He refines the original definition of Nutraceutical by its originator, DeFelice (Brower, 1998), "..a food (or part of a food) that provides medical or health benefits, including the prevention and/or treatment of a disease", but would run into opposition on Functional Foods as defined by Just-Food (2006) as "..those foods or food components (whole,fortified or enriched with functional food components) that are scientifically recognized as having physiological benefits beyond those of basic nutrition". The current market for such foods is estimated by this company to be US$ 7-63 billion and is expected to grow to US$ 167 billion by 2010. Its advantage is in being seen as "natural" medicine and requires no testing as a pharmaceutical product. If it is possible to test each individual (and charge heavily for it; US$ 150-1500) and then "prescribe" a diet, which, as it is tailored, you may supply at a special price, there is considerable business to be done.

Many health professionals see Nutrigenomics at present as a supplementary tool and not an individualized replacement for general dietary guidelines (SNE, 2004). Indeed there is much room for sowing confusion and frustration amongst consumers by providing conflicting advice when their confidence in nutritional advice itself is not high. However, the European Nutrigenomics Organization (NuGO) and its outreach arm, The Nutrigenomics Society, are dedicated ".. to develop, integrate and facilitate genomic technologies and research for nutritional science, to train a new generation of nutrigenomics scientists, in order to improve the impact of nutrition in health promotion and disease prevention."(NuGO, 2006). Their motivation and funding will enable rapid developments in this field to be pursued vigorously.

Not all scientists are convinced of the readiness of Nutrigenomics to play a significant role in the prevention and treatment of major chronic diseases. In 2003 The Nuffield Trust invited the Public Health Genetics Unit of Cambridge University to organize and host a workshop to explore "..this gene environment interaction in the important area of diet and susceptibility to chronic conditions such as heart disease, cancer, obesity and diabetes." Eminent national researchers were brought together with representatives of key policy making bodies to examine the evidence for nutrient gene interaction, to consider clinical applications and to make recommendations for policy implications. The workshop was held in 2004 and the report published in the following year (PGHU, 2005).

In 2004 and 2005 the Department of Health in the UK published two sequential White Papers developing the idea of more individual responsibility for dietary choice and resulting consumer health (DOH, 2004; 2005). The relationship between the two areas was not stated but cannot be discounted.

Results

Although there are many companies offering their Nutrigenomic products and services on the internet, only a token sample has been selected. Salugen (2005) "..pioneered nutritional gene therapy by analyzing a person's DNA to customize their nutritional supplement pill ingredients".

Nutrigen (2006) offers Nutrigenetic testing but specifies that it is only to indicate risk factors associated with Heart Health, Bone Health, Insulin Resistance, Antioxidant/Detoxification and Inflammation. However the company's founder will offer consultations with customers/patients based on their DNA analysis to help them build " a wellness plan that may help you to avoid health problems". A report of a more skeptical nature (Redherring, 2006) quotes a new company, Scionia, that has financing of US$ 8.3 million, that expects to make its CEO,James Bruce, a profit by tailoring perfect diets for customers based on a US$ 250 DNA test. This expectation is challenged by nutrigenomics researcher Dr Richard Krauss of the Childrens' Hospita[in Oakland, California. Another skeptic is Bruce Grierson (Moodfoods, 2006) who reckoned the concept was fine but the technology too distant.

The Expert Group at Cambridge (PGHU, 2005) in its recommendations, "..urged caution in promising too much from this very new science too soon and emphasized the need not to confuse population public health messages." The consensus of their meeting was that it was much too soon to develop any new " healthy eating" messages derived from nutrigenomics research. At the individual level there is a danger of scaring people about increased risks. The group noted that genotyping as a basis for delivery of personalized dietary advice is available over the internet. It was not thought that the level of such activity in the UK was sufficient at present to justify the issuing of public advice.

Two White Papers of the Department of Health, UK, appeared to move the responsibility for consumer health away from government and onto the population. The Food Ethics Council (2005), criticized the papers on the basis that good food choices cannot be made in ignorance or when suitable produce is unavailable for either economic or geographic reasons. A subsequent launch symposium discussed their own report and debated how responsibilities for diet and health should be shared amongst the state, companies and citizens (Food Ethics Council, 2006). One of its three conclusions asked how relevant personalized nutrition was within science and the food industry and will such an approach impact on public health policy. There was general skepticism that nutrigenomics was likely to deliver widespread health benefits, even in the long run.

Discussion and conclusions

There are many projects underway to advance the science and applications of Nutrigenomics for the improvement of human health and wellbeing, especially in organizations such as NuGO. The food industry is keen to launch products and services that will utilize the techniques and yield suitable profits. Many clinical and nutritional experts believe that the field is not well enough developed as yet and that its engagement too early will fail to impress, may frighten and could confuse consumers.

The Food Ethics Council and many followers believe that the UK government's new approach to personalize health will push health responsibilities away from government and onto both consumers and the food industry.

There will be an ongoing need to observe and scrutinize the developments of Nutrigenetics and Nutrigenomics and their applications to be confident of their place in public nutrition and the ethical shift of the responsibility for the nation's health.

References

Astley,S. (2006). *The European Nutrigenomics Organisation – NuGO.* Available on http://www.ifr.bbsrc.ac.uk/Science/ScienceBriefs/nugo.html [date of consultation: 10/01/2006].

Brower,V. (1998).Nutraceuticals: poised for a healthy slice of the healthcare market? *National Biotechnology,*16:728-731.

DOH (Department of Health, UK). (2004). *Choosing health: making healthier choices easier.* Department of Health, London. 128pp.

DOH (2005). *Choosing a Better Diet: a food and health action plan.* Department of Health, London. 52pp.

Food Ethics Council. (2005). *GETTING PERSONAL: shifting responsibilities for dietary health.* ISBN 0-9549218-1-X. Available on http://www.foodethicscouncil.org [date of consultation: 20/12/2005].

Food Ethics Council. (2006). *GETTING PERSONAL: shifting responsibilities for dietary health. Launch symposium report.* Available on http://www.foodethicscouncil.org [date of consultation: 21/04/2006].

Gibney,M.J. (2005). Nutrigenomics in human nutrition – an overview. *South African Journal of Clinical Nutrition,*18: 115-118.

Just-Food (2006). *Global market review of functional foods – forecasts to 2010.* available on http://www.just-food.com/store/product [date of consultation: 24/04/2006].

Karla, E.K. (2003). Nutraceutical – Definition and Introduction. *AAPS Pharmaceutical Science,* 5:article 25.

Moodfoods, (2006). *Eat right for your genotype.* Available on http://www.moodfoods.com/nutrigenomics/ [date consulted: 20/04/2006].

NuGO,(2006). *The Nutrigenomics Society, the Outreach of the European Nutrigenomics Organisation – NuGO.* Available on http://www.nugo.org/everyone [date consulted: 10/01/2006].

Nutrigen,(2006). *Nutrigenetics: The Art and Science of your Biological Code.* Available on http://www.1-888-nutrigen.com [date of consultation: 24/04/2006].

PHGU (Public Health Genetics Unit). (2005). *NUTRIGENOMICS. Report of a workshop hosted by The Nuffield Trust and PHGU on 5 February 2004.* The Nuffield Trust, London.26pp.

Redherring. (2006). *Top 10 Trends: Nutrigenomics.*Available on http://www.redherring.com/Article [date of consultation: 24/04/2006].

Salugen,(20060. *Innovations in Genetics.* Available on http://www.salugen.com/innovations.html [date of consultation: 24/04/2006].

SNE (Society for Nutrition Education) (2004). *Nutrition and Genes Report. Comments – Society for Nutrition Education, September 17, 2004.* Available on http://www.sne.org [date of consultation: 20/12/2005].

The transatlantic rift in genetically modified food policy

Celina Ramjoué
University of Zurich, Department of Political Science, Seilergraben 53, 8001 Zurich, Switzerland, cramjoue@pwi.unizh.ch

Abstract

The regulatory structures underlying United States and European Union policies regarding genetically modified food (GM) and crops are fundamentally different. The US regulates GM foods and crops as end *products*, applying the same regulatory framework that it would to non GM foods or crops. The EU, on the other hand, regulates agricultural biotechnology products as the result of a specific production *process* and has developed a wide network of rules that regulate GM foods and crops specifically. The result is a relatively permissive US regulation of GM foods, and a relatively restrictive regulation in the EU. Why are genetically modified food policies in the United States and the European Union different? This paper argues that differences between GM food policies in the US and the EU can be explained by three main elements. First, US and EU policies of the 1970s and 1980s on recombinant DNA research and events leading up to early GM food and crop regulation allow a deeper understanding of current policy. Second, underlying beliefs and values condition the content of current GM food and crop policy. Third, actors for and against agricultural biotechnology have achieved different levels of success in influencing policy in the US and the EU.

Keywords: genetically modified food, genetically modified crops, agricultural biotechnology, regulation, public policy

Introduction

Genetically modified (GM) crops are produced by genetic transfers *between different species*, whereas traditional crossbreeding is possible only *within the same species*. GM foods and crops are types of genetically modified organisms (GMOs). GM crops can mimic aspects of the metabolisms of unrelated species. Common GM crops are herbicide tolerant and/or insect resistant soy, corn, cotton, and oilseed rape. These crops are used in animal feed and processed foods. Main GM crop producers are the US, Argentina, Canada and Brazil. First shipments of GM grain to Europe in the mid-1990s led to first protests. A decade later, GM food is still very controversial, and the US and the EU are engaged in a World Trade Organization dispute over GM crops.

Central to this dispute is the question of how to regulate GM foods and crops. US and EU policies address the same types of products, and yet they are strikingly different. US policy is relatively permissive, while EU policy is relatively restrictive. The US regulates GM foods just as it would any other food, while the EU has regulations specifically for GM foods and crops. More GM crops and foods are approved more quickly in the US than in the EU.

In the light of these regulatory differences and the resulting transatlantic dispute, this paper addresses the question: *Why are GM food policies in the United States and in the European Union different?* It first gives a comparative overview of the differences between GM food and crop policy in the US and the EU, and then presents three elements explaining the transatlantic rift in GM food policy.

US and EU GM food and crop policy

US policy is based on the premise that GM foods and crops should be regulated like any other food and regardless of their method of production. This is often referred to as a "product-based approach". In contrast, EU policy takes a "process-based approach", which posits that the process of production is what should trigger a specific kind of regulation. Accordingly, the US does not have a regulatory regime specific to agricultural biotechnology, while the EU does.

In the US, three agencies and three statutes that make up the regulatory framework for GM foods and crops. The US Department of Agriculture (USDA) is responsible for protecting agriculture from pests and diseases under the Plant Protection Act. The Food and Drug Administration (FDA) ensures that food, feed, and food additives are properly labeled and safe to eat for humans and animals under the Federal Food, Drug, and Cosmetic Act. The Environmental Protection Agency (EPA) ensures that pesticides used in plants are safe for the environment under the Federal Insecticide, Fungicide and Rodenticide Act. Following the product-based logic, US agencies have interpreted these existing statutes to accommodate GM foods and crops. In 1992, FDA established that it would treat GM foods just like non-GM foods. USDA has made oversight of GM crops possible by defining GM plants as potential plant pests. EPA fulfills its mandate by regulating the insecticidal substances produced by certain GM crops as pesticides.

The EU's regulation of GMOs is divided into horizontal legislation, which covers GMOs in general, and vertical or sector-related regulations, which deal with specific types of GMOs (e.g. food, feed, seed) and individual GMO-related issues (labelling, traceability). EU rules cover two main types of authorizations: the deliberate release of GMOs into the environment, and the authorization of GM food and feed. Directive 2001/18/EC is a horizontal piece of legislation covering deliberate release into the environment for experimental purposes and for placing on the market. Main vertical instruments within the EU are Regulation 1829/2003 on genetically modified food and feed, and Regulation 1830/2003 on traceability and labelling.

With regard to labelling, in the US, FDA has a mandate to prohibit the entry of foods bearing a false or misleading label. Labelling to provide information on a food's method of production is not foreseen unless there is something tangibly different about a food, such that the common name can no longer be used. In practice, GM foods and foods containing GM food ingredients are not labelled in the US.

In contrast, in accordance with the process-based principle, European consumers are viewed as having a right to know how food is produced and whether it contains GM ingredients. Regulation 258/97 on novel foods first made the labelling of GM foods mandatory in 1997. Under Regulation 1829/2003, in force since 18 April 2004, GM food labelling is required above a threshold of 0.9% of GM content per ingredient. Moreover, Regulation 1830/2003 introduced the concept of traceability: all GM foods must be traceable along the entire food chain.

Explaining the transatlantic rift in GM food and crop policy

Why are GM food policies so different? In order to answer this research question, I derive three explanatory elements from theories coming from the wider fields of policy process theory (Sabatier 1999), comparative historical analysis (Mahoney/Rueschemeyer 2003) and science and technology studies (Jasanoff 1995). These elements are "policy precedents", "paradigms", and "actors and actor coalitions".

Preceding policies: building on rDNA research policy in the 1970s and 1980s

I define "preceding policies" as earlier policy or policies relevant for and/or leading to a current policy area. The policy preceding GM food regulation is recombinant deoxyribonucleic acid (rDNA) research policy. Current US and EU GM food policies build on distinct regulatory and socio-political experiences. Surprisingly, a "policy switch" seems to have taken place. The US started out with restrictive policies on rDNA research that paved the way for permissive GM food policy, whereas the EU began with a permissive situation for rDNA research that led to restrictive GMO policy.

The US developed restrictive guidelines on rDNA research in the 1970s in midst of an intense and wide-ranging public and media debate on the risks and benefits of genetic engineering. With the passage of time and increasing confidence that rDNA research was not an unacceptable hazard, these rules were relaxed, and the public debate subsided. By the early 1980s, a scientific and societal consensus had emerged, which led to later policy developments. This new era was dominated by neoconservative economic and research policies favouring relatively permissive regulation of biotechnology applications such as GM foods. The US government built on previous regulatory experience with restrictive rDNA policy in developing a permissive regulatory framework for rDNA applications.

A comparable experience did not exist at the European level, and this can be seen as having provided the necessary space for the development of restrictive regulation in the late 1980s and early 1990s. When rDNA research became possible in the 1970s, Europe was struggling to catch up with the US in terms of science and technology excellence, and therefore avoided strict oversight of rDNA research. In the mid to late 1980s, however, regulating GMOs had become necessary because first applications of modern biotechnology were in the pipeline. In Europe, this process coincided with the rise of the environmental movement and green parties, which in turn galvanized anti-GMO sentiment in Europe. This constellation resulted in the process-based and relatively restrictive policies which have remained the basis of the EU's GM food policies into the present.

Paradigms: the importance of underlying beliefs

A second explanatory element is based on the assumption that policy makers' collective ideas and beliefs or "paradigms" influence policy outcomes. I define a paradigm as an overarching framework of fundamental beliefs about a policy area, which often reaches into other policy domains. Indeed, empirical study indicates that GM food policy in the US and the EU is built on different underlying paradigms.

Although both the US and the EU view biotechnology as a driver of economic growth, US policy on GM food is dominated more than EU policy by a paradigm of economic growth in agriculture that privileges industrialization and cost reduction in order to increase profits ("agribusiness"). GM crops are a useful tool within this model, and thus are supported through permissive regulation in the US. In contrast, EU policy on GM food builds on a model of growth that is less cost efficient than the US model. Instead, it values concepts such as agricultural and environmental sustainability, and views agriculture and the environment as being indivisible. These very different models of economic growth in agriculture can in part be explained by geography: while the US has the land and the topography to dedicate vast areas exclusively to agribusiness, Europe's agricultural land is scarcer and less homogenous. It is usually located nearer to cities and towns, and is more closely intertwined with relatively untouched nature and areas dedicated to leisure.

A further fundamental paradigm difference lies in US and EU perceptions of science, and of how to deal with scientific uncertainty. The United States bases its GM food policy on the so-called "sound science principle", which supposes that science can and does deliver the incontestable truths necessary to make sound policy decisions. The EU implicitly also abides by the sound science principle, but makes an important addition by explicitly basing its GM food policy on the precautionary principle. This principle states that a lack of scientific information and certainty shall not stop measures to prevent potential hazards from being taken. The US and the EU define potential risks of GM foods in accordance with their sound science and precautionary stances. The US definition of risk is relatively narrow, including only direct and immediate effects on health and the environment. In contrast, the EU's risk definition includes indirect and delayed effects.

Actors and actor coalitions: different levels of influence

A third explanation of the transatlantic rift in GM food and crop policy consists of the different levels of success that US and EU actors and actor coalitions supporting and criticizing GM food have been able to achieve.

The US pro-GM food coalition is composed of the biotechnology industry, large and mainstream farmers, food producers, and retailers. These actors form a strong and successful coalition by engaging in coordinated activity revolving around maintaining a relatively permissive regulation. The fact that mainstream US farmers are a crucial part of the pro-GM food coalition is fundamental for its success in the US. Equally important is the observation that food producers and retailers did not fully reject GM foods after consumers started to question the presence of GM ingredients in their food, but merely became more cautious supporters. Moreover, pro-GM food actors in the US are bound by a powerful common vision of genetic engineering as a crucial and promising new tool in the agricultural toolbox. Within this discourse, GM technology is an environmentally friendly technology that can substitute pesticides and that may help feed the world's poor.

In the EU, support for GM food is less consolidated than is the case in the US. The only consistent supporter of GM food is the biotechnology industry. Mainstream European farmers stand to gain less from agricultural biotechnology than US farmers, and are therefore not strong supporters. Food producers and retailers were supportive when first GM crops came out in the mid-1990s, but the fierce consumer backlash of the late 1990s made them revise their positions. Many supermarket chains went "GM free" during this time, thus exerting backwards pressure on food chain actors supporting GM food.

Turning to GM food opponents, in the EU, critics consist of small-scale and organic farmers, consumers and environmentalists. These actors have successfully joined forces around the policy goal of rendering GM food policy as restrictive as possible. They have organized informational resources and events, as well as protest campaigns. They are also bound by a discourse about genetic engineering as a profit-oriented, unnatural and potentially dangerous technology. Consumers add to this the request for the right to choose between GM foods and non GM foods through labelling. In the EU, this anti-GM food discourse has gained mainstream acceptance and support. Extremely important for the EU anti-GM food coalition's success are a series of food and health-related shock events (especially mad cow disease) which fomented great public distrust of European regulatory institutions. The anti-GM food coalition was able to effectively utilize this mood to push for increasingly restrictive GM food policies.

In the US, GM food critics are fewer and less cohesive than in the EU. Environmentalists and some consumer groups pursue the goal of a more restrictive policy, although they have different end goals in mind. While environmental organizations would ideally like to achieve a process-based and restrictive regulation of GM foods, more moderate groups primarily aim for a more consistent, inclusive and transparent decision-making process. Moreover, food processors, producers or retailers may be only reserved supporters of agricultural biotechnology, but they are not prepared to be part of a coalition against GM foods.

Conclusion

US GM food and crop policy is relatively permissive, while EU policy is relatively restrictive. This paper has argued that this can be explained by three main elements: "policy precedents" (different experiences with rDNA research regulation), "policy paradigms" (dissimilar fundamental beliefs regarding GM crops and food), and "actors and actor coalitions" (disparate strength and cohesion of relevant actors and actor coalitions).

Acknowledgements

This paper is drawn from my dissertation (Ramjoué, forthcoming). I owe special thanks to my advisers Daniel Kübler, Frédéric Varone, and the late Ulrich Klöti for their constructive comments and consistent support. I would also like to acknowledge the financial support I received from the Swiss Federal Institute of Technology and the Swiss National Science Foundation.

References

Jasanoff, S., Markle, G.E., Petersen, J.C., and Pinch, T. (1995). *Handbook of Science and Technology Studies*. Sage, London.

Mahoney, J. and Rueschemeyer, D. (2003). *Comparative Historical Analysis in the Social Sciences*. Cambridge University Press, Cambridge.

Ramjoué, Celina. (forthcoming). *The Transatlantic Rift in Genetically Modified Food Policy*.

Sabatier, P.A. (ed.). 1999. *Theories of the Policy Process*. Westview Press, Boulder.

The ethical landscape of non surgical embryo-transfer in pigs: An explorative study of public concerns

Frans Stafleu[1] Dorothé Ducro-Steverink[1] and Jan Merks[2]
[1]Ethics Institute, Utrecht University, Heidelberlaan 2, postbus 80103, 3508 TC Utrecht, the Netherlands, F.R.Stafleu@ethics.uu.nl
[2]IPG, Institute for Pig Genetics B.V, the Netherlands

Abstract

Non surgical embryo transfer and cryopreservation (nsET) is a new technology that enables the development of a technique in pig production to improve genetic preservation and international trade. In the modern society however, the public is concerned about ethical aspects of animal production. To get an impression of these "public concerns" we have asked four groups an opinion about the technique: agricultural professionals involved in nsET, professionals of the Eurogroup for Animal Welfare, volunteers of the Dutch Society for the Protection of Animals and a group of members of the public. We used an ethical analytical tool (the ethical Matrix) to collect these opinions in a standardised way. From the results it emerges that, seen from inside the practice of pig breeding, it is a skilful technique with clear economic and animal welfare advantages. But many of the contra arguments come from an outside perspective and are aimed, among others, at the "unnaturalness" of the procedure and the negative impact that the technique may have on the biodiversity. Many respondents also associated the technique with intensive farming which has for many a negative connotation. It is concluded that the technique is morally disputable.

Keywords: pig, embryo transfer (ET), ethics

Introduction

Non-surgical embryo-transfer combined with cryopreservation is a combination of techniques to improve genetic preservation and international trade. These techniques have obvious advantages for the field of pig breeding, but in the modern society, the public/consumer is concerned about non-technical aspects of animal production. The implications of the nsET techniques have no direct influence on consumers in the strict sense. Consumers neither do directly "consume" the products of these techniques, nor are their lives influenced in another direct way by these techniques. Possible concerns are focussed on general implications about the "good society" and are better called "public concerns" as they address persons more as citizens and less as consumers (Brom, 2000, Meijboom and Brom, 2003). These "good society" aspects cover, generally speaking, ethical issues concerning environment, nature and animal welfare.
Part of the EU-funded project "Porcine embryo Cryopreservation and non-surgical transfer to improve genetic preservation and international trade"(QLK5-CT-2002-70983) was to address these issues by picturing an "ethical landscape" of the technical procedures.
The analysis of this "landscape" was rooted in an explorative survey of how professionals in agriculture, animal welfare, NGO's and members of the public review the ethical aspects of the nsET.

Material and methods

The opinion of four groups have been collected in different sessions: a session with agricultural professionals involved in nsET (n = 8), a session with professionals of Eurogroup for Animal Welfare (the European confederation of animal protection groups) (n = 2), a session with volunteers of the Dutch society for the protection of animals (N = 15) and one with a group of members of the public (n = 8). The professionals of the nsET are a well-defined group, as are the professionals of Eurogroup and the Dutch animal protectionists. The group "members of the public" is less well defined. The group used in this survey was a class of a school for geriatric attendants. It is clear that this group is not representative for "the European public", but as such "the public" does not exist and cannot be caught in a certain group. The group in our survey is expected to give an impression of the opinion of not especially interested citizens.
All the sessions were organized in the same way. In the first part information was given about the project. In the second part the ethical evaluation took place. First we asked a initial impression of the ethical acceptability of the project. The respondents were asked to give a short pro and contra argumentation. Then we used the "ethical matrix" as developed by Mepham (Mepham, 2000) for a more reflective discussion. The ethical matrix represents in twelve different cells the interpretation of respect for the principles of wellbeing, choice and fairness in terms appropriate to the interests of farmers, consumers, farm animals and the environment, respectively (Table 1).
The respondents were asked to discuss shortly the interests in each cell and indicate on an answering form whether or not the technique was positively or negatively influencing that interest. A short argumentation could be given.
In the analysis of the discussion a distinction was made into direct and indirect arguments in combination with intrinsic and extrinsic arguments. Direct arguments are the ones that concern directly the technique of nsET itself. Indirect arguments focus on the context of the technique. Intrinsic arguments concern the property of technique, extrinsic arguments concern the consequences of the technique, for example that the technique prevents diseases.

Table 1. The ethical matrix of Mepham.

Respect for	Wellbeing	Choice	Fairness
Farmers	Satisfactory income/workplace	Managerial Freedom	Fair trade rules
Consumers	Safety and quality	Choice and democracy	Affordability
Animals	Animal welfare	Behavioural freedom	Intrinsic value
Environment	Conservation	Biodiversity	Sustainability

Results and analysis

In Table 2 the most important intrinsic and extrinsic arguments given by all the respondents are subdivided per type of argument.

Direct and intrinsic arguments

The technique is unnatural.
From a theoretical point of view, an appeal on (un)naturalness as a moral argument is problematic because it is too vague. First, the border between "natural" and "unnatural" cannot be drawn,

Table 2. Arguments sorted by type of argument.

	Intrinsic	Extrinsic
Direct	The technique is unnatural The technique does not respect animals The technique is technically beautiful and efficient	Economic, breeding and diseases improvements Biodiversity Sustainability Poor farmers The technique harms animal welfare
Indirect	Intensive farming is unnatural	Intensive farming harms animal welfare Intensive farming causes pollution Intensive farming gives less choice to the consumer

Secondly, most people would, after some reflection, admit that "unnatural" does not necessarily mean "morally wrong". They would say that keeping domesticated animals is not wrong in itself, but that e.g. keeping them under certain circumstances is wrong. This vague argument must be taken as a sign that there may be something wrong. Our respondents express this with a question like: Are there not limits to what we are allowed to do to animals? Such a question can, from an analytical point of view, better be addressed with more precise terms like "welfare", "sustainability and "intrinsic value".

The technique does not respect animals

"Respecting" animals may mean two things. First "respect" can be explained as taking into account that animals are "sentient beings". This means that animals may suffer and that taking care of the welfare of animals is morally important. this is an extrinsic argument. The second interpretation is that people have to respect animals because they are living beings, with there own goals in life, and have a value of their own and may not be used in an pure instrumental way. This is often called the "intrinsic value" of an animal. Interpreted in this way non-surgical embryo transplantation may not respect the animals because it uses animals in an instrumental way in two different respects. First of all the animals are treated as instrument for better breeding. However, the use is also "instrumental" in the sense that the animals are treated with lots of instruments, i.e. techniques. The normal/natural procreation of pigs is replaced by a technical procreation, such as artificial insemination, super-ovulating etc.

Not all participants in the public debate recognize the intrinsic value as valid moral argument. Farmers for example think of their animals as being instrumental for the higher aim of producing good quality food for the world. Domesticated animals are in this world to do so and are by nature domesticated (Stafleu *et al.* 2004). For farmers "naturalness" and "intrinsic value" have no or little meaning in relation to their animals (Lund *et al.* 2004). From a farmers' point of view "good care for animals" is what morally matters (de Greef *et al.* 2005). Utilitarianism denies that anything else than the welfare of the animals is morally relevant.

The respect argument was not recognized by all of the members of Eurogroup. Those members would specify the moral relevance of "unnatural" using terms like "welfare" or "sustainability'. On the other hand, as is shown by our survey, there is a strong intuition of many people that treating animals in an "unnatural way" conceptualised as "respecting the intrinsic value" of animals, is indeed of moral relevance. Furthermore the intrinsic value argument can reasonably be defended on (deontological) ethical-theoretical grounds (Heeger and Brom 2000, Bovenkerk *et al.* 2002).

The technique is technically beautiful and efficient
This argument was not mentioned in the discussions but is clearly visible in the video from that was used to show the respondents how the technique works. Much emphasis is laid on the high hygienic standards, and the refinement of the specially designed "swinlet" a special developed catheter which is used to transplant the embryo's. For farmers an important value is to be proud on their work and to have pleasure in it. Technical skills are for many farmers and professionals active in this area part of the "sport" to steer nature in such a way that it serves the aim of producing good quality food in efficient way (Stafleu *et al.* 2004). From the video it is clear that this technique serves this aim and gives a reason to be proud of.

Direct and extrinsic argument
Most pro arguments that were given are direct and extrinsic. They concern the positive consequences of the technique: producing clean embryo's of high quality in such a way that the welfare of the animals is enhanced and the embryo's can be safely transported to third countries. For the donor breeders there is an enlargement of their markets, for the farmers in the third countries there is a chance for better participation in the (world) market. We see these advantages also is recognized by the non-professional respondents. Much of the negative critique, however, is aimed at the nature of the practice itself (non-natural, not respecting etc) and does not directly respond to the given pro's.

Biodiversity
This is an interesting argument, mentioned by all groups. The biodiversity argument is about the fear that local (e.g. China) breeds will disappear and be replaced by the modern European breed. This is morally negative in two ways. First a sustainability way: local pig genes could disappear and will be lost for future generations of pig breeders. The second reason is a cultural one. Local breeds may be an important part of the local (agricultural) culture. If those breeds are lost, the local culture will be harmed.

Sustainability
The professionals of Eurogroup mentioned a sustainability argument. They reasoned that the unnatural and technical character of the procedure and the resulting "designer pig" could harm the sustainability of agriculture, because these pigs may only function well under specific conditions. If in the future an other kind of housing would be preferred, then these designer pigs could not adapt anymore, which would be a loss for future generations of farmers.

Poor farmers
The poor farmers argument holds that poor farmers will not get the opportunity to buy the better breeding material and the fattening pigs. In this way they will not be competitive in the market and have to give up their vocation. The status of the argument is not really clear. First of all the nsET professionals were not certain this would happen and for some it is not clear that the disappearance of small farmers is a moral problem. The poor farmers argument seems to touch the general discussion about globalisation and scaling-up which is present in the society as well as in the agricultural sector.

The technique causes animal bad welfare
The animals protectionists were not convinced that welfare of the animals was enhanced, even in the context of the present practice. First of all the super ovulation of the donor sows was mentioned as possible harming the sows. Some of the animals protectionists were not sure that the recipient sow was not harmed by the "insemination". On the whole it seemed that this argument was not the most important one in the context of the technique.

Indirect intrinsic arguments

There was one important argument of this type. Some respondents took this technique as typically for intensive farming and for that reason they had moral problems with the technique. Interesting enough, the professionals denied this intrinsic connection between technique and intensive farming and argued that the technique was neutral in this sense and could also be used to breeding pigs in an extensive way. Although this is logically true, the development of this technique fits most easily in the common intensive practice farming.

Indirect extrinsic arguments

These are arguments that connect the technique of non-surgical embryo-transplantation with intensive farming and its supposed negative moral consequences like bad animal welfare, pollution of the environment, less consumers choice and the like. Moreover, the technique is seen as promoting intensive farming methods in places where most farming is still extensive. As mentioned above, there are good reasons to think that the technique will be used as part of intensive farming, although a pure logical necessary link is not there . It is not right to concentrate on this logical argument and isolate a technique from its use and only judge it from its intrinsic properties. It is also not wise, because in the public debate the connection with intensive farming will be made, irrespective of these possible logical contra arguments.

Conclusion and discussion

From an inside perspective, the technique of nsET is a skilful technique with clear economic and animal welfare advantages. A problem is that this point of view leans heavily on farmers morals and most of the critique comes from outside these morals. In farmers morals "care for the welfare of the animal" is the most relevant value. Arguments concerning "unnaturalness" of the technique and concerns about "not respecting the intrinsic value of animals" come from outside, the animal protectionist on the first place. Our survey shows however that also not especially interested citizens mention the unnaturalness of the technique as a possible disadvantage. A way to make a link between the "outside" argument of unnaturalness and an "inside" argument which is also valid in farmers morals may be the sustainability argument.
A disadvantage that was recognized by all respondents was the biodiversity argument, which holds that the technique may cause loss of local breeds which is morally problematic from sustainability and a cultural point of view. However, it is also recognized that nsET may help to preservation of biodiversity by freezing embryos.
For the above mentioned reasons it can be reasonably defended that the technique is from a moral point of view disputable. It may become a focus point for critique directly aimed the technique itself and indirectly as paradigmatic for intensive farming.

References

Bovenkerk, B., Brom, F.W.A., van den Berg, B.J., (2002). Brave new birds. The use of "animal integrity" in animal ethics. Hastings Centre Report Jan-Feb 32(1), 16-22.

Brom, F.W.A., (2000). Food, consumer concerns and trust: food ethics for a globalizing market, Journal of Agricultural and Environmental Ethics 12, 127-139.

De Greef, K.H., Stafleu, F., de Lauwere C., (2005). A simple value distinction as an aid for transparency in the farm animal welfare debate, Journal of Agricultural and Environmental Ethics (in press).

Heeger, R., Brom, F.W.A., (2001). Intrinsic Value and direct duties: From animal ethics towards environmental ethics? Journal of Agricultural and Environmental Ethics 14, 241-252.

Lund, V., S Hemlin, J. White, (2004). Natural behavior, animal rights, or making money - A study of Swedish organic farmers' view of animal issues. Journal of Agricultural and Environmental Ethics 17, 157-179.

Meijboom, F., Brom, F.W.A., (2003). Intransigent or Reconcilable: The complex relation between public morals. The WTO and consumers in: Vedder, A. (red.), The WTO and Concerns Regarding Animals and Nature, Nijmegen: Wolf Legal Productions, p. 89-99.

Mepham, T.B., (2000). A framework for the ethical evaluation of novel foods: the Ethical Matrix. Journal of Agricultural and Environmental Ethics 12, 165-176.

Stafleu, F.R., de Lauwere, C.C., de Greef, K.H., Sollie, P., Dudink, S., (2004). Boerenethiek, eigen waarden als basis voor een 'nieuwe' ethiek. Een inventarisatie. NWO januari 2004.

Agricultural animals as targets of genetic progress: Engaging with animal scientists about the impact of genomics

Richard Twine
ESRC Centre for Economic and Social Aspects of Genomics (CESAGen), IAS Building, Lancaster University, LA1 4YD Lancaster, United Kingdom, r.twine@lancaster.ac.uk

Abstract

Within social science analyses of genomics the field of *animal* genomics remains an under researched area. In an anthropocentric sense the field intersects with the substantial domains of food, health and environment, as well as their related economic, social and ethical elaborations. Moreover, animals are being given a resurgent role in the emergent cast of the 'bio-economy', as new sources of bio-capital. Beyond this animals are ethically and socially significant in their own right, having an important place in the social life of many people. Research that involves collaboration and dialogue between social scientists and animal genomics scientists is rarer still. Here I present results from semi-structured interviews with scientists working in the agricultural context of animal genomics. These include geneticists, molecular biologists and animal welfare scientists. The focus is to allow scientists in these fields to speak for themselves about the ethical, social and economic aspects of exploiting genomics knowledge for the enhancement of agricultural animals. How are such issues articulated? Is there a consensus between animal scientists when it comes to applying genomic knowledge? To what extent is a boundary placed around animal genomics vis-a-vis other technologies such as genetic modification and transgenics? The discourses articulated therein are examined for modes of framing the social and ethical relevance of their own work as well as for intimations of future impacts upon human/animal relations.

Keywords: genomics, animals, agriculture, ethics, bioeconomy

Introduction

In this paper I situate Animal Genomics within the emerging discourse of the bioeconomy, where post-genomic agricultural animals are cast as significant players in what they may bring in terms of added value to the EU economy. This paper presents initial findings from semi-structured interviews carried out with animal scientists in the UK concerning the ethical and social import of their work. Several preliminary themes are discussed, and tentative conclusionary remarks about the current state of animal genomics are made.

Social scientists and philosophers are drawn specifically to genomics and biotechnology more broadly due largely in part to the ambivalence intrinsic to it. Although genomics is constructed as one of the 'progress narratives' of our time especially in the hope of what it might contribute to both human health and wealth creation it is also clouded by potential imagined dangers (Franklin 2002). This ambivalence is brought more clearly into relief when we think about nonhuman animals. On the one hand genomics ushers in a significant consolidation to previous Darwinism in that it stresses animal-human kinship in terms of common genes, shared evolutionary histories, yet it also opens up new opportunities for instrumentalising animal life.

Ideas of the role of genomics and biotechnology in a narrative of progress have begun to coalesce around the notion of the bioeconomy. One can see this in the EU drive to produce a wide array of so-called technology platforms, essentially scoping documents for how different areas of science and technology may evolve over the next twenty years[80]. The term is also being used in China and most recently at the Biotechnology Industry Organisation in Chicago (BIO2006, 9th-1th April). The initial reaction from asocial scientist may be one of curiosity since historically the economy may have always been thought of as a *bio*economy. One may ask why the term is now being made explicity when historically the role of the bios in capital accumulation has been traditoionally denied. Now it is being used to give voice to the potential of biotechnology and related areas to glean new value from nature be that in industry, medicine or agriculture. The most relevant technology platform to this discussion is entitled 'Farm Animal Breeding and Reproduction – A Vision for 2025'[81] (FABRE-TP 2006). It is clear from this document and from listening to animal scientists in the UK and the US that the uptake of genomics in animal agriculture is premised upon a forecast increase in global demand for animal food products. The rise which is projected to sustain for twenty years is primarily located within developing countries, notably China and India. Yet it is not clear how this forcast rise in demand has been constructed nor whether it is sensitive to cultural differences, for example, in meat consumption between differing parts of the world.

There are various ways in which animal genomics and other biological techniques might be employed to add value – biocapital - to traditional products of animal agriculture. It is hoped that genomics will enhance disease resistance and lead to improvements in meat quality. The shift away from simple productivism is highlighted in a new focus on *qualitative* improvement such as the conversion of meat into a functional food. Actual genetic modification and cloning techniques are under research in the US to achieve similar ends - irrespective of a less conducive regulatory environment in the EU for such biotechnologies. In such cases of what we might broadly term 'genomic modification' the surplus value of the animal is yielded by the new knowledge available from sequencing and new modes of manipulation as in silencing or adding in genetic 'data'. Other examples of animal biocapital help foster a novel convergence of the agricultural and the medical where traditionally conceived 'agricultural' animals such as chickens or pigs are used to yield bio-pharmaceuticals or perhaps eventually organs for human transplant. Other modifications can produce more environmentally friendly animals as in the well known case of Enviropig that prodices more ecologically benign waste. Notably in all these examples the surplus value is not especially restricted to a whole animal but is exchanged speedily and transnationally in the form of enhanced embryos, gametes, meat or milk.

Methodological points

Given this context I am in the process of interviewing animal scientists in the UK about the ethical, social and economic aspects of such work. These new forms of biocapital may potentially introduce novel human/animal relations. Moreover they are altering the power that humankind can exert in order to contol other species. Yet there is little indication in the UK of institutional learning from the experience of GM crops in that animal genomics and other techniques are pursued with little or no public engagement. There is however now an emerging culture of science and society discourse within animal science communities. My aim is then to pursue the extent of that and to critically engage with the work of animal genomics scientists from my particular disciplinary perspectives, namely the socoiology of human/animal relations and the field of animal ethics. A further aim is to on the one hand give animal scientists the opportunity to discuss theses issues in their own words and to analyse this discourse for themes and points of interest. Here a present a brief extract of interview data and the emergence of four initial themes – ethics, scientific disagreements in conceptualisations of animals and selection issues. I have

[80] These two web-sites are especially relevant: http://cordis.europa.eu.int/technology-platforms/home_en.html and http://www.bio-economy.net/ (These were both last accesssed in April 2006).

[81] One can download the draft document from http://www.fabretp.org (last accessed April 2006).

not restricted my interviews to animal genomics scientists alone but recognising how areas of animal science are inter-related in terms of joint research I have also interviewed, for example, animal welfare scientists. This has the added benefit of looking at animal genomics through the eyes of another science community, one that ranges from complimentarity to antagonism in its relation to genomics. Interviews are recorded and identities of scientists are anonymised.

Ethics

Scientists interviewed had mixed views toward issues such as GM, xenotransplantation, and the ethics of patenting transgenic animals. Most were keen to point out boundaries that should not be transgressed in the case of GM. However for the purposes of this section I am going to focus on a different question; the issue of the extent to which genomics is continuous or not with traditional quantitative genetics methods of selection. From previous participation in multidisciplinary events on animal genomics I was aware of a discursive tendency amongst animal genomics scientists to employ what I have called a 'continuity argument' (Twine 2006) to suggest that the selection made possible by genomic knowledge or even practised in genetic modification is essentially no different from selective breeding. This may be criticised as something of a naturalising strategy so as to avoid critical examination of the potential newness of emerging technologies of food production. Consequently I have been interested to probe just what might be new about genomics in animal science. One animal welfare scientist was already very familiar with the continuity argument:

> *"Yes they always say that. ...I don't think it's fair to say well you know we've been doing it already in the past. You know there are ethicists that deal with that and I know that it's not an ethically sound argument because you know you are perfectly entitled to put cut off points and say this far and not further. You don't have to once you approve of one thing, you don't have to approve of everything that follows. I'm one of those people that says 'look, it's the way of thinking behind it that matters'"*

This unpacks some of the problems with a continuity position and also points to the sort of slippery slope argument (Williams 1995) familiar in bioethics. In this case for example one might see genomics as an intermediary step between selective breeding and GM. Responses from animal genomics scientists did reveal a consistent difference with genomics approaches:

> A - *"It just means that you're getting to where you want to get quicker and you may get where you want to get where you couldn't by conventional means"*

> B – *"I think the detail is still important, if you're applying genomics as such that you're essentially identifying variation that exists in your population as it currently exists and utilising existing variation I really don't think there's an new issue there, we're essentially doing what's done anyway but possibly more precisely...I think it allows you to basically home in on underlying genes much more quickly....It makes it easier, it doesn't make it straightforward, there's still a lot of work but with the sequenced genome you know what the genes are, you can then flit between species much more easily and that enables you to refine comparative genomics to get it a lot more quickly"*

Here we can see that genomics whilst not a straightforward solution does afford a significant improvement to animal scientists in gene identification and then functional genomics. Sociologically this is significant as genomic knowledge can afford us more control over animals in varying contexts and the potential changes in human-animal relationalities that could follow from this.

Scientific Disagreements

Animal Welfare scientists were concerned that the shift to genomics and molecular biology could be reductionist in how it figures the animal. There is certainly a striking difference evident when talking about animals to welfare and genomics scientists. This speaks to fundamental philosophical framings of the value of nonhuman animals as evidenced in this extract from an animal welfare scientist:

> *"I wanted to try and develop an approach to conceptualising animals as subjects rather than objects, and to see if that could hold within biological science... When I started off issues like consciousness, you know it was all just taboo. They're (Animal Genomics Scientists) in the first place conceptualising animals as complex production systems and then they're talking about the health of that system. An animal isn't just a complex system it's a being, a living animal"*

There is a marked contrast when in Genomics we hear about the potential contributions that may be made to 'genetic progress', the phrase from this paper's title. After first hearing this phrase I wanted to probe further. Since, as I have argued, genomics is a significant part of a new narrative of progress revolving around biotechnology and the emergent discourse of the bioeconomy it is interesting to find this phrase in animal science as it is a further framing of progress; one which is significantly reductive of 'progress' to the biological and economic level rather than social or moral measures. One interviewee defined 'genetic progress' as:

> *"The difference in say the mean performance of your flock this year compared to last year. So let's say you have a flock of sheep as an example and your aim is to improve growth rate a little bit. So this year the mean growth rate is 100g per day, next year is 110g, the following year is 120g a day you're making genetic progress of 10g per year"*

Performance here is clearly a commercially relative value determined by a concept of what is seen as valuable in a given context. For example 'marbled' meat is more desirable in some parts of the world and genomics has already been used to create a genetic test for marbling. But the most significant difference signalled by the last two extracts is a shift from an individual animal being to a general homogenous flock. This is of ethical import for whether one is able individualise and grant subjective, experiential existence to animals. The importance of progress defined in genetic terms in animal genetics and genomics inserts animals - the efficiency of their bodies - into what is called bioeconomic modelling where a sense of their being is obviously abstracted in favour of hopes for further optimisation based upon instrumentalisation. This is underlined by the term 'genetic progress' as it funtions in a similar way to 'selctive breeding' in the way that it demarcates a moral boundary between the human and the animal. If such terms were applied in the human case they would jar against the taboo of eugenics not hegemonically present in the animal case.

Selection Issues

In conversation with animal scientists it was clear there are a wide range of selection issues which arise to make the whole issue of enhancing animals and agricultural gain a complicated business. Certain selections run the risk of unintended consequences. For example over focus on selecting for milk yield has led to a reduction in fertility, and an increase in mastitis and lameness. Other issues include trying to understand genotype-environment interactions, how different selected lines respond to different environments, and how genes interact with each other. Moreover epigenetics also provides a further challenge to the degree of control animal genetics can exert. But I want to finish with the specific example of lameness, the condition where the internal structure of the hoof comes apart. Some UK farms have 50% lame cows and it is a significant welfare and economic problem. There now follows an extract of a dialogue between myself and an animal welfare scientist who works with animal genomics scientists:

> *RT – "What causes that in most cases?"*
> *Scientist: "It's a production disease, it's an intrinsic part of modern dairy production"*
> *RT: And so even though that's a production disease, presumably there are moves to try and address that through genetic selection?*
> *Scientist: Yes there are, you can select bulls who's daughters should have a better locomotion score. Several issues there, one is we believe on our side of the fence, the welfare side, they don't really measure this very well. And even if they did what we wanted them to do there's an enormous amount of difficulty at this end and just in measuring our farms. Consequently the heritability is poor. So they can't make much progress with it anyway because the heritability is slow.*
> *RT: But it's using the same method that's produced it to try and address it isn't it?*
> *Scientist: You mean breeding?*
> *RT: Yes*
> *Scientist: Yes which is curious.*
> *RT: Because if you tried to solve it through an environmental way it wouldn't be cost efficient?*
> *Scientist: Well people, very good question I mean, yes its not just that they've been bred to produce lots of milk, they're also kept indoors and the longer they're indoors, the bigger the risk they get lame. So they could be kept out more often, they could be fed at a lower rate and probably milked less often, that would probably reduce the risk. They could be in better quality cubicles. But that's investment. They could have their feet trimmed more often..*
> *RT: It's labour cost?*
> *Scientist: ..But it costs, this is the thing it costs a lot, you know lame cows don't produce milk, they often have to be replaced. And that's one of the things that we've been trying to get across as you know if you actually add up the hidden cost.*

I think what is most relevant from this extract is the choice of genetics or genomics as the problem solving methodology for lameness. It suggests that it may be being used partly because it fits into the cost-efficient model and so is cheaper than environmental solutions. Also it animates a long standing argument from the sociology of science that technoscience colonises and perpetuates by being able to cast itself as the solution to the problem that it was directly involved in creating (see Beck 1992). This is akin to contemporary debates over global warming where there is argument over whether technological solutions or lifestyle changes are the best methodology.

Some concluding remarks

Although space restrictions have prevented me from including more than just a preliminary highlighting of some of the emergent themes in my research one can already point to some significant conclusions and questions. The animal scientists I have interviewed have been very interested to engage on these issues and see it as a worthwhile activity to collaborate with social scientists.

Set within the wider context of animal science, animal genomics remains a new science with something to prove both scientifically and ethically. Animal genomics has had to adapt itself to new funding priorities including animal welfare and functional foods, due in part to the partial decline of a productivist paradigm in UK agriculture (see Buller & Morris 2003). Issues remain over what sort of conceptualisations of animals it relies upon and perpetuates – are these acceptable to the wider culture? Does animal genomics encourage a geneticisation of the animal; specifically of animal health, animal behaviour and animal welfare? Finally and rather fundamentally is it out of step firstly with other areas of science - most notably in this research animal welfare science - that point to an alternative trajectory of re-enchanting animal lives, and secondly, from a human health point of view, with a reduction in the consumption of animal products contained within much contemporary advice from nutritional research. These are further research questions which I hope that I and others will be able to pursue.

References

Beck, U. (1992) Risk Society – Towards a New Modernity Sage, London.

Buller, H. & Morris, C. (2003) Farm animal welfare: a new repertoire of nature-society relations or modernism re-embedded? *Sociologia Ruralis* 43 (3): 216-237.

FABRE-TP (2006) 'Farm Animal Breeding and Reproduction – A Vision for 2025' FABRE Technology Platform, Brussels.

Franklin, S., Lury, C. & Stacey, J. (2000) *Global Nature, Global Culture* Sage, London.

Williams, B. (1995) *Making Sense of Humanityand Other Philosophical Papers 1982–1993* Cambridge University Press, Cambridge.

Part 9
Nutrition, equality and global food security

Free vs. paid school fruit: Big difference with respect to social inequality

Elling Bere and Knut-Inge Klepp
University of Oslo, Department of Nutrition, P.O.Box 1046 Blindern, 0316 Oslo, Norway, ellingb@medisin.uio.no

Abstract

A paid School Fruit Programme (parents pay NOK 2.50 per day = € 0.30) is offered all elementary schools in Norway in order to increase the low consumption of fruits and vegetables. The programme is subsidised by the Norwegian Government with 1.00 NOK per pupil per day. The effect of the programme is limited due to low participation of schools and low subscription rates. The present study shows that those who choose to subscribe are healthier than the non-subscribers, e.g. they have a higher fruit and vegetable intake already before the programme starts, and the programme therefore tend to increase the inequality present. A free pilot version of the programme, on the other hand, have been reported to increase FV intake in all groups of children, also in at risk groups such as boys, low SES status groups, those eating little FV and those saying that they do not like FV. The present results show that a free school fruit programme, contradictory to the paid programme, decreases the initial differences present in FV intake.

Keywords: social inequalities, school-children, intervention, fruit and vegetables

Introduction

Fruits and vegetables (FV) are prerequisites for a healthy diet. Compared to the energy they contain, FV are expensive foods (Drewnowski *et al.*, 2004). This contributes to the reason why social inequalities are apparent in FV intake (i.e. that people of high social economic status (SES) eat more FV than people with low SES). Because a diet containing less FV will decrease the cost of the diet (Darmon *et al.*, 2003), the message from the health authorities about eating more FV therefore conflicts the wish from the same authorities about decreasing social inequalities (The Norwegian Ministry of Health, 2003). Despite a rather egalitarian society, there are great social inequalities in health in Norway (Directorate for Health and Social Affairs, 2005).
A paid School Fruit programme (parents pay NOK 2.50 per day = € 0.30) is offered all elementary schools in Norway (www.skolefrukt.no) in order to increase the low consumption of fruits and vegetables. A thoroughly description of the programme in English can be found in Bere *et al.* (2005). The programme is subsidised by the Norwegian Government with 1.00 NOK per pupil per day. The effect of the programme is limited due to low participation, and it has been reported that it may increase social inequalities in FV intake since subscribers eat more FV than non-subscribers already before the programme starts (Bere *et al.*, 2005). Contradictionary, a pilot project of the same School Fruit Programme offered for free did increase FV intake in all groups of children (Bere *et al.*, 2005; Bere *et al.*, 2006).
The present study investigate potential differences between those who subscribe in the national School Fruit programme and those who do not, regarding socio demographic factors, food intake, and other health related behaviours. In addition, it shows how FV intake between subscribers and non-subscribers (in the school year 2004/05) changed over time from September 2001 (when

they were not provided with any school fruit programme), to May 2002 (when they all got it for free) and to May 05 (when they had to choose to subscribe or not).

Method

Design

This study is part of the Fruits and Vegetables Make the Marks (FVMM) intervention project. The FVMM cohort has been followed from the pupils were in 6th and 7th grade in 38 elementary schools (September 2001) to May 05 (when they were in 9th and 10th grade in 33 junior high schools). As part of the intervention, nine of the 38 elementary schools participated for free in the Norwegian School Fruit Programme from October 2001 to June 2002 (Bere *et al.*, 2005). Pupils in three of these nine schools were in May 2005 in two different junior high schools that provided the paid School Fruit Programme. These two schools contain the study sample of the present study. Three questionnaire surveys have been conducted: In Sept. 01 when no school fruit programme was provided, in May 02 when all pupils participated for free, and in May 05 when the pupils had to decide to subscribe or not to the paid programme. At baseline, one parent of each child was invited to take part in a parent survey.

Sample

All pupils in 9th and 10th grade in the two junior high schools (n = 515) were invited to participate in the May 05 survey, and 441 pupils (86%) did; 229 boys and 212 girls. A total of 117 (27%) of these pupils did subscribe to the School Fruit Programme in May 2005. Of the 441 pupils in the study sample, 239 (77 subscribers, 32%) were part of the FVMM cohort, and had previously participated in both the Sept. 01 and May 02 surveys. Mean age of the study sample was 15.5 years in May 05. Of the study sample, 242 pupils had a parent that participated in the baseline survey.

Instrument

A written 24-h FV recall was used to assess pupils' FV intake (portions/day). The 24-h recall was read aloud to the pupils by a project worker (only in elementary schools). FV intake the previous day was recorded for school days (i.e., the surveys was conducted on weekdays, Tuesday through Friday). FV intake at school and all day are presented. In addition, habitual FV intake was measured by four FFQ items (times/week) and unhealthy snacks with three (soda/candy/chips, times/week). Both methods are previously presented, and validity and reliability properties have been reported (Andersen *et al.* 2004).

In addition, pupils reported whether they subscribed to the school fruit programme or not, gender, class level (initial 6th or 7th), educational plans (College/university or not), weight, height, dieting (yes or no), TV watching (hours/day), physical activity (times/week), smoking (yes or no), and whether they ate breakfast, lunch, dinner and evening meal yesterday (as part of the 24-h recall described above). Their parents recorded their own gender, dieting, FV intake, soda/candy/chips consumption, TV watching, physical activity, smoking, and 24-h meal frequency (all as described for the pupils), in addition to their own age (continuous), education (lower: no college or university education/ higher: having attended college or university), and household income (dichotomised).

Statistics

Analysis of variance and chi-square statistics has been conducted when comparing subscribes to non subscribers cross-sectionally. The longitudinal relationships between subscribers and non-subscribers are assessed using the repeated measures procedure. All analyses have been conducted using SPSS version 13.

Results

Subscribers are different from non-subscribers (Table 1). More girls than boys do subscribe to the programme, more subscribers have plans of higher education, they eat more FV, but less unhealthy snacks, they watch less TV, they are more physical active, they eat more often breakfast, lunch and evening meal, and they are leaner and fewer are on a diet. Their parents are also different from parents of non-subscribers; subscriber's parents are older, have more education and also a higher household income, they eat more FV, watch less TV, and they smoke less.

Table 2 show that the subscribers ate more FV than the non-subscribers already in September 2001 (3.5 years before they had to choose to subscribe or not). This difference decreased while the pupils got free school fruit in May 02, but increased, and were greater in May 05 than in September 01, when the pupils subscribed or not. The same pattern are seen for all FV outcomes; FV at school, FV all day and habitual FV intake.

Table 1 Socio demographic and health related differences between subscribers and non-subscribers and between their parents (t-test or chi-square).

	PUPILS (May 05)			PARENTS (Sept. 01)		
	Non-subsc n=321	Subscribers n=116	p-value	Non-subsc n=165	Subscribers n=77	p-value
Socio demographic factors						
Gender (% girls)	46%	59%	0.048	82%	82%	0.942
Grade level (% in 10th grade)	50%	48%	0.773			
Age				39.9	42.0	0.002
Education plans (% with plans of college/university edu.)	49%	68%	0.001			
Education (% with college/ university edu.)				50%	62%	0.074
Household income (% in upper half)				47%	65%	0.022
Weight aspects						
BMI	21.3	20.5	0.047			
% overweight	16%	10%	0.132			
% on diet (slimming)	18%	7%	0.007	20%	12%	0.123
Other health related behaviours						
FV all day (portions/day, 24-h recall)	1.9	3.2	0.000	2.6	3.3	0.009
Soda/candy/chips (times/week)	6.9	5.2	0.002	3.7	3.7	0.910
TV watching (hours/day)	3.0	2.5	0.008	1.7	1.2	0.006
Physical act. (times/week)	3.6	4.2	0.013	1.9	1.9	0.852
% smokers	11%	8%	0.270	38%	23%	0.028
Meal frequency (% that ate ... yesterday)						
Breakfast	75%	92%	0.000	91%	92%	0.720
Lunch	86%	94%	0.028	84%	88%	0.405
Dinner	93%	94%	0.614	94%	95%	0.917
Evening meal	74%	84%	0.042	71%	69%	0.726

Table 2. Difference between subscribers' and non-subscribers' FV intake in Sept 01 (NO school fruit), May 02 (FREE school fruit) and May 05 (PAID school fruit). Subscribers are the pubils subscribing to the paid School Fruit programme in May 05.

		Non-subscribers (n=160)			Subscribers (n=77)				
		Mean	95% CI		Mean	95% CI		difference	p-values
FV at school (portions/day)	Base	0.36	0.26	0.46	0.60	0.46	0.75	0.24	0.006
	May 02	0.96	0.84	1.08	1.06	0.89	1.24	0.10	0.333
	May 05	0.32	0.21	0.43	1.06	0.90	1.21	0.74	0.000
FV all day (portions/day)	Base	2.14	1.84	2.44	2.58	2.14	3.02	0.44	0.106
	May 02	2.45	2.12	2.77	2.74	2.26	3.21	0.29	0.385
	May 05	1.98	1.65	2.30	3.40	2.92	3.88	1.43	0.000
FV habit (times/week)	Base	13.0	12.0	14.1	16.1	14.6	17.7	3.1	0.001
	May 02	14.3	13.2	15.3	16.2	14.6	17.7	1.9	0.025
	May 05	13.6	12.4	14.7	19.6	18.0	21.3	6.0	0.000

Discussion

Previously we have reported that the paid Norwegian School Fruit Programme has low effect in increasing school children's FV intake, mainly due to low participation by schools and low subscription rates (Bere *et al.*, 2005). The present results indicate that the paid programme also increases the initial differences seen in FV intake between the subscribers and non-subscribers. The subscribers are clearly healthier than the non-subscribers, and the paid programme do not attract the needed. Those that would have benefited the most, and therefore given the greatest public health effect, do not subscribe. As the programme is subsidised by the Norwegian Government, the Government attributes to an increased social difference in FV intake, an inequality the health authorities very much want to reduce (Directorate for Health and Social Affairs, 2005; The Norwegian Ministry of Health, 2003).

A free school fruit programme, on the other hand, clearly increases FV intake in all groups of children, also in at risk groups such as boys, low SES status groups, those eating little FV and those saying that they do not like FV (Bere *et al.*, 2005). The present results show that a free school fruit programme, contradictory to the paid programme, decreases the initial differences present in FV intake.

Several studies show social inequalities in children's FV intake (Vereecken *et al.*, 2005; Rasmussen *et al.*, manuscript under review). Interventions aimed at increasing children's FV intake usually report low overall effects (Knai *et al.*, 2006), and in addition they tend to work better among the low risk groups; i.e. girls and high SES groups, such as the paid School Fruit programme. Free school fruit is an efficient strategy to increase all children's FV intake, and it is also a strategy for reducing the social inequalities present.

A limitation of the present study is the rather long time lag between the first and the last survey (the mean age of the sample increased from 11.8 to 15.5 years). A study from Northern Ireland found that social differences in FV intake seemed to increase from age 12 to age 15 (Van Lenthe *et al.*, 2001). The study by Van Lenthe and colleagues (2001) reported fruit and vegetables separately, and as percentage of energy intake. Therefore it is difficult to compare the results to the present study. However, the increase over time in the difference between subscribers and non-subscribers in the present study appears to be larger than the increase over time between non-manual and manual SES groups in the study by Van Lenthe and colleagues (2001). Therefore, it is reasonable to believe that the paid School Fruit Programme do contribute to the larger social differences seen in May 05 than in Sept. 01.

Conclusion

The Norwegian School Fruit Programme increases the social inequality seen in FV intake. A free version of the same programme reduces the inequality.

Reference List

Andersen, L. F., Bere, E., Kolbjornsen, N. and Klepp, K. I. (2004). Validity and reproducibility of self-reported intake of fruit and vegetable among 6th graders. *European Journal of Clinical Nutrition*, 58, 771-777.

Bere, E., Veierød, M.B. and Klepp, K.-I. (2005). The Norwegian School Fruit Programme: evaluating paid vs. no-cost subscriptions. *Preventive Medicine,* 41, 463-470.

Bere, E., Veierod, M.B., Bjelland, M. and Klepp, K.-I. (2006). Free school fruit--sustained effect 1 year later. *Health Education Research*, 21, 268-275.

Darmon, N., Ferguson, E. and Briend, A. (2003). Do economic constraints encourage the selection of energy dense diets? *Appetite*, 41, 315-322.

Directorate for Health and Social Affairs (2005). *Directorate for Health and Social Affairs' action plan against social inequalities in health* [Nor], Directorate for Health and Social Affairs, Oslo.

Drewnowski, A., Darmon, N. and Briend, A. (2004). Replacing fats and sweets with vegetables and fruits--a question of cost. *American Journal of Public Health*, 94, 1555-1559.

Knai, C., Pomerleau, J., Lock, K. and McKee, M. (2006). Getting children to eat more fruit and vegetables: a systematic review. *Preventive Medicine*, 42, 85-95.

Rasmussen, M., Krølner, R., Klepp, K.-I., Lytle, L., Brug, J., Bere, E. *et al.* (manuscript under review). Determinants of fruit and vegetable consumption among children and adolescents: systematic review of the literature. Part I: quantitative studies.

The Norwegian Ministry of Health (2003). *White paper 16 (2002-2003): Prescription for a healthier Norway* [Nor]. The Norwegian Ministry of Health. Oslo.

Van Lenthe, F.J., Boreham, C.A., Twisk, J.W., Strain, J.J., Savage, J.M. and Smith, G.D. (2001). Socio-economic position and coronary heart disease risk factors in youth. Findings from the Young Hearts Project in Northern Ireland. *European Journal of Public Health*, 11, 43-50.

Vereecken, C.A., Inchley, J., Subramanian, S.V., Hublet, A. and Maes, L. (2005) The relative influence of individual and contextual socio-economic status on consumption of fruit and soft drinks among adolescents in Europe. *European Journal of Public Health*, 15, 224-232.

Who's responsible for school meals?

Astrid Dahl
Technical University of Denmark, Department of Manufacturing, Engineering and Management, Innovation and Sustainability, Building 424, 2nd floor, DK-2800 Kgs. Lyngby, Denmark, ad@ipl.dtu.dk

Abstract

In this paper the current debate on school meals in Denmark is presented, looking at seeming dilemmas and disagreements. The idea of providing a meal at school is new in Denmark and, maybe due to the lack of tradition, a lot of different motives can be identified within the debate. The main arguments for introducing school meals are related to health promotion, learning ability, food education and school meals as a service, however the emphasis seems to be on the health aspect. In terms of responsibility for school meals it is discussed whether this responsibility should be public or private, in other words whether school meals is a public welfare task or not. Some examples is given on how different societal levels have responded to the question of school meals. It is argued that the current perceptions of school meals are characterized by a top-down approach, assuming that the implementation of school meals will follow a simple 'demand-response model'. A few school meal schemes have already started up, e.g. in the municipality of Copenhagen, and within this implementation process some barriers can be identified. Drawing on institutional theory an initial analysis of the social aspects of the implementation proces is presented from a school perspective. There seems to be reason to emphasize the local sensemaking processes and the general issue of acceptance of school meals. Finally it is concluded that school meals in Denmark are still subject of a discoursive battle and that no kind of institutionalization has happened yet.

Keywords: school meals, responsibility, implementation processes

Introduction

School meals are on the agenda in Denmark. This is something new, since Denmark has a long tradition of school children bringing a packed lunch from home. Within the last couple of years, quite a few reports and guidelines about school meals have been published by the national health authorities and recently a number of conferences and seminars have dealt with the issue. In practice some municipalities and single schools have started to provide school meals.
Looking at the debate and the early-stage practice, there seems to be a range of different perceptions of what a 'school meal concept' should include and which problems to address. I will outline the different arguments for school meals in order to illustrate the complexity and some of the dilemmas within this field.
The financial aspect of school meals is also part of the school meal debate. Should school meals be paid via tax like in Sweden, should the parents pay the full cost or how should this be arranged? This leads to the question of distribution of responsibility, which I will discuss in relation to the implementation of school meal schemes.
With inspiration from neo-institutional theory I will suggest an analytical approach to analyze the implementation process, looking at how a school meal scheme is expected to become part of the school setting. This will draw on an example of a school meal scheme in the municipality of Copenhagen.

I will relate the discussions to the initiate findings in my PhD work about public catering, which focus especially on implementation of school meal schemes.

The debate on school meals

School meals in 'the Danish Folkeskole' (primary and lower secondary school) are currently debated. In general there is a rising interest in offering children a meal at school, a practice which is a long tradition in many other countries. Both politicians, 'interest organizations', different expert groups and authorities take part in the debate. They all seem to agree that a school meal scheme in general might be a good idea, but when it comes to the question of which initiatives to take and by whom, there is no consensus.

In the following section I will outline the essence of the main arguments for establishing school meal schemes, concerning 'health', 'learning ability', 'food education' and 'food service'. This is based on different sources, debates in the media, internet, parliament, conferences and recently published reports and guidelines (e.g. www.folketinget.dk, Biltoft-Jensen *et al.*, 2005)

The introduction of healthy school meal schemes are by many actors seen as part of a health promotion strategy. As in many other western countries Denmark are facing problems of increasing obesity and overweight among children and adolescents. It is well known that overweight increases the risk of a range of diseases. Both the current and the future financial costs of treating diet-related diseases are discussed with a concern for the national economy. A 'fat nation' will be very expensive.

'Health' seems to be the main reason for showing a political interest in school meals. But also aspects of learning ability are for some actor groups a focal issue. It is questioned whether the children eat a "decent" lunch which enable them to pay attention and concentrate on their work in school. Improving the learning ability is central for those who would like to see Denmark as a leading 'knowledge society'.

Another argument for introducing school meals has to do with educational aspects. Since the school is an arena for learning, school meals should be seen as an invitation to learn about food, meals, food production etc. School meals are also seen as a tool to enhance the general food education (german: bildung) which are believed to be quite poor in Denmark.

Furthermore establishing a school canteen is said to be an extra service for the busy families of today, so that they don't have to use time for preparing the packed lunch or maybe, if all of the family have a hot meal in the middle of the day in the different canteens either at school or at work, they can spare some time in the evening.

Discussion

Both in debates and practice, the above arguments are combined and of course much more nuanced than this presentation. In the following part I will broaden the perspectives and outline some dilemmas within the school meal debate.

Healthy school meals are perceived as a preventive tool. However, among nutritional experts there are still some uncertainty about the actual 'health effect' of school meals, e.g. to which extent school meals can influence the eating habits of children and push their daily intake into a more healthy direction. So far there has been no scientific evidence to say that school meals can be a preventive factor. Even though menus are developed on the basis of nutritious principles of a balanced diet, a lot of other factors will determine the actual intake.

The concept of health seems to be dominated by a nutritional perspective. However, WHO has defined health as "a state of complete physical, mental and social well-being and not merely the absence of disease or infirmity" (http://www.who.int). The more general concept of well-being is related to the basic care for children. In the Danish welfare society it has been a 'natural' concern to take care of the well-being of citizens, however care in terms of food has mainly been considered to be the parent's responsibility.

The debate of school meals is to some extent reflecting the current political discussion about the welfare society. To which extent should the state be responsible for the welfare of the citizens?
I will come back to this in the discussion of distribution of responsibility.
Learning ability has become an issue since Danish children are said to have problems with e.g. the ability to read, compared to other countries (Andersen *et al.*, 2001). It has been stated that the criteria used in the PISA investigation are not sufficient and that the Danish school system provides other skills and qualifications, which are not assessed within the framework of PISA. However the PISA report puts a pressure on the school system and within the last years the focus on 'academic subjects' has been strengthened, e.g. sacrificing the more practical subjects such as home economics.
The educational aspect of school meals is based on a wish to change the food preferences of children. In this respect it can be discussed whether the focus should be on facilitating the children's ability of choosing for them selves or if they should be adjusted to certain norms. There seems to be a difference between the northern and the southern part of Europe in relation to this question. Both in U.K. and Denmark the school meal schemes are based on various products to choose between and to combine. In Italy a whole menu is typically provided based on the Italian food culture, the children are not meant to choose for themselves (Morgan, 2003).
Summing up, the debate about school meals seems to be dominated by a concern of health, however health is mainly perceived rather narrow in terms of nutrition. The 'nutritional effect' of school meals are questionable, where as the broader sence of childrens well being might be affected. Learning ability is another concern, but some will question whether there actually is a problem or not. In terms of 'food education' school meals might be a relevant issue, but the method of approach is an educational research question per se.

Responsibility and implementation

The most general question of responsibility is whether school meals are a private or a public responsibility. Since there has been a long tradition of 'the packed lunch', the responsibility of what children eat, have primarily belonged to the parents. With the introduction of school meal schemes it is argued that this responsibility should no longer be private. But as a 'classical' dilemma in discussions about the welfare society, it might be problematic to 'nurse' the citizens too much and undermine the responsibility of in this case the parents (Haastrup, 2003).
Regarding the distribution of the 'public responsibility' for school meals this can be arranged in many different ways. In Sweden there is a national law of providing 'free' school meals for all, financed via the tax system. The local authorities are responsible for running the service. In Denmark a majority in the parliament will not support the idea of 'free' school meals, in stead there might be a national support scheme. The role of the local authorities will depend on the local political majority and what is considered as part of the welfare obligation. As mentioned earlier, some municipalities have taken up school meals as a strategic element in fulfilling their policy, but also some single schools have started to provide school meals. This latter case is often based on a local 'fiery soul' who might be considered responsible for the success of the scheme.
In considering the implementation of school meal schemes, a very functional and 'instrumental' line of approach seems to be dominating the agenda, both in debates and in the initial practices. There seems to be an understanding that if well-balanced nutritious school meals are introduced, the children will be healthier and fit, their learning ability will be improved, etc. But it should be obvious that a lot of social and cultural factors will be interrupting and moderating this scenario and that a top-down implementation approach will face some barriers (Winter, 1990).
In the next section I will briefly present an analytical framework, as a suggestion to how the implementation of school meal schemes can be analyzed. The interest is to analyze the social

processes of implementation, with contributions from the institutional theory of organizations. Due to the length of this paper I will here focus on schools only.

The debate about school meals can be seen as a change in societal expectations towards those actors who in some way influence the everyday of young people. In this aspect the school is seen as a focal actor. Drawing on Scott (2001) we can understand the school as an organization, which relates to their environment in a specific way. Scott (2001) is pointing out that "...whether, when, and how organizations respond depends on their individual characteristics or connections" (Scott, 2001). In other words, schools are different! Any school meal scheme has to be open for local adjustment.

A new law or proposal can, from the perspective of an organization, be understood "as an occasion for sense making and collective interpretation, relying more on cultural-cognitive and normative than coercive elements for its effects" (Scott, 2001). This means that instead of assuming a 'demand-response model', the picture is much more nuanced and complex.

To illustrate some of the dilemmas when it comes to the implementation of a school meal scheme I will reflect upon an example from Copenhagen. In the municipality of Copenhagen it was decided by the city government in 2002 to run a school meal scheme with centrally produced food for sale in a canteen at each of the 69 public schools (www.skolemad.kk.dk). The schools are obliged to implement this new element in their 'everyday life', a canteen with certain features. The food for sale is healthy, partly organic and developed after what is believed to be the preferences of children. In order to establish a local 'ownership' the project is framed as a forum for education and not only a functional duty of supply. This means that students from 6th grade and a teacher are in charge of running the canteen. The school are expected to support the idea of having this kind of canteen and a 'food educational forum', letting the children and one or more teachers be part of the project. It has shown that the schools relate to the project in very different ways. For instance some are very annoyed at not having any influence on the supply of food, some are not willing to integrate the canteen in the administration of the school (e.g. in the over-all timetable planning), some are not informing the parents about the canteen as they are supposed to do, etc. In other words, quite a few schools have not come to see the project as 'their own', but still something belonging to the municipality and they do not take any responsibility for the canteen. While there are other examples of schools being very enthusiastic about the canteen and the concept as such, feeding back to the municipal administrators both on the function of the canteen and the further development of different aspects. The attitude of the school is one determinant for the 'outcome' of the school meal scheme. If the school doesn't focus very much on the canteen it might be the case that very few children use the canteen and an eventually 'health effect' (or other effects) will be very small. In Glasgow a very big school meal project has been focusing especially on this issue of acceptance, to actually get the children and the schools engaged. This was seen as fundamental for success and as a precondition to achieve the targets related to health promotion www.fuelzone.co.uk

Conclusion

'School meals' are part of a 'discursive battlefield' in Denmark. Many different issues are put on the agenda, but a concern for health seems to be the main argument for the introduction of school meal schemes. Also the question of responsibility is debated. This discussion is related to a broader and very ideological welfare debate where the distribution of responsibility for welfare duties is a main issue.

The first school meal schemes are running but are not overwhelmingly successful. This points to the fact that no matter what effects one wish to achieve by the introduction of school meals, the food need to be accepted! Therefore it is essential to focus on the implementation process and the conditions for implementation on the different levels.

We have still not reached a point where school meals are institutionalized in terms of regulation, norms or a socio-cultural framework.

References

Andersen, A.M., Egelund, N., Pilegaard, T.J., Krone, M., Lindenskov, L. and Mejding, J. (2001). *Forventninger og færdigheder - danske unge i en international sammenligning,* AKF, DPU & SFI-SURVEY, available on: www.sfi.dk/graphics/SFI/SFI-SURVEY/Danske_PISA_rapport.pdf [date of consultation: 19/04/06]

Biltoft-Jensen, A., Ygil, K.H., Christensen, L.M., Christensen, S.M. and Christensen, T. *Forslag til retningslinjer for sund kost i skoler og institutioner,* Report DFVF Publikation nr.14, available at www.dfvf.dk

Haastrup, L. (2003). Mad og måltider i skolen. In: *Mad, Mennesker og Måltider,* Holm, L. (ed), Munksgaard Danmark. p 247-263.

Morgan, K. and Morley, A. (2003). *School meals: Healthy Eating & Sustainable Food Chains.*The Regeneration Institute, Cardiff University. 58 p. www.cardiff.ac.uk/cplan/ri

Scott, W.R. (2001). *Institutions and Organizations*. Second Edition. Sage. 255 p.

Winter, S. (1990). Integrating Implementation Research. In: *Implementation and the Policy Process,* Palumbo, D.J. and Calista, D.J. (eds), Greenwood Press. p 19-38.

www.folketinget.dk/Samling/20051/beslutningsforslag/B38/som_fremsat.htm [date of consultation: 19/04/06]

www.fuelzone.co.uk [date of consultation: 19/04/06]

www.skolemad.kk.dk [date of consultation: 19/04/06]

www.who.int/about/en/ [date of consultation: 19/04/06]

Technological interventions for global food security

Albert E.J. McGill
James Martin Institute for Science and Civilization, Said Business School, Oxford University, Park End Street, Oxford OX1 1HP, United Kingdom, albert.mcgill@sbs.ox.ac.uk

Abstract

At its World Congress in 1995, the International Union of Food Science and Technology (IUFoST) pledged on behalf of its members to use its skills to alleviate world hunger. By 2003 there was no evidence of any work having been done. Following an inspiring plenary paper by Dr Seragelden, a project was formulated to engage the grass-roots membership in local technical support for existing programmes. This project has the endorsement of the organization and initial stages of the project development are presented. A major challenge has been the lack of recognition by both UN and national agency programmes of the value and essential contributions that food science and technology can and must play if the provision of edible, nutritious food is to be the real target and not just the availability of crops in the field. Details of key programmes in food security are described and the paucity of attention to the food supply chain noted. Equally notable is the failure of IUFoST to be well enough informed about other programmes in food security to be able to make effective, complementary contributions.

Keywords: food security, hunger, nutrition, food science

Introduction

The International Union of Food Science and Technology (IUFoST), a country-membership organization, is the sole global food science and technology organization. It is a voluntary, non-profit association of national food science organizations linking the world's best food scientists and technologists. The present membership comprises 67 countries, each of which is represented through its Adhering Body (AB). The food professions of each country contribute to their national AB as best fits their individual circumstances. IUFoST is best known for its Congresses, Conferences and Symposia and is a full Scientific Member of the International Council of Scientific Unions (ICSU).

At the Congress in Budapest in 1995, a declaration was approved by the delegates that extended the intentions for the organization put forward initially by Lord Rank., and drew upon the outcomes of the International Conference on Nutrition (ICN, 1992) The key points of the declaration emphasize the delegates' " ...determination to work for the elimination of hunger and the reduction of all forms of malnutrition throughout the world ...ensure sustained nutritional well-being for all people in a peaceful, just and environmentally safe world ....recognizing that access to nutritionally adequate and safe food is the right of each individual." (IUFoST, 1995).

In a plenary lecture in Seoul at the 2001 Congress, Owen Fennema surprised many delegates by criticizing the abject failure of IUFoST to make any progress towards meeting the obligations of the Budapest Declaration (IUFoST, 2001). At the Chicago Congress the same failure to make progress was highlighted by Ismael Serageldin (IUFoST, 2003) who drew an analogy between the challenge of tackling hunger and that of tackling slavery. Discussions with colleagues at the Congress generated the idea of not attempting to tackle hunger as an international problem on a grand scale which can be seen as too overwhelming, but rather to break it down into local components that might be influenced by the voluntary work of individuals, "the new abolitionis ts",complementing, through their professions and Adhering Body, the existing work of the major agencies such as WHO/FAO/WFP and other voluntary groups such as OXFAM. But how?

Methodology

Mobilizing the profession

It soon became clear that the optimal way to involve the largest number of qualified food professionals on any project was to use the organizational structure of IUFoST and its Adhering Bodies. This would require the authority and approval of the Governing Council (GC) and the support of the secretariat. Accordingly a proposal entitled, "Intervention and Mobilization for Food Security" was put before this body at its meeting in Kuala Lumpur in March, 2005.

Definitions

Although the initial focus and concerns of the project were on world hunger, examination of both the Budapest Declaration and the initial President's address, the terms food security/insecurity are considered more appropriate and parallel the approach of the major food agencies:
"Food Security exists when all people, at all times, have physical and economic access to sufficient, safe and nutritious food to meet their dietary needs and food preference for an active and healthy life" (FAO, 1996). This definition is in keeping with the principle that everyone has a right to adequate food, to be free from hunger, and to enjoy general human dignity, enshrined in the International Bill of Human Rights (ECOSOC, 1999). The definition also makes clear that food production does not equal food security.

Communications

The presentations by Fennema and Serageldin at consecutive Congresses bemoaning the paucity of activity by IUFoST as a whole on the Budapest Declaration presupposed that a survey of such activities had been carried out and/or a recording system was available to track projects or other events that could be seen to be contributing to the relief of hunger or the improvement in food security. It was discovered that no significant action had been taken in this area as there was no related project, apart from that of Joseph Hulse as Chair of an Integrated Food Systems Task Force, which had produced only a draft outline report. Clearly a more robust system of communication and record would be required.

Complementary programmes

Both the size and the technical focus of IUFoST and the professions accommodated within its Adhering Bodies, limit the inputs and the impact that can be directed onto the issue of global food security. Accordingly much of what may be achieved will depend on a considerable knowledge of the activities and programmes of the major players, namely the UN agencies of WHO and FAO, supported by the voluntary agencies, such as OXFAM. Identifying the active agencies, understanding their programmes and negotiating supporting activities are the key factors for this project.

Results

Mobilizing the profession

After considerable discussion and constructive criticism the meeting of the Governing Council of IUFoST in Kuala Lumpur, March 2005, endorsed the proposed project and undertook to advise their Adhering Bodies ie. country members that the project should be supported.

Definitions

The use of the terms food security/insecurity has made the identification of existing and complementary programmes much more effective.

Communications

Contact with the IUFoST secretariat has been effective but has highlighted how little hard information exists on the composition of many adhering bodies with respect to the professions they have contained within their representation. No attempts have been made to correct this until a clear plan has been formulated and a clear message can be sent.

Complementary programmes

An extremely important and useful source of data and commentary on current programmes is "The State of Food Insecurity in the World", which is published by FAO each year in November and initial data were taken from this source (FAO, 2004; 2005) in making the presentation to IUFoST in Kuala Lumpur in March, 2005. The impact of all aspects of hunger on the populations of all countries for which data can be found (>80 in 2004) is presented and could therefore provide a starting point for the Adhering Body of any country listed. The first task would be to validate the FAO findings for their own country and, if necessary and possible, correct the picture. A summary of some of the key progress factors was made by Sanchez and Swaminathan, (2005), noting that although the hunger target of the MDGs will require unprecedented levels of effort, the cost of additional assistance amounts to 60 U.S. cents per month for every person living in a developed country during 2010-2015.

An initiative of the FAO, the FAO Special Programme for Food Security (SPFS) was established in 1994 to help Low Income Food Deficit Countries (LIFDC) to improve their national food security. Regional programmes have been developed with the appropriate Regional Economic Group, FAO Member Governments and other key players. Regional Programmes boost regional integration among developing countries so the impact of national programmes is maximized. Regional Programmes also work to strengthen national programmes by holding capacity building and training sessions with country representatives (FAO, 2005a). The SPFS activities have not been without criticism having been reviewed and reported (FAO, 2003). Although comprehensive in their work at regional and national level, there is little evidence in any programme of any significant work directed at post-harvest and supply-chain food processing for either improved local usage and nutrition or for the value adding necessary to bring surplus export income to reduce poverty among the primary producers.

Another invaluable source of data on Food Security across the world is the Food Insecurity and Vulnerability Information and Mapping Systems (FIVIMS, 2005), which provides direct country by country information that the IUFoST project can access and employ.

A number of papers prepared for the Hunger Task Force and the Millennium Project emanate from the International Food Policy Research Institute (IFPRI), so the updating of their strategic plan (IFPRI, 2005) is of significant interest. This 5-10 year projection discusses policy at length but pays little attention to processes that convert crops to food.

Another major consideration that impacts upon the world's food production and processing is the effect on the environment and associated climate change. In 2002 an international interdisciplinary research project was launched "..to better understand the relationship between Global Environmental Change (GEC) and food security". A joint project of the International Geosphere-Biosphere Programme (IGBP), the International Human Dimensions Programme on Global Environmental Change (IHDP) and the World Climate Research Programme (WCRP) was established as the Global Environmental Change and Food Systems (GECAFS) project. It also has formal research partnerships with the Consultative Group on International Agricultural Research (CGIAR), the UN Food and Agriculture Organization (FAO) and the UN World Meterological

Organization (WMO). More importantly, from the perspective of this project, it has origins with, and strong support from the International Council for Science (ICSU) of which IUFoST is a member. In fact the GECAFS project is described in the ICSU Series publication, "Global Environmental Change and Food Provision", (ICSU, 2002). Throughout both the ICSU publication cited and the GECAFS Science Plan and Implementation Strategy (GECAFS, 2005) there is a confusion of crops with the whole food system and supply chain which is surprising given the energy demands of that system and its critical needs, if food surpluses are to be marketed to provide much needed income for the poor in those countries. That neither ICSU nor GECAFS have made use of the resources of IUFoST or its members is surprising, particularly in field work. Such confusion and oversight is not new and can be found in key contributions on food security, (Rosegrant and Cline, 2003).

Discussion and conclusions

The images of hunger and starvation around the world, especially amongst children move all of us to want to make a difference. The global efforts of United Nations' agencies, particularly FAO, together with very many NGOs and voluntary aid agencies has been and is enormous and yet the overall effects have been modest when viewed against the size of the problem that remains. Initial research has shown that, almost without exception, little cognizance has been shown of the differences between crops, food and nutrition. The potential role of IUFoST in this regard has been ignored, sidelined or merely overlooked by the major agencies or assumed to be covered by advice from overworked staff of FAO. In parallel it must be noted that few, if any, members of the food professions represented by IUFoST are aware of the breadth and depth of work already accomplished in the agricultural, social, economic and political aspects of the Global Food Security issues. Much education and understanding of the comparative roles and activities of the many participants in this field must be undertaken if the resources of substantial technical expertise of the IUFoST members is not to be disengaged.

In conclusion, this project is still at an early stage of engagement but has produced some useful insights:

The Budapest Declaration (IUFoST, 1995), whilst having strength of pledge, had no specific stages, timelines or clear outcomes to measure success.

The means of communication with the Adhering Body (AB) members are not yet responsive or synchronized. The composition of ABs in terms of professions or capacities is not recorded nor easily determined.

The content and range of existing research and implementation programmes on Food Security has not been appreciated by IUFoST nor have the many agencies with whom IUFoST is familiar, and indeed a member, factored in the resources available in the food science and technology professions in contributing to their projects. Much of this appears to be from ignorance and oversight rather than deliberate exclusion and may be subject to revision.

This project must continue if the ethical stance of IUFoST is not to be undermined through failure to honour their own Declaration.

References

ECOSOC (United Nations Economic and Social Council).(1999). *"International Bill of Human Rights."* New York.

FAO (Food and Agriculture Organization of the United Nations). (1996). *"Rome Declaration on World Food Security and World Food Summit Plan of Action."* Adopted at the World Food Summit, November 13-17, Rome.

FAO. (2003). *"The Evaluation of the Special Programme for Food Security: Report on Follow-up Action." Summary of Actions taken to Respond to Recommendations of the Independent External Evaluation* May, 5-9 Rome. Available on http://www.fao.org/DOCREP/MEETING/006/Y8966E.HTM [date of consultation:10/11/2005].

FAO. (2004). *"The State of Food Insecurity in the World 2004. Monitoring progress towards the World Food Summit and Millennium Development Goals."* FAO, Rome. ISBN 92-5-1051X

FAO. (2005). *"The State of Food Insecurity in the World 2005. Eradicating world hunger – key to achieving the Millennium Development Goals."* FAO, Rome. ISBN 92-5-105384-7.

FAO. (2005a). *The Special Programme for Food Security. Support to national programmes.* Available on http://www.fao.org/tc/spfs/support [date of consultation: 01/12/2005].

FIVIMS (Food Insecurity and Vulnerability Information and Mapping Systems). (2005). *"Focus on Food Insecurity and Vulnerability."* Available on http://www.fivims.net/ [date of consultation: 20/10/2005].

GECAFS. (Global Environmental Change and Food Systems). (2005). *"Science Plan and Implementation Strategy."* Available on http://www.gecafs.org/ [date of consultation:20/02/2006].

ICN. (International Conference on Nutrition). (1992). *Final Report of the Conference on Nutrition."* ICN/92/3 and PREPCOM2.ICN/92/FINAL REPORT. FAO and WHO, Rome.

ICSU. (International Council for Science). (2002). *"Global Environmental Change and Food Provision: A new Role for Science."* ICSU Series on Science for Sustainable Development No.7. ICSU, France. ISSN 1683-3686, 20pp.

IFPRI. (International Food Policy Research Institute). (2005). *"IFPRI's Strategy: Toward food and nutrition security: Food Policy research, capacity strengthening, and policy communications"* IFPRI, USA, 47pp. Document may be downloaded at http://www.ifpri.org/about/ifpristrategy.pdf.

IUFoST. (International Union of Food Science and Technology). (1995). *"World Food Congress, Budapest, Hungary."* Information available on http://www.iufost.org/ [date of consultation: 05/01/2005].

IUFoST. (2001). *"World Food Congress, Seoul, South Korea."* Information available on http://www.iufost.org/ [date of consultation: 05/01/2005].

IUFoST. (2003). *"World Food Congress, Chicago, USA."* Information available on http://www.iufost.org/ [date of consultation: 05/01/2005].

Rosegrant, M.W and Cline, S.A. (2003). Global Food Security: Challenges and Policies. *Science*, 302: 1917-1919.

Sanchez, P.A. and Swaminathan, M.S. (2005). Cutting World Hunger in Half. *Science*, 307: 357-359.

Urban challenges and solutions for ethical eating

Albert E.J. McGill
James Martin Institute for Science and Civilization, Said Business School, Oxford University, Park End Street, Oxford OX1 1HP, United Kingdom, albert.mcgill@sbs.ox.ac.uk

Abstract

With the move of people from rural to urban locations across the world, so that the majority now live in cities, the challenge to provide food security for all, with its emphasis on choice as well as access and nutritional satisfaction, is much greater. To do so in a way that respects environmental as well as social and economic drivers is even more difficult. Two case studies are described, one in the UK, the other in Japan, that have and are succeeding in meeting the challenges while allowing their consumers diverse and ethical food choices. There would appear to be hope that discussion, collaboration and mutual respect will afford better health through good food choice and less damage to the associated environments.

Keywords: food strategy, health, sustainability, cities

Introduction

The Food and Agriculture Organization of the United Nations (FAO) refers, in its definition of Food Security, (FAO, 2004), not only to "...physical and economic access to sufficient, safe and nutritious food to meet dietary needs...for an active and healthy life.." but also "..food preference..." or indeed food choice. The seemingly inexorable movement of people from rural to urban environments has resulted, during this year, in more people living in cities than in the countryside. The populations at risk from hunger still live predominately in rural environments, but even in Africa, this situation is likely to reverse by 2020 (FAO, 2005). Consequently, the battle for food security will be fought as much in cities as in the countryside. Efforts are being made to ensure that, even in cities, farming and food production should be developed and expanded (RUAF, 2006). For example, "organic" produce generated A$28 million in sales in 1990 and A$300 million by 2004, in Australia (The Weekend Australian, 2006). This development has little impact on lower socioeconomic groups, whose local food supply is more likely to be dominated by fast food outlets.
How can affordable food from environmentally responsible suppliers be made available to a general, urban population? The following two "case studies" may indicate challenging solutions to the problem.

Methodology

The London food strategy

This is known as the Mayor's Food Strategy for London, is still in draft form and is based on his observation: "that the food system in the capital is not functioning in a way that is consistent with the ambition that London should be a world-class, sustainable city."(LDA, 2005).
Many features of the London food system are positive:

- Every day, millions of Londoners are able to access food they want and can afford.
- World-class retailers contribute to the commercial prosperity of the city, and provide employment for many thousands of people.

- Hundreds of businesses throughout London and the surrounding regions prosper by processing and manufacturing the food the city needs.
- The city's extraordinary social and cultural diversity is reflected in a restaurant culture that has seen it recently crowned the "gastronomic capital of the world".

Many features are much less positive:

- A rising number of Londoners, particularly children are becoming obese.
- In some parts of London, people struggle to access affordable, nutritious food.
- The safe preparation of food, both in the home and in London's plethora of restaurants and cafes, remains a key issue.
- Many small or independent enterprises, such as retailers, farmers, food manufacturers, marketers, struggle to survive and their (business) fragility may affect both the diversity and resilience of London's food system.
- The environmental consequences of the way London's food is grown, processed, transported and disposed of are profound and extensive. (LDA, 2005a).

As with all other cities in the world, London operates within a globalized setting and is equally affected by the negotiations of the World Trade Organization (WTO) and the corporate strategies of global enterprises such as Wal-Mart, Tesco or Nestle. The effect of the power of general market forces will also impact upon
consumer information and choice. Both market forces and consumer preferences are dynamic and are subject to change. London has the means to do this and a responsibility to act in the interests of its citizens. However, the capital's food system is inextricably linked to bodies, individuals and places outside London and often abroad. This is why the Mayor will work with and through other partner organizations, as often as may be necessary, to achieve his vision for London.

Daichi-o-Mamoru-Kai (Association to Preserve the Earth) (iNSnet, 2005)

In Japan, most foods are derived from imported crops and the calorie-based self-sufficiency ratio is only about 40 percent, the lowest of all OECD countries. Land available for agriculture has been declining annually, dropping to 12.8 percent of the total land area in 2002. The primary industry workforce (agriculture, forestry and fisheries) also declined to 4.7 percent of the total. Under these circumstances, various groups, large and small, have been formed to protect and develop local food traditions and primary industries, by establishing closer relationships among producers who farm organically, processors who maintain traditional manufacturing methods and their supporting consumers. One of these groups, Daichi-o-Mamoru-Kai, was established in 1975 as a result of an encounter with farmers disillusioned with agrichemicals, which they believed were damaging both the eco-system and consumers of their produce, to seek ways of growing agrichemical-free vegetables and rice and delivering them to consumers in urban areas. As the number of participating producers and consumers increased, the association established a distribution company, moving from a group-purchasing system to a door-to-door delivery service for individual households. It now handles up to 3,500 items, including meat, fish and processed foods.

Results

The London food strategy

Few attempts have been made to consider food as part of an integrated and interdependent system. Diet, health and wellbeing are inextricably linked and food lies at the heart of the Mayor's cross-cutting themes of health, equalities and sustainable development. Food waste requires new composting facilities and also offers potential for emerging renewable energy initiatives.

The importance of food within London is best summarized through the five principal themes of the Strategy framework:
Health: The relationship between diet/food and health is now regularly acknowledged (Rayner and Scarborough, 2005). The improvement of Londoners' diets could deliver significant benefits ranging from the incidence of cancer and coronary heart disease to type-2 diabetes. While many in cities around the world suffer from under-nutrition, in London diseases related to over-consumption are increasingly common. Air pollution associated with road freight in London has a number of major health implications, and the incidence of food-borne disease remains a key concern.
Environment: The food system has significant environmental impacts. In the case of London, it is responsible for 41 percent of the ecological footprint, while food preparation, storage and consumption account for between 10-20 percent of the environmental impact of the average household. More than one third of domestic waste in London is food related. In total, it has been estimated that close to half of the human impact on the environment is directly related to the operation of the food system. Consumer choice and more efficient food transportation could reduce this impost.
Economics: In London the agri-food sector is significant, employing nearly 500,000 people; the food and drink sector is the capital's second largest and fastest growing manufacturing sector; food retail businesses in the city employ tens of thousands of people; and London's thriving restaurant culture accounts for around one quarter of total UK activity in the sector and is a major attraction for tourists to the city. Londoners spend GBP 8.8 billion each year in retail food outlets.
Social & Cultural: Although food provides the essentials for survival, this is surpassed by its central role in socializing, providing pleasure and sustaining health and lifestyles. Many Londoners now prefer to eat out using the 12,000 restaurants (half the nation's total), 6,000 cafes and 5,000 pubs and bars.
Food Security: Unlike the accepted FAO definition of food security (FAO, 2004), this strategy reserves the term for issues associated with the ability of the food system to withstand an emergency or crisis event, such as flooding, disruption of energy supplies or terrorist attack.
The five principal themes are presented in the columns of the framework (Table 1).
Matched with these principal themes are the eight stages of the food chain comprising:
Primary production; processing and manufacture; distribution and transport; retailing: food purchase; food preparation and cooking; consumption; and disposal of waste.
The stages are presented in each row of the framework (Table 1).
Acting upon the food system and involved throughout it are the owners, providers and power brokers. These are presented in the second column of the framework (Table 1).
The framework is used throughout the strategy to structure the detailed plans and allocate timelines and responsibilities.

Daichi-o-Mamoru-Kai (Association to Preserve the Earth)

Having encouraged its primary producers to collaborate and share their own particular production methods, the association ran into difficulties with its customers. Consumers who took for granted attractive, uniform, unblemished vegetables complained immediately about misshapen, varied produce which was frequently marked or worm damaged and sometimes in short supply. The association turned these complaints into a marketing strategy through the education of its customers in the advantages of chemical-free foods from known producers. It now holds about 100 events each year when consumers visit farms, farmers hear complaints and explain the processes necessary to supply their vegetables. This process re-unites consumers with the point of production of their food and helps to minimize ignorance of food quality.

Table 1. The food strategy framework for the mayor's draft food strategy, London development agency, 2005 (from Table 3.1, www.lda.gov.uk/londonfood).

Stages of the Food Chain	Key organisations	Heath	Environment	Economic	Social & Cultural	Food Security
Stage 1: Primary production	Abattoirs, allotments, city farmers, community growers, cooperatives, fisheries, market gardens.	Public access to nature (health benefits), labour standards & pesticides exposure, health & safety, farmer welfare, public health & antibiotic use, nutrient content, animal feed (quality & sourcing of).	Biodiversity, energy/water use, climate change, agri-environmental schemes, pest-icides, GM crops, flooding, soil erosion, environmental management, soil fertility, quality assurance standards, pollution (air, water & soil), fishing by-catch, fish stocks, bush meat trade.	Income, employment, labour skills, access to markets, farming methods, diversification, non-food crops, crime (inc. vandalism, fly-tipping, theft), subsidies, economies of scale & farming intensity, quality assurance standards.	Public access to nature (educational benefits), labour standards, animal welfare, migrant labour & gang masters, quality assurance standard, ethnic food production, skills.	Biodiversity & genetic crop diversity, energy & water scarcity, GM crops, climate change, flooding, soil erosion, skills, potential for self sufficiency, animal & human disease, food scares.
Stage 2: Processing & manufacturing	BME processors, farmers, large processors, packaging companies, SME processors.	Health & safety, public health (nutrition, additives & flavouring), labour standards.	Energy use (include. Heating & cooling), climate change, air quality, water use, packaging, waste & recycling.	Employment, skills, income, access to markets.	Labour standards.	Disruption to fuel supplies, human & animal diseases, food scares.
Stage 3: Transport, storage & distribution	Distribution companies, farmers, logistics companies, retailers & supermarkets.	Mode of transport impacts, vehicle design, schedule, pollution (noise & air), congestion, infrastructure maintenance, nutrition.	Mode of transport, vehicle design, load profile, driver training, fuel type, air quality, food miles & CO2 emissions/climate change, energy use, packaging.	Mode of transport & costs, employment, vehicle design, load profile, information & communication technology, refrigeration, storage & warehousing.	Labour standards, skills & training.	Emergency/ disruption, oil dependency, 'Just-in-Time' delivery, mode of transport, infrastructure maintenance, international relations, climate change.

Stages of the Food Chain	Key organisations	Heath	Environment	Economic	Social & Cultural	Food Security
Stage 4: Food retail	BME retailers, catering companies, convenience retailers, direct selling (box schemes, internet, markets), importers/ exporters, target groups, markets (street & farm), off licences, public sectors, restaurant, SME retailers, cooperatives, social enterprises.	Transport impacts, food safety & hygiene.	Transport impacts, congestion, climate change, production methods.	Price, employment, pricing system (e.g. farm gate price) & contract criteria, Quality, reliability, WTO rules, import/ export duty, quality assurance standards.	Eating out & 'on the go'.	Emergency disruption to supplies, price, quantity, reliability, food access, nutritional value, transport infrastructure, climate change, biodiversity, energy supply, diversity of supply.
Stage 5: Purchasing food	Consumer, public procurement.	Nutrition, consumer preference, labelling.	Transport mode, vehicle efficiency, journey profile, air pollution, congestion, energy, consumer demand for organic food.	Household incomes, food price, consumer demand & preferences, emerging markets (e.g. ethical goods, internet shopping).	Lifestyles/habits, income, convenience & physical access, work patterns, cooking skills, nutrition/food knowledge, education, consumer preference, labelling.	Skills, facilities, disruption to utilities.

Stages of the Food Chain	Key organisations	Heath	Environment	Economic	Social & Cultural	Food Security
Stage 6: Food preparation, storage & cooking	Catering companies, community groups, individuals, public sector, restaurants, take-away outlets.	Lifestyle/habits, nutrition/vitamins, skills, ethnic food & ethnic food skills, health & safety, food safety & hygiene, target groups (age, ethnicity, pregnant mothers).	Energy & water use, climate change, air quality, cooking skills & shopping preferences.	Skills, cooking equipment, employment.	Lifestyles/habits, skills, cooking clubs, ethnic food & ethnic food skills, work patterns, target groups, cultural/special events.	Skills, facilities, distribution to utilities.
Stage 7: Eating & consumption	Business, care homes, community groups, individuals, public sector, Restaurants.	Lifestyle habits, nutrition/vitamins, health/well-being, breast feeding, dieting, nutrition standards.	Climate change, food related litter & disposable packaging.	Eating out, tourists, ethnic food, corporate procurement, public procurement, taste/quality, take-away, employment.	Lifestyle habits, family groups, breast-feeding, recipes, work patterns, cultural/special events, maslow, dieting, books & magazines, recipes, work patterns, take-away, cultural/special events.	Contamination of food & water supplies.
Stage 8: Disposal	Community groups, households, individuals, local authorities, markets, manufacturers, public sector, retailers, restaurants, waste companies.	Possible health impacts from landfill as well as visual pollution & smell: possible health impacts of incineration.	Loss of land to accommodation landfill, leachates from landfill, methane & CO_2 emissions from incineration, congestion & air quality issues from the transport of waste.	Transport costs of collection/infrastructure, increasing cost of waste management, need for investment in new facilities, job creation through recycling.	Waste recycling & composting collections, home composting, lifestyle/habits (e.g. convenience food & eating out), propensity to compost influenced by lifestyle/habits.	Threat of distribution to the collection & disposal of waste.

The next stage in the associations plans is to raise the awareness of consumers to the energy consumption of providing their food. This is the "food miles" concept, multiplying the volume of food by the distance it travels. According to the Ministry of Agriculture, Forestry and Fisheries, Japan's total food mileage for 2001 was about 900 billion ton-kilometres. This is 8.6 times that of France, three times that of the USA and 2.8 times that of South Korea. Because of the large volume of food imports and the great distances covered, huge amounts of energy are consumed and carbon dioxide (CO2) produced.

Another critical problem with raising consumer demand for domestically-produced foods that are fresher and healthier is that they are more expensive than imported foods. With interest in trying to reduce global warming since the ratification of the Kyoto protocol, the association decided to enhance their image by showing that purchasing their produce helped to reduce CO2 emissions. They estimated the CO2 amounts from the transport of various types of food coming from different production areas to compare domestic with imported foods. Tokyo was used as the destination for all produce and they developed the concept of a unit called the "poco", where 1 poco = 100 grams CO2. So choosing asparagus from Hokkaido means a reduction of 4 pocos of CO2 compared with the same produce from Australia. Daichi-o-Mamoru-Kai disseminates this kind of information through its public relations magazines and website as a way of advocating domestically-grown foods to reduce CO2 emissions. It estimates that eating 100% domestically grown food would reduce per capita CO2 emissions by 20,000 tons during a one year campaign involving its 70,000 members.

Discussion and conclusions

Both case studies have changed, or are attempting to change, both their respective food markets and local consumer behaviour.

Daichi-o-Mamoru-Kai has been established since 1975 and has developed not only a range of 3,500 food items for home delivery, but also an active, consultative 70,000 membership. It has extended its activities from giving its consumers a choice in purchasing locally sourced foods virtually free from chemicals and pesticides to making an ethical choice that connects production, processing and their environmental impacts on CO2 production through its concept of "pocos".

The Mayor of London, Mr Ken Livingstone, has had developed a Draft Food Strategy for London that considers food as an integrated and interdependent system and sits alongside, and is expected to network with, other similar strategies on Spatial Development, Transport, Municipal Waste Management and Economic Development. It is supported and enhanced by the detailed study on providing a Food Hub for the storage and distribution of food throughout London (LDA, 2005b). Both are a complex and detailed proposals that have had extensive consultation through both public forums and thematic specialized committees. The Food Strategy has been out for web based comment until December, 2005, and will be issued in final form sometime in May, 2006. It is not expected that there will be an easy consensus on its implementation but its creators are committed to consultation and compromise. This strategy, when implemented should give many Londoners access, diversity of products and informed choice that can allow them to make ethical eating their norm. It will also allow them to take greater responsibility for the health outcomes of their food choices and goes some way towards answering a major criticism of the Food Ethics Council's report (2005) on UK government policy on health and diet, (DOH, 2004; 2005) namely , that good food choices cannot be made in ignorance or when suitable produce is unavailable. Other aspects of the debate on more "personalized" diets have been discussed elsewhere (Food Ethics Council, 2006).

Both case studies show that the developing commitment of consumers to be able to exercise their choice of food on ethical as well as health, safety or economic grounds can be harnessed to support, and indeed drive, changes in environmental policy more effectively than might have been expected.

References

DOH (Department of Health, UK). (2004). *Choosing health: making healthier choices easier.* Department of Health, London. 128pp.

DOH (2005). *Choosing a Better Diet: a food and health action plan.* Department of Health, London. 52pp.

FAO. (Food and Agriculture organization of the United Nations). (2004). *The State of Food Insecurity in the World. Monitoring progress towards the World Food Summit and Millennium Development Goals.* FAO, Rome. ISBN 92-5-1051X.

FAO. (2005). *The State of Food Insecurity in the World. Eradicating world hunger – key to achieving the Millennium Development Goals.* FAO, Rome. ISBN 92-5-105384-7.

Food Ethics Council. (2005). *GETTING PERSONAL: shifting responsibilities for dietary health.* ISBN 0-9549218-1-X. Available on http://www.foodethicscouncil.org [date of consultation: 20/12/2005].

Food Ethics Council. (2006). *GETTING PERSONAL: shifting responsibilities for dietary health. Launch symposium report.* Available on http://www.foodethicscouncil.org [date of consultation: 21/04/2006].

iNSnet. (2005).*Unique NGOs in Japan. Daichi-o-Mamoru-Kai.* Available on http://www.insnet.org/ins_spoton [date of consultation: 25/10/2005].

LDA (London Development Agency). (2005). *Better food for London: the Mayor's Draft Food Strategy.* London Development Agency. London. Available on http://www.lda.gov.uk/lpndonfood [date of consultation: 25/10/2005].

LDA. (2005a). *ibid.* page 1.

LDA.(2005b). *London Sustainable Food Hub. Opportunities for a sustainable food logistics centre in London.* Available on http://www.lda/gov.uk [date of consultation: 30/10/2005].

Rayner,M. and Scarborough,P. (2005).The burden of food related ill health in the UK. *Journal of Epidemiology and Community Health,* 59:1054-1057.

RUAF. (Resource Centres on Urban Agriculture & food security). (2006). *What is urban agriculture?* Available on http://www.ruaf.org/node/512. [date of consultation: 22/04/2006].

The Weekend Australian. (2006).*Jury out on chemicals v. organic battle.* 21-22-January, p 22.

Governance of healthy eating: The changing role of governments in promoting healthy dietary habits and its implications for consumers and civil society

Bent Egberg Mikkelsen
Danish Institute for Food and Veterinary Research, Mørkhøj Bygade 19, DK-2860 Søborg, Denmark, bem@dfvf.dk

Abstract

Public health strategies that can be used to fight the increase in obesity and overweight are strongly needed. There is a general acceptance of the idea that such strategies should focus on the promotion of healthy eating and increased physical activity. Such strategies have traditionally focused on downstream strategies putting the individual in the centre, but there is a growing acceptance of the fact that wider socioeconomic and environmental factors need to be taken into account if effective promotion of healthy eating is to be achieved. This paper addresses these wider aspects and places the promotion of healthy eating into a governance framework. It hypothesizes that we face a modernisation of governing of society in which traditional decision making processes and governing structures are changing from primarily politician and bureaucrat driven to network decision making in which a wider range of stakeholders are involved. It argues that this will influence public health policy aimed at promoting healthy eating. As a result a reorientation and rethinking of healthy eating strategies is necessary. The paper assumes that a reorientation must involve a broader stakeholder perspective than normally included. The papers argue that such reorientation must include NGO's, consumer organisations as well as food industry and sets out for a discussion of the questions that this development gives rise to. In conclusion it discusses the implications of new type of nutritional governance. It argues that modernisation of our governing structures is taking place in the food and nutrition area and that this will have consequences for public health nutrition policy.

Introduction

The incidence of overweight and obesity are increasing in most countries (WHO, 2004) due to lack of physical activity and unhealthy eating habits and the World Health Organization (WHO) has identified the obesity epidemic as one of the biggest threats to public health worldwide and that there are now more overweight people than underweight (WHO, 2003). Public health strategies that can be used to fight the increase in obesity and overweight are therefore strongly needed. Such strategies have traditionally focused on downstream strategies putting the individual in the centre, but the idea that wider socioeconomic and environmental factors needs to be taken into account is gaining momentum. Such up stream strategies involve the commercial food sector including manufacturing, retailing and out of home eating. According to the WHO (2003), the high intake of energy-dense foods, heavy marketing of energy-dense foods and fast foods outlets, large portion sizes, are among the main causes for overweight and obesity and the high proportion of food prepared outside the home (WHO, 2003) contributes to the increased incidence of overweight and obesity. A number of studies have shown that food eaten outside the home contribute to unhealthy diets and larger food intake (Edwards *et al.*, 2005, Roos, 2004, Pereira *et al.*2005). The contribution of food industry to the health problems resulting from unhealthy eating has lead to an increased awareness of the responsibility of

different stakeholders to take part in preventive strategies, including commercial stakeholders and NGO´s.
This paper addresses the development in society in which up stream factors is taken into account wider. It hypothesizes that we experience a modernisation of governing structures in society in which traditional decision making processes and governing structures are changing from primarily politician and bureaucrat driven towards network decision making in which a wider range of stakeholders are involved. It argues that this must influence public health policy aimed at promoting healthy eating and that as a result a reorientation and rethinking of healthy eating strategies is needed.

Theoretical foundation

According to contemporary political science modern societies face a development towards more networked kinds of decision-making in which governments, NGOs and industry interact and "governs" thereby practising governance. This seems to coincidence with a development in which traditional kind of representative democracy are replaced by new "networked" types of decision making. Where elected governments formerly were expected - on behalf of the citizens - to adopt policies for the benefit of citizens and the society as a whole and to make sure that they were implemented, modern governance involve non public partners in a new way. In neo-liberal societies applying the principles of new public management the political rationale is in favour new types of service provision including slimmer public services and more market. The notion of governmentality (Foucault 1999) plays a central role in explaining the modernization of society, which calls for a re-orientation of public health nutrition policy.
According to governmentality studies government is there not only to serve the sovereign of the state but also has a role in protecting the wellbeing of the self i.e. the health of the individual. Governing in the new context is about managing a whole population and society so that both prosper. Foucault describe the new meaning of government as the ``the conduct of conduct'' and thus as a term which ranges from "governing the self" to "governing others". This implies that governing is about shaping and guiding citizens so they do what is best for them and for society as a whole (Foucault 1991). Taking a governmentality approach term also implies that there has been a shift in governance so that "looking of the good of the citizens" i.e. promoting healthy eating is not just a responsibility of the government but is regarded also as a shared responsibility of a range of stakeholders including government, market and civil society. The point is that governmentality is a more accurate description of the range of rules, programs, and institutions involved in running a modern country than do standard concepts of ``the government'' or ``the state''. This means that governance also includes the action and governing carried out by non governmental and private agencies.

Methods

The paper is empirically based on three sources. Analyses of the Danish Public Private Partnership Conference held in October 2003, the Danish Obesity summit for industry government and NGO´s held in march 2005 and the Nordic Out of Home healthy eating seminar held in May 2005 (Mikkelsen and Trolle, 2004, Mikkelsen, 2005, Mikkelsen *et al.* 2005).

Who can influence our dietary habits?

At least four important actors have strong influential roles in the shaping of our eating habits and thus have role in promoting healthy eating: the individual, government, industry and civil society.

The individual

Where as the public traditionally has been playing an important role in protecting health and providing health care the individual responsibility for healthy eating has come back into focus in the shift from welfarism to neoliberal governance. Thus within the health care sector, the concept of the active consumer, has come to replace the notion of the passive patient and citizens are to an increasing extent expected to maintain a lifestyle that minimises their contribution to health care costs by. This means individuals are expected to practice their agency in line with public health recommendations i.e. nutritional guidelines. But in most theoretical models of health behaviour the individual plays an important role but not the only one. In Bandura´social cognition theory (1999) the environment plays an important role as well and in the context of healthy eating this means that healthy options should be available.

Government

Traditional it is assumed that governments through the mandate given to them in elections should govern societal condition so as to facilitate healthy eating. It is assumed that the public has to instruments at hand: service provision and regulatory instruments and that these regulatory instruments include legal instruments (laws), campaigns and taxations. For a long time governmental agencies, have been in the frontline in searching for ways of enabling citizens to make the right choices and promoting healthy eating patterns through campaigns. Thus nutrition has since long been a central arena for public engagement and regulation in most European countries (Kjærnes 1993).

Food industry

To an increasing extent food industry is engaging in corporate nutritional or diet related schemes (Lang and Heasman 2004, Mikkelsen and Trolle, 2004) and seems to co-exist with public regulations in the same area (Marsden *et al.* 2000). This kind of private regulation has been named self-regulation, and is the dominant way of EU regulation in a number of adjacent areas, e.g. food safety and environmental management (Majone 1999). In some cases, these private initiatives, take on responsibilities that public authorities otherwise would cover.

Civil society

Public interest NGO´s as well as trade organisations are engaging in regulation in this field Lauber and Sheard, 2001 and there is a long tradition for private health associations and NGO's engaging in nutrition and healthy eating related work. National and international associations that supports prevention of cancer, diabetes and CHD are obvious examples

What is new in relation to promoting healthy eating?

Results of the analysis of healthy eating initiatives taken by commercial stakeholders, non governmental organisations alone or in partnerships shows that non governmental stakeholders to an increasing extent engage in a number of activities related to the promotion of healthy eating. The three events referred to in the paper shows that the problems related to overweight and obesity now has reached a level where action needs to be taken by other than government agencies. Although companies does not necessarily regard themselves as being part of the problem they do increasingly recognise that the might be a part of a solution.
Government on their sides has com to realise that stand alone campaigns has failed to bring about sustainable change in dietary habits and that they must engage in partnerships with the stakeholders that control the supply chain, in this case food industry. Most markedly this has shown its viability in the case of the 5 a day fruit and vegetable promotion schemes. Against this background government bodies engage in networks and partnerships to an extent that

was not known before and at the same time attempts to promote healthy eating through legal measures have decreased.
Non-governmental organisations on their side do also engage in healthy eating schemes and increasingly they do so in cooperation with governmental bodies and companies. This is the case both for public interest NGO´s such as health promoting/patients organisations and for trade industry organisations representing different sectors of industry.
The development underpin the assumption argue for in this paper that a change in governance of healthy eating is taking place. As Governmentality studies show governance is not anymore in the hands of governments alone, rather power has shifted to be in the hands of a number of stakeholder's i.e. corporate actors and NGO's. But although according to Governmentality studies emphasises that the responsibility for health matters has shifted from a patient/client perspective to a consumer perspective where individual responsibility is expected, the problem of obesity has now reached a level where other stakeholder need to take action and where not only individual behaviour but also availability and environmental change has to be addressed. These findings are supported in recent papers and commentaries. As argued by Eves *et al.* (1996) caterers and Njor (2004) food service operators are generally confident in the belief that they can influence consumer choice and that they play a role in providing a healthy diet. Also in other areas of the food sector examples can be found. Trade associations within food industry (Mikkelsen, 2005, NAOS, 2005) have launched initiatives that aim at increasing corporate social responsibility in relation to healthy eating.

Discussion

Involvement of commercial actors is called for in modern policy papers on nutrition and healthy eating from the EU and WHO (WHO, 2004, EU, Raben 2003). In particular, involvement of the private food sector is sought for in recent EU policy statements (Telicka, 2004, Kyprianeou, 2005). But companies within the food sector seem to react not only because they feel forced to it but also as a result to "business" forces. For modern businesses the term Corporate Social Responsibility (CSR) has become the preferred term to describe the phenomena that businesses, besides their obligation to earn money and comply with legislation, should in addition take responsibility for societal issues (Caroll, 1980). These new responsibilities are in contrast to traditional business thinking in which companies usually have measured their success solely in terms of sales, profit and market share. In this way focus has shifted from the perspective of shareholders towards the perspective of stakeholders. The increased industry participation in the governance of healthy eating is subject to continuing and ongoing debate. Some commentators regard industry participation as problematic (Nestle, 2003; Critser, 2004) while others see industry participation in governance as positive and thus two competing views on the development are found.
According to *one viewpoints* industry participation in governance of society is unwanted because a large part of the food industry is dependant on making a profit on marketing of unhealthy products and thus in general influence our dietary habits in a negative way. Thus traditionally there have been reservations concerning the role that industry could play in relation to promotion of healthy eating. In general there has been concern that the shaping of our dietary habits seems to be in the hands of food industry rather than in the hands of legal and democratic bodies such as governments as one would like to see it. In fact many commentators see food industry as one of the main barriers to effective promotion of healthy eating.
According to the other viewpoint such governance involving industry in decision making is positive (or at least unavoidable) since it is assumed a priory that it is not possible to influence

dietary habits radically without being able to control the supply chain[82] and thus the availability. In such control food industry (namely food manufacturer, retailer and food service operators) participation is unavoidable. Although successful examples of partnerships in which industry and government (and NGO's) has been able to make a difference are found predominantly among 5 a day schemes (in which food industry supply has an obvious interest in promoting the marketing of their products) the belief that involvement of industry can be useful is gaining momentum. According to this viewpoint governance should not be seen as a positive phenomena, which could raise synergy in society. This argument lies behind the concept of public private partnerships, which has influenced much of the contemporary New Public Management and neo-liberal thinking.
But regardless of viewpoint the development calls for more work to be carried out. There is a need to identify best practices in industry, trade organisations and civil society and for developing a framework for evaluating the appropriateness of initiatives, developing managerial framework and benchmarking for the food industry to manage nutritional and healthy eating issues. Also the development give rise to important questions regarding the for public health nutrition policy. In some cases i.e. promotion of fruit and vegetable/five a day schemes there are obvious advantages in involving commercial partners and thus there is a need to formulate ideas to industry, governments and civil society on partnership working.

References

Bandura, A. Social Cognitive Theory: An Agentic Perspective. *Asian Journal of Social Psychology* **2** (1), 21-41: 1999

Bessant, J. Kaplinsky, R. Lamming, R. Putting supply chain learning into practice, *International Journal of Operations & Production Management*, **23,** 2, 167 – 184: 2003

Caroll, A.B. The Four Faces of Corporate Citizenship. *Business and Society Review*. 1998;100: 1-7: 1998.

Critser, Greg Fat land : how Americans became the fattest people in the world, 2004. Penguin

Edwards J.S.A., Engström, K. and Hartwell H.J. Overweight, obesity and the food service industry, In Conference Proceedings of Fifth International Conference on Culinary Arts and Sciences, *Global and National Perspectives*, Bournemouth: University, 161-171: 2005

Eves, A., Corney, M., Kipps, M., Noble, C., Lumbers, M., and Pice, M., The nutritional implications of food choices from catering outlets. *Nutrition and Food Science,* September/October, 26-29: 1996

Foucault, M (1991) Governmentatilty in *The Foucault effect –studies in governmentality.* Burchell, G Gordon C and Miller P (eds) London: Harvester Wheatsheaf

Kyprianou, M. Interview with the Commisioner for Health & Consumer Protection DG (www.ft.com): 2005.

Lang, T. and Heasman, M. *Food Wars, The Global Battle for MoutMinds and Markets*, London: *Earthscan*: 2004

Lauber, R.P, and Sheard, N.F. Nutrition, Science & Policy, The American Heart Association Dietary Guidelines for 2000: A Summary Report, *Nutrition Reviews* 2001 - 59 - 9 - 298

Majone, G.: The Regulatory State and its Legitimacy Problems. *West European Politics*, 22: 1-24: 1999

Marsden, T., Flynn, A. and M. Harrison: Consuming Interest. *The Social Provision of Foods.* London UCL Press: 2000

Matten, D., Grane, A., Chapple, W., Behind the Mask: Revealing the Mask: Revealing the The True Face of Corporate Citizenship. *Journal of Business Ethics* 45: 109-120: 2003

Mikkelsen, B.E. The declining role of governments in promoting healthy eating – time to rethink the role of food industry? *Scandinavian Journal of Nutrition*, vol 49, no 3., 127-130: 2005

[82] Here it is assumed that the public only has three regulatory instruments avialible , laws, campaigns and tax (Winter, 1994) and that the these are insufficient in this case

Mikkelsen, B.E., and Trolle, E., Partnerships for better nutrition – an analysis of how Danish authorities, companies, organisations and practitioners are networking to promote healthy eating, *Scandinavian Journal of Nutrition*; 48, 61-69: 2004

Mikkelsen BE, Jørgensen MS, and Bruselius Jensen M Danish Food Sector's Involvement In Strategies

Towards, Overweight And Obesity - Findings From A Survey. IUNS 2005 Durban S.A. Nestle, M. *Food politics : how the food industry influences nutrition and health*, 2003. University of California Press

NAOS (2005)Spanish Food Industry NAOS strategy, www.naos.es

Njor, L. The cutting edge. [På forkant med udviklingen]. Editorial in *Visitor*, no 4: 2004

Pereira, M.A., Kartashov, A.I., Ebbeling, C.B. van Horn, L, Slattery, M.L., Jacobs, D.R., Ludwig, D.S. (2005) Fast-food habits, weight gain, and insulin resistance (the CARDIA study): 15-year prospective analysis, Lancet, 365, 9453, 36

Raben, A. Jumbosize Europe? (2003) European Union conference on obesity calls for immediate action, *Scandinavian Journal of Nutrition*; 47 (1): 29-38

Roos, E, Sarlio-Lähteenkorva, S and Lallukka T. (2004) Having lunch at a staff canteen is associated with recommended food habits Source: *Public Health Nutrition* Volume: 7 Number: 1, 53 – 61:

Telicka, P. (2004), Speech of the commissioner. *European Commission*, October

WHO (2003): Diet, Nutrition and the Prevention of Chronic Diseases. WHO Technical report series 916. http://www.who.int/dietphysicalactivity/publications/trs916/en/

WHO (2004): Global strategy on diet, physical activity and health. Fifty-seventh Health Assembly, WHA 57.17, 22 May 2004, http://www.who.int/dietphysicalactivity/goals/en/

Food system policies in rich countries and consequences in poor ones: Ethical considerations

Per Pinstrup-Andersen
H.E. Babcock Professor, 305 Savage Hall, Cornell University, Ithaca, NY 14853-6301, USA, pp94@cornell.edu

Abstract

Food and Agricultural policies in OECD countries are likely to have spillover effects on developing countries and their low-income people. The importance of such effects grows with increasing globalization. OECD policies that result in increasing poverty, hunger and related human suffering in developing countries present a serious ethical dilemma for OECD countries, whether such effects are intended or not. Existing agricultural trade, technology and food safety policies are particularly likely to cause harm to poor people. The ethical aspects of these policies are analyzed in this paper. Failure to explicitly consider such international spillovers from OECD policies in policy design and implementation creates serious ethical problems and contributes to social injustice, poverty and food insecurity.

Keywords: food policy, ethics, nutrition, hunger

Introduction

Should ethical considerations regarding poverty and hunger in developing countries influence the design and implementation of policies for the food system in OECD countries? This question will be addressed here for three policy areas, where rich countries' policies may influence the welfare of poor people in poor countries. These are: trade and subsidy policies, technology policies and food safety policies.

Trade and subsidy policies

The principal ethical considerations of agricultural trade and subsidy policies in OECD countries are related to the impact of these policies on market access and export subsidies, both resulting in trade distortions and negative effects on the poor in developing countries. Between 70 and 75 percent of the world's absolute poor live in rural areas of developing countries. The majority depends on agriculture directly or indirectly. In most of these countries, export opportunities are critical for agricultural and broader economic growth. High tariffs as well as non-tariff barriers imposed by the OECD countries on imports of agricultural commodities and products place severe constraints on the opportunities for developing countries to compete on the OECD markets.

Export subsidies and similar measures in OECD countries facilitate exports of agricultural commodities at prices below cost of production, making competition in the international market almost impossible for developing countries, most of which cannot afford outright export subsidies. Furthermore, some OECD countries, notably the United States, dump agricultural commodities, such as maize and sugar, in selected developing countries either as free food aid or at prices significantly below cost of production, thus undercutting farmers in their national markets.

While consumers in importing countries may benefit from low import prices, at least in the short-run, opportunities for growth in the agricultural sector and related income increases among the rural poor, will be bypassed. Because of the very large multiplier effect resulting from agricultural growth in which an additional dollar of agricultural income may, as a rule of thumb, result in a general economic growth of 2.5 dollars in low-income developing countries (Pinstrup-Andersen, Lundberg, and Garrett 1995), stagnation in the agricultural sector has very serious ramifications for general economic growth and poverty alleviation in those countries.

Low-cost producers, including those in many poor countries, of textiles, sugar, cotton, dairy products, and a variety of other commodities and products are prohibited from competing on rich country markets and they are faced with competition from highly subsidized commodities in the international market. The economic costs are very high (OECD 2003) and the ethical aspects are worrisome. There are no reliable estimates of the humanitarian costs of OECD policies in terms of poverty, hunger, and child death, but there is no doubt that they are high.

The low international prices resulting from OECD agricultural trade and subsidy policies make investments in agricultural and rural areas of developing countries very difficult to justify financially. This is an important reason for under-investment in rural infrastructure, markets, and agricultural research. Thus, compared to a free-trading system, the agricultural trade and subsidy policies in rich countries result in more poverty, hunger, and human misery in developing countries. While this is clearly not the intent, it is the result rather than the intent that counts—if we subscribe to utilitarianism. As the evidence of such negative effects mounts, it is difficult to justify the policies on deontological grounds – that the intent is good.

From an ethical point of view, I believe that the most devious trade-distorting tool used by OECD countries is the tariff escalations applied to processed commodities. It is devious because it permits the tariff rate to increase in response to the increasing rate of processing. Tariff escalation helps to maintain low-income countries as a source of raw materials for processing, value addition, and employment creation in high-income countries, a policy that also characterized colonial times. It places severe constraints on opportunities by low-income developing countries to add value to staple food commodities and generate additional employment in rural areas. It is also inconsistent with the same OECD countries' development assistance to promote rural employment through agricultural processing for export.

OECD countries have repeatedly pretended to commit themselves to opening their markets for export from the least developed countries, as exemplified by EU's "Everything but Arms" and the U.S.' "Africa Growth and Opportunity Act". While these initiatives pretend to offer duty-free access for exports from the poorest countries to the EU or U.S. markets, respectively, many of the commodities and products of greatest importance for poor countries' export are excluded. Furthermore, safeguards that allow rich countries to cancel imports if domestic suppliers are threatened are typically built into such initiatives. Safeguards were used to limit imports of textiles and clothing into the European Union and the United States when the Agreement on Textiles and Clothing was discontinued. Current trade negotiations within the Doha framework permit countries to identify so-called "sensitive" products, which would be protected in agreements towards freer market access

Agricultural research and technology

While agricultural research and technology has played an essential role in expanding agricultural productivity and reducing poverty in both high- and low-income countries, modern agricultural technology is opposed by some. Perceived risks and uncertainty, along with ethical concerns have contributed to the opposition

The ethical issue of concern in this analysis is how the behavior by high-income countries affects low-income countries and poor people within these countries. Do rich societies have the right to influence the technology agenda for poor societies? Does it matter whether the influence is intentional or not? There are several ways in which the behavior by high-income countries may affect the welfare of low-income countries and their poor people, only some are intentional (Pinstrup-Andersen and Schiøler 2001). The Nuffield Council on Bioethics, a world renowned institution on the subject, concludes that the moral imperative for making genetically engineered crops available to developing countries that want them is compelling (Nuffield Council on Bioethics 1999).

If we believe in utilitarianism, it is outcome that counts. Outlawing or restricting genetically engineered food on the EU market, which would cause more hunger and child death in developing countries, would be equally bad whether the intent was to protect European consumers from risks and uncertainty or the intent was to cause harm to poor people in developing countries. But the intent does matter. Policies with the intent to harm poor people in developing countries would not be acceptable in OECD countries even if such policies would bestow large benefits on OECD citizens, or would they?

Positions regarding the use of genetic engineering for food and agriculture taken in Europe and other high-income countries may be adopted by developing countries. Such transmission of values, if imposed on developing countries through advocacy, trade policy, or conditions related to development assistance, may lead to ethically questionable outcomes, such as pressure on developing country policymakers not to approve the production of genetically engineered food, in cases where it may be the most appropriate solution to a food-related problem. Poor countries and poor people face food-related problems that are very different from those faced by non-poor. Therefore, benefit/cost ratios and the most appropriate solutions are likely to differ as well. Rich countries' very positive attitudes towards the use of genetic engineering in human medicine and the negative attitudes towards genetic engineering in food and agriculture, illustrate the point. OECD citizens get sick but they do not go hungry. Furthermore, the budget share spent on food is much lower in rich countries than in poor ones. Productivity increases in agriculture, including those that could be brought about through genetic engineering, resulting in lower unit costs and consumer prices are a powerful tool to help expand poor consumers' purchasing power but have little impact on high-income consumers.

With 50 to 80 percent of the population living in rural areas, and primary agriculture being the principal income source, poor countries are much more likely to be interested in productivity increases in agriculture than are rich countries, where only 2 to 5 percent depends on primary agriculture and where subsidies are an important source of income.

But what if genetic engineering offers large productivity and income gains for poor producers and consumers and food for export cannot be kept separate from food for domestic consumption? The issue here is one of choice, not ethics. Producing something the consumers do not want, makes no economic sense. However, suppose that European opposition to genetically engineered food, whether by consumers, supermarkets, or governments is a result of advocacy with ulterior motives and that there is no evidence of any negative health or environmental effects of approved genetically engineered food. Suppose it is, to use a term used by the U.S. government, not substantially different from food that is not genetically engineered? Without questioning the right of consumers to choose, does government action to prohibit import of genetically engineered food, which has been properly tested, not present an ethical problem because of its negative effect on poor farmers? Efforts to understand and resolve the controversy surrounding the use of genetic engineering in food and agriculture should address the underlying values, ideologies, and paradigms including ethical issues (Omamo and von Grebmer 2005).

Property rights to agricultural technology present another potential ethical issue. Should it apply to all potential users or would it make more sense from an ethical perspective to give free access by poor farmers? As a minimum, poor farmers, who buy seeds protected by patents or plant breeders' rights, should have the right to use their own production as seed in future seasons.

Food safety

Uniform international rules and regulations regarding food safety would facilitate international food trade and reduce the opportunities for the above mentioned misuse of sanitary and phytosanitary regulations. Efforts to establish such uniform rules and regulations are undertaken within the framework of Codex Alimentarius, under the guidance of the United Nations Food and Agricultural Organization (FAO) and the World Health Organization (WHO). However, the level of food safety demanded by consumers and societies tends to be a function of the income level. Thus, national food safety requirements become more stringent with increasing incomes. Similarly, within a given country, food safety requirements are likely to be at a much higher level among high-income consumers. This raises the question of whose food safety standard should be adopted as the universal level. The problem arises because increasing food safety requirements are likely to increase food prices.

Low-income people in many developing countries spend 60-80 percent of their total income on food while the average budget share spent on food in OECD countries is 10-15 percent. Therefore, any increase in the price of food brought about by higher safety standards is likely to have a greater negative impact on real incomes of the poor, than on real incomes of the non-poor

Similarly, elevation of food safety to high levels may—intentionally or not— function as a barrier to export by low-income countries. One could argue that the decision to elevate the standards contradicts utilitarian ethics and may have been driven by a virtue ethics on behalf of the policymakers to be seen to do the right thing for European consumers, even though no significant health effect was expected. Such government behavior is facilitated by the finding that perceptions of potential food safety risks appear to affect consumer preferences more then technical information does (Thilmany and Barrett 1997)

Do the OECD countries have an ethical obligation to restrain their food safety requirements in deference to the potential negative impact on the poor through higher food prices or export opportunities foregone or, at least, consider the impact on the poor before setting new and higher standards? I agree with Caswell and Friis-Bach (2004) that "There is an important role for a shared vision of moral rights and duties to play in international discussion and action on food safety".

References

Caswell, J. and C. Friis Bach. 2004. "Food Safety Standards in Rich and Poor Countries." Paper presented at Ethics, Globalization, and Hunger Workshop, Cornell University, Ithaca, NY, 17-19 November.

Nuffield Council on Bioethics. 1999. *Genetically Modified Crops: The Ethical and Social Issues*. London: Nuffield Council on Bioethics.

Omamo, S.W. and K. von Grebmer. 2005. "Chapter 8: Lessons and Recommendations". *Biotechnology, Agriculture, and Food Security in Southern Africa*. Washington, DC: International Food Policy Research Institute.

Organisation for Economic Co-Operation and Development. 2003. *Agricultural Policies in OECD Countries: Monitoring and Evaluation*. OECD Publications.

Pinstrup-Andersen. 2005. "Ethics and Economic Policy for the Food System" American Journal of Agricultural Economics, Vol 87, #5, pp 1097-1112.

Pinstrup-Andersen, P. and E. Schiøler. 2001. *Seeds of Contention: World Hunger and the Global Controversy Over GM Crops*. Baltimore, MD: Johns Hopkins University Press.

Pinstrup-Andersen, P. , M. Lundberg and J.L. Garrett, 1995. "Foreign Assistance to Agriculture: A Win-Win Proposition". IFPRI, Washington, D.C.

Thilmany, D.D. and C.B. Barrett. 1997. "Regulatory Barriers in an Integrating World Food Market". *Review of Agricultural Economics* 19(1): 91-107.

Advertising, social responsibility and moral myopia: An overview on the fast food industry

Athanasios Vlachos and Asterios Tsioumanis
School of Natural Resources and Enterprises Management, University of Ioannina, G. Seferi 2, 54100, Greece, nasosvla2003@yahoo.gr

Abstract

Advertising has been the centre of much critique during the last decades focusing on the sophisticated techniques being used. These techniques render the initial scope of conveying information more and more distant. Criticism has been accusing marketers of creating new needs, mixing dreams with reality while advertising practitioners often *have difficulty seeing ethical issues or have a distortion of moral vision that has been referred to as "moral myopia" by researchers.* All these elements become even more important as today, advertising has a continuously more profound impact on everyday life. The paper analyses the history and status quo of the fast food industry, its advertising techniques and targets. Fast food manufacturers have been obliged to deal with pressure at least as far as promotional tactics towards children are concerned while food companies together with advertising agencies have tried, often aggressively, to protect their right to choose their advertising target groups. The analysis focuses on fast food companies' initiatives to behave in a social responsible way, including their promotional strategies aimed at children. The paper concludes that fast-food companies seem to have little intent to evolve the content of its communications to children and thus private initiative shall not prove adequate. Approaching the notion of consumer socialization, a lot more than a growing epidemic of childhood overweight and obesity may be at stake.

Keywords: children advertising, fast-food companies, advertising ethics

Introduction

The emergence of children as a specific age group with increasing spending power over the last decades could not pass unnoticed by marketers. The main reason behind that is that children direct spending has been growing exponentially as children aged 4-12 spent in the U.S. $2.2 billion in 1968 and over $40 million in 2002 (Dotson and Hyatt, 2005). These numbers probably devaluate the role of children as a single agent in the marketplace as their indirect influence of their parents' purchasing behaviour may not be easily quantified.

Advertising has been the centre of much critique during the last decades focusing on the sophisticated techniques being used. These techniques render the initial scope of conveying information more and more distant. Criticism has been accusing marketers of creating new needs, mixing dreams with reality while advertising practitioners often have difficulty seeing ethical issues, which has been described by the term "moral myopia".

Advertisers who target children have three main objectives: to directly seek children as consumers, to work indirectly on parents through children's "pester power" and to imprint the younger generation with positive brand associations (Paul, 2002). Although the acknowledgment that the human brain can make brand associations even before it develops the ability to speak is relatively recent, the fast food industry has long ago identified the importance of young consumers.

A tiny bit of marketing history reconsidered

The emergence of children as consumers with increasing purchasing power led to a rapid increase in volume of companies focusing their promotional tactics on children during the 1980s and ever since. However, even before that and depending upon the economic branch, a variety of companies (toy manufacturers, fast and snack food industries, movie and entertainment companies), which identified children as an important segment of their potential customers had been trying to understand their behavior as consumers.

Marketers soon came to realize that designing appropriate promotional tactics for children was not an easy task as their behavior proved in many ways unpredictable and their likes and wants change faster than any consumer group. Despite the obstacles the efforts to understand children consumptive behavior did not cease as the reward is great; *such behaviors tend to remain relatively unchanged throughout a person's life, provided the person remains within the specific subculture experienced during his or her formative years* (Moschis, 1987).

Fast food industry has utilized a variety of approaches in order to attract the junior consumer and create an international children's market. Their ideas were not confined in aspects of the product but rather shifted towards the evocation of a series of pleasing images. Thus, toy-like wrapped food items and bright colors came along with the allocation of thousand of playgrounds in fast food restaurants. In addition, the mutual agreement on promotional deals with toy manufacturers, lead from the inclusion of a free toy in the price of a children's meal to elaborated global marketing agreements. Some of the most successful short-term promotional tactics involved giving away simple toys, including the spring 1997 almost ten-fold increase in McDonald's volume of "happy meals" sales accompanied by a baby toy. Children's clubs have also played an important role in effective targeting and the popularity of kid-based web sites largely removed all existing technical obstacles.

Cartoon characters advertising potential in attracting young consumers has been evident by numerous examples of successful campaigns even beyond the fast food industry. Joe Camel before the discontinuity of the ad campaign was very familiar to children, Taco Bell's talking chihuahua proved a big success while Ronald McDonald comes second among most recognizable figures for American schoolchildren only to Santa Claus (Atlanta Journal Constitution, 8/8/03). Together with linking to movie studios, sports leagues and athletes also became business associates of the fast food industry. Finally, more aggressive tactics include the incursion of fast food chains into the education system at least within the U.S., where more than 20% of US high schools offer brand-name fast foods, such as Pizza Hut, Taco Bell, or Subway (Wechsler *et al.*, 2001). Direct advertising in schools is a common theme and one may find *corporate logos on athletic scoreboards, sponsorship banners in gyms, ads in school newspapers and yearbooks, free textbook covers with ads, and screen-saver ads on school computers for branded foods and beverages* (Story and French, 2004).

On fast-food advertising focusing on kids

Despite the diversity of promotion tactics, the bulk of fast-food restaurant budgets are allocated to television. Although the increasing number of hours spent watching in TV for children provides a good lead, the background behind this choice is not central to the line of argument and it will thus be omitted with a brief quotation. *Children believe a majority of the health messages communicated via advertisements regardless of the truthfulness of the message* (Lewis and Lewis, 1975).

Trying to approach the relationship between purchase requests by children of various food types and the advertising of different foods the everlasting question over advertising haunts the quest. Is the stated behavior a consequence of the child responding to advertising or the

advertiser responding to the wants of the child market? In an effort to clarify the issues some central notions will be briefly discussed.
Comprehension of the nature of advertising is an important pre-condition. John (1999) in his review of relating research, points out that young children present limited comprehension of the persuasive intent of advertising. Prior to 7 or 8 years of age, children tend to view advertising as fun, entertaining, and unbiased information. Preteens, from ages 8-10 years, possess the cognitive ability to process advertisements but do not necessarily do so (Strasburger, 2001). The same stands for the rest of the adult population. Given than the lack of cognitive ability to fully comprehend or evaluate the advertising message is present, young children should not be exposed to such stimuli as this deceitful process constitutes a blatant violation of rights.
Focusing on the fast food industry and its advertising methods, it should be noted that most of the food items promoted to children are predominantly high in sugar and fat and in general contradict the healthy food patterns. Taking into consideration that the fast food industry is only partially responsible as one of the operating agents in food choice, it should be underlined that *diet-related illnesses can be prevented if a healthy lifestyle and diet are established in childhood* (Caroline Walker Trust, 1992). A recent FAO/WHO report concluded that while the evidence that the heavy marketing of fast food outlets to children causes obesity is not unequivocal, sufficient indirect evidence exists to place this practice in the "probable" category for increasing risk of obesity (World Health Organization, 2003).While the aesthetics concerning human body have shifted considerably throughout history, it is evident that good health is recognized as desirable and its pursuit may be considered a concrete stated objective in a social sense. The mutual agreement on a common goal is prominent as, given the complexity of the issues, the debate is doomed to failure if a common goal can not be mutually agreed upon.
The context under which the decisions in the social arena are made is equally important. Under a strict utilitarian framework, the social optimum should be approached on a cost-benefit structure. This would include the calculation of cost of illness presupposing that a causal relationship would be able to be attributed between health problems and fast food consumption. In addition, the internalization of all related costs and benefits is a pre-requisite while some of them are clearly not quantifiable. Luckily, such an enormous obstacle does not have to be tackled as specific children rights have been widely recognized and institutionalized.

Corporate social responsibility: Is nowadays fast-food marketing any different?

Ethics and social responsibility are two of the key challenges faced by management in recent years. Social responsibility refers to the obligation of a firm, beyond that required by legal or financial considerations, to pursue long – term goals that are good for society. Defining corporate social responsibility (CSR) is not an easy task. In general, definitions may refer to ethical behavior, sustainable development, the environment and philanthropic ideas. However, the idea that the right conduct takes into account the welfare of the larger society is central. CSR is not a recently invented concept, but as customers' demands and social norms are changing rapidly, the adoption and implementation of ethical trading principles is integral to an organization's overall marketing strategy (Omar, 1999).
Jones *et al.*, (2005), examined the role of CSR in the UK's leading food retailers. They found that each company has its own approach to CSR and there are substantial variations in the nature and extent of the reporting process. While some, and not always the smallest of the leading retailers provide relatively limited CSR information and other offer detailed and comprehensive reports they all make the case for locating CSR as an integral element of the core business. Piacentini *et al.*, (2000) assessed motivation to engage in CSR behavior amongst food retailers in the UK. They found that customers' pressure is a very significant reason for companies to introduce the

CSR notion. Depending on they way each company see CSR, they have different commercial strategies overall. Customer's satisfaction is the first main motive for the companies. The second motivating factor is profitability.
As the market for ethical consumption expands, companies need to continually monitor the strategic role of corporate ethics in their respective sectors. In a study by Schroder and Mc Eachern (2005), the examples of Mc Donald's and KFC have been utilized in order to show that consumers' valuation has changed. No longer can either company rely solely on convenience and product consistency as unique selling points. A strong corporate emphasis on consumer health, quality and socially responsible initiatives must be incorporated. New advertising campaigns or new product lines alone are not sufficient to communicate changing brand values. There is a variety of CSR initiatives available for fast food companies, analogous to the number of the related ethical issues. Thus, animal welfare, environmental protection, workers rights or the littering of public spaces need to be considered together with the nutritional value of food and relating health issues.
It is a fact that recently, there has been some alteration, which is often identified as progress, in fast food companies' way of conduct. A number of them have reviewed the fat and sugar contents of their product ranges, and reconsidered the size of the portions they offer. *Another recent innovative strategy by McDonald's and KFC enables dietary information for each meal to be accessed via nutritional calculator tools on each company's web site* (Schroder and Mc Eachern, 2005). McDonald's issued its first Corporate Responsibility Report in 2002 and a second, more substantive one in 2004 and were the first, followed by the Yum! Brands (KFC, Pizza Hut, Taco Bell) to develop animal-welfare principles. Responsible purchasing programs designed to protect fishing stocks and forest resources, waste management schemes as well as community support through cash and in-kind donations also have been reported as initiatives of the fast food multinationals.
However, relatively less attention, in terms of CSR, has been drawn upon policies on the general issues of advertising and marketing. In a review by Lang *et al.* (2006) on reported commitments and practice of 25 of the world's largest food companies only four of them had any policies on advertising and all of them were manufacturers.

Should we leave kids alone?

Advertising and promotional tactics in general constitute one of many sociocultural variables, such as parents, peers, schools, and other media sources, that can influence food choice in children. However skepticism may be justified as advertising has a continuously more profound impact on everyday life.
Food marketers spend billions each year on advertising and highly sophisticated techniques are being used in consumer research, including those targeting to children. A large share of the achievements reported by the fast-food industry, which seem to take into account the welfare of the society, has largely been shaped in reaction to public controversy. Specifically fast-food companies seem to have little intent to evolve the content of its communications to children. The obvious conclusion that private initiative will not prove enough to achieve the social objectives is supported by a number of economic and philosophical arguments. In addition, from an historic perspective, ethical concerns have not been an issue taken into account when designing the promotion strategy of fast-food companies. Given the relatively poor record that many of these companies hold in other CSR issues (e.g. labour and environmental standards) and their unwillingness to voluntarily address issues surrounding their promotion and advertising tactics towards children, there is little space for optimism.
In many European countries, there have been bans, proposals to ban or restrictions to television advertising to children while similar legislation covers fast food marketing at schools. However,

whether the growing epidemic of childhood overweight and obesity is a major public health concern or not, a lot more may be at stake. Consumer socialization may be defined as *the process by which young people acquire skills, knowledge and attitudes relevant to their effective functioning as consumers* (Ward, 1974). As long as children are exposed to excessive commercial messages this process will be imbalanced. Creating new generations of consumption-driven citizens in the developed world, overweight or not, is not an action inherently good or evil. The fact that in economic terms, depending upon the point of view, it may easily make sense to create a new generation of shopping maniacs has little to do with the value judgment as the latter is related to the aspirations of our society as a whole. However, what should be advocated is that by exposing children to more and more commercial hype, a massive projection of our world to them takes place living thus limited space for unbiased choices.

References

Atlanta Journal Constitution (2003). Article on August 8th.

Caroline Walker Trust (1992). *Nutritional guidelines for school meals,* Report of an expert working group, the Caroline Walker Trust, London.

Dotson, M.J. and Hyatt, E.M. (2005). Major influence factors in chlidren's consumer socialization. *Journal of Consumer marketing,* 22/1:35-42.

Gallo, A.E. (1999). Food Advertising in the United States. In: *America's Eating Habits: Changes and Consequences.* USDA/Economic Research Service (ed). USDA, Washington, DC. 173–180.

John, D.R. (1999). Consumer socialization of children: A retrospective look at twenty-five years of research. *Journal of Consumer Research,* 26:183–213.

Jones, P., Comfort, D., Hillier, D., and Eastwood, I. (2005). Corporate social responsibility: a case study of the UK's leading food retailers. *British Food Journal,* 107 (6): 423-435.

Lang, T., Rayner G. and Kaelin E. (2006). *The food industry, diet, physical activity and health: a review of reported commitments and practice of 25 of the world's largest food companies,* Centre for Food Police, City University, London

Lewis, C.E. and Lewis M.A. (1975). The impact of television commercials on health-related beliefs and behaviors of children. *Pediatrics,* 53-3: 431-435.

Moschis, G.P. (1987). *Consumer socialization: A Life-cycle Perspective,* D.D. Heath and Company, Lexington, MA.

Omar, O. (1999). *Retail Marketing,* Prentice Hall, London.

Paul, N.C. (2002). Branded for life?. *Christian Science Monitor,* April 1.

Piacentini, M., McFadyen, L. and Eddie, D., (2000). Corporate social responsibility in food retailing. *International Journal of Retail & Distribution Management,* 26 (11): 459-469.

Schroder, M., J., A., and McEachern, M., G., (2005). Fast foods and ethical consumer value: a focus on McDonald's and KFC, *British Food Journal,* 107 (4): 212-224.

Story, M. and French S. (2004). Food Advertising and Marketing Directed at Children and Adolescents in the US. *International Journal of Behavioral Nutrition and Physical Activity,* 1: 1-17.

Strasburger, V.C. (2001). Children and TV advertising: nowhere to run, nowhere to hide. *Journal of Development and Behavioral Pediatrics* ,22:185–187.

Ward, S. (1974). Consumer socialization. *Journal of Consumer Research,* 1: 1-14

Wechsler, H., Brener, N.D., Kuester, S., Miller, C. (2001). Food service and foods and beverages available at school: results from the School Health Policies and Programs Study 2000. *Journal of School Health,*71:313–324.

World Health Organization, (2003). *Nutrition and the Prevention of Chronic Disease.* Joint WHO/FAO Expert Consultation on Diet, World Health Organization, Geneva.

Part 10

Who is responsible? Corporate responsibility and political governance

Re-doing responsibilities: Re-doing the state politics of food and the political market-place

Kristin Asdal
Center for Technology, Innovation and Culture (TIK), University of Oslo, Box 1108 - Blindern, 0317 Oslo, Norway, Kristin.Asdal@tik.uio.no

Abstract

Institutional responsibilities within the Norwegian politics of food are changing. Responsibilities are shifting from state administration and national ministries to the market place and the consumer. How is this going on in practice? Ministries and state-bureaucracies are crucial actors in this process, hence the practices and technologies of government are pertinent to study, also when it comes to grasping consumer-roles and markets. In doing this science and technology studies (STS) in combination with economic sociology and the governmentality-literature in the tradition of Michel Foucault, are viable resources. This is as they draw the attention to the practices of government: the technologies of politics. Interestingly, the displacement of politics to new sites and arenas does imply that markets and consumers are in themselves enacted as crucial political technologies. Through politics and administration markets are enacted as political sites only by other means. The consumer of neo-classical economics, the 'homo economicus', is enacted side by side with 'homo-ethicus'. Is this a viable vision for food-politics?

Keywords: consumers, economics, governmentality, markets, politics and administration

Over the last few years the political and administrative domain of food-matters has been subject to profound changes. Politics of food simply doesn't mean what it used to. Before politics of food referred to a fairly specific set of problems within a particular set of arenas. Now it has now moved or been extended in novel directions. Before food was, at least predominantly, seen as a source of nutrients and the handling of this issue took place within the domain of state bureaucracy. Now food-matters are tied up with all sorts of connections and relates to food production, distribution and consumption.
The introductory store above is not genuinely mine. It is a version of the story told by Marianne E. Lien in her introduction to the edited volume *The Politics of Food* (Lien and Nerlich 2004). A crucial argument of this story coincides with a point which has also been underlined within the field of science and technology studies (STS): Neither politics is what it used to be. Politics has expanded to fields and arenas not traditionally thought of as political. One way of framing this is to say that new objects; such as food, are being politicized. Another way is to say that what has been going on for some time is a displacement of politics (Marres 2005).
Through the displacement of politics to new sites and arenas, responsibilities are shifting as well. Who's responsible? Until recently food matters were predominantly political, however political in another sense than 'contested'. They were political in the sense that food matters were political decisions. Those responsible for decisions and strategies were national ministries. This applies not the least to The Ministry of Agriculture to which the Chief Veterinary Officer used to be an integrated part. Within the last 10-year period responsibilities have to an increasing extent shifted from national ministries, to the level of international institutions, and, at a national level, to scientific expertise on the one hand, markets and market agents, consumers, on the other.

In sum what is becoming apparent within the politics of food is a new structure. Those in the authority to act, those ascribed the authority to make decisions independently, are no longer ministries, at least not to the extent they used to be, but new agents and institutions; namely science and market-agents (Asdal 2005). So maybe there are two processes going on simultaneously? On the one hand there is a general process of politicization, in the sense that food has become a highly contested object and affair. On the other hand however, there is a process we could, to simplify somewhat, label de-politicization. This in the sense that what used to be political decisions, decisions for which ministries were responsible, to an increasing extent are left over to expert bodies and the market. Elsewhere I have labeled this the abdication of politics (Asdal 2005).

There is more to the notion of politics and the political than conflict and contestation. It is also a matter of the capacity of drawing things together and about closing off (Asdal 2004a). Exploring the different notions and implications of politics and the political is crucial - to politics of food. The question 'who's responsible?' is one way of attending to this issue. The next step however, would be to ask 'What are they made responsible *for*?' That is as the interesting question is not only *who* are the responsible actors and institutions, but just as much 'What are the politics of food made to be through the ongoing shift in responsibilities?' Thus I attend to what has been labeled the politics of what (Mol 2002). First however I will attend to the question of materials and method, the question of which sites and materials to study.

Taken together, what might seem to follow from the stories above is that studies of state bureaucracies and administration are no longer crucial to food-matters. Hence these are no longer the relevant sites to study. In stead, we should go directly to the markets, consumer-behavior, every-day practices as well as expert bodies.

Even though I agree, of course, that these are all crucially important sites to study, the argument of this paper is different. Precisely in this process of an on-going shift in responsibilities ministries and state bureaucracies are crucially important sites and actors. The various displacements of politics are not emerging from "nowhere" neither from "below" or from the outside of ministries, exclusively. On the contrary, ministries and state administration are themselves crucial actors in the re-doing of the responsibilities of the state. One site through which this re-ordering of responsibilities is done is within white papers to Parliament. Thus drafts for as well as the final outcome of white papers provide a rich material to processes of politics in the making – or politics of displacing. Let me give only one little example.

As pointed to by Lien (2004) politics of food used to imply seeing food as a source of nutrients. However this has changed. How did this happen? It takes work to displace politics to new sites and issues. In addition it takes a lot of fumbling and uncertainty, and sometimes reluctance. The way in which The Ministry of Agriculture through the mid -90ies re-worked its politics of food is interesting in this respect.

To begin with The Ministry of Agriculture was not particularly interested in re-composing the food-domain. To a certain extent the Ministry wanted to keep the situation as it used to be. This was a situation in which political commitments to the right composition of nutrition complied with a protectionist or at least protective agricultural policy. One of the early drafts for the white-paper later to become 'Stortingsmelding nr 40 (1996-97)' on food-quality and consumer-safety demonstrates the way in which the Ministry tried to combine both considerations: The Ministry suggested that tariffs on cheese with a high level of fat should *not* be reduced. The problem however was that the political and economical situation of which agricultural politics and politics of food and trade-issues took part was changing. Agreements both with The World Trade Organization (WTO) and the European Union (EU) implied a less protective trade and import-policy. In accordance with the WTO-agreement nation states could not, on political grounds exclusively, stop import. This had, eventually, to be backed up with science. In principle the situation could be handled and compensated through increasing tariffs on selected products.

The problem was that other ministries did not share this version of a new trade-policy. The draft-proposal was immediately rejected by the Ministry of Trade.

Hence, the next draft to the white paper was different. Now, the issue of a potential increase in trade was not any longer defined as a problem. On the contrary it was turned into an opportunity – even when it came to nutritional issues: "Lower prices of fruit and vegetables throughout the year will increase their consumption. We need to encourage this development" (all precise references to documents can be found in Asdal 2005). Thus the politics of food was still linked up with nutrition, however in a different way than before. And it was linked up with a situation the Ministry of Agriculture was not to the same extent as before in command of. National agricultural interests were losing ground and status to the international: "Import conditions" said one of the early drafts, "are not as they were before. (They are no longer determined nationally). We must adapt to an international set of laws and regulations, legislation has become international." Not surprisingly another notion; 'the market' started to circulate more often within the draft white papers. For instance: "Things are developing in a direction where more and more are handed over to players in the market".

The transformation process above could be analyzed as the rolling back of the nation-state. But already it is clear. While the nation state is being rolled back, it is also "rolling-up", actively constructing, new sites and subjects. As a consequence responsibilities are shifting too. As trade-conditions are not any longer determined nationally, i.e. through state directives or policies, new agencies and actors are being made responsible for the politics of food. One of these agencies is a new expert-body, *The Food Safety Authority*. Another is a market-agent, namely the consumer. It is to this figure; the consumer and the market place the next part of this paper will attend. In doing this I engage with and discuss theoretical resources within the field of economic sociology and STS. There is nothing natural to market transactions (Zelizer 1992, Miller 1994).Under which conditions are calculativeness possible? Under what conditions do calculative agents emerge? Michel Callon (1998) who is the one asking, points to an answer: Calculating is a complex collective practice which requires that the calculative agencies or agents are equipped. Calculativeness requires a complex material reality or a complex set of technologies in order to take place. Crucially to STS-researchers engaged in economic sociology is the point that these processes of calculation are practices that economists themselves and economic theory take part in. Neither economic theory nor economists are outsiders to 'the economy'. On the contrary, they take actively part in producing markets and market-actors (Callon 1998, Barry and Slater 2002).

As I have already indicated, also ministries and state-administrations take actively part in producing markets and market-actors. Markets are coming into existence also through the administrative practices of bureaucracies and ministries. They enable calculative agencies. I will give a few examples. As I pointed to already, the re-doing of responsibilities for food-matters did, to begin with, happen rather reluctantly. In early drafts for white papers in the mid -90ies the consumer was a rather suspicious, emotional, more or less ignorant figure in no capacity of taking on responsibilities. She was however reworked into a new and more trustworthy figure. Or rather new figures, in the plural, as what emerged were competing and conflicting versions of the consumer, in the same white paper. I will use these as the starting point for my concluding discussion.

"The consumer is the employer of agriculture, industry and trade". And the consumer, we are told, is "expected to be [even] more influential in the future". What was the Ministry of Agriculture doing? On the one hand they went "by the book": The consumer is enacted in line with neo-classical economics: the rational actor of economic theory. She is the 100% informed, the all-knowing economic subject; able to calculate and to act in line with her preferences. So the consumer is a market-actor. Perhaps she is not *born* to be free, but she is being *enacted* (Law

2004) as free. She chooses and she calculates and within this neo-classical model any choice is legitimate.

So the consumer is an economic actor, a real homo-economicus. But at the same time she is being enacted as the perfect, omnipresent political actor, as 'homo-ethicus'. In the market she is not just made into a calculator of prices, she is also enacted as being competent and responsible for calculating qualities as well. Moreover, she is the responsible consumer who is less interested in prices than in ethics; in norms and values. "We are probably about to enter an era in which consumer-demand will to an increasing extent be related documented production, to animal welfare and to ethics". This is not like the homo-economicus to whom all choices are legitimate. This is homo-ethicus; ascribed the responsibility of moving the politics of food in a specified and politically viable direction. Or if not the responsibility; so a trust, a hope on behalf of the Ministry that this is what will come through. Interestingly, without this engaged and responsible consumer politics and administration do not seem capable or to hold the authority of prescribing policy.

What I have wanted to demonstrate is the work it takes and has taken to displace politics to new sites and subjects. Through this I have, as I have also done elsewhere (Asdal 2004a), argued for a relational approach which take into consideration the ways in which politics of the state are linked to other sites and practices. This is done through a reading and re-working of the governmentality-tradition stemming from Foucault (1991) concerned with the art and practices of government; the technologies of politics (Asdal 2004b).

The question remains; to what extent is the process of displacement I have pointed to also an anti-political move, a process of de-politicization? In the sense that the state does not any longer hold the same authority to act independently as before, this can be said to be the case. However, simply jumping to this conclusion would be too easy. What my material show is also a profound effort to re-inscribe politics into new sites and actors. The consumer-figure as she, sometimes, is configured through the documents of the state is the chief political figure enacting the market-place as *the* political space. Thus; the market is understood as a site for politics by other means.

To this I will conclude by turning the conventional sociological critique of economics on its head. Whereas the conventional critique is that the 'economic man' of economics is too 'thin' to be true, the calculating homo-ethicus of the governmental white papers is too 'thick' to come through. Too many responsibilities on her shoulders - and not properly equipped to act - or transact. What we might be left with is a politics of food of which can be said that the only real matter of concern is what passes through the consuming (human) body.

References

Asdal, K.(2005). *Grensetrafikk. Nedslag i matpolitikkens og veterinærvesenets historie.* Unipub Forlag, Oslo.105 p.

Asdal, K.(2004a). *Re-inventing politics of the state. Science and the politics of contestation.* [on line]. Available on http://www.tik.uio.no/natureculture/papers/re-inventingpolitics.pdf

Asdal, K. (2004b). *Politikkens teknologier. Produksjoner av regjerlig natur.* Unipub Forlag, Oslo. 317 p.

Barry, A. and Slater, D. (2002). Introduction: The technological Economy. *Economy and Society*, volume 31: 175-193.

Callon, M. (ed). (1998). *The Laws of the Markets.* Blackwell Publishers, Oxford 1998. 278 p.

Foucault, M. (1991). Governmentality. In: *The Foucault Effect.* Burchell, B., Gordon, G. and Miller, P. (eds). The University of Chicago Press, Chicago. 87-104.

Law, J. (2004). *After Method.* Routledge, London. 188 p.

Lien, M. E. and Nerlich, B. (eds) (2004). *The politics of food.* Berg. Oxford/New York. 224 p.

Marres, N. (2005). *No issue, No Public. Democratic Deficits after the Displacement of Politics*, Academisch Proefschrift, Ipskamp Prinpartners, Amsterdam. 175 p.

Miller, P. The invention of calculating selves and calculating spaces. In: *Rethinking Objectivity*. Megill, A. (ed). Duke University Press, Durham. 239-264.

Mol, A., (2002). *The Body Multiple*. Duke University Press, Durham. 196 p.

Zelizer, V. (1992). Human Values and the Market. In: *The Sociology of Economic Life.* Granovetter M. and Swedberg R. Westview Press, Oxford. 285-304.

The responsibility of food corporations for their customers' health

Tal Caspi and Yotam Lurie
School of Management, Ben-Gurion University of the Negev, P.O. Box 653, Beer-Sheva, 84105 Israel, kaspit@bgu.ac.il

Abstract

Food experts tell us that the abundance of food, its easy availability and its chemical ingredients can have a negative effect on our health. The main focus in the academic literature regarding the responsibility of the food industry for the health of their customers is on the marketing of rotten or damaged food products, which may have a negative affect on our health that is immediate. However, an issue which is neglected is the responsibility of the food industry for food products which affect the consumers' health in the long run. Their influence on our health in the long run is not one dimensional, and not always known to the marketer, even after long periods of production. This paper examines the responsibility of the food industry for the customers' long term health. It assesses the relevance of the classical positions that regarding the responsibility of manufactures for consumers' safety. It shows the weaknesses of this model for the food industry since our eating habits are not just commodities we choose to consume. Subsequently it asks about the proper ethical norms in the food industry. Establishing ethical norms in the food industry requires that we choose a responsible risk management policy. In this sense, the desired norms should be built on the constitutive principle according to which those who have the power (ability) have the responsibility.

Keywords: responsibility, risk management, food safety, eating habits, autonomy

Introduction

In a certain respect the average person today has less vitality than people had fifty years ago. In the past, people did physically strenuous work .Today people sit in an office, in front of a computer or talk on the phone for eight hours a day. At the end of the day these people get into an elevator that drops them off right next to their cars. On their way home, they park their cars two steps away from their homes, and they once again take the elevator to get up to their apartments. When people decide to go the mall on their way home from work they aren't exactly doing work that strains their muscles. Escalators in the malls take us from floor to floor, allowing us to stroll peaceful without having to strain our muscles. In the United States, even the small amount of exercise one gets from stepping out of their cars is eliminated with the creation of the drive thru window.

This cultural fact should be supplemented with the tremendous increase in the availability of food in the developed world. Affluent and developed countries do not suffer from food deficiency, but rather offer their citizens a variety of high calorie foods while using much less physical energy in the in their lives in general and in preparing food in particular. The food that we eat today is made from more carbohydrates than before. This destructive process begins with attacking organs associated with transporting blood (veins, arteries) which increases over the years. This causes diseases, such as heart attacks, strokes, and diabetic related illness to develop, and it also increases the chances of developing various types of cancer. Many physicians and food experts hold the opinion just described above. The Research Staff at the American Organization for Heart Diseases shows the connection between a fat diet and the development of heart diseases (Klein *et al.* 2004).

The food we eat damages not only our health but it also has negative effects on the economy. In contrast to what economics claim, that we have solved the problems of hunger and scarcity of food, we are rather in a situation that due to excess of food we are suffering new economic problems with the high rise in the cost of health care and other negative economical affects. A research conducted by The Food and Drug Administration concludes that the recent economical gains have a price tag attached to them; they were attained due to the damages that were caused as a result of diseases developed. The damages and diseases caused as a result of Americans eating habits have been assessed at 70.9 billion dollars (Frazao, 1999) This cost includes the cost of treatment for sick patients as well the loss work days because of sick days. The damage done to the economy as a result of this new attitude towards food, in Europe, is related to diseases. The damage caused by Coronary Heart Disease alone costs Europe 169 Billion Euros (Statistical Data, 2005).
Though there is no doubt that our eating habits are leading us down a dangerous road, this in itself is not sufficient to point a blaming finger against the food industry. The logic of consumption with respect to food is dangerous: more is not necessarily better. Hence, even if the food industry is not to blame for the present situation, it continues to fuel this destructive logic of consumption and in this, maybe, lies their responsibility. The logic of economics, i.e., consumption and growth, controls the food industry. This logic is controlled by the simple rule that the more they sell, the more money they will make. In order to sell a lot, they have to find a way to make the consumers buy more products so that they have these products in their possession. What will the customers do with the products that they buy? In terms of food there is only possible use that is worth it which is to put it in their mouths and eat it!. The destructive results of this trivial process have been explained and shown above. The fact that it is preferable to eat less is something that we have known for more than thirty years, and despite this, the food production is still stuck in processes of trying to create products that fit the products that are made from real utensils, not like the ones that are staying in our bodies nowadays.
What this means is that the profit of the food corporations is made through negative health effects and a huge economic cost to society. The economic burden on our health system might be viewed as a negative externality which results from the high profit of food producers. Hence, we need a new normative model that will cut-down on the negative health effects and the huge economic cost to society. This is an issue which bears not only upon the relations between customers and producers of food, but has to do with safety and transparency in the food industry.

Responsibility of manufacturer for the safety of their consumers

Velasquez (1992) present three theoretical positions regarding consumer risk: the contractarian view; the due care position and the social cost position. According to Velasquez, the contractarian position divides the responsibility equally between the manufacturer and the customer who needs the product; both are autonomous agents. The due care theory transfers more of the responsibility for the quality of the product to the producer of the product. The third position, the " The Social Cost View", lays the blame and the full responsibility on the producer for any possible future damages. We argue that these three positions might be relevant in extreme unique circumstances for corporations in general; but they definitely do not create a solution for the food industry.
The contractarian, which uses the model of a rational contract to understand the relations between consumer and producer, fails to show that both the client and the marketer are both signing a contract out of their own free will. First of all, there is no balance in the relationship between the marketer and the customer. In other words, the customer does not have the ability to attain by himself valuable and accurate information about the product. Moreover, products go through a process where they go through "endless handy work conducted by passing from hand to hand" until they arrive to the person ultimately in need of the product, and it is impossible to start counting the products from start to finish because this process of counting will never

end. What is more is that most of the marketed products are very complex and it is not really permitted to separate contracts on the basis of the different ways that they are built and or made. Also the idea about the product (as if it were on sale) that both sides came into the agreement of the contract out of free will, does not pass the test of reality, so to speak. This specifically applies to many products and especially food. People do not even have the ability, even hypothetically, not to be part of the consumers who buy food necessities.
The Due Care Theory with its "Caveat Vendor" doctrine witnesses the fact that the responsibility for the injuries to the customer is a subject that could not be foreseen and is a burdened on the marketer and the maker of the product. The responsibility of the manufacturer requires that he will take steps worth taking in order to prevent any damage that he is able to prevent. According to this view, the maker of the product has to deal with more than just quite a few difficulties. First of all there is no way of checking that the maker of this product acted in a way that obeyed the "Due Care Theory", and the idea behind this theory itself is applied differently to different products and from customer to customer. Similarly, there is an existing problem with the maker of this product dismissing his responsibility for damages that he could not foresee. One of the major cons for the maker of this product is that this itself is actually a disadvantage for the maker of this product in terms of the aspect of food.
The view of the Social Cost View broadens the responsibility of the maker of the product. It claims him responsible for any damage that could possibly occur including those damages which could not be foreshadowed, all the more so if the customer of the product had been subject to damage while he was not at all using the product irresponsibly. This view puts the burden of the responsibility of the damage on the maker of the product. He is burdened with this responsibility whether or not he was able to have avoided it beforehand, and even if he warned the customer about possible injuries that could occur to him from using the product beforehand. The basis of this view is that it aims to understand what the societal cost is to the fullest, in every way that will broaden the products uses, and cause much more of a worry for the company executives. In our case, all of the cost for the future damages is burdened on the maker of the product. The main problem is that for the maker's work and for the marketing of the product, they create a larger potential cost for damage done to the customers in need in relation to how much the product costs. Something that is clearly understood is that all of this matter discussed above cannot be related to the food branch.
The opposition to "The strict product liability attitudes came also from the concept of fairness; it is "unfair" to put the blame and to create outcomes that will lead to damages on people that aren't necessarily guilty or irresponsible in terms of the specific cause for the damage and in terms of the ability to avoid the damage. In order to understand the theory of definite responsibility while making sure those involved are sincerely honest and straightforward, there must be opposition to the matter. In regards to opposition to the theory's logic, a lot of researchers withhold their opinions about the matter. Coleman says that straightforwardness is not applicable in this case at all. To understand the attitude that lacks straightforwardness and placing the blame on the client will create a situation where he is to blame which is completely unfair because he is not to blame (Coleman.1976).

Beyond simple manufacturer: Consumer relations

What's special about the food industry? An important distinction should be made at this stage. It seems that one of the main reasons why the concept of responsibility with reference to consumer safety seems futile in this context is that food is much more than a commodity or an article of trade. Food is deeply connected to different life styles; and when considering life styles, individuals are not autonomous to choose a life style in the same way that they choose what shirt to wear. This insight serves as the basis for Korthals' (2001) discussion of consumer autonomy in the context of food. He concludes that food should not be treated merely as a dangerous and risky product; when considering products that are not dangerous in and of themselves, responsibility should be divided between consumer and producer. However, since the producers have more power, they also have the ability to lead changes.

If food consumption has to do with a certain life style, let's see how the food industry tries to affect and impact the formation of this life style by political means. Marion Nestle (Nestle 2002) demonstrates this. She does not only blame the food industry for causing people to eat unhealthy food, but it also shows that the management of food industry are busy trying to advance this situation further through political channels. Yes, it is true that fatty food and sweet food taste better, it is easy for us to get used to it, and it is cheaper to create. It is also true that the usage of chemicals and other unnatural ingredients allows those who produce the food to create whatever comes to their minds, and genetic engineering changes the main ingredients so that there won't be anything missing. The customers won't even be able to develop an opinion about these unhealthy ingredients that are placed in their food. In fact, they will buy these new products with the new ingredients without even thinking twice. When someone tries to speak out about the situation, lobbyists make sure that any changes do not occur that will affect and cause those creating the goods to change their minds. This is exactly what happened in the American Congress when Edward R. Medigan decided to speak out. He stopped the printing of the food pyramid which illustrates the so called correct way to eat, on the basis of him saying that it would confuse children. However, the real reason for him stopping the printing was that he was really worried about raising cows that were capable of hurting people. He made this statement on the basis of a recommendation he received to decrease the amount of milk and meat needed for each person to eat in order to remain healthy.

Conclusion

Dealing with the challenges presented by the food industry, with its long term negative health and economic effects, cannot be done with the traditional tools for considering acceptable risk in consumer – manufacture relations. Food has to do with a certain life style and thus the consumer, no matter how autonomous he might be, does not have the ability to make a difference. Hence, responsibility lays on shoulders of food corporations that have the ability to change the way the industry is managed and open new spaces for discourse and communication regarding food safety. Instead of using their power and influence positively, they continue to promote consumption, over eating and even obesity.

References

Coleman, J (1976). The Morality of Strict Liability Theory. *William and Mary Law Review 18 (1976): 259-286*

Frazao, Elizabeth. (1999). High Cost of Poor Eating Patterns in The United State, America Eating Habits: Changes in Consensus. *Department of Agriculture, Economic research Service Agriculture Information Bulletin* No. 750 Chp 1. 1999

Klein, S., Burke, E.L., Bray, A.G., Blair, S., Alison, D., Pi-Sunyer, X., Hong, Y., and Eckel, H.R., AHA Scientific Statement: Clinical Implementation of obesity With Specific Focus on Cardiovascular Disease. *Circulation*, 110: 2952-2967. America Heart Association Inc.

Korthals, Michiel. (2001). Taking Consumers Seriously: Two concept of Consumer Sovereignty. *Journal of Agriculture Environmental Ethic,* 14: 201-215.Kluwer academic publisher. Nederland.

Velasquez, Manuel G. (1993). The Ethics of Consumer Production. *Ethical Theory and Business*. Bowie and Beauchamp (eds.). 4th Edition, 1993

Nestle, Marion,.(2002). *Safe Food,* University of California Press

Peterson, S., Peto, V. and Rayner, M..(2005).*European Cardiovascular Disease Statistics.* [on line].Available on http://www.ehnheart.org/content/sectionintro.asp?level0=1457 [01/04/06].

Governing sustainable food and farming production futures using integrated risk assessment approaches

A.H. Gesche[1] and A. Haslberger[2]
[1]School of Humanities and Human Services, Queensland University of Technology, GPO Box 2434, Brisbane, Qld. 4001, Australia, a.gesche@qut.edu.au
[2]Vienna Ecology Center and Department for Nutritional Science, Althanstrasse 14, A-1090 Vienna, Austria, alexander.haslberger@univie.ac.at

Abstract

Nanofoods, 'functional foods' and biopharming are three production futures that are strongly developing despite being characterised by significant gaps in knowledge and understanding, and a peculiar scarcity of proactive processes with which to seize opportunities and minimise and manage potential risks and public concerns which could negatively impact on the industry. In order to better assess benefits and risks and to build public trust, the paper suggests the establishment of an integrated health/food and environmental risk assessment regime that also incorporates and is responsive to the ethical concerns, socio-economic realities and local demands of various stakeholders – right from the beginning of a development. In order to have a global as well as a national practical effect, the assessment regime needs to conform to national and accepted international regulations and observe fundamental principles in bioethics and public sector ethics, such as integrity, access, autonomy and choice. Such a pro-active approach might lead to improved collaborations, to constructive communication channels and to enriched and more mutually acceptable futures.

Keywords: integrated risk assessment, ethics, nanotechnology, GM foods, biotechnology

Introduction

Second generation biotechnology crops, so called 'functional foods' are beginning to enter the market promising benefits to consumers, such as foods with enhanced nutrient content and bioavailability, improved quality and taste, reduced allergenicity, or a higher satiety index. Another emerging production future uses plant and animal systems as living 'bioreactors' to produce pharmacologically active substances (biopharming), which may be more cost-effective than employing conventional bacteria, yeast or cultured mammalian cell systems. Straddling both, and still largely at the research stage, are the 'nanofoods' - food technologies that are associated with nanoscience and nanotechnology, a science that functions at the molecular and atomic level. In the food sector, one possibility being explored in this area is the development of nanosensors and nano-scale diagnostic devices that monitor the safety and quality of foods. Another area of applied research investigates how medicines or health supplements can be embedded in nanoparticles in order to deliver nanoscale nutrients or drugs directly to targeted cells, thereby maximising their health benefits and minimising their adverse effects. Currently more than 200 companies worldwide, including Unilever, Nestlé, and Kraft Foods, are reportedly involved in a rapidly developing nanotechnology industry. The value of its applications for foods alone is forecast to surge to US$20.4bn in 2010 (Helmut Kaiser Consultancy, 2004).
While the number of scientific papers addressing aspects of these emerging technologies is climbing, as with all emerging technologies, a lack of empirical data makes it difficult to accurately forecast their actual benefits and risks to human and environmental health. There

is also a paucity of publications assessing their ethical and social impacts. Indeed, there is little evidence to suggest that their implications are being seriously discussed by either the scientific or civil community. Public knowledge about these novel production futures is generally low to almost non-existent. This is particularly the case for nanotechnology (Mnyusiwalla *et al.*, 2003; International Risk Governance Council, 2005). Specific laws and regulations in this area are also largely absent. It is becoming apparent, however, that change is underway. For example, a newly funded program by the European Commission Research Directorate-General called "NanoBio-RAISE", aims to anticipate and then respond to ethical and social issues and concerns that might arise from nanotechnology (NanoBio-RAISE, 2006-2007) and in the US, the "21st Century Nanotechnology Research and Development Act", section 10 requires that "ethical, legal, environmental and other appropriate societal concerns" be considered and integrated into nanotechnology research and development programs (21st Century Nanotechnology Research and Development Act, 2003). Although change might slowly be underway, uncertainty and a lack of public debate makes solid risk assessments difficult to achieve.

Current issues with conventional risk assessment methodologies

Traditionally, risk assessment methodologies in the agroenvironmental sciences have focussed on scientific and technical multidisciplinary data and science-based risk assessment methods. For example, Codex Alimentarius principles (2003) guide the assessment of the safety and nutritional aspects of genetically modified (GM) foods, while the Cartagena Protocol on Biosafety (1992/2003) guides an assessment of genetically modified organisms (GMOs) on the environment. Both take a case-by-case empirical approach. Both refer to direct as well as indirect effects. The Codex includes an assessment of indirect effects of novel foods on human health and the environment (Codex 2003); and the Cartagena Protocol recommends an inclusion of "uncertain" effects with regards to the "conservation and sustainable use of biological diversity...taking also account of risks to human health" (Cartagena Protocol, 1992, Annex III).
When considering risk assessment for emerging agroenvironmental bionanotechnologies, a number of problems present themselves. Firstly, uncertainty has no numerical value. The existence and likely long-term persistence of scientific uncertainty, which would call Article 15 of the Rio Declaration on Environment and Development (1992) into action (defined as "the precautionary principle") demanding protecting the environment and human health. However, scientific uncertainty has no numerical value that is measurable and reproducible. Secondly, empirical scientific data are open to interpretation. While scientific methods of data gathering may be objective and reproducible, empirical data are still open to interpretation with alternative explanations possible. In addition, there are often no universally accepted and reliable testing methods or standards yet available. At other times, any risk may not become apparent immediately or may be indirect, especially where the gathering of data is incomplete and still evolving. Furthermore, interpretations of data may be effected by social parameters. As discussed in detail by Kimmelman (2004), the interpretation of scientific data – even among scientific experts – is influenced by an individual's assumptions about risk acceptability, social affiliation, and political and social worldviews. Alternative interpretations could even lead to conflict until more data becomes available or a temporary consensus can be established. Thirdly, current conventional risk assessments largely ignore the human dimension and do not embrace ethical, cultural or social issues. Government-sponsored surveys in Australia indicate that a better informed public assesses risks arising from biotechnologies not so much by evaluating scientific data, but rather on outcomes and processes. For processes, not only do the various publics demand participatory decision-making, they also expect that technology providers present proof of benefits that trusted information sources can verify (Biotechnology Australia, 2003). Indeed, recent research tends to suggest that public disquiet and distrust regarding GM

foods and feeds has become more differentiated in that a perception of distrust and unethical behaviour of researchers and companies connected to the industry also rank high in the public mind. Surveys conducted by Cormick and Ding (2005) indicate that for biotechnology, many policy decisions that were made by governments and industry in the past were based on an erroneous assumption that the public needed scientific facts to evaluate risks, when in reality the public was more interested in who would ensure their safety. Education campaigns designed to convince a sceptical public that a technology is safe does not appear to guarantee acceptance and quell public disquiet, indeed, a recent study investigating the acceptance of genetically modified foods has shown that exposure to information can alert people to potential dangers, even making people more distrustful of existing regulations (Poortinga and Pidgeon, 2004). According to Cormick and Ding (2003), the public is more likely to accept a new technology application if:

- The information about an application is provided from a credible source.
- It is morally acceptable and not harmful to people or the environment.
- The application is regulated by government and not by industry.
- The public has had some meaningful input in the development of a technology.
- Consultations address major public concerns and public feedback is seen as having some effect.
- Consultations occur before an application is developed and not afterwards.
- Consumers can choose to accept or reject an application.
- The application has clear individual or societal benefits.
- The largest beneficiary is the public and not a multinational company.

Adapted from Cormick and Ding (2005).

The aforementioned problems indicate that conventional risk assessments are insufficient in addressing the needs of the various publics who are asked to embrace emerging technologies and to trust researchers and governments although scientific uncertainty is common and risks and benefits difficult to ascertain.

Adopting an integrated risk assessment approach for emerging technologies

It has recently been recommended to adopt an integrative approach when assessing genetically modified organisms and GM foods and link health and food safety, and environmental considerations (World Health Organization, 2005; Haslberger, 2006). In the following, we suggest that an integrated approach is also taken for emerging food and biopharming technologies in order to better assess benefits and risks and to build public trust. Such integrative methodology would be holistic and include deliberate, multi-stakeholder public participation and an assessment of normative values. The approach would also not so much concentrate on endpoints but on various points of intersections, preferably right from the beginning of a given research and development action (Kapuscinski *et al.*, 2003) until the post-marketing phase.

When allowing normative values to enter risk assessment protocols, an immediate hurdle presents itself, namely how to measure and standardise them. Normative values cannot easily be measured, because they are multidimensional and affected by different variables, such as social forces, cultural practices and by moral codes, vulnerabilities and impact. While they might not lend themselves to scientific analysis, they can nevertheless be discussed and evaluated in a similar analytical and rational process. One possible "tool" with which to facilitate such assessment is the ethical matrix, introduced by Mepham (2000) and first applied by Kaiser and Forsberg (2001). The ethical matrix does not provide a numerical value. In its mechanics, the matrix is a grid of cells headed by a set of ethical principles first proposed in the 1970s by Beauchamp and Childress (2001) for the field of medical ethics. With respect to each ethical principle, the matrix identifies the most important impact of a particular technology for a given stakeholder group and transcribes it into its corresponding cell. Once every cell has been filled in, stakeholders consider and deliberate on all impacts, if necessary ranking them, so as to arrive at an ethically acceptable position. Furthermore, since each ethical matrix is constructed for a particular case/scenario, local as well as regional, national or international particularities and complexities could be addressed as soon as they arise and in a feedback loop inform subsequent development steps. When working under the guidance of an ethical matrix for assessing normative risks, the sharing of information could make complexities and interdependencies transparent, could foster a sense of co-responsibility for future outcomes and could bring about a sense of ownership of the technology.

However, this method is not optimal either. Similar to the shortcomings discussed above with regards to interpreting scientific data, bias and power relationships can influence processes and outcomes. One possible way to minimise bias and the effect of unequal power relationships is by agreeing on which ethical principles to use and by balancing stakeholder participation. Such standardisation might be difficult to achieve. For example, unlike in medicine, where the ethical principles first suggested by Beauchamp and Childress (2001) have long guided medico-ethical decision-making, their concept of 'autonomy' is not universally applicable across all cultures. However, there is some flexibility in which set of ethical principles is chosen for a given circumstance. For modern food biotechnologies with a potential global reach and impact, the following four ethical principles have already been proposed for a global code of ethics for modern food biotechnologies: the principle of beneficence (doing good), the principle of non-maleficence (avoiding harm), the principle of justice and fairness (which includes integrity and access), and the principle of choice and self-determination (Gesche *et al.*, 2004). The latter replaces the principle of autonomy, which is contrary to the communal decision-making practices in many cultures. The four ethical principles could be extended by a fifth, such as 'respect for the law and the system of government' (Queensland Government, 2001), which would link the ethical framework to national and accepted international regulations.

It needs to be acknowledged that an integrated risk assessment would place an additional inconvenience on industry, which would embrace ethical and social issues as elements of their corporate strategic planning. Companies driving the evolution of emerging technologies expect to be constrained by a country's business norms and legal system. However, understandably, they try to negotiate minimal operational conditions to contain their costs and to maximise shareholder benefits. Already having to comply with at times stringent scientific regulatory requirements, they make a strategic decision where to locate their operations and to what extent they commit to corporate social responsibility (CSR as defined by McWilliams and Siegel (2001)). Their voluntary commitment may depend on the existence or absence of a competitive market (McWilliams *et al.* (2006), shareholder expectations and stakeholder demands. For the latter, it can be anticipated that in capacity-strong countries with vocal stakeholders and a tradition of strong regulatory oversight, companies will be more mindful of their ethical and social responsibility than in capacity-poor countries, where public objections and regulatory measures may be minimal. Two points stand out: regardless of capacity, without integration, ethical and social issues are likely to remain marginalised until public disquiet or fear brings them to the surface, potentially starting and repeating a cycle of conflict and distrust similar to the one that has long plagued the GM biotechnology sector. Furthermore, without integration of other, equally valid non-scientific knowledges, scientific risk assessment processes are likely to result in uni-dimensional views regarding risk, with long-held beliefs and inferences becoming further entrenched and enforced by like-minded peers. If those beliefs and inferences are contrary to public opinion, rejection of the new technologies is a real possibility. Worse is to be expected if those interpretations turn out to be erroneous and the public comes to a conclusion that it has been misled.

A paradigm shift is needed from industry. In a recent survey conducted by the International Risk Governance Council (2005), most scientific nanotechnology experts on the panel from different economies in different continents saw no need for public engagement at this stage. Only a few recognised that civil society would make judgements on societal values rather than scientific facts. Only one expert went further and recommended "early citizen and stakeholder collaboration...as an essential step in preventing risk in the longer term". Whatever the misgivings or inconveniences, if industry aspires to develop these technologies further and to turn substantial investments into profits and if governments and civil society aspire to move towards a sustainable, equitable future, a more engaged, holistic, and integrated approach to risk assessment is necessary. By following an integrated risk assessment regime, the scope of assessment is widened and responsibility for developments, including adverse effects, shared between the scientific and public community.

Conclusion

Nanofoods, 'functional foods' and biopharming are three production futures which are strongly developing despite being characterised by significant gaps in knowledge and understanding. They have the potential to significantly contribute to feeding a rapidly growing world population and to respond to people's escalating medico-pharmacological needs in an affordable manner. However, while scientific discoveries and economic application leap ahead, the ethics and social science discourse lags dramatically behind. Remaining silent and waiting for the scale and economic force of the emerging technology to make public criticism ineffectual would be an unsatisfactory development. The preferred option is a proactive one that creates dialoguous relationships and shares knowledge and responsibilities. Reaching this goal will require synergy between scientists, industry, the government and the public. It requires international agencies to take the lead in developing new standards and to function as a repository and disseminator for analysis and exchange of data and experiences. It requires as its foundation one unifying global ethical framework whose principles guide different levels of conduct and interactions. It requires an integrated risk assessment approach that is systematic on the one hand and dynamic on the other, where the kinetics of scientific knowledge, technical change, difference in capacity and values, ecological, economical and political particularities and realities are being reconciled as best as possible for enriched and mutually accepted outcomes.

References

Beauchamp, T.L. and Childress, J.F. (2001). *Principles of Biomedical Ethics*. 5th ed., Oxford University, Oxford, 454 p.

Cartagena Protocol on Biosafety. (1992). [on line]. Available on http://www.biodiv.org/convention/articles.asp [date of consultation 18 April 2006].

Biotechnology Australia. 2003. *Biotechnology Public Attitudes Towards Gene Technology Study*. [on line]. Available on http://www.biotechnology.gov.au/index.cfm?event=object.showContent&objectID=F771AD9E-BCD6-81AC-15E8E2C315F1B4DA [date of consultation 17 April 2006].

Codex Alimentarius Commission. (2003). *Codex Work on Foods Derived from Biotechnology* [on line]. Available on http://www.who.int/foodsafety/biotech/codex_taskforce/en/ [date of consultation 18 April 2006].

Cormick, C. and Ding, S. (2005). *Understanding Drivers of Community Concerns About Gene Technologies*. [on line]. Available on http://www.biotechnology.gov.au/assets/documents/bainternet/Beijing200520060209145752%2Epdf. [date of consultation 5 April 2006].

Gesche, A., Haslberger, A. and Entsua-Mensah, RE. M. (2004). Towards A Global Code of Ethics for Modern Foods and Agricultural Biotechnology. *Preprints of the 5th congress of the European Society for Agricultural and Food Ethics, 2-4 September* (Catholic University of Leuven, Belgium, pp.125-128.

Haslberger A.G. (2006). Need for an "Integrated Safety Assessment" of GMOs, Linking Food Safety and Environmental Considerations. *Journal of Agricultural and Food Chemistry*. Accepted for publication.

Helmut Kaiser Consultancy. (2004). *Study: Nanotechnology in Food and Food Processing Industry Worldwide: 2003-2006-2010-2015.* [on line]. Available on http://www.hkc22.com/nanofood.html [date of consultation 7 April 2006].

International Risk Governance Council. (2005). *Survey on Nanotechnology Governance: Volume A. The Role of Government*). [on line]. Available on http://www.irgc.org/irgc/projects/nanotechnology/_b/contentFiles/Survey_on_Nanotechnology_Governance_-_Part_A_The_Role_of_Government.pdf [date of consultation 12 April 2006].

Kapuscinski, A.R., Goodman, R.M., Hann, S.D., Jacobs, L.R., Pullins, E.E., Johnson, C.S. *et al.* (2003). Making 'safety first' a reality for biotechnology products. *Nature Biotechnology,* 21(6): 599-601.

Kaiser, M. and Forsberg, EM. (2001). Assessing Fisheries – Using an ethical matrix in a participatory process, *Journal of Agricultural and Environmental Ethics*, 14: 191-200.

Kimmelman, J. (2004). The Ethical Review of Clinical Trial Safety. Kennedy Institute of Ethics Journal, vol. 14 (4), p. 369ff.

McWilliams, A. and Siegel, D. (2001). Corporate social responsibility: a theory of the firm's perspective. *Academy of Management Review*, 26: 117–27.

McWilliams A., Siegel D.S., Wright P.M. 2006. Corporate Social Responsibility: Strategic Implications. *Journal of Management Studies*, 43: 1. [on line]. Available on http://www.blackwellsynergy.com/doi/full/10.1111/j.1467-6486.2006.00580.x [date of consultation 5 April 2006].

Mepham, B. (2000). A framework for the ethical analysis of novel foods. *Journal of Agricultural and Environmental Ethics* 12: 165-176.

Mnyusiwalla, A., Daar, A.S. and Singer, P.A. (2003). 'Mind the gap': science and ethics in nanotechnology, *Nanotechnology*, 14: R9-R13.

NanoBio-RAISE. (2006-2007). *Nanobiotechnology: Responsible Action on Issues in Society and Ethics*. [on line]. Available on http://nanobio-raise.org/ [date of consultation: 26 March 2006).

Poortinga, W. and Pidgeon, N.F. (2004). Trust, the Asymmetry Principle, and the Role of Prior Beliefs. *Risk Analysis*, 24(6): 1475-1486.

Representatives of the United States of America in Congress. (2003). *21st Century Nanotechnology Research and Development Act.2003. Public Law 108-153.* [on line]. Available on http://frwebgate.access.gpo.gov/cgi-bin/getdoc.cgi?dbname=108_cong_public_laws&docid=f:publ153.108 [date of consultation 7 April 2006].

World Health Organization. (2005). *Modern food biotechnology, human health and development: an evidence-based study*. [on line]. Available on http://www.who.int/foodsafety/publications/biotech/biotech_en.pdf [date of consultation: 21 April 2006].

Queensland Government. (2001). *Code of Ethical Practice for Biotechnology in Queensland.* Available from: www.biotech.qld.gov.au. Accessed 22 March 2004.

Ethical principles and practice in organic processing: Governance and corporate responsibility

Mette Weinreich Hansen, Thorkild Nielsen and Niels Heine Kristensen
Technical University of Denmark, Department of Manufacturing, Engineering and Management, Innovation and Sustainability, Building 424,2nd floor, DK-2800 Kgs. Lyngby, Denmark, mwh@ipl.dtu.dk

Abstract

This paper will give a short introduction to the historical roots of the organic production in Denmark. The organic production has gone from a pioneer phase with few dedicated actors, to an institutionalized and standardized food production system. The institutionalization has opened the organic market for new actors - especially this paper will concentrate on the processing companies who, for a large part, took interest in the organic production after the ending of the pioneer phase. The authors emphasize the importance of reflecting values in relation to organic food production especially because both the standards for organic food processing and the value discussion are almost non-existing related to processing. There are two main claims in this paper. One claim is that there could be reluctance from the companies in discussing organic values because they fear to be associated with the organic pioneer movement and subsequently exclusion from mainstream. Another claim is that there is a need for upgrading the existing networks or maybe invite to creation of new networks around reflection on values since this issue need qualification in order to be used as an active for the companies. One way into discussing values without leading to exclusion could be to use the concept Corporate Social Responsibility (CSR), which is getting an increasing attention also in Denmark. Some tendencies show that consumers are still more critical and therefore it can be an attractive way of improving credibility. A possible combination between CSR and organic food is presented and the preconditions for this combination are briefly touched.

Keywords: organic values, companies, CSR, ethics, Denmark

Introduction

The organic production as it is put in practice is different from other food production in the sense that the driving forces originally have been a critique of the consequences of modernity. Throughout the western world organic production has to a varying extend been defined by a social movement with some ethical and environmental visions for the concept of food production in opposition to the conventional agricultural production. According to Jacobsen (2005) the organic movement in Denmark has its recent roots partly in the biodynamic anthroposophical farming, partly in political left wing groups wishing a different ownership to farming land and partly in a general concern about the biological surroundings and the way they were (and are) treated. The first attempts in Denmark to characterize the alternative way of growing food as an "organic movement" started in the 1970'ies with inspiration from movements in USA, Germany and UK (Bramwell, 1989, Jacobsen, 2005).
The Danish organic movement has undergone a great deal of change since the early days in the 1970'ies and the most recent discussion seen in newspapers and magazines is now whether the organic actors can be seen as part of a social movement at all. The development shows more diverse reasons for growing and producing organic, and now the concept of growing and

producing organically has a political and institutional suspension. This development has been analysed by different authors as a development towards more "professional" and less idealistic organic actors – more professional in the sense more oriented towards the economic and standardized aspects of the production and less idealistic in the sense less oriented towards ideas and visions from a socially founded movement (Jacobsen, 2005; Christensen, 1998; Kaltoft, 1997). This development is supported by the institutionalization of the organic principles and ideas, which happened especially through the late 1980'ies and 1990'ies in Denmark. In 1987 the state was involved in defining organic production in the sense that Denmark got its eco-label (a red capital "Ø"). Together with this state-label the organic production got subsidised from the Danish state and thereby became economically dependent. This naturally led to sort of trade of between the critique and the possibility of getting economic support, and the organic production took the first step to be put into measurable standards in order to make production controllable. Hereby the grass-root organic social movement was starting to get "mainstream", or to get included in the mainstream way of acting. Or maybe more correctly the idealistic and political pioneer actors were loosing influence on the definition and direction to where the ideas, visions and also production standards should go, and subsequently the ideas and visions were fading slowly. In 1991 EU formulated their organic standards and in that way paved the road to an even more mainstream organic production. Now it was difficult to produce and sell anything organically without following the Danish State and the EU. This development naturally led to a loss of idealistic visions and many – or most – of the vision formulated by the first pioneers were lost in the translation to institutionalised and standardized organic agriculture. This development is interestingly the same that opened the organic production to new actors who were not idealistic or organic "by heart". In this period the Danish market for organic products increased. Until late 1990'ies there was a constant increase of the organic sale in supermarkets and this has led to a constitution of organic products as a normal choice in the supermarket among all other choices. Today organic products are a part of almost all supermarkets and discount shops in Denmark with quite high market shares for some products. These two things together – the standardization and the market increase - has led to a development where many new actors are contributing in characterizing the market of organic products. Amongst others the processing industry and companies in secondary production are relatively new actors on the organic market, with the exception of a few companies.

Loss of values – loss of identity

The question is now whether this development is problematic. From the perspective of the authors there are some aspects of the development, which can be predicted to give the organic sector a problem in the future. The standardization and the fading of reflected ideas coupled with the new actors who – generally speaking - are less oriented towards any values beside the requirements connected to the organic label, can from the authors point of view lead to a scenario where the organic products are just variations of conventional products with no other qualities or distinctions attached to them. They will loose their identity and peculiarity. Consequently it can be hard to tell the difference between an organic and a conventional product. Looking at the standards for processing organically there is almost no directions. Basically the two limits are the number of additives, which is constantly being put under pressure from powerful food industries and others, in order to allow more additives. Recently the nitrite has been allowed after many years of discussion and lobby activity. The other focus for the organic standards are the traceability of the products back to primary produce in terms of organic guarantee certificates on every primary product used in the processing industry.

If producers are not actively reflecting what organic production imply for them one consequence could be that organic and conventional products are getting still more similar and subsequently the price difference can be hard to explain and convince consumers to pay.

After this introduction to the problems of processed organic products we will examine the possibility of using this production as a potential for promoting responsibility.
Also the lack of potential of the existing organic networks are touched – and the need of an empowerment of these networks in order to be able to focus on the value discussion, which in relation to processing companies is almost absent today.
Firstly we will discuss the value concept. Is it reasonable to view organic values as a naturally given set of directions?

What are the organic values?

An important question looking at the organic production today is: are there any specific organic values that all organic producers are directed towards? The description above shows that the more common values of the pioneers early in the organic history were taken over by standardized rules defined by the state and the EU. What happened to the value basis in this standardization? Following the debate in different organic magazines and newspapers, values have to a limited extend been a continuously debated issue in relation to organic production, although almost only in relation to primary production.
The standards of the processing of organic production are – as earlier mentioned - not very detailed. Only the questions about additives and the documentation and traceability of the primary produce are required. How the products are made and the issues on how environment, packaging, skills and craft connected to the production and the quality are all individual areas to the company to decide. As mentioned most processing companies seems to organize their organic production primarily by incorporating the standardized directions.
Looking at the voluntary organizations for organic production, values are discussed also to some degree related to processing although limited compared to the ideas and values of the primary production. The international umbrella organization for national organic movements IFOAM (International Federation of Organic Agricultural Movements) recently approved a new set of recommendations and guidelines for organic production. Here the ethical dimensions of the production are explicated also to some degree of relevance for the processing of food. Also the Danish organization for organic farming and food production, Organic Denmark, has a set of visions or values – some of them reminiscences of old value wordings from the early pioneer days.

Organic values in the companies

The first results from a doctoral (Ph.D.) work with a case study perspective on food processing companies indicates that there seems to be a weak connection between the organic ideas as they are represented in the above mentioned organizations and the production in the processing companies. It seems that the companies are in many cases not familiar with any values defined by the organic organizations.
This analysis points at different aspects in trying to explain this seemingly decoupling between the organizations who invokes the organic values and the processing companies who produce the organic food products.
Looking at the actors involved in formulating the values in the organizations it seems that companies are almost absent, and the claim is here that this leads to a lack of ownership in the companies towards the explicit ethical thoughts.
In Organic Denmark, an organization where the farmers are the dominating member group, there is a board of companies where about 60% of all Danish organic producing companies are members according to the organization themselves, but only a few are more or less active in the arrangements directed towards companies and even fewer are active in the work at the board

(Nielsen, 2005). On top of that three of the biggest market actors, who have the dominating part of the market share of organic food products, are not members of the organization.

Marketing and sale recognition

Another aspect of the work of this board is that there is no specific value-oriented focus. The secretariat in Organic Denmark, who have the primary contact with the companies are marketing educated and have no personal or organizational "roots" in the organic circles. Most of the work in relation to companies are directed towards sale and marketing of the products. It is clear that this is not promoting the value-reflection in the companies if they are not themselves reflected on an individual basis.

The claim here is that there is recognition between the marketing educated secretariat and the companies, because in spite of many different company cultures and different organizational traditions there is a common focus on sale and marketing in all companies – it is a simple survival necessity. Thereby the relationship confirms a certain way of looking at the organic production where marketing initiatives are prioritized and for example reflection of values are almost totally absent.

The fear of exclusion

A hypothesis of our analysis is that another important aspect of the reluctance in taking the organic values into ownership is the fear of exclusion. As history shows the first critical voices in the debate about organic production had a critical view on mainstream society and therefore in many ways excluded from this. This is not the image that the mainstream companies want to have today. They are reluctant in getting associated with the excluded idealistic part of the organic production. In order to get values on the agenda in the companies today it is therefore necessary to give some new tools which are not associated with the "old" pioneers of the organic movement. This is where CSR is may be useful as a future tool for reflecting organic values.

CSR

The concept of Corporate Social Responsibility (CSR) or Business Ethics refers to an understanding of companies as having a responsibility – socially, but later expanded to cover also broader ethical and environmental aspects. The notion "the Triple Bottom Line" covers the meaning that there is more to a company than economic rationale and moneymaking (Mac, 2003). In relation to the value discussion of corporations it is important to underline that values are not new. To consider a company as a "strategic money-machine" is also loaded with value. The focus on values are therefore here more considered as a change of values in relation to organizations/ companies rather than moving from a place with no values into a new place loaded with values (Mac and Rendtdorff, 2001).

CSR is based on a moral ethical responsibility and thereby not connected with any authority control and this can been seen as a weakness because the company therefore in principle can use the ethical communication without having any reality behind the words. This is of course also where the consumer can be critical towards the concept – is it only image? The believers of this responsibility trend correspond that any cheating will be seen through sooner or later and only the trustworthy companies will survive (See Mac and Rendtdorff for a discussion of these issues).

CSR and organic food production

In Demark it is our impression that this focus on explicit values has been gaining attention for the last years, although the food industry has not yet been so visible in this sense. It is an important aspect of legitimizing the value discussion that it can be acknowledged in mainstream

circles, and it is our impression that the concept of CSR gains a lot of attention and respect in the company environment.
Here the potential in focusing the organic food production to the concept of CSR is addressed. By reflecting and expressing the social, environmental, quality and responsibility issues of organic food production the companies can strengthen their image towards consumers and maybe see new potentials for the organic food products, potentials which are not seen as long as the reflection on why the choice in producing organically is tacit. One challenge is to create the basis for trustworthy reflection where also e.g. inevitable dilemmas of the daily production can be communicated and discussed.

Conclusion

This paper addresses a need for a renewed focus on organic values in relation to food production. In order to overcome the reluctance in discussing values in food processing companies CSR can be a tool into reflecting the organic production, and at the same time getting acknowledged in mainstream circles. In order to upgrade the discussion it is the author's opinion that it is necessary to pay attention to value networks, either by upgrading exiting networks or creating new networks where values can be reflected.

References

Bramwell, A. (1989). Ecology in the 20th Century. A History. Yale University Press. New Haven and London. 292 p.

Christensen, J. (1998). Alternativer. Natur. Landbrug. Akademisk Forlag. Doktor-afhandling. 463 p.

Jacobsen, N.K.B. (2005). Den økologiske jordbrugsbevægelses historie – fra 1970'ernes græsrødder til 1990'ernes professionelle. Saxo Instituttet – Afdeling for Historie. Københavns Universitet. 222 p.

Kaltoft, P. (1997): Naturetik som praksisbegreb. En undersøgelse af naturforståelse, praksis og viden i økologisk jordbrug. Institut for Teknologi og Samfund, skriftserie nr. 5. Danmarks Tekniske Universitet, Lyngby. 302 p.

Mac, A. and Rendtorff, J.D. (2001): Værdier og socialt ansvar i virksomheder. Fra profitmaksimering til interessentdialog. GRUS nr.65. p 69-87.

Mac, A. (2003): Værdibaseret virksomhedsteori. Workingpaper 9/2003. Center for Værdier i Virksomheder. Institut for

Samfundsvidenskab og Erhvervsøkonomi. 14 p.

Nielsen, T.K. (2005). Interview with then Head of Marketing Department at Organic Denmark. August 2005.

Scott, W.R. (2001): Institutions and Organizations. Second Edition. Sage. 255 p.

Corporate responsibility

Karsten Klint Jensen
Danish Centre for Bioethics og Risk Assessment, Rolighedsvej 25, DK-1958 Frederiksberg C., Denmark, kkj@foi.dk

Abstract

Appeals to corporate responsibility often simply take for granted that businesses have ethical responsibilities that go beyond just respecting the law. This paper addresses arguments to the effect that businesses have no such responsibilities. The interesting claim is not that businesses have no ethical responsibility at all but that their primal responsibility is to increase their profits. The extent to which there is reason to take such arguments seriously delineates the limits of corporate responsibility. It is shown that Milton Friedman's famous right-based argument fails, because it assumes social responsibility to imply that the corporate executive acts against the interests of the share holders. But why should not share holders be ethical responsible? However, a more pragmatic but better argument refers to the division labour between the market and the political level: the market works most efficiently if consumers and businesses make decisions based on their preferences for consumption and profit, respectively. The flipside of this coin makes it the job of the political level to ensure that the market works within an ethically fair framework. The implications of this argument are demonstrated. Most importantly, the argument presupposes that there is a democratic regulation of the market. If there is not – which at least to some extent appears to be the case with multi national corporations – then the ethical responsibility stays with the business itself. Interestingly, the cases where people tend to level ethical requirements directly at businesses seem to be exactly this kind of cases.

Keywords: democracy, efficiency, ethical responsibility, free rider, rights

Introduction

Appeals to corporate responsibility build on the implicit premise that businesses have ethical responsibilities that go beyond just respecting the law. However, this premise is often simply taken for granted without justification. But is it evident that businesses do have ethical responsibilities? On the face of it, it seems a plausible assumption. Why should businesses as opposed to other agents be exempted from ethical responsibility? However, the interesting claim is not that businesses have no responsibilities at all. It is rather that it is the responsibility of businesses to increase their profits. Thus, Milton Friedman (1970) famously claimed: The social responsibility of business is to increase its profits. In other words, there could be important ethical reasons supporting the claim that businesses ought to concentrate on increasing their profits.
This paper uses Friedman's article as point of departure for a closer scrutiny of arguments to the effect that the ethical responsibility of businesses is to increase their profits. The objective is to clarify the extent to which there is reason to take them seriously. Since we end up with the default assumption – that businesses have ethical responsibilities other than increasing their profits – to the extent these arguments fail, it thereby at the same time becomes clarified what the ethical responsibility of a business is and how far it extends.

Preliminary remarks

Does it make sense at all to ascribe responsibility to a collective entity? Is it not, as Friedman claims, only individuals who can have ethical responsibilities? I shall follow Friedman here in only committing myself to a minimal ontology; that is, I assume that all statements about the acts and responsibilities of a business can be reduced to statements about the acts and responsibilities of individuals.
Friedman acknowledges that it may well be in the interest of a corporation to act in ways that can be described as exercising ethical responsibility. And many people in fact claim that it pays off for a corporation in the long term to act ethically responsible. For Friedman, the attempt to rationalize such acts as an exercise of ethical responsibility is just hypocritical window-dressing. I do not need to follow him in this. The important point is, however, that when we discuss whether or not a business has an ethical responsibility that goes beyond to the duty to increase its profits, this discussion becomes empty if the responsibility does not imply that sometimes the business will have a duty to act in ways contrary to its interest in increasing profits.

The argument from rights

What then, according to Friedman, is wrong with corporations acting ethically responsible? Suppose a corporate executive make expenditure on reducing pollution beyond the amount that is in the best interest of the corporation or that is required by the law. He would then be spending someone else' money without their consent for the sake of a general interest. If his actions reduce the return to shareholders, he is spending their money. If his actions raise the price to consumers, he is spending the costumers' money.
Each of these parties could have spent their own money on this action if they wished. The executive is exercising a distinct ethical responsibility only if he spends the money in a different way than they would have spent it. In effect, he is imposing taxes and deciding how to spend them. Thereby he is violating the right of property and acting as a self-appointed government.
In a democracy, the imposition of taxes and expenditure of tax proceeds is exclusively a governmental function. This function is controlled by the constitution, the parliament and a range of laws to ensure that taxes are imposed and expenditures enacted after democratically legitimate procedures and in accordance with the preferences of a majority of the population. The immediate consequence of Friedman's objection is that ethical responsibility, in so far as it goes beyond increasing profits and involves taxation for the sake of the public good, should be dealt with through the political system and not by arbitrary managers acting as self-appointed legislative and executive power.
However, Friedman is furthermore sceptical about letting the political mechanism act for the sake of the public good. The reason is that the political mechanism involves coercion of the minority. An individual has a vote, but if he is overruled, he is coerced into spending money in ways he would not do on his own. By contrast, the market mechanism does not coerce anyone into spending money in ways he does not want to. All transactions are made voluntarily.
According to Friedman, the political mechanism is a threat to freedom. As individuals, each of us has an ethical responsibility. But this responsibility should be exercised voluntarily. It is wrong if a majority coerces individuals into moral acts they do not consent to (Friedman call it 'socialism'). Hence, the political mechanism should be used as little as possible.
We might well agree with Friedman that it is wrong if an executive acts against the interests of the shareholders. We might even grant him the conclusion about the political mechanism's inherent tendency of coercion. None of this suffices to show that the ethical responsibility of a business exclusively is to increase profits. The point is that the owners of a business as individuals have

ethical responsibilities. Event though they are granted the right to spend their money as they see fit, they are not exempted from ethical responsibility in this regard.
Friedman implicitly admits this himself when he says of an individual proprietor exercising an ethical responsibility that he is spending his own money, and that there cannot be any objection to his doing so. If we add the general premise that individuals do have ethical responsibilities, we get the conclusion that the proprietor as individual has an ethical responsibility. Shareholders will similarly have ethical responsibilities as individuals. Of course, they may be overruled by other shareholders, but they are free to invest in other companies in accordance with their ethical views.

The argument from efficiency

Some of Friedman's remarks point in the direction of another, more pragmatic argument. Interestingly, he notes in passing that it takes monopolistic power for a business to be able to make expenditure on ethical responsibility. In other words, on a competitive market a business taking on higher costs than its competitors for the sake of ethical responsibility would soon run into trouble. Since a free market – reasonably constrained – is in the interest of consumers, it should be the responsibility of the political sphere to ensure an ethically responsible framework for the market. I shall now present this argument in more detail.
One production plan for a business is more efficient than another just in case, for the same input, it results in more output. According to economic theory, a production plan, which is profit maximizing on a free market under the given prices, cannot be made efficient. In other words, on a free market profit maximizing on a free market is an indicator of efficient production.
Moreover, a market equilibrium where producers are profit maximizing and consumers are utility maximizing (i.e. they choose the most preferred consumption plan given prices and budget constraint) results in a Pareto-optimal state. One state of the economy Pareto dominates another if and only if every consumer either prefers the first to the second or is indifferent between them, and at least on consumer prefers the first. A state is Pareto-optimal if and only if no other state Pareto dominates it.
If two agents in an economy both see a benefit in some exchange of goods, the state after this transaction would Pareto dominate the state before the transaction was made. A Pareto-optimal state is an indicator of efficiency in the sense that all such possibilities for mutually beneficial transactions are exhausted. In this state, a consumer may very well want more goods, but none of these goods are for sale at a price he wants to pay. For this reason, the market mechanism is by many considered at least as good as any alternative in determining how scare resources should be allocated.
The next step of the argument is to point out the fact that a free market by itself is unable to produce public goods. A public good is characterised as a good which is non-rival in consumption; i.e. the consumption of it by one person does not reduce the amount available for others. A second property is non-excludability: Once the good is made available for one person, it is available for all.
The difficulty for a free market in providing public goods stems from the fact that most people are likely to have the preferences of a free rider. Consider a situation where each individual has to consider whether or not to contribute to the provision of a public good, depending on whether or not others will contribute. It is understood that contributing involves a cost for the individual. A free rider will have the following preference order:

1. I do not contribute, others contribute.
2. I contribute, others contribute.
3. I do not contribute, others do not contribute.
4. I contribute, others do not contribute.

If others have the same preference order, we get the following picture:

	Others contribute	Others do not contribute
I contribute	2, 2	4, 1
I do not contribute	1, 4	3, 3

The numbers in each box indicate the place in the preference order, for me and for the others. Even though all prefer the provision of the good (2, 2) to its not being provided (3, 3), it is a dominant strategy not to contribute: regardless of what others do. I get the more preferred outcome if I do not contribute (1 compared with 2, if others contribute, and 3 compared with 4 if others do not contribute). Since the situation is symmetric, this will be case for others as well. Consequently, no one is motivated to contribute and the good will not be provided.
One way to solve this problem is that the state, by enforcing positive and/or negative incentives, can change the preference order of the individuals. Suppose it becomes more costly both for me not to contribute than to contribute, regardless of what the others do, and similarly for all others. Then we get the following picture:

	Others contribute	Others do not contribute
I contribute	1, 1	2, 3
I do not contribute	3, 2	3, 3

In this situation, the good will be provided, because it is better for all to contribute rather than not, regardless of what others do.
The free market is known to have some deficiencies. First, a Pareto-optimal distribution is not necessarily a just distribution. It is indicator of an efficient use of resources, given the initial distribution of income. But the initial distribution of income may well be unjust. Moreover, the market is unlikely by itself to provide protection of fundamental rights.
Second, if there are externalities, the market will result in a Parcto-optimal outcome. An externality occurs when there is difference between marginal private costs and marginal social costs in some activity; in other words, when the price does not reflect the social costs of the activity. Similarly, if there are monopolies, the market will not produce a Pareto-optimal outcome.
The correction of these deficiencies can be considered public goods. Hence, if we sum up the argument from efficiency, it looks like this. If all businesses profit maximize under free market conditions, a Pareto-optimal outcome will emerge. This is clearly in the interest of consumers. However, in a free market, businesses are unlikely to act ethically responsible without the state enforcing incentives for them to do so. Hence, there is good reason to have a division of labour, such that businesses concentrate on profit, while the state should set up a framework for the market, such that it works in an ethically acceptable way.

What is then the ethical responsibility of a business?

Note that the argument from efficiency does not exempt a business from ethical responsibility. It only claims that the responsibility is better fulfilled if the state exercises it through the legal and economic framework it sets up for the market. To the extent that this argument is not tenable, the responsibility stays with the individual business. Let us now consider the strength and extension of the argument.
The most important restriction of the argument concerns areas where the market there is not democratically regulated. To some extent, the international trade is regulated by the WTO. However, on the international market, there can still be many problems in countries governed by

dictatorships or otherwise democratically deficient states. In these situations, a business cannot exempt itself from ethical responsibility with reference to the argument from efficiency.
Interestingly, it is very often multi national corporations that are met with requirements of ethical responsibility. Many people appear to believe that multi national corporations are not working under sufficient democratical control. Clearly, if individual businesses are to exercise ethical responsibility by themselves, the free rider problem will pop up again. This is notoriously difficult to handle. The only solution seems to be that corporations engage in voluntary agreements on ethical standards and actively try to keep free riders out of the good company.
Another restriction of the argument concerns questions which are difficult to legislate about. An example would be the treatment of employees in cases of individual crises. In such cases, where the legislation cannot go into detail, it seems reasonable to assume that the business has an ethical responsibility beyond just living up to legal requirements.
More generally, it should finally be noted that is not necessarily true in all cases that the state is better than individuals in exercising an ethical responsibility. Liberal minded people like Friedman will worry that, because the free rider problem, the demand for public goods will be overestimated. Another problem stems from the fact that the political sphere often act within a rather narrow time horizon. Multi national corporations often work with a much longer time horizon. This could be another reason for levelling ethical requirements directly to them.

References

Friedman, M (1970). The social responsibility of business is to increase its profits, *New York Times Magazine* September 13.

For public prevention or on private plates? The ethical evaluation of personal diets and the shift towards market applications

Franck L.B. Meijboom, Marcel F. Verweij and Frans W.A. Brom
Ethiek Instituut, Utrecht University, Heidelberglaan 2, NL-3584 CS Utrecht, the Netherlands, f.l.b.meijboom@ethics.uu.nl

Abstract

Most of the nutrigenomics research in the Netherlands has been initiated from the idea that the future applications will be beneficial on a public health level or for special groups of citizens that have clear genetically induced increased risk to a certain ailment. However, especially the effects on overall public health are believed to be small. Nevertheless, personalised dietary advices are still considered to make sense for individuals, but now as a market instrument rather than as a tool for public health policy. In this paper we show that this shift in the focus on future areas of application of the scientific developments in nutrigenomics has consequences for an ethical evaluation of this development. Various ethical issues surface in the context of the introduction of genetically tailor-made dietary advices. Most of them can be accurately addressed by the interpretations of values and principles that are widely shared within a specific context of application (e.g., the market or public health). However, the interpretation of such values is not identical in all contexts. This is problematic when contexts are overlapping, which is the case when market parties start direct-to-consumer marketing of personalised diets. Since citizens are also consumers that operate in different contexts at the same time they are confronted with a situation in which it is unclear what they can expect of other agents. Hence it is important to systematically reflect on what one may expect of the market and of government when predictive genetic testing and personalised diets are introduced.

Keywords: nutrigenomics, diets, corporate responsibility, political governance

Introduction

In our research project on genetically tailor-made dietary advices (TMD), it has become evident that the added value of such advices for prevention and health promotion at the level of general public health is considered as marginal. In interviews with researchers and during an expert meeting, there was a consensus that increasing compliance to the general, already existing advices for diet and lifestyle is a more promising strategy for the prevention of mass diseases than the introduction of genetically tailored diets.

Although the effects on public health are believed to be small, TMD are still considered to make sense for individuals, either because they belong to high-risk groups or since they have a personal interest in disease prevention. Therefore, the improvements in the field of nutrigenomics with the aim to develop personalised diets still seem to have a future. However, it is likely that it becomes a market instrument rather than a tool for public health promotion (cf. Meyer, 2005).

This entails not only a shift in the areas of application of the scientific development; it has consequences for an ethical evaluation of this development too. In this paper we argue that the most interesting ethical issues arise when market parties offer dietary products with preventive or health claims. We differentiate between three contexts of application: (a) public health (b) high-risk groups and (c) market. We discuss ethical issues that arise in each of these contexts.

From this we conclude that TMD do not face us with complete new or insurmountable ethical issues. However, when the contexts overlap the questions become more problematic. This will be the case when market parties start direct-to-consumer marketing of personalised diets.

Public health & high-risk groups

Most of the nutrigenomics research in the Netherlands has been initiated from the idea that the future applications will be beneficial on a public health level or for special groups that have clear genetically induced increased risk to a certain ailment. In both contexts it is possible to formulate ethical issues (Meijboom, *et al.*, 2003).

Public health

With regard to the public health applications we can identify ethical issues on three fields: (1) screening, (2) information and consent, and (3) the problem of the added value. First, genetic testing is necessary in order to tailor diets to one's genes. The test result can be valuable in the understanding of the causes and course of certain ailments, and hence in raising the possibilities of preventive medicine. When TMD aim to be relevant at a public health level it is important that testing is not restricted to an individual level, but includes a screening programme. This raises issues with respect the justification of the worries and uncertainties about one's health that will be caused by the introduction of such a programme. The question is whether we have good reasons that can justify this. We will return to this point later on. Furthermore, screening faces us with questions of the justification of the criteria of in- and exclusion. In the cases such as the national screening programmes for prevention of breast and cervical cancer these criteria are relatively clear: women with the age of about 40[83]-50[84] years and older are included since they have evidently an increased risk. With respect to TMD these criteria are more subject to discussion. In principle everyone may benefit from these new dietary advices since they are not restricted to one (type of) ailment. However, it will not be equally beneficial for each individual, since the tests only indicate risk factors not whether one has a disease or that it will emerge necessarily. Hence the decision of governmental health bodies to invite certain groups to participate in a screening programme will depend on risk analyses that will be subject to discussion and are inherently value-based.

This surfaces the second range of ethical issues: those related to information and consent. First, individuals should be assisted in making informed decisions whether one wanted to be tested. Second, there are problems with regard to communicating the results. Genetic risk information is very complex and full of uncertainties about the relation between polymorphisms, a specific test result, the effectiveness of the proposed diet and the emergence of certain diseases. The results of genetic testing do not automatically enable the individual to make autonomous choices about his health. On the one hand, genetic testing contributes to a more individual choice with regard to health promotion strategies, since the test provides information about your own health status. On the other hand, this individual information is rather vague since it is about relative risks. Thus, individuals have more information, yet it is questionable whether they are better informed.

Finally, the question of the added value of TMD underlies both above-mentioned issues. The discomfort of screening and the change of diet need a moral justification. Therefore, the question surfaces whether the added value is substantial enough to do this job. This question is not limited to the direct effects of new dietary advices, but also concerns the added value in comparison to other health risks, e.g., one's own lifestyle (e.g., smoking) or external factors (e.g., the influence

---

[83] Age used by the Centers for Disease Control and Prevention, Atlanta USA (http://www.cdc.gov/)

[84] Age used by the Dutch National Institute for Public Health, RIVM (www.rivm.nl)

of particulate matter). According to various experts it is questionable whether the effect of TMD is substantial enough to justify a public offer of screening and dietary advice (Ethics Institute, 2005).

High-risk groups

For high-risk groups personalised dietary advices seems to be relevant. This concerns individuals that have a substantially increased risk for a disease. In this context most of the issues from the public health context can be addressed more easily. For instance, the issue of the added value is less complicated. These persons obviously have an increased risk that can be reduced with the help of TMD. Moreover, the issues of information and consent are easier to deal with since the individuals are relatively well-informed, mostly because of the incidence of that ailment within their family. Consequently the offer of a genetic test and dietary advices will not come as a surprise. Furthermore, the question whether to start a public screening programme is irrelevant. Only specific, well-identified groups are tested. Nevertheless, the criteria for in- and exclusion are not completely beyond discussion. There is the problem of the identification of individuals as a member of a high-risk group. This is especially problematic with respect to family members. When a son is diagnosed as having a substantial risk for a certain type of cardiovascular disease it might be legitimate to offer genetic tests to his direct family. However, should one also contact other members of family, like aunts or nephews? Especially when you know that the risk is smaller in their case, yet they may have an increased risk. At this point the problem of the legitimacy of testing given a small added value returns. Especially since the impact of the testing results on one's diet will be in this context more profound than in that of public health. Therefore, involving a broader group of members of family entails a further medicalisation of their life, which is not morally unacceptable, yet needs arguments to be justified.

From this overview we conclude that applications in the context of public health and high-risk groups raise important ethical questions, yet each of them can be addressed with the help the common values of as well-being and autonomy. These can put sufficiently moral constraints to the development in order to result in a morally justifiable practice and to prevent the introduction of such dietary advice if they were not to have any added value.

Nutrigenomics and the market

Industry is an interested agent in the context of nutrigenomics. It may be an attractive way to produce food products with added value that also has better profit margins. Nevertheless, it confronts industry with a challenge since it requires new marketing and operational behaviours (Ethics Institute, 2005).

In this context many of the above-mentioned issues pop up in a slightly revised form, however, some issues that are relevant in the context of public health are hardly interesting with respect to the market. For instance, criteria of in- and exclusion are not relevant since the decision whether to purchase a genetic test-kit will be completely up to the individual consumer. It would even be remarkable to exclude consumers from the purchase of products in a market context. The consumer is considered as a free and autonomous agent that should have the opportunity to buy what is available at the market. Even when this would entail that the products are only accessible for the affluent this is not directly a problem within a market context, since market mechanisms determine the distribution of goods.

Nevertheless, the offer of TMD via the market raises specific moral questions, e.g., with regard to the responsibility of companies. Regarding health professionals it is clear what one may expect of them when they offer tests and advices, yet these patterns of expectation are less clear in the case of the market. For instance, although the results from the genetic test are not always easy to communicate it is relevant information for other commercial parties. Hence a company

might have a commercial interest to sell this information to other market parties (e.g., insurance companies). This obviously leads to a conflict between the protection of individual privacy and commercial interests, which requires regulation in order to protect consumers. Furthermore, some well-known are defined in another way because of the market structure. Consumers have the freedom to choose whether to buy TMD and from whom, yet their choice is limited because of a monopoly of big companies: only large market parties are able to develop such high-tech products like genetic tests and dietary advices.
As we have concluded for the other contexts, we can also claim for the market that TMD raise no new issues if we perceive them from the market perspective. However, there is a difference between the market and the other contexts with respect to the interpretation of standard (medical)-ethical norms and procedures.

Implications of a changing relationship

The above-mentioned moral issues require substantial attention, yet as long as they are addressed within one of the contexts they do not lead to serious problems. However, the contexts are overlapping when (a) the market actively addresses high-risk groups and (b) the extent of the offer of TMD-becomes close to that of a public health context.

High-risk groups

An interesting group for the market are consumers with a high risk. These individuals can really benefit from personalised dietary advices. However, this also entails that they have a vulnerable and dependent position. They need these products. Hence, emphasising this need or even "sowing worries" is an effective marketing strategy in order to keep them buying your products. In the health sector, the threat of abuse of this vulnerability is considered as incompatible with widely-used principles of non-malificence and beneficence. For a market party this dependency is certainly something that will be addressed with care, yet it is also has another dimension: it is highly important from the perspective of market strategy to keep them dependent. An exclusive product that consumers are depending on is a unique selling opportunity that has interesting profit margins. Hence although companies have clear responsibilities, there is a tension between their interests and the interpretation of principles such as not harming consumers.
This illustrates that although all agents share certain values and principles it is not directly clear how they will interpret them when TMD are offered to high-risk groups by the market.

Public health perspective revisited

The possibility to tailor advices to the genetic risk profile of individuals makes nutrigenomics interesting for the market. However, the personalised scope seems to be opposed to market's interest to create large groups of potential consumers. Especially since the development and production of TMD is a capital-intensive investment it is preferable to distribute these costs over a larger group of consumers. Thus TMD will be especially profitable when as much as possible consumers purchase the products. Hence a revisited public health approach is an attractive one and consequently the problem of added value returns again. From the perspective of public health policy TMD is not relevant, yet for the market the public health perspective is certainly interesting even when the added value is small. This faces us with the question whether it is justified to introduce advices that are not considered to have enough value to be relevant as a tool for public health policy to as much as possible consumers. This issue cannot be simply addressed by the common argument of the market that it is up to the individual consumers. The complexity of TMD makes us wonder whether consumers have the opportunity to make individual assessments and evaluations. Hence adequate and independent advice is necessary.

In the health care context this information comes standard, yet for the market it asks for some serious modifications in how products are offered.
From this it possible to conclude that, although all parties will admit that TMD need to be beneficial to individuals and that the autonomy of consumers has to be respected, it is not directly clear whether it is morally preferable that TMD are offered by market parties as a preventive tool to address mass diseases.

Conclusion

The ethical issues that surface as the result of genetically tailor-made dietary advices ask for reflection and deliberation. As long as we deal with the issues within one context of application the moral questions can be accurately addressed by the common interpretation of values and principles such as autonomy and well-being. The interpretation, however, can be different in each of the contexts. This is relevant, since citizens are also consumers that operate in different contexts. Hence it is important to systematically reflect on what one may expect of all stakeholders when TMD are introduced.
This reflection is not only relevant at the stage of introduction, but also during the process of research and development. Ethical questions are already (implicitly) present in the research stage. Hence we have to differentiate between an ethics of development and ethics of application. This leads to two types of recommendation. First, with regard to moral issues that surface in the stage of research we have to reckon with the assumption that TMD will be mostly offered via the market. Second, with regard to applications, it is important to discuss the need of regulations. This does not imply governmental regulation only. In line with existing independent organisations that advise people with mortgages, it can be helpful to launch independent institutions that advice individuals with regard to personalised dietary advices.

Acknowledgement

This paper is part of the project "Food, prevention, and ethics." financed by The Health Research and Development Council of the Netherlands (ZON/Mw, 2300.0025).

References

Ethics Institute. (2005). *Report of the International Seminar 'Genetically tailor-made diets: future prospects and ethical challenges'*, Ethics Institute, Utrecht

Meijboom, F.L.B., M.F. Verweij, F.W.A. Brom. (2003). 'You eat what you are. Moral dimensions of diets tailored to one's genes', *Journal of Agricultural and Environmental Ethics*, 16/6, 557-568.

Meyer, G. (2005). 'Pharma-food: Are tailor-made, individual diets relevant for problems of public health?', *EurSafe News*, 7/1 pp. 1-3.

Can the freedom to choose between gm-free and gm-products be guaranteed by voluntary GM-free regional co-operations in the long-run?

Frauke Pirscher
Institute for Agricultural Economy and Rural Landscape Planning, Martin-Luther-University Halle-Wittenberg, Emil-Abderhalden Str. 20, D-06099 Halle (Saale), Germany,
frauke.pirscher@landw.uni-halle.de

Abstract

By legalising the cultivation of genetically modified (GM) crops the EU wants to increase producers' and consumers' freedom to choose between different production systems. However, GM-crops can affect non-GM-production on adjacent fields by cross-pollination. Co-existence" of different production systems in a highly fragmented cultivated area like Europe seems to be an illusion despite all legal obligations of GM-farmers to reduce pollen dispersal. This means that after a few years of GM-cultivation no freedom of choice for those who reject this production system would exist. In Germany farmers reacted with the spontaneous foundation of GM-free regions because spatial segregation is viewed as one decisive means to reduce the probability of cross-pollination considerably. To guarantee a GM-free agriculture farmers commit themselves in written form to produce without applying GM-crops. The maintenance of the freedom to choose between different production systems mainly depends on the stability of the co-operations. It can be shown that the governance structure in the current design of voluntary agreement is insufficient to meet the challenge of distributional conflicts within the region. The danger that the regions will collapse very soon after their establishment can be reduced by inducing changes in the sharing-rules between farms that are bound by a contract not to exceed the legal threshold and those that are not and on the other hand farms at the edge of the region compared with those in the centre.

Keywords: freedom of choice, genetically modified free regions, voluntary co-operation, governance structure

Introduction

In 2001 the EU has legalised the cultivation of genetically modified (GM) crops by passing the directive 2001/18/EG. After having tightened up the licensing procedure for GM-crops now the Commission considers the commercial cultivation as safe and wants to offer farmers the possibility to choose between GM-, conventional and ecological production systems. However, the majority of European consumers reject the consumption of food that includes GM-ingrediences (European Commission 2000). Therefore, as part of the amendment of the European food legislation, the EU issued an obligation to label all products exceeding 0.9 percent of GM-share (Regulation No 1830/2003). This should guarantee the freedom to choose for proponents and opponents within the group of producers and consumers. However, both production systems cannot co-exist independently. Pollen dispersal from GM-fields to non-GM-fields affects the possibility of non-GM-farms to produce GM-free products. Experimental results on cross-pollination and seed translocation lead to the conclusion that gene-flow between adjacent fields cannot be avoided (Belcher *et al.* 2005). Thus, co-existence has to be viewed as illusion especially in Europe with highly fragmented cultivated areas. Consequently, a few years after introducing the option to grow GM-crops, possibility of consumers to choose non-GM-products will no longer exist. However, the freedom of choice can be view as key element of the principle of autonomy (Korthals 2004).

Spatial segregation of the production systems is viewed as a decisive means to reduce cross-pollination to an extent that can guarantee GM-free crop production. Therefore, farmers in one area have to decide cohesively to do without GM-crops. This renunciation has to be voluntary. A mandatory designation of GM-free regions is not consistent with the EU-law. Therefore, the maintenance of consumers' and producers' freedom of choice particularly depends on the present ability of farmers to organise voluntary GM-free regional co-operations that will be stable in the long-run. In Germany farmers reacted with the spontaneous foundation of GM-free regions on the basis of voluntary agreements. These associations included ecological as well as conventional farms.

It is the aim of this paper to analyses weather the incentives to co-operation within the association will be sufficient to guarantee for long-lasting GM-free region. In theory on collective action it is argued that co-operation will only be stable if it creates a collective gain and this surplus is distributed between the members on the basis of sharing rules that allow all of them to improve themselves (Ostrom and Gardner1993). Thus the paper wants to identify the main determinants of the collective surplus and its distribution between different groups of farmers within the GM-free regions. Subsequently it wants to analyse whether the current sharing-rules of the voluntary agreement are able to avoid distributional conflicts and cope with the danger that the regions will collapse soon after their establishment. With respect to the incentive for co-operation I will distinguish four different groups: On the one hand farms that are bound by a contract not to exceed the legal threshold or even produce beyond it in contrast to those that are not and on the other hand farms located at the edge of the region compared with those in the centre.

In the next chapter I will shortly characterise the existing voluntary GM-free regional co-operations, Afterwards incentives for co-operation under different sharing-rules will be analysed and proposals for improvements presented. The paper closes with some conclusions.

Voluntary GM-free regional co-operations – key elements

Voluntary agreements are an instrument within environmental politics where a group of producers mutual commits to reduce or refrain from some kind of environmental damage whereas the obligation is legally non-binding. The same is true for GM-free regional co-operations. Here farmers commit themselves in written form to produce without applying GM-crops. The life of the contract is one year extending automatically. Furthermore, farmers commit to take samples of all inputs used and products sold and keep them for five years. Farmers shall act on their private contractor to clean their machines carefully after having worked for a GM-farmer to reduce contamination.

With respect to exertion of influence by entrepreneurs on the level of the standard voluntary agreements on GM-free production differ from other voluntary agreements on environmental standards. In case of the GM-free regions purity standards are regulated by law. Any tightening or relaxation is not provided by the agreement. In typical voluntary agreements the participating entrepreneurs can negotiate with the government on the level of purification standard that should be achieved. The enterprises commit voluntarily to fulfil this standard while the government relinquishes any mandatory regulation. Thus, the entrepreneurs have a strong incentive to meet the obligations, although they are not justiciable. A further remarkable difference is that in case of the GM-free regions not the persons causing a pollution or here better pollination commit themselves not to do so, but the injured party, those who refuse GM-cultivation. Consequently, their incentive to co-operate is different from typical voluntary agreements that try to achieve a low purification standard.

Therefore, the question arises what kind of benefit farmers expect by signing a voluntary agreement not to grow GM-crops.

Analysis of the stability of co-operation

The incentive for co-operation depends on the potential damage that can be caused by a neighbouring GM-farmer. Here two groups of farmers can be distinguished. First, there is a group of farmers who will be sactioned in case their products exceed the threshold for obligatory GM-labeling. They can no longer sell their products at a higher price or sell them at all. This group includes mainly ecological farms but also all other farmers that have guarantee by contract to stay below the threshold. In the following I will generally call this group as eco-farms. The second group includes all conventional farms that prefer to continue producing without GM-crops, but will not have any economic disadvantage from exceeding the threshold. Their produce can be sold at the same price even it is labeled. While for eco-farmers the existence of GM-production can deteriorate their economic result this is not true for conventional farms. For the latter in any case their benefit of production π_i is total revenue, that is price p_i times quantity $q_{i,}$ less production costs c_i.

$$\pi_i^k = p_i^k q_i^k - c_i^k$$

For the ecological farms the existence of GM-farms increase their production costs. Information costs i_i arise when farmers have to find out the production decisions of their neighbours. Control costs s_i to analyse samples of their products have to be added as well as cleaning costs r_i in case a private contractor will be hired. Finally, the possibility that the farmer no longer receives the higher price for his or her product compared to the conventional farmer can be expressed by damage costs (Beckmann and Wesseler 2005). The probability that this happens depends on the quantity Q_i of GM-fields in the neighbourhood and the diffusion coefficient . This coefficient is crop and farm specific. It depends on the distance to the GM field, the cultivated crop, the existence of buffer zones or other management practices to reduce pollen dispersal. The damage costs are zero if the GM-share of the crops stays below the threshold for obligatory labelling. If the threshold is exceeded, eco-farmers get the lower conventional price. Thus the damage costs amount the price difference between the eco- and conventional products times the harvested quantity of eco-products. While for eco-farms without co-existence the benefit can be considered as:

$$\pi_i^e = p_i^e q_i^e - c_i^e \quad \text{with } p_i^e > p_i^k;\ q_i^e < q_i^k \text{ and } c_i^e > c_i^k$$

with GM-production in the neighbourhood it is

$$\pi_i^e = p_i^e q_i^e - c_i^e - i_i^e - s_i^e - r_i^e - d_i^e$$

with

$$d_i^e = \begin{cases} 0 & \text{if } \dfrac{\alpha_i Q_i}{q_i} < T \\ (p_i^e - p_i^k)\, q_i^e & \text{if } \dfrac{\alpha_i Q_i}{q_i} \geq T \end{cases}$$

This can be viewed as a benchmark for the decision to co-operate.
Generally, co-operation will only take place if for the whole group a surplus is generated. The surplus of aligning with the neighbouring farms to GM-free regions results in reduced of the

diffusion coefficient and thus a low probability of a price reduction, reduced information costs on the production decision of the neighbours, reduced costs for sample analyses because of economics of scale and reduced or no clearing costs, in case several members of the GM-free region hired the same private contractor. The surplus increases with the number of participants as the reliability on the neighbours' behaviour increases. It also depends on the shape of the region. The diffusion coefficient decreases the less the percentage of borderline in relation to the whole area of the region. That is the more the region follows the form of a circle. Consequently, to increase the overall benefit it is not only important how many farms participate but also their location within the region and with respect to the GM-farms. Farms at the edge of a region are very important to guarantee to overall surplus of the co-operation.

The incentive of a single farm to co-operate depends on the share it receives form the overall surplus compared to the situation without co-operation. The sharing rules have to be found that guarantee all four groups a benefit from co-operation. Otherwise the GM-free regions will not be stable. Under the institutional arrangement of voluntary agreements the share each farm receives differs with respect to the production system and the location within the region. It can be deduced from the composition of the overall surplus that only eco-farms profit from co-operations while conventional do not and farms in the centre of the region receive a higher share of the surplus than those at the edge. The reduction of costs caused by co-existence affects only eco-farmers because conventional farmers do not have to bear these costs and consequently have no benefit from their reduction. Not only that they do not improve themselves by co-operation their production costs now increase compared to the situation without co-operation. From the decrease of the diffusion coefficient farms in the centre of the region mainly benefit because the diffusion coefficient decreases from border to centre. Surrounding farms serve for the centrally located as a protection against cross pollination. Consequently, under the current sharing-rules conventional farmers and those located at the edge of the region do not profit from co-operation.

The governance structure of voluntary agreements does not guarantee for a stable co-operation in the long-run. The fact that currently GM-free regions already exist cannot be interpreted as rational in economic terms. Here, others factors than economical ones have lead to their foundation. Probably conventional farmers and farms at the edge will not prolong their agreement, although they do not switch to the GM-production system. But by contracting out they will reduce the surplus of the whole co-operation.

To stabilise the regions changes in the sharing rules have to be made. To compensate for disadvantages caused by the location, a payment depending on the distance to the border of the region would be conceivable. A contribution of centrally located farms to the costs of farms at the edge of the region can increase the incentives of these farmers to continue co-operating. This can be done by introducing a fee where the amount each farmer has to pay depends on the location of the farm within the region or the percentage of borderline the farm posses. The more centrally located a farm is the higher the fee will be, or the less plots at the border of the region the farm posses the higher the fee will be. Prerequisite would be that the co-operation will chose a more binding institutional form like e.g. a club. However, one has to keep in mind that with this the danger of free riding for centrally located ecological farms increases in case the payment would be fixed too high so that it would over compensate the costs advantages of a membership.

Further stabilisation decisively depends on the processors of agricultural products mainly the mills. Their behaviour influences the incentive of conventional farmers to produce without exceeding the threshold and thus continue to be a member of a GM-free region. Mills can accept crops from conventional farms containing a GM-share beyond the threshold without any changes in the prices. Here, conventional farmers would have no incentive to control the GM share of their products and thus to co-operate. The mill accepts products containing a GM-contamination beyond 0.9% but only pays the low price for feed grain as they fear the consumer do not accept products that have to be labeled. This assumes that mill can segregate GM-products form non-GM-products. Then, for conventional farms it would be as attractive as for eco-farms to co-operate in a GM-free region to partake in the cost reduction of e.g. the sample analysis. Finally, mill can reject grain exceeding the threshold if they do not have the possibility to segregate both production lines. On the one hand this would induce the highest incentive for co-operation on the other hand is it unlikely that any other farmers would decide to grow GM-crops. This would make the necessity of GM-free regions obsolete. The decision of the mills very much depends on the presumed attitude of consumers towards GM-products. As up to now no real market for GM-products exists their assumed paying behaviour has only hypothetical character. Currently, the fear that consumer will reject theses products is great enough to guarantee GM-free production. But this can change very rapidly and might lead to the situation that a mixture has taken place before consumers recognised that they lost their option to choose. EU law allows for labeling products as "without GM" if the suppliers can guarantee GM-freedom along the whole food chain and bears all control costs for proofing it. Up to now this costs appear to be an insuperable barrier to market entrance. Certifying products as "produced in a GM-free region" as a label of origin is not compatible with is EU law. However, this would allow stabilising the association of farmers to a GM-free-region at a point of development that can guarantee for long-run segregation of the production systems and thus consumers' and producers' freedom of choice. It appears to be incomprehensible why EU law do not allow consumers of offer their willingness to pay for a process standard but only for a product standard. .

Conclusions

Farmers established GM-free regions spontaneously to express their political and ethical concerns against GM-production systems. In the long-run these regions will only continue to exist if all participants achieve a positive share of the collective surplus by co-operation. Under the current sharing-rules this is not the case. The participation of conventional farmers and farmers at the edge of a GM-free region is altruistic when assuming a utility maximising behaviour. For farmers located at the edge production costs increase while the probability of contamination remains the same than without co-operation. For conventional farmers the production costs increase as well but in contrast to eco-farmer they do not get any price supplement for GM-free production. Thus, the current institutional arrangement does not stabilize the regions. A great danger exists that they will collapse soon and by this the option to produce without GM-contamination will vanish. Institutional changes can reduce this danger. The contribution of centrally located farms to the costs of farms at the edge of the region can increase the incentives of these farmers to continue co-operating. Further stabilisation depends on the behaviour of the processors and consumers. If the processors require a certificate about the GM-share of the products from all farmers co-operation would reduce production costs for conventional farmers as well. This they will only do if a considerable part of the consumers continues to reject GM-food. Finally, it is worth to consider whether the institutional barriers established by the EU that prohibit offering the willingness to pay for products originating from GM-free regions are justifiable. The possibility to express ones preference for GM-free regions as a process standard would increase the incentive to co-operation and thus contribute to the maintenance of producers' and consumers' freedom of choice.

References

Beckmann, V. and Wesseler, J. (2005). *Governance of Genetically Modified Crops in the EU*. Unpublished Manuscript.

Belcher, K.,Nolan, J. and Phillips, P.W.B. (2005). Genetically Modified Crops and Agricultural Landscape: Spatial Patters of Contamination. *Ecological Economics*, Vol. 53: 387-401.
European Commission (2000). *The Europeans and Biotechnology*. Eurobarometer 52.1.Brussels.
Korthals, M. (2004). *Before Dinner – Philosophy and Ethics of Food*. Springer, Dordrecht.
Ostrom, E. and Gardner, R. (1992). Coping with Asymmetries in the Commons: Self-Governing Irrigation Systems can Work. *The Journal of Economic Perspectives*, Vol7, No.4: 93-112.

Part 11
Critical issues in aquatic production systems

Turning cheap fish into expensive fish? The ethical examination of an argument about feed conversion rates

Matthias Kaiser
The National Committees for Research Ethics, Prinsensgate 18, Box 522 Sentrum, 0105 Oslo, Norway, matthias.kaiser@etikkom.no

Abstract

This paper examines one particular debate about the long term sustainability of intensive aquaculture of carnivorous species (e.g. salmon). It asks whether improved knowledge will help us resolve this debate. However, given the uncertainties in the knowledge base and the complexity of the ecosystems in questions chances are that the debate cannot be resolved by science alone. Rather it is pointed out that the debate rests on a number of different framings that are deeply value dependent and raise different ethical questions. Several such framings are discussed and ethical dilemmas are pointed out. Two strategies are proposed to deal with this situation: a wide use of participatory measures in technology assessment, and strengthening the ethical awareness of experts by soft-law as e.g. ethical codes.

Keywords: aquaculture, feed conversion, sustainability, framing problem, ethics

The logic of aquaculture

Aquaculture relates to fisheries roughly the same way as agriculture relates to hunting. Surely this is a positive feature of aquaculture. Furthermore, it is beyond doubt that aquaculture production is in principle better adapted to the demands of modern markets than traditional capture fisheries. It provides for a much more stable and targeted delivery of food products. Aquaculture is currently the fastest growing food sector world wide, with the main growth happening in Asia, especially China. In Europe aquaculture is struggling with a negative image that holds its grip on consumers and the media.
Obviously there are a number of critical issues in aquaculture. I shall here only discuss one major issue that relates to the sustainability of aquaculture. What are we to think about production fish being caught along the shores of, say Chile, being processed somewhere in Europe, say Finland, then being fed to aquaculture salmon in, say Norway, for this fish then being finally flown to, say Japan where consumers can enjoy high quality / price products? Can the energy use along the way be considered to be sustainable in the long run? And what are our alternatives? Is this the unavoidable prize of globalization, a prize that all food production has to pay? I surmise that this then is at least a type of logic that very few ordinary people find easy to reconcile with their values, especially when they are expected to reduce their consumption and save energy. And who is responsible? From an ethical point of view, if the "system" is wrong, then all those who "follow" the system take on co-responsibility.

Fishing down and farming up the food chain?

In this connection, aquaculture sometimes makes the claim that fishing down and farming up the food chain must be the only sustainable and reasonable way to go. Among others they cite arguments that the feed conversion rate in salmon is much better than in other animals. Feed conversion for salmon looks impressive when given as e.g. 1:0.8 (i.e. 1 kg of salmon produced for every 0.8 kg of fish feed fed), but one needs to remember that this is in relation to *dry* feed, i.e. the actual biomass needed to produce the feed will usually be much higher. For salmon it would be something like 1 kg of farmed salmon for 4-5 kg of biomass that is used as feed. But this is assumedly still better than for most agricultural animals. For instance, Åsgård and Austreng, comparing energy and protein retention in salmon, poultry, pigs and sheep, give their digestible energy rate at 34%, 17%, 20% and 1.3% respectively, while protein retention is given at 30%, 18%, 13% and 2.1% respectively. Furthermore, we get much more edible energy out of 10 kg of capelin if used as feed for farmed salmon (~28 MJ), than if eaten by cod that then is captured by fisheries and sold on the markets (~3 MJ) (Åsgård and Austreng 1995). In other words, where nature left alone is wasteful, modern farming techniques appear highly efficient. This looks like a strong argument in favour of intensive aquaculture (cf. Figure 1).

However, marine biologists and ecologists typically oppose this line of argument (Naylor *et al.* 2000, Naylor *et al.* 2001). For one thing, they point to scientific uncertainties about the "energy flows between marine fish at different trophic levels" (Naylor *et al.* 2000, p.1019). They also point to the risk of unsettling complex eco-systems if "ever increasing amounts of small pelagic fish should be caught for use in aquaculture feeds" (ibid.). In this connection they also point out that aquaculture already today is a major contributing factor in the over-exploitation of our ocean resources (Naylor *et al.* 2001; Poseidon *et al.* 2004). A further factor they want us to consider is that the typical inclusion rate of fish meal in agricultural feed of 2-10% requires an order of magnitude less fish meal than for fish products (Poseidon *et al.* 2004, p.63). Feed providers, however, argue that significant reduction of the inclusion rates can also be expected for fish feed in the near future. But given the growth of aquaculture a further increase of fish meals and oils can be expected in the future (cf. Figure 2.).

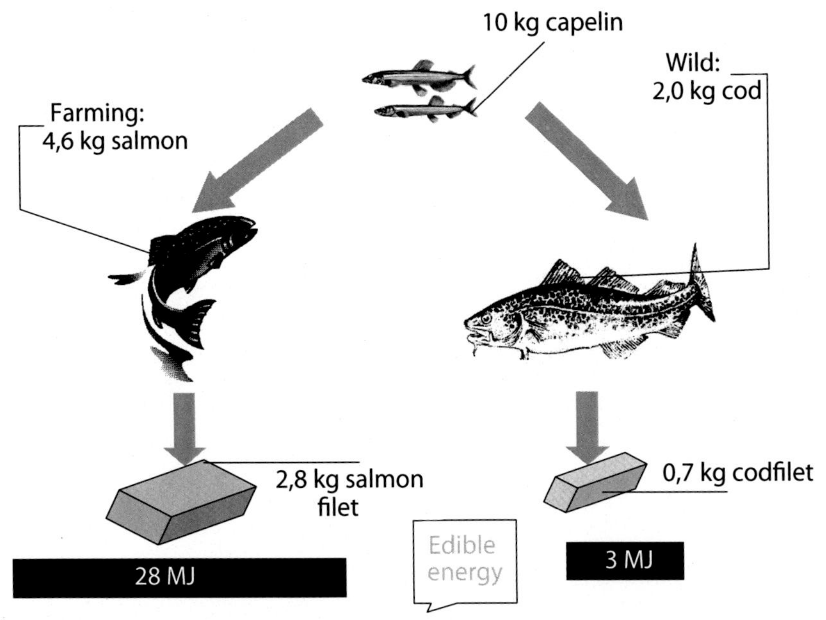

Figure 1. Amount of farmed salmon and wild cod produced from 10 kg of capelin (adopted from Åsgård and Austreng, 1995).

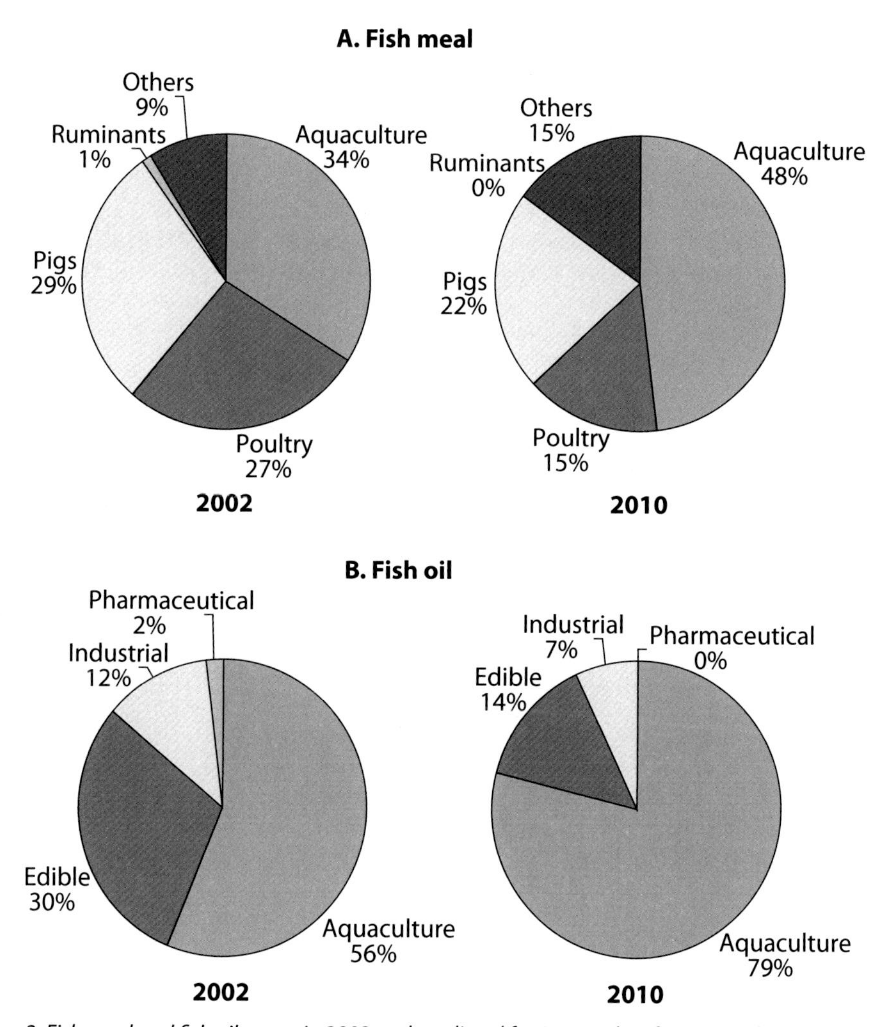

Figure 2. Fish meal and fish oil usage in 2002 and predicted for 2010; taken from Poseidon et al. *2004.*

Ethical issues

The pro and con-arguments of this debate are interesting, albeit unresolved. The most tempting rejoinder is obviously to reply something along the following lines: Let us first resolve the scientific issues and remove all remaining uncertainties. Once we know all we need to know, we will be able to define a sustainable level of fishing for production purposes.

However, the problem is that this rejoinder is probably based on a fallacious conception of scientific knowledge and (un)certainty. Recent contributions to this topic suggest that uncertainties do not simply vanish when we gain more knowledge, rather they may increase (Walker *et al.* 2003; Van der Sluijs *et al.* 2005). Our sciences typically reveal ever increasing complexities that do not support simple policy solutions.

We seem left with the necessity to make management and policy decisions under conditions of major prevailing uncertainties. There are several theoretical ways to go about this task. For one thing, the Precautionary Principle can be said to be tailormade for this kind of situation where we have to make decisions now even if major uncertainties remain (UNESCO / COMEST 2005). The Precautionary Principle is itself based upon an ethical commitment. However, one of the important insights from clarifications of the Precautionary Principle is that even if one is certain that the principle needs to be applied in a certain case, this does not necessarily prescribe a unique precautionary measure (see also Kaiser 1997). What measure to apply to a case, depends always in some sense on values. Precautionary measures that are suitable to either prevent harm or limit the scope of it can be directed towards multiple potential leverage points like e.g. regulatory measures, technological measures, or market mechanisms. Another (supplementary) possibility to tackle complexity and uncertainty is to increase the resilience of the system (cf. Gunderson and Pritchard 2002). But also here it seems that several measures will be possible and the choice is largely value-dependent.

At this point it might be important to step back and ask some more fundamental questions. One is the question what precisely is at stake here? This refers to the well-known framing problem. Obviously there are a number of possible questions involved here. A narrow framing of the problem would be to limit it merely to the issues of sustainable management of limited aquatic resources. That is to say that we need to find a way so that the amount of small pelagic fish taken out of the ocean by fisheries and used for the production of fish-meal and fish-oil needs to be such that it does not threaten the fish populations. Framed like this, we indirectly accept a number of other premises, like e.g. the premise that it is largely up to market mechanisms to determine how much of that resource we have available for the farming of carnivorous (or to some extent omnivorous) marine species. Therefore this framing of the issues does not question the very logic of intensive production of aquatic or agricultural food animals. Yet there may be good reasons to doubt that ethical worries will be answered by markets.

But we may also frame the issue differently. We may ask whether we can afford utilizing scarce marine resources in feed for the production of animals that are largely ineffective converters of energy and protein. In other words, we may say that the real issue is to choose between fish or meat or poultry on our plates. Can it be sustainable in the long run to increase our diet with more meat rather than with more fish? And should we not reduce intake of meat and increase intake of fish for health reasons in order to have a more healthy balance between saturated and unsaturated fatty acids? This is perhaps a framing of the issues that the aquaculture industry would like us to accept, but at the same time a framing that brings it into conflict with the agricultural sector. Note that there are a number of possible ethical dilemmas associated with this framing. One is the problem of paternalism. Once we lift the question of the composition of the diet from the individual level to the level of policy making, we seem to automatically restrict individual freedoms that others may count as essential. Another unresolved issue is the question of the extent to which we – from an ethical point of view – are at liberty to regard all animal species as on par when it comes to using them in industrial food production. For some animals, like cows or pigs, the ensuing living conditions in industrial food production may be such that they conflict with ideas of animal welfare. But on the other hand, the same might possibly be said even for fish species used in aquaculture. And where does that leave us?

We may frame the issue even wider. We may e.g. be of the opinion that the global population growth is such that we need to utilize all possibilities for growth in the food sector in order to feed future generations that soon will experience food shortages given present productions. A similar line of argument has been utilized by proponents of e.g. biotechnology. Obviously this is a very complex matter. But on the other hand one should not overlook that this line of argument corresponds well with our ethical responsibility for future generations. Given that food shortages might be a real possibility of the not too distant future, one may be inclined to stretch our range of acceptable actions to secure food production to its limit.

Other framings of the issue are obviously possible. One such framing would be to look only at options that are politically realistic. We ask where states and international governmental and non-governmental organisations have a potential for action that is politically achievable and at the same time a step towards more sustainability. This is the "Realpolitik"-viewpoint. We may have to compromise some ethical ideals, but at the same time we may hope for some Pareto-optimality that provides positive net results.
At this point, ethicists may ask us to step back once more and consider whether we really have addressed the issue properly. One important factor is whether we have accounted for all possible options. What could be considered a "possible" option is again dependant upon some very fundamental convictions. For instance, one might wonder whether we could "educate" food consumers in such a way that we can avoid the whole problem in the first place. What if we all would be vegetarians half of the week? This would obviously relieve us from a lot of pressure on the natural marine resources. Can or should we systematically work for a change of our consumption pattern? Again, questions of paternalism enter this picture. Wide spread defection from high goals may not only be the problem of the individuals who defect, but also a problem of the goals themselves. Here cultural factors will play in, and the range of possible options will be linked to our different cultures.
Other possible options may be easier to handle. This applies first of all to technological options that we have not satisfactorily considered so far. For instance, one might argue that we indeed should strive for growth in aquaculture, but that this growth should be achieved not by high-priced carnivorous species, but by low-price omnivorous or herbivorous species. Carp and Tilapia may be good examples here, given also that we do not restrict ourselves to intensive production forms, but explore further the potentials of semi-intensive, extensive and integrated fish farming. Some NGO's seem to argue this point (Naylor *et al.* 2001). But then obviously "Realpolitik" comes into the picture. Who can steer and control such a development on a global scale, without restricting the rights of states to produce food products that under present market conditions secure high profits?

Conclusion

We are bound to address rather complex and uncertain issues. Ethical issues are often sacrificed in expert dominated discourses about technology. We need to ask who frames the issue, and how stakeholders would frame it. How can we avoid that an initial focus on limited questions soon brings us to big, global, controversial and perhaps unsolvable questions of the good and morally right life on earth?
The strategy I would propose as achievable and ethically justified is of a procedural kind. Two points seem vitally important to me. First, we need to employ participatory techniques on the widest scale possible. The whole register from local hearings to national consensus conferences (Danish style) should be utilized as a means to address these issues. Second, we need to strengthen the ethical awareness of those experts that are part of the innovative forces of the relevant sector. This cannot be done by law or regulation alone, but should be done by "soft-law", i.e. by ethical guidelines that codify moral responsibility towards the common good. Those at the forefront of technological innovation are possible leverage points for changing the course of development. Ethicists, philosophers, lawyers, and other social scientists are among the specialists that need to engage in these issues. We see much too little of that in our current academic environment.

References

Åsgård, T and Austreng, E. (1995), "Optimal utilization of marine proteins and lipids for human interest", in: H. Reinertsen and H. Haaland (eds.), *Sustainable Fish Farming*, A.A. Balkema: Rotterdam / Brookefield, pp. 79-87.

Gunderson, L. and Pritchard, L. Jr. (eds.) (2002), *Resilience and the Behaviour of Large-Scale Systems*, Island Press: Washington/Covelo/ London.

Kaiser, M. (1997), "Fish-farming and the precautionary principle: Context and values in environmental science for policy", *Foundations of Science* 2, 307-341.

Naylor, R.L., Goldburg,R.J., Primavera, J.H., Kautsky, N., Beveridge, M.C.M., Clay, J., Folke, C., Lubchenko, J., Mooney, H. and Troell, M. (2000), "Effect of aquaculture on world fish supplies", *Nature*, 405 (29 June 2000), 1017-1024.

Naylor, R.L., Goldburg,R.J., Primavera, J.H., Kautsky, N., Beveridge, M.C.M., Clay, J., Folke, C., Lubchenko, J., Mooney, H. and Troell, M. (2001), "Effects of aquaculture on world fish supplies", Issues in Ecology, 8 (Winter 2001).

Poseidon Aquatic Resource Management Ltd & The University of Newcastle-Upon-Tyne (2004), Assessment of the sustainability of industrial fisheries producing fish meal and fish oil, Report to the Royal Society for the Protection of Birds, available electronically: http://www.rspb.org.uk/Images/fishmeal_tcm5-58613.pdf

UNESCO / COMEST (2005), The Precautionary Principle. World Commission on the Ethics of Scientific Knowledge and Technology, Paris; available electronically: http://unesdoc.unesco.org/images/0013/001395/139578e.pdf

Van der Sluijs, J.P., Craye, M., Funtowicz, S., Kloprogge, P., Ravetz and Risbey, J. (2005), "Combining Quantitative and Qualitative Measures of Uncertainty in Model-Based Environmental Assessment: The NUSAP System", *Risk Analysis*, 25 (2), 481-492.

Walker W.E., Harremoës P., Rotmans J., Van der Sluijs J.P., Van Asselt M.B.A., Janssen P. and Krayer von Krauss M.P. (2003). "Defining uncertainty a conceptual basis for uncertainty management in model-based decision support", Integrated Assessment, 4 (1), 5-17.

A warm heart for a cold fish: Moral obligations and welfare considerations in fish farming

Vonne Lund and Cecilie M. Mejdell
National Veterinary Institute, PO Box 8156 Dep., N-0033 Oslo, Norway, vonne.lund@vetinst.no

Abstract

"Fish welfare" is a rather new issue, given steadily increasing attention by authorities, the public and fish farmers. Fish has usually not been given considerations comparable to that given to higher vertebrates like mammals and birds. The reasons for this difference may be found in history and tradition as well as communication difficulties among species and the lack of factors making the basis for human empathy. In the western world sentience is commonly considered to be the single most important criteria for moral status, and hence moral concern and moral obligations towards animals. Current scientific knowledge, based on neuroanatomy, neurophysiology and behaviour, indicates that pain perception is likely to be present at least in bony fishes (teleosts). This is setting the stage for a change of moral obligation towards fish. The understanding of the welfare concept then becomes an issue, since it determines how to judge welfare and what welfare indicators to select: obviously it is necessary to know what to measure. Substantial research efforts are needed to shed light on these issues and to unveil even basic fish needs in relation to fish welfare.

Keywords: animal welfare, fish farming, moral community, pain perception

Introduction

At least in the Western culture there is a general agreement that animals are entitled to certain animal welfare considerations and standards. England got its first animal welfare legislation as early as in 1822, and the comprehensive Cruelty to Animals Act was passed in 1876. The European Union acknowledged in the 1997 Treaty of Amsterdam that "Member states shall pay full regard to the welfare of animals ...". However, fish have not been given welfare concerns comparable to other vertebrates.

This seems to be about to change. "Fish welfare" is given steadily increasing attention by the public, researchers, authorities, as well as the fish farmers. The Holmenkollen Guidelines (Sundli 1999), adopted in 1998, were among the first international protocols suggesting that ethical principles ensuring health and welfare of fish, including humane slaughter, should govern the industry. In 2002 the "world health organization" for animals, the OIE, pointed out animal welfare as a priority area (Håstein *et al.*, 2005), aiming at developing standards for animal welfare in agriculture as well as in aquaculture, to be incorporated in the OIE Terrestrial and Aquatic Animal Health Codes. These are generally the basis of standards and guidelines for animals in international trade. In 2005 the Council of Europe passed recommendations for the keeping of fish for farming purposes under the Convention for the protection of animals for farming purposes. In Norway, the forthcoming new Animal Welfare ordinance will regulate the keeping and treatment of fish in similar way as other vertebrates, departing from that animals have intrinsic value and therefore should be handled with *"care and respect for the animals' distinctive character"*.

Fish has not traditionally been part of the moral community

The reasons why fish welfare until now hardly has been considered may be found in history and tradition, the difficulty of communication across species boundaries, as well as in lack of knowledge. This is particularly apparent in countries where animal welfare traditionally has a strong position, such as in the Scandinavian countries.

Fish farming in Scandinavia has its origins in commercial fishing rather than in traditional farming. Commercial fisheries may be considered rooted in an ancient tradition of hunting and catching the prey, incorporating no elements of care or husbandry. The first salmon farmers were often fishermen, bringing along their "hunting" experience and related underlying values and attitudes into their new profession. These fish farmers did not possess the approach of "ordinary" farmers who keep and care for individual animals, ensuring their health, well-being and production during their entire life. The development of fish farming into big business, run by multi-national corporations owing duty to shareholders distant from the production, has underlined the view of fish as a "crop" to be harvested rather than individual animals to be cared for.

Fish certainly do not possess factors arousing human empathy in the way mammals do. Fish have no facial mimic and they do not scream or make sounds which humans can perceive, urging for human concern. They have an unfamiliar body language, they live in an element strange to us, and we know little about their natural lives. It is hard for us to imagine what it is like to be a fish, let alone a happy fish. Perhaps as a result of this lack of communication among species, fish have not been considered as sentient beings. The consequences have been profound, since in western society sentience is commonly considered the single most important criteria for moral status. Hence, fish have not been included in the moral concern or moral obligations towards animals.

Today, the picture is changing. Not only has animal welfare become a major concern in western society. For example, the Council of Europe has stated that "thc humane treatment of animals is one of the hallmarks of Western civilization." Different kinds of animals are also increasingly included in this concern – the moral circle seems to be expanding. In addition, human knowledge is advancing. Based on findings in neuroanatomy, neurophysiology, and ethology, an increasing number of scientists now conclude that pain perception is very likely to be present at least in bony fish (teleosts) (*e.g.* Sneddon *et al.*, 2003; Chandroo *et al.*., 2004; Sohlberg *et al.*, 2004; Huntingford *et al.*, 2006).

Fish are likely to feel pain

Fish has many neuroanatomical similarities to mammals in regards to nociceptors and nerve fibres. The most obvious difference is that the relative brain size of most fish species is smaller, and most important, the fish brain does not have any neocortex, responsible for human cognitive abilities and involved in the conscious perception of pain (Bermond, 1997). However, the neocortex is also absent in birds, that still show convincing signs of pain perception, and many bird species also have impressing cognitive abilities (*e.g.*, Pepperberg, 2005). McLean has shown that human emotions originate from the phylogenetically oldest part of the brain, the "reptilian brain" (McLean, 1990). Thus, the neocortex modulates and refines emotions, but does not initiate them. 24 out of 25 characteristics of the "reptilian brain" are also found in fish (Verheijen and Flight, 1997). Also the neurophysiology and -biochemistry are very similar in all vertebrates. Most neuropeptides, neurotransmittors, opioids, and their receptors, involved in nociception and pain modulation in mammals, are also found in fish. Behavioural studies show that fish display behaviours indicating pain in situations that are painful for mammals, while this behaviour does

not appear when painkillers are administered (*e.g.*, Sneddon, 2003). By experience fish also learn to avoid situations which have caused painful consequences (Beukema, 1970).
Some authors claim the capacity to feel pain depends on the animal's cognitive ability and is related to its position in the phylogenetic hierarchy. Thus, a higher order consciousness, an awareness of "self", is a prerequisite for pain perception, making it dubious that other than primates may experience pain (*e.g.*, Rose, 2002). In spite of intensive research, it has not been possible to localise the centre for consciousness to a specific region of the human brain. Brain structures shown to be involved in conscious thinking are not exclusively human. It is interesting that even some invertebrates show behaviours indicative of planning and conscious communication. It has been claimed that consciousness is the most effective way to deal with complex sensory input, especially if the brain size is small. Thus consciousness is likely to be present in species such as honey bees and spiders (Griffin and Speck, 2004). Broom (1998) argues that pain is a relatively simple feeling like thirst and hunger, which does not need extensive brain processing, in contrast to more complex feelings like guilt and jealousy.
We argue that the feeling of pain is essential for creatures, because it makes them motivated to avoid injury and thus increases fitness, *i.e.* chance of survival. Therefore, pain perception is conserved through evolution. Indeed, some physical reactions to nociceptive stimuli are reflexive, like the retraction of a hand from a burning hot item. That is, the muscular reaction is faster than the conscious awareness of the situation. However, this does not mean that there is no conscious experience of the pain. The conscious experience is very important for the learning process, teaching the animal to be more careful next time. Reflexive reactions are adaptive to limit damage in acute situations, but conscious memory of aversive events will help the animal to modify behaviour and thus minimize future risk. Long lived animals are believed to have special benefits of learning by experience. Some of the 28,500 species of fin fish may live even longer than human beings. Among the farmed species the halibut can be more than 50 years old.

Fish should be included in the moral concern

One may conclude that scientific findings indicate that fish are sentient beings with a complex behaviour. Even if proofs are not absolute, the consequences for not taking notice are so big that the precautionary principle should be applied.
From a utilitarian view, pain perception is a definite argument why fish also should be included in human moral obligations. Even for non-utilitarians arguments for this can easily be found. In the case of fish farming, it is lies near at hand to view farmed fish as part of the moral community, and thus count this fish among those whose interest should receive the same consideration as our similar interests, as discussed *e.g.* by Mary Midgley (1983). The discovery of fish sentience has uncovered some essential fish interests. The latter view emphasises the importance of taking a husbandry approach to fish farming, rather than one of hunting and harvesting.

What is welfare for fish?

Once it is clear that fish welfare should be considered, the question becomes: what *is* it? Again, here we must deal with the problems related to interspecies communication, relying on more or less good interpretations. There are commonly three different concepts of animal welfare, defining it in terms of biological functioning, the affective state of the animal, or emphasizing the importance of a natural life (Duncan and Fraser, 1997). The understanding of the welfare concept has more than an academic interest, since this obviously influences what welfare parameters to measure and eventually, the overall conclusion drawn.

The biological function view says that the animal has good welfare when its biological coping systems are not overloaded, and the animal is healthy, grows, and reproduces normally. Important parameters include health data, physiological parameters, behaviour and production data.
The natural life view is based on the fact that different animal species have evolved in interplay with certain surroundings, thereby developing the need for certain physiological and behavioural feedback. For example, even if a farmed salmon does not need to migrate to find nutrients or breed, it may have an internal drive to do so as part of "the salmon nature", or bright coloured tanks may induce stress in species adapted to dark surroundings.
The affective state view postulates that the important issue is how the animal feels about its situation. However, positive and negative emotions of other species are not easily accessed, especially not those very different from ours, and how measurable parameters correlate to internal emotions. While the biological functions are relatively straight forward to register, current knowledge on natural living and motivations of the cultivated fish species is very limited. Preferably, when assessing welfare, one should use a selection of welfare indicators covering the three approaches. However, the natural living approach may give the best clues of what a good fish life entails. It is therefore important to study the farmed species in natural or semi-natural environments to learn about their behavioural repertoire, preferences and time budgets, e.g. to know whether a halibut is inactive most of its time or swims around also when not searching for food. This will help interpreting whether the quiet halibut is showing contentedness or if it is apathetic. In this work it is useful to look to work developing terrestrial animal welfare, to avoid some pitfalls. Methodology used in terrestrial animal welfare research may be useful for fish, and some is already in use. This includes preference studies for temperature, light, salinity, oxygen content, and fish density. It is important to study and understand the basic behavioural mechanisms when setting welfare standards (Damsgård *et al.*, 2006). Interviewing experienced stockpersons may also reveal candidates for welfare indicators. Developing welfare concerns at fish slaughter, important comparative aspects may be found in experiences from terrestrial animal slaughter plants, based on the use of natural behaviour to minimize stress (see Grandin). In terrestrial animal production positive human animal contact is shown to improve animal welfare and the attitudes and beliefs of the stockperson are crucial. This should be taken into consideration also in aquaculture.

Conclusive remarks

Fish are traditionally not included in the moral community or given welfare concerns. There are however arguments for changing this. There is strong and increasing evidence that fin fish species are sentient. There is still lack of knowledge on emotions and emotional expressions in fish, which calls for the use of the precautionary principle. However, as science advances the interspecies communication becomes better, improving preconditions for including fish in human moral obligations, because they are sentient or since they are part of the moral community. A professional "mental conversion" is demanded from the fish farmer (and the fish farming industry), to change approach from hunting and harvesting to one of husbandry, where the welfare of the individual animal is in focus. Obviously, in order to make this a reality it must be supported by the consumers as well as by national and international legislation. However, in order to know what fish welfare really entails and how this can be measured, substantial research efforts are needed.

References

Bermond B. (1997).The myth of animal suffering. In: *Animal consciousness and animal ethics: perpectives from the Netherlands.* Dol, M., Kasamoentalib, S., Lijmbach S., Rivas, E., Vandenbos, R., eds. Assen: Van Gorcum. 125-143.

Beukema, J.J. (1970). Angling experiments with carp (*Cyprinus carpio* L.) II. Decreasing catchability through one-trial learning. *Neth J Zool.* 20: 81-92.

Broom, D.M. (1998). Welfare, stress, and the evolution of feelings. *Adv. Stud. Behav.* 27: 371-403.

Chandroo, K.P., Moccia, R.D. and Duncan, I.J.H. (2004). Can fish suffer? – Perspectives on sentience, pain, fear and stress. *Appl. Anim .Behav. Sci.* 86: 225-250

Damsgård, B., Juell, J. and Braastad B. (2006). Welfare in farmed fish. *Fiskeriforskning, Report* 5/2006, 100 pp.

Duncan, I.J.H. and Fraser, D. (1997). Understanding animal welfare. In: Appleby, M.C. and Hughes, B.O. (eds). *Animal welfare.* CAB International: Wallingford, 19-31.

Grandin T. *www.grandin.com*

Griffin, R.G. and Speck, G.B. (2004). New evidence of animal consciousness. *Animal cognition* 7: 5-18

Huntingford, F.A., Adams, C., Braithwaite, V.A., Kadri, S., Pottinger, T.G., Sandøe, P.and Turnbull, J.F. (2006) Current issues in fish welfare. Review paper. *J. Fish Biol.* 68: 332-372

Håstein, T., Scarfe, D. and Lund, V.L. (2006). Science-based assessment of welfare: aquatic animals. *Rev. Sci. Tech. Off. Int. Epiz.* 24: 529-547.

MacLean, P.D. (1990). *The triune brain in evolution, role in paleocerebral functions.* Plenum Press: New York.

Midgley, M. (1983). *Animals and why they matter.* University of Georgia Press: Athens.

Pepperberg, I.M. (2005) An Avian Perspective on Language Evolution: Implications of simultaneous development of vocal and physical object combinations by a Grey parrot (*Psittacus erithacus*). In: *Language origins: Perspectives on Evolution.* Tallerman, M. (ed.) Oxford University Press.

Rose, J.D. (2002). The neurobehavioral nature of fishes and the question of awareness and pain. *Rev. Fish Sci.*; 10:1-38.

Sneddon, L.U., Braithwaite, V.A. and Gentle M.J. (2003). Do fishes have nociceptors? Evidence for the evolution of a vertebrate sensory system. *Proc R Soc Lond B Biol Sci.* 270: 1115-21.

Sohlberg, S., Mejdell, C., Ranheim, B. and Søli, N.E. (2004). Oppfatter fisk smerte, frykt og ubehag? - en litteraturgjennomgang. *Norsk veterinærtidsskrift* 116: 429-438 (In Norwegian)

Sundli A. (1999). – Holmenkollen guidelines for sustainable aquaculture (adopted 1998). In *Sustainable aquaculture, Food for the future.* Svennevig, N., Reinertsen, H. and New, M. (eds). 343-347.

Verheijen, F.J. and Flight, W.F.G. (1997). Decapitation and brining: experimental tests show that after these commercial methods for slaughtering eel *Anguilla anguilla* (L.), death is not instantaneous. *Aquacult. Res.* 28: 361-6.

The implications of the use of GM in aquaculture: Issues for international development and trade

Kate Millar and Sandy Tomkins
Centre for Applied Bioethics, School of Biosciences, University of Nottingham, Sutton Bonington Campus, Loughborough, Leics, LE12 5RD, United Kingdom, kate.millar@nottingham.ac.uk

Keywords: aquaculture; GMOs; international development; biotechnology

Introduction

The major source of fish is still capture fishing. However, current practice is proving increasingly unsustainable. Global marine stocks are diminishing at alarming rates due to over exploitation, threatening marine biodiversity and creating economic hardship for traditional fishing communities across the globe. Recent statistics indicate that fish stocks for a number of major species are at the point of collapse (FAO, 2003; 2005).

In response to these pressures, technological innovation has stimulated increased production from aquaculture. Aquaculture production now accounts for over 30% of overall marine products (FAO, 2003). Commercial aquaculture as seen today has only developed over the last three decades, increasing from 5.3% to 32.2% of the total world fisheries landing between 1970 and 2000 (FAO 2003). Although significant advances have been made in aquaculture production, gains have not been as significant as those seen from the domestication of crops or livestock. However, enhanced aquaculture production has increased the supply of high quality fish protein for rural and urban populations in developing countries. Uptake of production technologies such as improved containment systems and selective breeding have played an important role in improving production, but the industry still faces other challenges in areas such as resource use, waste management, infrastructure investment, and product quality / traceability. For a number of developing nations increased aquaculture production has been characterised as a force for social development, providing local and global food security, alleviating poverty, improving rural livelihoods, creating employment and generating income in some of the poorest regions.

However, a number of environmental and social concerns have been raised regarding the rapid development of the aquaculture industry. These include environmental impacts from the increased use of pesticides and antibiotics, discharge of production wastes and pollutants, and reliance on high energy inputs, such as aquafeeds. Intensive aquaculture production systems are also being scrutinised for their animal welfare standards and the overall nutritional quality of the products.

As a result of biotechnology innovation in a number of key aquaculture producing countries, such as China, there is growing pressure to move aquaculture development into a new phase by applying genetic modification (GM) technologies. GM technologies have been demonstrated for Atlantic salmon, tilapia and other potentially important aquaculture species, bringing with them the potential for improving production efficiency, resource use and competitiveness. However, the use of these technologies in aquaculture raises a number of unique questions, including issues relating to environmental impacts, fish welfare, food safety and distributive justice.

Examining the implications of GM aquaculture development in the broader framework of international development and trade is a challenging task. A number of prominent issues and questions have been identified, these include:

i. What are the comparative issues raised by using different technologies in aquaculture development in terms of environmental impacts, food security and safety, production efficiency, distributive justice and wider international development implications.
ii. How might international competition driven by technology change in key aquaculture producing countries influence current regulatory frameworks and impact on international trade in fish products.
iii. What is the extent to which key aquaculture producing countries with scientific strengths (e.g. China, India, Cuba) may proceed with developing and applying GM technologies, and what are the potential ethical issues raised by such decisions.

In order to explore these issues and other key questions that arise, an ethical framework is being used to map potential impacts. A preliminary analysis is presented below.

Ethical analysis

In order to explore the key ethical dimensions, policy implications and areas for further discussion, the Ethical Matrix method was used to map the issues for the defined interest groups affected by the use of these technologies (Mepham *et al.*, 2006). The Ethical Matrix (EM) framework is based on a set of commonly acknowledged *prima facie* ethical principles that encapsulate traditional ethical theories in the form of a 'common morality'. Application of the Ethical Matrix facilitates the assessment of biotechnology use in terms of respect (or lack of respect) for the ethical principles - wellbeing, autonomy and fairness as applied to the defined interest groups. For the analysis of the use of GM in aquaculture systems four interest groups have been defined. The interest groups are defined as Treated Organisms (e.g. fish); Aquaculture Producers (including producers from Developed and Developing Countries and their related communities; commercial producers; technology providers) Consumers (e.g. consumers, affected citizens), and the Environment (e.g. biota, water quality).

The use of GM technology is assessed to determine whether it respects or infringes each principle (i.e. whether there is a positive or negative ethical impact). The weight assigned to particular principles in specific cases is determined by the evaluation and appeals to several forms of evidence. 'Evidence' is defined here as "anything that provides material or information on which a conclusion or proof is based". Such forms of evidence include, for example:

- Scientific and economic data.
- Assessments of the consequences of risk and uncertainty (e.g. reflected in the different ways people apply the Precautionary Principle).
- Assessments of the intrinsic value of different forms of life (which may reflect people's differing world views).

Qualitative or quantitative assessments of impacts recorded in the different cells of the Ethical Matrix (ethical analysis) provide a road map of salient ethical concerns, the different weightings of which underpin the various ethical judgements made. This form of analysis facilitates an assessment of the potential technology trajectories for the case that is being assessed.

Potential ethical impacts and areas of uncertainty

Studies have indicated that the use of GM fish in aquaculture will improve productivity by increasing feed conversation ratios and reducing the time needed for a single production cycle for on-shore facilities (e.g. Aerni, 2004; Goos *et al.*, 2001). At present no GM fish have been licensed for use in commercial production. A number of patents have been granted for GM fish, but only one commercial company, Aqua Bounty, has applied for a commercial production licence. This is currently under review with the US FDA, with a decision pending. GM fish technological developments are focused at present on enhancing production efficiency, however future innovations are likely to focus on production areas such as improving the nutritional quality of the product or increasing disease resistance in the modified fish. These new breeds of GM fish have been proposed as a technology option for increasing productivity and production output in a number of developing countries. As well as the commitment made by a number of aquaculture companies, countries such as China, India and Cuba have made `significant investments in GM technologies through national research programmes.

The global market has already shown resistance to the use of GM technologies in food production and initial analysis indicates that this resistance will be extended and possibly amplified if GM fish and shellfish were licensed for use in aquaculture production. The present cultural symbology of fish will modulate the acceptability of any GM products and this is likely to be a significant factor. Within the premier market sector a number of fish farming companies (e.g. the Canadian Aquaculture Industry Alliance; Interior Alaska Fish Processors Inc) have issued statements indicating that they will not use GM technology directly or source GM fish products (CFS, 2002).

From a market perspective the number of regulatory uncertainties and consumer perception issues surrounding the use of GM fish is likely to impact on the potential uptake of this technology within the fish industry in Europe and, possibly to a lesser extent, in North America. For those sectors of the industry in developing countries that are focused on export markets as well as increasing production for local markets, economic uncertainties and potential retailer aversion to GM products are likely to act as a barrier to GM use.

As well as the market dynamics and regulatory issues, there are also a number of issues raised by the use of GM technology in aquaculture that need further analysis and consideration. These ethical issues will feedback on the regulatory process, but they also need to be further considered as important issues for development of the industry, particularly in relation to the proposed use of these technologies in developing countries. The use of GM fish raises a number of welfare concerns that require further assessment. The potential increases in productivity from the use of GM fish with accelerated growth could make significant contributions to local access to high quality protein in a number of developing countries. However significant ethical concerns such as economic access to the technology, the IPR conditions affecting use and possible market restrictions - all raise key issues of distributive justice for producers in developing countries. Recent reports analysing the role of innovation and IPR in public health provision for developing countries highlights the need to consider these issues comprehensively (WHO, 2006).

Further development of GM aquaculture, focusing on disease resistance and reduced external inputs, might be more acceptable to producers, regulators and consumers than current growth enhancement models. However stakeholders' perspectives of ethical acceptability will not only be modulated by distribution of risks and benefits, but also by the impacts on stakeholder autonomy and notions of fairness. There is need to stimulate an informed dialogue between stakeholders in the industry to further explore the perceived opportunities for the aquaculture industry and the ethical issues raised by various options. As part of this process, strenuous efforts should be made to ensure the participation of producers and their representatives from developing nations.

This represents an initial characterisation of issues, further work is needed particularly on aspects of distributive justice and the use of the precautionary principle in the assessment of environmental risk and food safety for this case.

References

Aerni P (2004) Risk, regulation and innovation: the case of aquaculture and transgenic fish Aquatic Science 66, 327–341

CFS (2002a) List of 25 fish distributors who took the "GE fish pledge" not to buy or sell GM fish http://www.centerforfoodsafety.org/pubs/Distributor%20List.pdf accessed 20 Mar 2006

FAO (2005) Ethical Issues in Fisheries. FAO Ethics Series 4. FAO, Rome, 39pp

FAO (2003) Review of the State of World Aquaculture. FAO Fisheries Circular. No. 886, Rev.2. FAO Inland Water Resources and Aquaculture Service, FAO, Rome, 95pp

Goos H, Rastogi R, Vaudry H, Pierantoni R (2001) Are transgenic fish for aquaculture doomed to extinction? Proceedings of the International Conference on Perspectives in Comparative Endocrinology: Unity and Diversity. 26-30 May 2001, Sorrento, Italy. ISBN: 88-323-1526-2

Mepham B, Kaiser M, Thorstensen E, Tomkins S and Millar K (2006) Ethical Matrix Manual. Agricultural Economics Research Institute (LEI), The Netherlands

WHO (2006) Public Heath, Innovation and Intellectual Property Rights. WHO Press 228pp

Economic values of fish welfare and application of market experiments

Ingrid Olesen[1], *Frode Alfnes*[2], *Mia Bencze Rørå*[1], *Ståle Navrud*[2], *Kari Kolstad*[1]
[1]*AKVAFORSK, Institute of Aquaculture Research, PO. Box 5010, N-1432 Ås, Norway, ingrid.olesen@akvaforsk.no*
[2]*Department of Economics and Resource Management, Norwegian University of Life Sciences, N-1432 Ås, Norway*

Abstract

A challenge in animal breeding and production is to value such intangibles as animal welfare in monetary terms, because market prices reveal the value of private goods, which are excludable and rival in consumption. Animal welfare, however, is a public good, that can be both non-excludable and non-rival in consumption, and that is usually not traded and valued in the market. Hence, market prices do not fully reflect the economic value of fish welfare. However, if fish farmed under a better animal welfare regime is labelled (e.g. as organic or Freedom food), the higher market price consumers are willing to pay (WTP) for this fish characteristic will capture parts of this value. In order to estimate a lower bound for the consumers' willingness to pay for improved welfare for farmed salmon, an experimental market study with eco-labelled salmon was carried out. 115 consumers made 30 choices between pairs of packages of either conventional, organic or Freedom food salmon fillets with different colours and prices. To impose real economic incentives, one out of 30 choices was randomly drawn as binding, and the consumer had to buy the package chosen. The average respondent preferred eco-labelled salmon to conventional salmon when the colour was the same, and was willing to pay additional 2 euros per kg fillet for eco-labelled salmon. For organic salmon with paler colour, due to the less artificially produced pigment in the feed, people were willing to pay a significantly lower price.

Keywords: salmon, animal welfare, organic farming, willingness to pay, breeding goal

Introduction

Animal breeding for sustainable production systems requires more focus on animal welfare and environmental concerns. A challenge in the definition of breeding goal is to value such intangibles as animal welfare in monetary terms. Market prices only reveal the value for private goods that are characterized by being excludable and rival in consumption. Rivalry in consumption implies that as you use or consume the good, nobody else can use it at the same time (e.g. bread and meat). Excludability in consumption means that purchasing the good implies access to consuming it, and no payment implies no access (e.g. cable TV). Animal welfare, however, is a public good, which is not, or only partly, traded and valued in the market. This characteristic of animal products can be both non-excludable and non-rival good. People not eating farmed fish may also benefit from improved fish welfare. Hence, market prices do not fully reflect the economic value people place on fish welfare. However, if the fish product can be labelled and marketed as produced with better welfare for the fish, parts of this value can be captured by the higher market price people that eat farmed fish is willing to pay extra. Organic salmon and Freedom food salmon are examples of such eco-labelled fish products.

The objective of this paper is to consider valuation of animal traits in breeding goals and to present results from a survey of consumers' attitudes to farmed salmon as well as a market experiment for valuing fish welfare.

Economic values of goods and services from animal traits

The traits' values in the aggregate genotype or breeding goal may be split into non-market values and market values in the money economy (Olesen *et al.*, 2000). The value of improvements, that reduce the costs per unit of product (e.g. feed costs due to improved feed efficiency), can be derived from market prices of feed and products. For some private good characteristics, e.g. colour of salmon fillet, it is not possible to reveal the market values directly due to e.g. biological and technological reasons. However, there may become a differentiation in future market prices that should be accounted for in breeding programs. For estimating the relative economic values of e.g. fillet colour, we can apply market experiments as demonstrated by Alfnes *et al.* (2006). The market values of the characteristics of eco-labelled products including e.g. animal welfare improvements, given the external conditions defined by the available technology and the present sociological and political situation, can also be obtained by such market studies and experiments. In market experiments we can elicit consumers' willingness to pay (WTP) for improvements in animal welfare. From consumers' WTP we can get an indication of a potential consumers' surplus. Parts of this consumers' surplus that we can observe for the improvement can be converted to producer surplus by eco-labelling and an excess price of the product.
For capturing more of the society's total valuation of fish welfare and the non-market value, other studies of people's attitudes and willingness to support economically alternative production systems and breeding programs are needed. A shift in political situation can also make tax financed or subsidized breeding programs emphasizing animal welfare profitable. Hence, a basis for an increased non-market economic value of animal welfare in the breeding goal is possible without a change in fish market prices.

Valuation methodology

Methods for deriving non-market values of traits in animal breeding goals may be based on either individual preferences (stated and revealed preference methods), opportunity costs principle (restoration costs), decision makers and politicians revealed preferences (implicit pricing) or experts' preferences, in addition to a method based on desired gains or restricted selection indices (Olesen *et al.*, 1999). Contingent valuation (CV) has been a popular method based on individual preferences (Mitchell and Carson, 1989). It can measure all parts of the total economic value including future and/or hypothetical changes. A problem with stated choice experiments and CV studies of hypothetical and non-observed behaviour is that people may not really pay what they have said. In order to apply real economic incentives, simulated market experiments have been implemented. A new type of experimental markets using posted prices has been used for eliciting WTP for private as well as public goods (Carlsson and Martinsson, 2001). In these real choice (RC) experiments, participants are asked to make choices between products with various quality traits in a series of shopping scenarios. After the participants have completed all shopping scenarios, one shopping scenario is randomly chosen as binding and the product chosen in that scenario is sold for the posted price. Hence, the incentives to reveal true preferences are relatively transparent, and the choice tasks are relatively similar to consumers' choices in grocery stores. This makes the RC experiment an incentive compatible method for investigating WTP for the attributes of both private goods as well as nonexclusive and non-rival goods.

Material and method

We conducted an RC experiment during four evenings in February 2004. In total, 115 participants were recruited through various local organizations in South-eastern Norway. Between 13 and 16 persons participated in each session. First, the respondents filled in a questionnaire with a total of 34 questions. In each session, three-times ten choice scenarios were presented. The organizations were given NOK 200 for each participant they recruited, and the participants were given NOK 300 to take part in the experiment.

Market experiment

A total of 115 participants were asked to make 30 choices between pairs of packages of salmon fillets of different colours, prices and labels. In this study, 20 of these choices were utilized: 10 choices between conventional salmon fillets with different colours and prices, and 10 choices between either conventional, organic or Freedom Food salmon fillets with different colours and prices (see Table 1 for descriptions and information to the respondents). The organic salmon was fed Scottish organic feed with only astaxanthin in natural feed sources, and were therefore paler (ranging from 20 to 22 on the Roche scale) than any of the regular (23-30) and Freedom Food salmon (22-28). The Freedom Food salmon on the other hand were slightly redder than the average regular salmons.
The fillets were grouped into five colour categories; hereafter referred to as alternatives R21, R23, R25, R27, and R29. The price attribute took the levels NOK 24, 30, 36, 42, and 48[85].
One of the 30 choices was randomly drawn as binding, and the respondent had to buy the package chosen.

Table 1. Information given to the participants about production form.

Regular salmon:	This is regular salmon farmed according to Norwegian laws and legislations. Similar salmon were last week sold for NOK 89 pr kilo at Maxi in Ski and for NOK 119 pr kilo at Mega in Ås. This corresponds to NOK 35.60 and NOK 47.60 per 400 grams, respectively.
Organic farmed salmon:	This is salmon that in addition to fulfilling the criteria for regular salmon production also fulfils the production criteria given in the rules and regulations of Debio, the Norwegian agency for control and certification of organic production. This implies among other things stricter environmental criteria for the treatment of sea cages, that the color added in the feed is from natural sources, and that the fish is fed feed that complies with the rules and regulations of Debio. As regular salmon, the organic salmon is vaccinated and medicated to avoid and treat diseases.
Freedom-food salmon:	The production of this salmon comply with the rules and legislations of regular farmed salmon, and in addition comply with the production criteria necessary to achieve the international Freedom Food label. This is an international label with stricter animal welfare production criteria than the standard criteria in the Norwegian regulations. The fish in each cage has e.g. more space available. The Freedom Food salmon follows conventional rules and regulations for feed and medication.

[85] In four of the eight sessions, the price attribute of the organic and eco-labelled salmon took the levels NOK 36, 42, 48, 54, and 60.

The RC data was analysed with the following random parameter logit model (also known as a mixed logit model):

$$U_{nis} = (\beta_0 Color_{nis} + \beta_1 Organic_{nis} + \beta_2 Freedom_{nis}) * Weight_{nis} + \beta_3 Price_{nis} + \beta_4 Price2_{nis} + \beta_5 Dummy24_{nis} + \sigma_s^{-1} * \varepsilon_{nis} \quad (1)$$

where U_{nis} is the utility of individual i from alternative i in choice s, $Colour_{nis}$ is a vector of colour-specific dummies; $Organic_{nis}$ is a dummy taking the value one if the salmon is an organic salmon, and zero otherwise; $Freedom_{nis}$ is a dummy taking the value one if the salmon is a Freedom salmon, and zero otherwise; $Weight_{nis}$ is the exact weight of the fillet; $Price_{nis}$ is the price of the alternative; $Price24_{nis}$ is a dummy taking the value one if the price is NOK 24, and zero otherwise; σ_j is the scaling parameter taking the value one if the choice was from scenario 11-20

The mean WTP per kilogram of alternative *i* can be calculated by dividing the utility of one kilogram of alternative *i* by the price sensitivity parameter, and multiplying the result with minus 100:

$$WTPi = -100 * \left[\frac{\beta_0 Color_i + \beta_1 Organic_i + \beta_2 Freedom_1}{\beta_3}\right] \quad (2)$$

where WTP_i is the estimated mean WTP per kilogram of alternative *i*; and all other variables and parameters are as described in equation 1.

Results and discussion

Table 2 shows the distribution of answers for the questions in the questionnaire that we consider most relevant for this study. The questionnaire responses show that a relatively large percentage agreed or partly agreed that the Norwegian animal welfare standards are sufficient strict (78.1%)

Table 2. Distribution of responses (%) to statements in questionnaire.

Statement \ Response %	Don't know	Totally disagree	Partly disagree	Neither nor	Partly agree	Totally agree
It is important for me that salmon's feed do not contain artificial pigments	1.74	0.87	5.22	9.57	28.70	53.91
I think the Norwegian standards for animal welfare are sufficiently strict	4.39	1.75	9.65	6.14	45.61	32.46
I think the fish welfare is sufficiently ensured in Norwegian fish farming	2.63	1.75	10.53	20.18	41.23	26.68
Organic food tastes much better than regular food	11.30	13.91	10.43	39.13	17.39	7.83
Organic food is much healthier than regular food	3.48	4.35	13.91	21.74	39.13	17.39
Organic food is much safer than regular food	3.48	6.09	12.17	17.39	40.87	20.00
Organic production is much better for the nature than regular food	7.89	0.75	1.75	14.04	30.70	43.86
Organic production ensures better the fish welfare than regular food	14.78	0.0	6.09	26.09	30.43	22.61

and that the fish welfare is sufficiently ensured in Norwegian fish farming (67.9%). However, a relatively large percentage also agreed that organic food is much healthier (56.5%), safer (60.9), better for the nature (74,6%) and ensures better the fish welfare (53%) than regular food.
Table 3 contains the mean WTP with standard errors and *p* values. On average the consumers preferred the redder alternatives to the palest alternatives R21. Furthermore, the marginal utility of increased redness was positive but decreasing up to R27. A more thorough discussion of these results on colour is given by Alfnes *et al.* (2006).

The average consumer preferred organic and Freedom Food salmon to salmon from conventional salmon farms when the colour was the same. The mean WTP for organic salmon with the colour R21 was NOK 123.2 and significantly less than the mean WTP for the Freedom Food salmon with the colour R27 (NOK 170.7) and regular salmon most commonly seen in the Norwegian market.
The result for organic salmon also corresponds with the price premiums found for organic salmon in Norway (35%) and Belgium (43%) (Reithe and Tveterås, 1998). Information from Norwegian exporter of organic salmon in Sweden and from Scottish producer of Freedom Food salmon in Scotland about recent price premiums obtained (17-30%) also correspond well with these results.
The information given about Freedom Food salmon (Table 1) emphasized the stricter fish welfare criteria for this label, so the higher WTP for this product is expected to express WTP for improved fish welfare. The information given for organic farmed salmon referred to the rules and regulations of Debio (2005). Fish welfare was not mentioned specifically. However, the results from the questionnaire indicate that many were aware of the emphasized fish welfare in organic farming.
Furthermore, the study shows that improving colour in organic salmon will provide large benefits. However, the new Norwegian Debio standards permit other additives such as yeast (Debio, 2005), and now makes it possible to produce organic feed with better pigmentation ability at an acceptable price.

Table 3. Mean Willingness to Pay (WTP), standard errors (Std.err) and probability value (p value).

Variable	Mean WTP	Std. err.	*p* value
Organic	16.25	7.89	0.04
Freedom food	14.79	7.74	0.06
R21	106.99	10.07	0.00
R23	145.44	9.04	0.00
R25	151.21	8.26	0.00
R27	155.89	9.72	0.00
R29	145.22	10.83	0.00
Summary statistics			
Number of choice observations			1676
Number of participants			84

Conclusion and implication

The average respondent preferred salmon from the eco-labelled salmon to salmon from conventional farms when the colour was the same. A higher WTP of 15-16 NOK (ca 2 Euro) per kilo filet for eco-labelled salmon was estimated from the data. However, the combined effect of organic and the pale colour due to the less intensive pigmentation feeding was significantly less than the utility of the regular salmon most commonly seen in the Norwegian market.
The WTP achieved for the product attributes can be utilized to value fish welfare traits in the breeding goal. Parts of the consumers' surplus that we can observe from WTP for improved fish welfare can be converted to producer surplus by eco-labelling and an excess price of the product. This market value of improving welfare may for example be correlated to the market value of improving fish health. This information may be included in estimation of income required for the profit equation of improved health.
For capturing more of the society's total valuation of fish welfare and the non-market value, other studies of people's attitudes and willingness to support economically alternative production systems and breeding programs are also needed.

Acknowledgements

This research was funded by the Research Council of Norway (project no. 143045/140).

References

Alfnes, F., Guttormsen, A., Steine, G and Kolstad, K. (2006). Consumers' willingness to pay for the color of salmon: A choice experiment with real economic incentives, Amer. J. Agr. Econ., forthcoming 89.

Carlsson, F. and Martinsson, P. (2001). Do hypothetical and actual marginal willingness to pay differ in choice experiments? J. Environ. Econ. Manag. 41: 179-192.

Debio (2005). Standards for Organic aquaculture. Debio, N-1940 Bjørkelangen, Norway. 19pp. Available on http://www.debio.no/_upl/okologisk_akvakultur_2005.pdf

Mitchell, R. and Carson, R. (1989). Using surveys to Value Public Goods. The Contigent Valuation Method. Resources for the future, Washington DC. 463 p.

Olesen, I., Gjerde, B. and Groen, A. (1999). Methodology for deriving non-market trait values in animal breeding goals for sustainable production systems. Interbull Bulletin No. 23(1999): 13-21.

Olesen, I., Groen, A and Gjerde, B. (2000). Definition of animal breeding goals for sustainable production systems. J. Anim. Sci. 78:570-582.

Reithe, S. and Tveterås, R. (1998) Demand for Organic Salmon in the European Union. SNF Report No. 50/1998, Stiftelsen for samfunns- og næringslivsforskning, Breiviksveien 40, N-5045 Bergen, Norway. 51 p.

A story of two fishes

Guro Ådnegard Skarstad
Norwegian Agricultural Economics Research Institute, Postboks 8024 Dep, N-0030 Oslo, Norway, guro.skarstad@nilf.no

Abstract

Environmental contaminants in farmed salmon have the last two years created a big stir in Norway and around the world. Particularly an article published in the prestigious journal Science in 2004, although highly criticised, contributed to putting environmental contaminants in fish on the agenda in a more powerful and general way than earlier. This paper takes the 2004-controversy as its starting point, and investigates its claims and responses with theoretical resources from the field of Science, Technology and Society which aims to understand how objects, such as fish, are produced through practices. The controversy will be interpreted as a clash between two fishes, "a healthy fish" and a "dangerous contaminated fish". The paper will show how the clash challenged existing boundaries between scientific disciplines, administrative fields, as well as established relations between fish and human health. The governmental handling of the situation that arose was a cry for more science. The ethical premises for governmental handling of the issue will be discussed.

Keywords: farmed salmon, environmental contaminants, nutrients, scientific productions

Introduction

In 2004 an article published in the journal Science claimed that farmed salmon is not as healthy as considered previously due to high level of environmental contaminants which again may cause cancer. The authors of the article had made a global assessment of organic contaminants in salmon collected from all over the world, and concluded that farmed Atlantic salmon was the most contaminated. The authors advised a maximum of approximately one meal per month of farmed salmon, and consequently branded farmed salmon as dangerous food. The article stirred up a big debate in Norway which is the world's largest exporter of farmed Atlantic salmon (NSEC 2006). Much was at stake for Norwegian fish farming.

This paper tells the story of how the farmed Atlantic salmon, as it appeared in the article in Science, travelled, and thereby challenged the established notion of fish as being healthy food by linking farmed salmon to environmental contaminants and health risk. The story will be developed with the use of theoretical resources from the field known as Science, Technology and Society (STS). A central tenet to what has been referred to as the "turn to ontics" [onto = to exist] in the field of STS is that objects are produced relationally through practices (Law 2004). Science is not only in the business of knowing the world, but contributes to its creation. The tradition of laboratory studies has long suggested that the laboratory is the place par excellence for object production (cf. Latour 1983). As will be argued, the scientific process of collecting, analysing and interpreting data in the making of the article in Science contributed to producing a general "dangerous contaminated fish". However, science is not only confined within the walls of the laboratory. First, scientific facts travel (Latour 1983). Second, a recent turn in the field of STS suggests that other places and non-scientific practices, such as administrative practices, can also be studied as sites for object production (cf. e.g. Barry 2001). The article's trip to Norway became highly turbulent, to extend the metaphor. Its conclusions became heavily criticized as it was in conflict with established administrative and scientific practices related to what a fish is. The article contributed to making visible an established two-fold situation reflected in the difference between the nutritional field producing a "healthy fish" on the one hand, and a toxicological – food control field, producing a "dangerous contaminated fish" on the other. The governmental response to the article was to ask for more science as a foundation for more correct dietary advice, taking into consideration both contaminants and nutrients. At the end of the paper, I will briefly discuss how the issue of environmental contaminants, if being treated by the government as solely a question of dietary advice, may imply an ethic of utilitarianism which may bar out other relevant issues. To sum up, the paper will illustrate how food objects might be produced, how they may change, and how the production may exclude other issues/objects.

The production of a "dangerous contaminated fish"

The journal article in question was published on the 9th of January 2004 in Science and had the title "Global Assessment of Organic Contaminants in Farmed Salmon". The article was written by Professor Ronald Hites at Indiana University and co-authors from other American universities. However, the article was not only an analysis of farmed Atlantic salmon, but also of wild salmon from the Pacific. More specifically, 700 salmons, both wild and farmed, as fillets and as whole salmon, had been collected from around the world, and brought to the laboratory of one of the co-authors of the article, a private Canadian laboratory in Sidney, which specializes in the analysis of dioxins and congener specific PCBs (Axys Analytical Services Ltd., 2006). The 700 salmon, some of them collected from Norway, had been analysed in the laboratory for 14 organochlorine contaminants, including dioxins and PCBs. As reported in the article, this analysis proved farmed Atlantic salmon to have a higher level of contaminants than wild Pacific salmon. Also, farmed salmon raised in European waters had higher levels than farmed salmon from American ones. The article concluded that people in general should restrict their consumption of farmed Atlantic salmon to approx. one time per month, unless they wanted to increase the risk of developing cancer.
This conclusion created headlines in news media all over the world on the 9th of January, well supported by a media campaign led by an international PR organization, promoting the article and its conclusions (cf. e.g. Times Online 2004). My initial meeting with the article in Science was through the eyes of Norwegian actors who portrayed the article as a "scientific curiosity" because of its dissident method and as an example of interest-driven knowledge production. This was by and large the interpretation of the article in Norway. The article was also criticized by other European governments, the US government, and later by FAO (2004:46-47). The criticism of the article may be said to have been two-fold. First, the article was criticised for not taking into considerations the many health benefits of eating fatty fish. Especially the beneficial effect of marine omega-3 fatty acids to cardiovascular health was pointed to (e.g. Tuomisto 2004). Second, the article was criticised for founding its conclusion on a model of risk that was not internationally accepted for the types of contaminants analysed. The article had lowered the threshold for what could be considered a risk, by employing a risk model for genotoxic substances, meaning that any dose was considered to indicate a certain degree of risk. The argument made by the criticising voices was that the level of contaminants found in farmed salmon would not lead to an intake that would exceed the internationally tolerable intake values established for these contaminants.

A clash between "two fishes"

From a natural scientific viewpoint the claims of article were considered problematic. From a social scientific perspective, and more specifically, from a theoretical perspective interested in how scientific and other types of practices contribute to object production, the article and the subsequent discussion is interesting. Studies of controversies are a well-known method as they stir up hidden assumptions or relations that go into the making of the established world. They make it easier to "trace the associations" (Latour 2005) or relations that are assembled in the making of an object, in this case a food object.

The salmon which were collected in Norway, brought to a laboratory in Sidney, analysed for environmental contaminants, interpreted according to a risk model, and published in Science, were returned to Norway through headlines in the newspapers as a totally "different fish" than the fish originally collected. The Norwegian health government recommends an increased consumption of fish. That fish is healthy food is a well-established fact (cf. Døving 1997). Farmed salmon, as other fatty fish, is considered as healthy food. The salmon, as it appeared in the article in Science on the other hand, was full of contaminants; and it could potentially cause cancer for all people who consumed more than one meal per month. By linking farmed fish to environmental contaminants and health risk, the article challenged the established relations between fish, nutrients and good health, or a "healthy fish".

Environmental contaminants in fish are not a new issue. The first dietary advice given due to environmental contaminants was given in Norway in the 1960/70s (Økland 2005). The first advices were typically advice given for fish in specific fiords, often close to industrial plants, and meant for the local population. Environmental contaminants in fish has traditionally been handled and treated as mainly a local problem (Skarstad, in progress), although in recent years more people (e.g. pregnant women) as well as more fish (e.g. large freshwater fish) have been covered by consumption advice. In this respect, the article in Science can be said to have made visible a more general development, despite its controversial nature. The article in Science and the subsequent discussion contributed to making environmental contaminants in salmon, and fish, into a general problem. The article can be said to have produced a general "dangerous contaminated fish" by being a *global* assessment of salmon; by lowering what is considered a risk; by giving consumption advice to all kind of people (not only vulnerable groups); and by analysing the salmon for contaminants exclusively.

By highlighting environmental contaminants in fish as a general problem, the article made visible a knowledge-political challenge, which was the starting point for the Norwegian Food Safety Authority's handling of the situation that arose in 2004. Which advice should be given when "the same fish" can be a source of both nutrients and contaminants? In June 2004 the Norwegian Food Safety Authority commissioned the experts in the Norwegian Scientific Committee for Food Safety to investigate the overall health effects of eating fish with regard to both contaminants and nutrients. The Norwegian Ministry of Fisheries invited the same month other fish farming countries to establish a co-operation in order to convey "independent documentation and information to consumers about fish and seafood as safe and healthy food" (Communiqué 2004). The governmental response of treating the issue of contamination and nutrients together is interesting in relation to how fish-as-food traditionally has been handled scientifically and administratively. The article in Science was based in toxicological analysis. Toxicological knowledge has been the foundation for the dietary advice and other actions actuated by the Norwegian food control authorities throughout the years to handle the issue of contamination. However, the general advice has been to eat more fish. This has been given by the health authorities and been based in nutritional scientific knowledge. Hence, fish-as-food has in the post war period by and large been handled in two different scientific and administrative fields of practices. This is reflected in the difference between the nutritional field producing a "healthy fish" on the one hand, and a toxicological – food control field, producing a "dangerous contaminated fish" on the other (Skarstad, in progress). The conflict following the article in Science may be described as an encounter between these "two fishes", an encounter which made them visible, and which challenged their existence as "separate fishes".

Environmental contamination in fish: An ethical issue?

The governmental response to the situation that arose in 2004 was a cry for more science. The article in Science had made it more evident that in order to give a "true" dietary advice to the consumers, the "whole fish" had to be taken into account, in order to assess the overall health effects of eating fish. The field of food politics, both the nutritional field as well as the field of food control, has traditionally been heavily based on scientific knowledge. The governmental responses, as well as the subsequent discussion which to a large extent centred on the correctness of the claims of the article, was no exception to this. When I refer to the knowledge-political challenge that the issue of environmental contamination pose, caused by the bisection of "food science" in a nutritional and food toxicological field, I may also seem to approach the issue of contamination as mainly a question of knowledge.

However, I have also tried to show how the science of the Science-article, not only "discovered" its conclusions, but produced them. Science is not only an epistemological, but also an ontological activity. This implies that science is not free of norms or ethics. However, the productions made through decision-making processes, whether based in scientific practices or not, also always imply exclusions of other possible productions. That is what decision-making is all about (cf. Skarstad 2004, ch.7).

This opens up the question of what kind of ethics the policy process, implied by the governmental response to the article in Science, may produce/exclude. The policy process is still in progress. Hence, the points I will put forward will not be based in a thorough analysis, but be of a more general and open character. My first annotation draws upon Lang and Heasman's (2004) analysis of what they refer to as the productionist paradigm. This is a paradigm in which the food's contribution to health comes from the "right" ingestion of the "correct" balance of ingredients. According to Lang and Heasman (2004:36), there are no good or bad foods in this paradigm, only good or bad diets. If dietary advice or questions of food consumption becomes the only means of governmental response to the issue of environmental contamination, another field of food politics, namely the production of food, may become ignored. This point is highly relevant as the issue of food consumption/health vs. food production/economy is a central conflict line within the field of food politics in Norway (Asdal 2005). Second, a governmental handling of the issue of environmental contamination which is solely based in the logic of a utilitarian "cost-benefit" analysis in relation to human health effects, may bar out issues that concerns responsibilities, duties and rights, as well as "actors" that are not part of the estimation, such as fish, oceans and human bodies made external. Examining the issue of environmental contamination with ethically-oriented approaches is important to highlight these exclusions.

References

Asdal, K. (2005): *Grensetrafikk. Nedslag i matpolitikkens og veterinærvesenets historie.* Unipub, Oslo. 105 p.

Axys Analytical Services Ltd. (2006). *History* [on line]. Available on http://www.axysanalytical.com. [Date of consultation 04/04/06]

Barry, A. (2001): *Political machines.* The Athlone Press, London-New York. 305 p.

Communiqué (2004): *The round table conference on consumer information concerning seafood as safe and healthy food* [on line]. Available on http://odin.dep.no/filarkiv/215023/Kommunike_Rundebord_23_6_04_(bearbeidet).pdf. [Date of consulation 04/04/06].

Døving, Runar (1997): *Fisk: en studie av holdninger, vurderinger og forbruk av fisk i Norge.* SIFO, Lysaker

FAO (2004): *The State of the World Fisheries and Aquaculture.* [on line] Available on ftp://ftp.fao.org/docrep/fao/007/y5600e/y5600e01.pdf [Date of consultation 05/04/06]

Hites, R. *et al.* (2004): Global Assessment of Organic Contaminants in Farmed Salmon. *Science*, Vol. 303, No. 5655: 226-229.

Lang, T. and Heasman M. (2004): *Food wars: the global battle for mouths, minds and markets.* London, Earthscan. 365 p.

Latour, B. (1983): 'Give me a laboratory and I will raise the world!' In: *Science Observed,* K. Knorr-Cetina and M. Mulkay (eds.): Sage, Beverly Hills: 141-170

Latour, B. (2005): *Reassembling the Social – an Introduction to Actor-Network-Theory.* Oxford University Press, Oxford. 301 p.

Law, J. (2004): *Enacting Naturecultures: a Note from STS.* Lancaster, Centre for Science Studies, Lancaster University. Available on http://www.lancs.ac.uk/fss/sociology/papers/law-enacting-naturecultures.pdf [Date of consultation 11/04/06]

NSEC (2006). *World Trade Atlantic Salmon 2004/2005.* Powerpoint presentation [e-mail communication].

Økland, T.E. (2005): *Kostholdsråd i norske havner og fjorder.* Mattilsynet, Vitenskapskomiteen for mattrygghet, Statens forurensningstilsyn, Oslo.

Skarstad, G.Å. (2004): *Naturens politikk. Ein diskursanalyse av den politiske avgjerdsprosessen i Øvre Otta-saka.* University of Oslo, Department of Political Science, Oslo.

Skarstad, G.Å. (in progress): *Fisk som farleg og sunn mat. Vitskaplege og administrative møte mellom to fiskar.*

Times Online (2004): *Thunderer: Answer this: who benefits from the salmon scare?*[on line]. Available on: http://www.timesonline.co.uk/article/0,,8122-964617,00.html [Date of consultation 11/04/06]

Tuomisto, J.T. *et al.* (2004): Risk-Benefit Analysis of Eating Farmed Salmon. *Science,* Vol 305, No. 5683: 476-477.

The ethical slaughter of farmed fish

Ambrose Tinarwo
Department of Clinical Veterinary Science, University of Bristol, Langford Bristol, BS40 7EU, United Kingdom, Ambrose.Tinarwo@bristol.ac.uk

Abstract

There has been increased interest by consumers and legislators on ethical aspects of food production systems including the welfare of animals kept for food. Fish farming is a new and fast growing industry that presents unique welfare challenges to farmers. Recent scientific debate on the capacity of fish to feel pain has raised interest in welfare aspects of fish farming. Some traditional methods of harvest, in particular the killing process, have been demonstrated to cause avoidable suffering. As consumers become more aware of fish farming procedures and the scientific knowledge on the welfare implications of fish harvest increases, alternative methods have to be developed. This paper discusses UK legislation concerning the welfare of farmed fish at slaughter. An exploration of current knowledge on the welfare implications of traditional fish killing methods leads to the conclusion that some current methods are ethically and sometimes legally unacceptable. Original research from the scientific development of stunning by electronarcosis for fish stunning is discussed as an example of alternative commercial and ethically acceptable means of killing farmed fish. The example illustrates how scientific inquiry can attempt to satisfy the animal welfare imperative and balance some critical considerations such as profitability, food safety, product quality, ease of operation and operator safety.

Keywords: electronarcosis, fish, legislation, slaughter, welfare

Introduction

Fish respond to stressors with rapid loss of body condition. Consequently, there is much information on potential stressors such as disease and stocking densities which affect productivity (Huntingford *et al.*, 2006). This concern for fish welfare has, however, not been extended to the low profile and potentially painful process of killing at harvest until recently. There is need for alternative means of killing fish that do not cause avoidable stress and pain.
Aquaculture produce in the United Kingdom (UK) grew by more that 400% between 1990 and 2004 (FAO Fisheries Department, 2000). With this growth, the debate on the welfare of farmed fish at killing is increasingly coming to the forefront of society's concern. Questions on the ethical killing of farmed fish include:

1. Are fish capable of suffering during the killing process?
2. How much concern should we have for any such suffering?
3. Can we do anything to limit any such suffering?

UK legislation on the protection of fish

The UK has a long history of protecting animals and boasts the world's first parliamentary legislation for animal welfare which was enacted in 1822. Despite this history, the welfare of farmed fish has little legal protection which is provided by the following general laws.

The protection of animals act (1911)

This Act makes it an offence to cause, by commission or omission, unnecessary suffering to any domestic or captive animal. It applies to fish farmed on land and at sea in cages. As knowledge of stress and pain in fish improves, farming practices are expected to improve inline with modern technology and scientific knowledge to prevent unnecessary suffering. This is a potentially useful legislation although no prosecutions concerning fish farming have been undertaken under it. This Act is due to be replaced by the Animal Welfare Bill which is more proactive and relates to potential rather than unnecessary suffering.

The agricultural (miscellaneous provisions) act 1968 (AMPA '68)

This legislation makes it an offence to cause or to allow unnecessary pain or distress to livestock held on agricultural land. It applies to fish held in tanks on land for commercial exploitation but is not applicable to fish farmed in sea cages or fish on land but kept for private consumption.

The welfare of farmed animals regulations 2000

The regulations require owners and keepers of animals kept for farming purposes to ensure that animals under their care are not caused any unnecessary pain, suffering or injury. The regulations are made under the AMPA '68 and are therefore limited to animals on agricultural land at the exclusion of fish kept in sea cages. Improvement notices may be served if an inspector thinks that animals may suffer unnecessarily if certain improvements are not made.

The welfare of animals (slaughter or killing) regulations 1995

These regulations make it an offence to cause or permit avoidable excitement, pain or suffering to any animal during stunning, slaughter or killing. It introduces a requirement for anyone killing any animal to have received sufficient training to do so humanely. Although a system exists for training and licensing slaughter men who kill terrestrial animals in abattoirs, there is no equivalent system for the killing of fish. It is difficult to qualify the pain or suffering caused during slaughter as avoidable therefore these regulations are rarely enforced.

Animal welfare law is based on the assumption that animals can suffer and therefore deserve protection. It is also a reflection of the level of empathy felt for a particular animal. Recent debate on the ability of fish to feel pain has resulted in calls for a review of legislation with regards to fish welfare.

Reasons behind the concern for fish welfare

Pain and stress in fish

Arguably, if fish are incapable of suffering during slaughter, it does not matter how we kill them. There is overwhelming evidence that fish subjected to severe conditions respond in a manner analogous to the mammalian stress response (Kestin, 1994) therefore fish welfare may be compromised by stress or pain.

Animal pain is defined as an aversive sensory experience representing an animal's awareness of damage or threat to the integrity of its tissues (Molony and Kent, 1997). Receptors and nerves similar to those involved in detection and transmission of pain signals in mammals have been found in fish (Sneddon, 2003). However, there is controversy on whether fish merely detect potentially harmful stimuli (i.e. nociception) or whether they can translate nociceptive signals to the unpleasant sensory experience that is pain.

Fish brains lack a cerebral cortex which, in humans, is used for processing emotions such as pain. It is therefore unlikely that fish experience pain in the same manner as humans (Rose, 2002). However, fish react adaptively to potentially painful stimuli in ways suggestive of pain perception. For example, carp that have been hooked once and released learn to avoid bait and remain difficult to catch for at least a year (Beukema, 1970). It is therefore reasonable to give fish the benefit of the doubt and not subject them to potentially painful experiences when possible. Given the paucity of scientific information on the matter, how we should treat fish remains largely an ethical issue.

Consumer attitudes towards fish ethics

There is less concern for the welfare of fish compared to that of other animals like lambs. People are unsure that fish are capable of emotions such as pain, stress or happiness (Frewer *et al.*, 2005). This may be because human empathy increases with resemblance between humans and an animal. Moreover, consumers are largely unaware of fish husbandry practices and their potential welfare implications. However, as fish farming grows, consumers will become more aware and demand that fish are reared in a welfare friendly manner.

Most systems of killing fish are designed for ease of operation, maintenance of carcass quality and cost control rather than animal welfare (Robb and Kestin, 2002). Many traditional methods do not cause a rapid or painless loss of sensibility and therefore impinge on animal welfare and the law.

Traditional methods of slaughter

Methods used to kill fish depend on local resources, expertise and fish characteristics. Common methods of killing fish and their effects on fish welfare are summarised in Table 1.

Table 1. Methods of killing fish and their effects of welfare.

Generic	Method	Typical species	Time to unconsciousness	Fish welfare*
Asphyxiation	In air	Small fish eg trout	slow	bad
	In ice	Trout, sea bass, turbot	slow	bad
	In inert gas	Experimental in small fish	slow	good
Mechanical	Exanguination	Salmon, halibut, cod, tuna	slow	bad
	Spiking	Tuna, salmon	immediate	good
	Concussion	Salmon, halibut, trout	immediate	very good
	Hypothermia	Eel, turbot, sea bass	slow/ intermediate	bad
	Eviceseration	Eel	slow	bad
Chemical	CO2 Narcosis	Trout, salmon	slow/ intermediate	bad
	Anesthetics	Salmon, trout	slow/ intermediate	very good
	Warm salt	Eel, turbot, sea bass	slow	bad
Electrical		Trout, salmon	immediate	very good

* assumes correct application

Asphyxiation

Asphyxiation by removal from water is the most common method of killing fish. It is inexpensive and easy to implement with minimal resources and expertise. Fish gill lamellae collapse when exposed to air, resulting in asphyxiation, cerebral ischemia and ultimately death.

Fish find removal from water highly aversive; they vigorously attempt to escape and exhibit a maximal stress response. Loss of brain function is slow in most species and is dependant on temperature. For example trout lose brain function in 9.6 minutes at 2°C and only 2.6 minutes at 20°C (Kestin *et al.*, 1991). Fish that evolved in low oxygen environments such as, crucian carp, tilapia, turbot and halibut have some tolerance to asphyxiation. Death by asphyxiation results in earlier entry into rigor and softer flesh. Asphyxiation is therefore not ideal for fish welfare or product quality.

Hypothermia

Highly priced fish such as salmon and eel may be rapidly chilled in brine or packed on ice at or below 1˚ for transport to markets as live fish where they receive a premium. Chilling results in reduced energy supply to the brain and eventually, unconsciousness. This method is more effective as the temperature differential between the fish and the cooling media increases.
Loss of sensibility can be slow and temporary. The common eel loses sensibility to pain after 12 minutes in water around 1˚C while temperatures around -18˚C result in irrecoverable loss of sensibility after 27 $\pm$ 17 seconds (Lambooij *et al.*, 2002). Live chilling causes stress in fish (Skjervold *et al.*, 1999) and may be painful. It should be discouraged as it causes avoidable suffering.

Exsanguination

Large fish such as salmon and halibut may be killed by cutting or manually pulling the gills. The resultant bleeding results in unconsciousness and eventually death as the brain is starved of oxygen and energy. A variation to exsanguination involves gutting the animals while they are still alive. Bleeding reduces blood spotting, improves colour and reduces blood induced off-flavours. However, there is no added quality benefits from bleeding without stunning (Gregory and Wilkins, 1989).
Exsanguination without stunning causes a slow and painful death; salmon die after bleeding for 4.5 minutes while turbot take 60-90 minutes. Exsanguinations should therefore be done after stunning in order to ensure good welfare. Live gutting subjects the animals to unnecessary pain and distress and is not ethically acceptable. It has been banned by law in Germany since April, 1999. Both exanguinations without stunning and live gutting may be illegal under UK legislation and should be discouraged.

Carbon dioxide narcosis

Trout and salmon may be stunned by immersion in carbon dioxide saturated water prior to exsanguination. CO_2 gas is bubbled continuously through sea or fresh water until saturation (pH 4.5-5). Fish exposed to CO_2 saturated water absorb the CO_2 into their blood resulting in a pH drop, hypercapnia and narcosis.
Fish find exposure to CO_2 water highly aversive and show a maximal stress response and increased mucous production indicating irritation. Induction of narcosis is slow. It takes 2 minutes for salmon and 110 minutes for eel to become anesthetised in CO_2. Slaughter by CO_2 results in stressed fish and an inferior end product and is thus not recommended. Anoxia created by inert gases (e.g. nitrogen and argon) may be more humane since the fish cannot detect the gases. However, inert gases are expensive and difficult to dissolve in water.

Spiking

Large fish may be killed by inserting a spike into the skull and destroying the brain. This may be followed by pithing with a rod or wire to destroy the anterior part of the spinal chord to reduce carcass convulsions. Accuracy of incision is important as fish have relatively small brains and will struggle when taken out of water.
Failure to correctly apply the method results in an injured and conscious fish. In salmon, loss of consciousness was between 0 and 27 seconds when the spiking was accurate but this increased to

102 seconds when inaccurate (Robb *et al.*, 2000). Moreover, fish have to be taken out of water and restrained by the gill which causes stress. When correctly applied spiking is good for welfare.

Chemical anaesthetics

A food grade clove oil extract, eugenol (2-methoxy-4-propenylphenol) has been used to stun fish by anaesthesia. The fish do not have to be taken out of water and the induction of insensibility appears to be without discomfort (van de Vis *et al.*, 2001). Death may be achieved by overdose of the anaesthetic.
The method is painless and prevents ante-mortem stress resulting in improved welfare and product quality. Residual odour and taste from absorption of the anaesthetic can be prevented by using the synthetic isomer, iso-eugenol. Food grade chemical anaesthetics have great potential for the humane slaughter of fish but are yet to be licensed in the EU.

Percussive (concussion) stunning

A blow of sufficient energy to the head results in the brain colliding with the skull, causing rapid depolarisation of the neurones and insensibility or death. The effectiveness of percussive stunning will depend on the velocity, shape, surface area, and positioning of the blow. Correct use results in immediate and irrecoverable insensibility.
Percussive stunning is not suitable for small fish like trout which are difficult to handle. Fish should not be kept out of water for long periods as this increases stress and may result in lower product quality. Correct percussive stunning guarantees good welfare at slaughter.

Electrical stunning

Applying electric currents of sufficient strengths and duration causes a rapid depolarisation of brain membranes resulting in a non responsive seizure like brain state. Electricity can render large numbers of fish immediately unconscious, with little loss of product quality.
It has the added advantage that the fish do not have to be isolated or taken out of the water thus reducing the pre-slaughter stress. Stun-kill systems that render fish immediately and permanently insensible have been developed to commercial production for trout (Lines and Kestin, 2004). Incorrect stunning may cause haemorrhaging and downgrading or cause paralysis without unconsciousness, resulting in pain and stress in the animals.

Designing an electrical stunner for turbot

Current methods of killing turbot include asphyxiation in ice, exsanguination, and percussive stunning. Turbot are tolerant to anoxia and show signs of consciousness and aversion for up to 15 minutes after exsanguination. There are currently no technologies for automated percussive stunning of flatfish.
Turbot farmers were consulted on their expectations of a turbot stunner. They were primarily concerned with:

1. product quality;
2. equipment purchasing and maintenance costs;
3. ease of use;
4. speed of harvests;
5. public approval;
6. mobility of equipment.

Electronarcosis had the greatest potential of fulfilling the farmers' expectations and provide a humane stunner for turbot.

Possible conflicts in design requirements

High frequency electricity is used to reduce haemorrhaging but it is less effective at inducing insensibility and killing animals. This may be offset by using higher voltages but this increases overall equipment cost and the maximum voltage is limited by operator safety considerations. More electrical power is required to effectively stun in sea water (Lines and Kestin, 2004) but there is higher risk of accidental shock because sea water is highly conductive.

The electrical stunner

Behavioural and physiological signs of insensibility were used to determine the effectiveness of all stuns. The result was a system whereby the fish were dewatered immediately prior stunning with a high voltage of Pulsed DC (PDC) electricity. Passing a PDC current across the fish brains resulted in immediate electronarcosis (insensibility). Pulsing the current allows the equipment to rest and recharge in between the pulses. This allows higher voltages to be used with smaller, less expensive circuitry than is required for continuous Alternating Currents (AC). Dewatering the fish meant less energy was wasted in powering the sea water. This reduced the overall power requirement and improved operator safety as there was less current leakage. Moreover, the stunner could be powered by a 12V battery allowing portability for use on other farms or on boats.

A quality test conducted by a commercial turbot processor revealed no significant loss of quality in experimentally killed fish when compared to those killed using the traditional method (asphyxiation in ice). PDC systems, which are fully automated and easy to operate at commercially acceptable speeds, are at an advanced design stage for commercial production with the support of turbot farmers.

Conclusion

There are concerns for the welfare of fish during slaughter as some traditional methods cause avoidable pain and stress to the fish. There is weak legislative imperative to change current systems as there are few viable alternatives and little scientific information on fish slaughter. Collaboration between farmers and scientists can result in systems that are humane and address the commercial interests of farmers. More work is needed to inform future legislation on the welfare of fish at slaughter.

Acknowledgements

This work was done as part of PhD studies jointly sponsored by the Humane Slaughter Association and the Biotechnology and Biological Sciences Research Council.

References

Beukema, J. J. 1970. Angling experiments with carp; decreased catchability through one trial learning. *Netherlands Journal of Zoology,* 20:81-92.

FAO Fisheries Department, D. a. S. U. Fishstat Plus: *Universal Software for Fishery Statistical Time Series.* [2.3]. 2000.

Frewer, L. J., A. Kole, S. M. A. van de Kroon, and C. de Lewere. 2005.

Consumer attitudes towards the development of animal-friendly systems. *Journal of Agricultural and Environmental Ethics,* 18:345-365.

Gregory, N. G. and L. J. Wilkins. 1989. Effect of Slaughter Method on Bleeding Efficiency in Chickens. *Journal of the Science of Food and Agriculture,* 47:13-20.

Huntingford, F. A., C. Adams, V. A. Braithwaite, S. Kadri, T. G. Pottinger, P. Sandoe, and J. F. Turnbull. 2006. Current issues in fish welfare. *Journal of Fish Biology,* 68:332-372.

Kestin, S. C. 1994. *Pain and stress in fish.* Royal Society for the Prevention of Cruelty to Animals.

Kestin, S. C., S. B. Wotton, and N. G. Gregory. 1991. Effect of slaughter by removal from water on visual evocked activity in the brain and reflex movement of rainbow trout (*Oncorhynchus mykiss*). *Veterinary Record,* 128:443-446.

Lambooij, E., J. W. van de Vis, R. J. Kloosterboer, and C. Pieterse. 2002. Welfare aspects of live chilling and freezing of farmed eel (Anguilla anguilla L.): neurological and behavioural assessment. *Aquaculture,* 210:159-169.

Lines, J. and S. Kestin. 2004. Electrical stunning of fish: the relationship between the electric field strength and water conductivity. *Aquaculture,* 241:219-234.

Molony, V. and J. E. Kent. 1997. Assessment of acute pain in farm animals using behavioral and physiological measurements. *Journal of Animal Science,* 75:266-272.

Robb, D. H. and S. C. Kestin. 2002. Methods used to kill fish: Field observations and literature reviewed. *Animal Welfare,* 11:269-282.

Robb, D. H., S. B. Wotton, J. L. McKinstry, N. K. Sorensen, and S. C. Kestin. 2000. Commercial slaughter methods used on Atlantic salmon: determination of the onset of brain failure by electroencephalography. *Veterinary Record,* 147:298-303.

Rose, J. D. 2002. The Neurobehavioral Nature of Fishes and the Question of Awareness and Pain. *Reviews in Fisheries Science,* 10:1-38.

Skjervold, P. O., S. O. Fjaera, and P. B. Ostby. 1999. Rigor in Atlantic salmon as affected by crowding stress prior to chilling before slaughter. *Aquaculture,* 175:93-101.

Sneddon, L. U. 2003. Trigeminal somatosensory innervation of the head of a teleost fish with particular reference to nociception. *Brain Research,* 972:44-52.

Van de Vis, J. W., J. Oehlenschlager, H. Kuhlmann, W. Munkner, D. H. Robb, and A. A. M. Schelvis-Smit. 2001. Effects of the commercial and experimental slaughter of eels (*Anguilla anguilla L*) on quality and welfare. In: *Farmed Fish Quality,* S. C. Kestin and P. D. Warriss (Eds.). pp. 234-248. Blackwell Science, Oxford.

Consumer perception about ethical and sustainability issues of fish

Filiep Vanhonacker, Wim Verbeke and Isabelle Sioen
Ghent University, Faculty of Bioscience Engineering, Department of Agricultural Economics, Coupure links 653, 9000 Gent, Belgium, Filiep.Vanhonacker@Ugent.be

Abstract

Ethical and sustainability issues in livestock production are of increasing public importance. Till now, little profound consumer research has been conducted to reveal consumers' perception about ethics and sustainability from fisheries and aquaculture. Cross-sectional data were collected from a sample of 381 Belgian women aged 25-50 years in June 2005 through a postal self-administered survey. Interest in sustainability issues was measured as the importance attached to the depletion of natural fish stocks and the environment. Also perceived importance of fish welfare was included. Finally, perception of farmed versus wild fish was measured. Consumers express higher interest in environmental issues compared to fish welfare when buying fish, meaning a higher concern about a healthy world (in favour of their own health) instead of fish welfare. Analyses also indicate a clear interest and concern in sustainability issues relating to aquaculture and fisheries, while consumers' knowledge appeared rather limited. Ten percent of the respondents claims to refuse farmed fish. A similar amount claims to refuse wild fish. The latter attach more importance to fish welfare and sustainability and are more convinced that aquaculture contributes to natural fish stock conservation. Hence, rejecting wild fish is associated with attaching importance to sustainability and fish welfare. Consumers rejecting farmed fish show no different interest in these matters as compared to those accepting it. Higher importance attached to ethical and sustainability issues correlates with a stronger interest in capture area and origin, and a stronger benefit expectation of more information, but it does neither result in a different total fish consumption frequency nor attitude towards eating fish. Higher interest in sustainability associates with lower consumption of cod. Apparently, consumers have some awareness of the challenge facing cod fisheries concerning the depletion of natural cod stocks.

Keywords: attitude, consumer, ethics, fish, perception, sustainability

Introduction

The consumption of fish and derived fish products has greatly increased over recent decades, due to fish's contribution to a healthy human diet, the increasing world population, higher living standards and the good overall image of fish among consumers (Cahu *et al.*, 2004). This increase in demand has led to an expansion of the fishing fleet. This, together with a higher efficiency of fish capture, has contributed to the depletion of several natural fish stocks (Van Delsen *et al.*, 2005). In response to the depleting wild fish stocks and the increasing consumer demand for fish, consumers are now being offered farmed fish as a valuable alternative (Cahu *et al.*, 2004). Aquaculture has been the fastest growing food production sector of the world (FAO, 2000), with an annual growth of about 9% or an evolution starting from a world production of 3.5 million ton in 1970 to 54.8 million ton in 2003 (Van Delsen *et al.*, 2005). This fast development, together with an increased attention to health and food safety, has made aquaculture of increased public importance. However, aquaculture and aquacultural products are subject of rather conflicting communications towards the public. Hence, consumer perception of farmed fish plays a crucial role in the success of aquaculture (Kole, 2003). Recently, several studies were conducted

opposing farmed fish to wild fish regarding consumer perception (Farmer *et al.*, 2000; Kole, 2003; Arvanitoyannis *et al.*, 2004; Verbeke *et al.*, 2006). In general, wild fish is perceived more positive than farmed fish by the consumer, while there is no scientific evidence to confirm this (Luten *et al.*, 2002; Cahu *et al.*, 2004; EFSA, 2005). The main cause of this seems to be the lack of knowledge concerning aquaculture, resulting in the use of emotions to judge aquaculture (Verbeke *et al.*, 2006). Additionally, little is known about how differences in perception influence consumer behaviour. Finally, although several studies show a strong increase of consumer interest in ethical and environmental issues (Ethical Consumer, 1997/1998; Shaw and Clarke, 1999; Follows and Jobber, 2000), only few research is devoted to reveal the role and impact of these ethical and sustainable concerns in consumer decision making (Shaw and Shiu, 2003).
The objective of this paper is to verify if ethical and sustainability matters contribute to consumers motives in choosing or rejecting either farmed or wild fish. In addition, characterisation is made of persons who attach more importance to ethical matters and sustainability when eating fish, based on behaviour, attitude towards fish and socio-demographics.

Method

Cross-sectional data were obtained through a consumer survey with data collection during May-June 2005. The sample was composed of 381 women and includes respondents from a variety of different age, educational and professional categories. Nevertheless, the specific respondent selection and recruiting procedures yield a statistically non-representative sample. Hence, findings apply within the characteristics of the sample, whereas generalisation to the overall population are rather speculative.
The questionnaire consisted of three main parts: first, questions measuring fish consumption frequency and attitude towards fish were incorporated. A second part intended to measure the perception of ethical and sustainability issues. Also the topic farmed versus wild was highlighted within this section. Finally, some socio-demographics were included.
With respect to ethical and sustainability issues, items were used from the scale of Lindeman and Väänänen (2000) dealing with ethical motives involved with food choice. These items sound out consumer consciousness about fish welfare, depletion of natural fish stocks and protection of the natural environment. Next, respondents were asked to which degree they agree that by choosing farmed fish, one can attribute to the saving natural fish stocks from depletion. Also, respondents were asked if they would not eat fish products if they knew that they had farmed (wild) origin. Also perceived importance attached to the capture area of the fish consumed was measured. Finally, several items investigating whether the consumer believes that more information can play a role in guaranteeing ethical issues like fish welfare and over-fishing, were included. Seven-point interval measurement scales were used to measure perceived importance and attitudes.

Results

Description of the mean values

Ethical and sustainability issues are measured by five items, which yield a satisfactory cronbach's alfa of 0.93. Nevertheless, not all the items are combined, because different issues are touched, namely fish welfare (2 items), over-fishing (1 item) and environmental issues (2 items). The latter was given the highest importance (5.77 on a 7-point scale), followed by over-fishing (5.60) and fish welfare (5.21), which indicate a higher concern for environmental care compared to fish welfare. The difference in means proved to be significant ($p=0.000$). With regard to the item measuring the extent of agreement that by choosing farmed fish, one can contribute to the saving fish stocks from depletion, a more neutral average was found (4.11). Hence, people seem to express interest and concern in ethics and sustainability, but with a rather limited conviction. Information and communication can play a crucial role here. People are slightly aware of this, given a mean score of 4.83 on the belief that more information can guarantee a better conservation of natural fish stocks. A similar question regarding the impact of more information on animal welfare resulted in a mean of 4.33. Regarding the items about the interest in origin (farmed versus wild) and capture area, a low mean score is obtained (respectively 3.49 and 2.83 on a 7-point scale).

Identification of consumers who refuse farmed or wild fish

Respectively 10.2 and 11.7 percent of the respondents declare to refuse either farmed or wild fish. Cross-tabulation indicates that these groups are not composed of the same persons. Characterisation of those two groups is made based on behaviour, attitude towards fish, importance attached to ethical and sustainability issues and socio-demographics. Regarding behaviour, the overall fish consumption frequency and the consumption frequency of different fish types are of interest. Salmon for instance is expected to be less consumed by respondents indicating not to eat farmed fish. Besides salmon, also more than half of the trout on the market has farmed origin. The other fish (cod, tuna, pollack and sole) are mainly or exclusively wild fish. With regard to overall fish consumption frequency, no significant difference was found, in spite of a clear difference in mean and direction (Table 1). Respondents refusing to eat farmed fish have a higher average fish consumption, while the opposite is true for respondents refusing wild fish, corresponding with the fact that consumers perceive wild fish as more tasty and healthy (Verbeke *et al.*, 2004). With regard to both farmed and wild fish types, no significant differences were found, neither between respondents who do or do not refuse farmed fish, nor between respondent who do or do not refuse wild fish. Apparently, consumers do not really know which fish types have either farmed or wild origin.

Regarding attitude, the items general attitude, health and nutritional value are expected to be in favour of wild fish, based on literature (Luten *et al.*, 2002; Kole, 2003; Arvanitoyannis *et al.*, 2004; Verbeke *et al.*, 2006). Absolute values confirm literature data when respondents refusing farmed fish or not are taken into consideration. However, only the item nutritional value yields significance ($p=0.040$). With respect to wild fish, none of the items showed significance ($p>0.05$).

Next, analyses are performed to investigate whether people refusing either farmed or wild fish act out of ethical or sustainability considerations (Table 2). Refusal of farmed fish does not seem to be based on ethical or sustainability considerations, given no differences in importance attached to fish welfare, over-fishing and environmental issues between both groups. Also, the group not refusing farmed fish does not claim a significant higher extent of agreement on the statement that by choosing farmed fish, one can attribute to the savings of natural fish-stocks. A totally different

Table 1. Typifying consumers who refuse farmed or wild fish based on behaviour and attitude (n=381).

	Refuse farmed fish			Refuse wild fish		
	Yes	No	p	Yes	No	p
Total consumption frequency#	5.43	4.48	0.126	3.88	4.67	0.063
General attitude	5.73	5.61	0.554	5.42	5.66	0.338
Perceived nutritional value	6.60	6.25	0.040	6.28	6.29	0.940

frequency per month

Table 2. Typifying consumers who refuse farmed or wild fish based on ethical and sustainability issues (all items are measured on a 7-point scale).

	Refuse farmed fish			Refuse wild fish		
	Yes	No	p	Yes	No	p
Importance of fish welfare	5.21	5.19	0.935	5.84	5.11	0.007
Importance of over-fishing	5.51	5.60	0.750	5.90	5.56	0.187
Importance of environmental care	5.71	5.77	0.824	6.23	5.71	0.034
Choosing farmed leads to less over-fishing	3.64	4.16	0.149	4.64	4.05	0.046

conclusion can be drawn when the respondents who do or do not refuse wild fish are taken into consideration. A significant higher importance attached to fish welfare and environmental care by the group indicating to refuse wild fish indicates that ethical and sustainability considerations contribute to this refusal. Those who refuse wild fish are also more convinced of the contribution of fish capture to the depletion of natural fish-stocks. Hence, rejecting wild fish seems to be partly based on sustainability issues. This is confirmed with a negative correlation between interest in sustainability and cod consumption, indicating consumer awareness of the challenge of cod fisheries concerning the depletion of natural cod stocks.
Finally, socio-demographic differences are examined between the different groups (refuse farmed versus not refuse farmed and refuse wild versus not refuse wild). No significant differences are found with regard to age and the presence of children. Concerning education level, it appeared that relatively more lower educated people are encountered in the group claiming to refuse wild fish.

Typification based on importance attached to ethical and sustainability issues when eating fish

For this discussion, scores on the three ethical and sustainability issues (fish welfare, over-fishing and environmental care) are combined into one construct, based on the satisfactory cronbach's alfa (0.93). Correlations are used to link this construct to behaviour, general attitude, interest in capture area and origin and perceived benefit of information in terms of guaranteeing better fish and environmental welfare (Table 3).

Table 3. Correlations with importance of ethical and sustainability issues.

	Importance of ethical and sustainability issues associated with fish consumption	
	Pearson r	p
Behaviour (fish consumption frequency)	0.099	0.064
General attitude	0.078	0.151
Interest in capture area	0.245	0.000
Interest in origin (farmed / wild)	0.223	0.000
Choosing farmed leads to less over-fishing	0.267	0.000
Importance of information	0.431	0.000

A slightly positive, but non-significant correlation is found between fish consumption frequency and general attitude towards fish consumption on the one hand and importance attached to ethical and sustainability issues associated with fish consumption on the other hand. Hence, neither total fish consumption frequency nor general attitude seems to be strongly shaped by ethical and sustainability matters.
With respect to interest in capture area and origin of the fish consumed, a positive and significant correlation is found. People who attach more importance to ethical and sustainability issues with respect to fish are also more interested in knowing the origin of fish. Hence, either capture area and origin are associated by the respondent with ethical and sustainability issues, or respondents attaching the largest importance to ethical and sustainability issues are the ones with the highest interest and concerns in general. The strongly positive correlation with the extent to which the respondent beliefs that more information can guarantee a higher degree of both fish welfare and saving of natural fish stocks, points in the direction of the latter possibility.
Finally, the positive and significant correlation with the statement that by choosing farmed fish, one can contribute to saving natural fish stocks from depletion, indicates a higher awareness of the present problems associated with capture fisheries among respondents with higher importance attached to ethics and sustainability.
With regard to socio-demographics, only age appeared to result in a different importance of ethical and sustainability issues ($r=0.14$; $p=0.010$), indicating that ethical and sustainability issues are valued somewhat higher among older consumers.

Conclusion

Based on a consumer survey performed in June 2005, this paper examined the importance of ethical and sustainability issues towards fish and its relationship with fish consumption frequency, attitude towards fish consumption, socio-demographics and the refusal of either farmed or wild fish. In general, ethical and sustainability issues were indicated as being important by the consumer in relation to fish consumption. This importance however, was neither translated in a significant correlation with total fish consumption frequency nor with general attitude towards eating fish. The choice not to eat wild fish seemed to have part of its origin in ethical and sustainability issues, given a significantly higher importance attached to these issues by consumers who refuse wild fish, and a negative correlation between interest in sustainability and cod consumption. This was not the case with respect to rejecting farmed fish. Finally, respondents with the highest importance attached to ethical and sustainability seemed to be the ones with the highest interest in information in general and with the highest belief in a potential benefit of receiving more information.

References

Arvanitoyannis, I.S., Krystallis, A., Panagiotiaki, P. and Theodorou, A.J. (2004). A marketing survey on Greek consumers' attitudes towards fish. *Aquaculture International*, 12: 259-279.

Cahu, C., Salen, P. and De Lorgeril, M. (2004). Farmed and wild fish in the prevention of cardiovascular diseases: Assessing possible differences in lipid nutritional values. *Nutrition Metabolism and Cardiovascular Diseases*, 14: 34-41.

EFSA, 2005. Opinion of the scientific panel on contaminants in the food chain on a request from the European Parliament related to the safety assessment of wild and farmed fish. *The EFSA Journal*, 236: 1-118.

Ethical consumer (1997/1998). The rise and fall of ethical consumerism. *Ethical Consumer*, 50: 22-25.

FAO, 2000. The state of world fisheries and aquaculture 2000, part 1 world review of fisheries and aquaculture, fisheries resources, trends in production utilisation and trade overview. FAO information division, Rome.

Follows, S.B. and Jobber, D. (2000). Environmentally responsible purchase behaviour: a test of a consumer model. *European Journal of Marketing*, 34: 17.

Kole, A. (2003). Consumer opinions towards farmed fish, accounting for relevance and individual knowledge. In: *Quality of fish from catch to consumer. Labelling, monitoring and traceability*. Luten, J.B., Oehlenschläger, J. and Ólafsdóttir, G. (Eds.). Wageningen Academic Publishers, Wageningen. 393-400.

Lindeman, M. and Väänänen, M.(2000). Measurement of ethical food choice motives. *Appetite*, 34: 55-59.

Luten, J., Kole, A., Schelvis, R., Veldman, M, Heide, M., Carlehög, M and Akse, L. (2002). Evaluation of wild cod versus wild caught, farmed raised cod from Norway by Dutch consumers. *Økonomisk Fiskeriforskning*, 12: 44-60.

Olsen, S.O. (2001). Consumer involvement in seafood as family meals in Norway: an application of the expectancy-value approach. *Appetite*, 36: 173-186.

Shaw, D. and Clarke, I. (1999). Belief formation in ethical consumer groups: an exploratory study. *Marketing Intelligence and Planning*, 17: 109-119.

Shaw, D. and Shiu, E. (2003). Ethics in consumer choice: a multivariate modelling approach. *European Journal of Marketing*, 37: 1485-1498.

Van Delsen, B., Dhont, J., Bossier, P. and Sorgeloos, P. (2005). Aquacultuur: sleutel tot de blauwe revolutie. *Het Ingenieursblad*, 11-12: 34-40.

Verbeke, W., Sioen, I., Pieniak, Z., Van Camp, J. and De Henauw, S. (2004). Consumer perception versus scientific evidence about health benefits and safety risks from fish consumption. *Public Health Nutrition*, 8: 422-429.

Verbeke, W., Sioen, I., Brunsø, K., Van Camp, J. and De Henauw, S. (2006). Consumer perception versus scientific evidence of farmed versus wild fish: exploratory insights from Belgium. *Aquaculture International*, submitted.

Part 12
Politics of consumption

The political-commercial framing of ethical consumption

Eivind Jacobsen and Arne Dulsrud
National institute for consumer research (SIFO), P.O.Box 4682, Nydalen, N-0405 Oslo, Norway, Eivind.jacobsen@sifo.no and arne.dulsrud@sifo.no

Abstract

In Europe as well as world wide, increasing political and commercial attention has been focused on ethical-political consumption. However, comparative studies display considerable differences between countries regarding peoples' attention and willingness to address ethical questions *as consumers*. Such differences have been explained with reference to peoples "national character", to countries different levels of modernity and to other culture relativistic factors. SIFO-studies have added to this with institutional explanations, pointing to differences in the historically defined legal and fiscal framing of national markets and to how this implies differences in the construction of *consumer roles*. In this paper we elaborate further on this, focusing on supply side strategies that may contribute to the framing of consumer choices, and potentially explain differences between different markets. On a theoretical basis we discuss companies' motivations for selling and promoting "ethical" products. We discuss these in relation to cases referring to organic food, animal friendly and environmentally sustainable products in different markets. Most cases are related to food. The paper concludes with a discussion of consumer agency and asks whether we can rely on consumers to solve the ethical challenges of modern consumption.

Introduction

In Europe as well as world wide, increasing political and commercial attention has been focused on ethical consumption (e.g. Micheletti *et al.*2003; Sørensen 2004; Nordic Council of Ministers 2004). Shopping is being moralized and politicized and certain product categories, like hardwood timber, tuna fish or fôis gras are turning into moral minefields. Invite you university friends for a glass of wine to admire your new teak-wood garden furniture and you now what we mean. You will probably already have thought out a morally defensible justification of your purchase and its impact on the rainforest.[86] Hence, heavy burdens are put on the shoulders of consumers these neo-liberalistic days. Increasingly they are supposed to take on the responsibility for a morally virtuous handling of technological change and the liberalization of world trade.
A lot of consumers tend to acknowledge this responsibility. They try to, or at least claim to try to consume in an ethically virtuous way. However, a lot of studies have shown a discrepancy between consumers' stated honourable attitudes and their pitiable behaviours in areas like e.g. recycling or purchasing of ecological and animal friendly foods (e.g. Stø *et al.*2005; Torjussen *et al.* 2004; Roex and Miele 2005). People tend to know what's the right thing to do, they claim to want to do it, but don't. This attitude-behaviour gap has been explained in various ways, e.g. with reference to the limited reliability of stated attitudes and behaviours (Swann 2002) or more generally with reference to the moral weaknesses of man (Von Wright 1993).[87] In this paper we

[86] Moreover, the moral inconveniences of your specific purchase, comes on top of the more general consumer society condemnation brought forth by the "greens", where almost every form of consumption is conceived of as a sin.

[87] The environmental movement is full of overtly contemptuous moralistic writings about consumers and consumer society.

will explore another approach focusing on the political-economic framing of consumer choices. Comparative studies have shown that there are systematic differences between countries regarding the scope of ethical consumption and the willingness of people to engage *as consumers.* These differences point to structural determinants and how consumer choices are differently framed in different countries. In this paper we elaborate on these dynamics and ask what this implies for our notion of consumer sovereignty and ethics of consumption.

A market for ethics

Through their *shopping* behaviour a lot of consumers demonstrate concerns for the environment, for the working conditions in producer countries, for the wellbeing of production animals and so on. They boycott and (so called) buycott[88] products and services by withholding or utilizing their purchasing power. But, consumers also show ethical and political concerns in their *use of* purchased goods. People recycle paper, tin cans and bottles in order to realize a sustainable lifestyle (e.g. Stø *et al.* 2005). The motivations for such behaviours may vary considerably. Follesdal (2003) lists several possibly overlapping motivations: People may just want to have clean hands (its dirty and I just don't want to be a part of it), they may seek to construct some kind of identity for themselves through their consumption practices (self understanding or respect for the other), or they may actively try to influence other consumers (to effect moral pressure on others by means of own virtuous behaviour) or producers/sellers practices. The latter has of course been most focused. This is also so because such clearly political behaviours have the potential to transcends the actions of individuals to form and be part of political movements with the potential to challenge political and economic powers (Merlucci 1996).
The motivations reflected in ethical-political consumption practices are outcomes of moral negotiations between different everyday considerations (Halkier 2004; Bugge 2005). In the socially and culturally complex realities of modern everyday life, ethical-political virtues often have to be compromised against other obligations and highly legitimate concerns, framing the actual options available for choices. As such purchasing decisions e.g. related to food, are often entangled in overlapping and often conflicting moral expectations, care and power relations, economic, time-use and nutritional concerns and taste preferences (e.g. Holm and Kildevang 1996). Hence, the often noticed discrepancies between honourable attitudes and pitiable behaviours may partly be explained with the complexity of everyday morals and practices.
Besides, these behaviours are highly entangled in morally consequential languages games, rhetorical tropes and classificatory schemes (Bowker and Star 2000; Jacobsen 2004). They are part of "ontological politics", more or less explicit negotiations over what is real (Mol 1999). Hence, as classification and language trope, the notion ethical-political consumption is interwoven with the grids of power in society and reflects or negates these powers, defining certain actions as more desirable than others. Naming and displaying the ethical virtues of certain consumption practices, is therefore also a contestation of power.
But, ethical-political consumption has powerful supply side drivers as well: Modern consumer markets are dominated by highly concentrated manufacturing industries and vertically integrated retailers basing their consumer communication on branding, TV-advertising and product placement. This is also so in the food sector, where the market shares of international branders and retailers are constantly rising across national borders (e.g. Dobson 2003). The branding producers and distributors react to consumers' concerns and protests to avoid blame, they actively appeal to consumers' consciousness to differentiate markets, and they take responsibility and demonstrate citizenship to avoid regulatory intervention from authorities (e.g. Marsden *et al.* 2000). They also need to "dress up" to motivate employees. Economic performance often

[88] actively choose a product for some ethical-political reason

depends on employees' motivation and belief in the virtues of their company. Due to these and other reasons, companies put up private quality standards, regulatory schemes and labelling programs to communicate more or less justified claims of ethical virtues (e.g. Ponte and Gibbon 2005). In the era of brands, CSR has become an indispensable part of any consumer oriented production and marketing (e.g. Kotler and Lee 2005). The internet and globalized media push producers of branded goods to be seen as ethically virtuous and consistent and to enforce their claimed virtues throughout the value chain (e.g. Ind 2003). This way ethical consumption is approached and developed just like any other market, by means of product development strategies, product placement, adverting and other marketing devices, and where producers when successful, may hope to build up a niche market where they can take out a price above the standard products.

National authorities are also of relevance for this development. In the post-cold war era a notion of a slim but strong state have grown in popularity (e.g. Majone 1996), where ambitions outside core functions increasingly are pursued by means of indirect policy instruments. Soft laws, private standards, codes of conduct and stakeholder involvement in different kinds of policy formulation and implementations, are encouraged (Jordana and Levi-Faur 2004). Even though the state "may carry a big stick", "the soft voice" is due to the economic and political limitations of modern states and their dependency on commercial actors and civic society to do their parts (Braithwaith 1997). This is even more the case when the actions and actors in question are outside the jurisdiction of the nation state, e.g. in far off exporting countries. In these cases private standards may supplement and extend "the arm of the law". But also at home, on national soil, governments may be hampered by international treaties and obligations in politically "unwanted" ways. In such cases private codes of conduct and labelling schemes, may accomplish what governments aren't allowed to (or don't want to or haven't the guts to do), and thereby release governments from domestic political pressures and nuisances.[89] Moreover, in a lot of cases such codes of conduct have a trade-protectionist effect, by indirectly coining most imported goods (from third world countries) unethical, e.g. with reference to poor working conditions in sweat shops, unsustainable resource management and so on. This way ethical labelling may accommodate protection where international trade agreements prevent governments from using fiscal tariffs and import quotas.

Even though ethical consumption seems to be on the rise in most European countries, there are some interesting variations to be notices. Comparative studies of e.g. animal welfare concerns and environmental sustainable consumption practices, display systematic differences between countries regarding peoples' attention and willingness to address ethical questions *as consumers* (Stø *et al.* 2005; Poppe and Kjarnes 2003; Terragni and Kjarnes 2005). As such the scope of questions relevant for consumers to consider seems to vary considerably between Europeans (Kjarnes and Lavik 2006). In some countries the care for e.g. the environment, the welfare of production animals or the livelihood of far off production workers, are more in focus in citizens' roles, pointing to more collective action approaches to these issues. This is not the place to elaborate and explain these differences.[90] Here it suffices to register that consumer roles and accompanying moral responsibilities, are differently constructed across Europe, entangled in local and foremost nationally networks of political traditions and power.

[89] This was for instance the case with GM-food in Norway, where the supermarket chains agreed not to import such foods.

[90] Regarding animal welfare concerns, see Kjarnes and Lavik 2006.

Discussion: Consumer agency?

Hence, the phenomenon of ethical consumption evolves in the dynamics between different consumer aspirations and everyday practices, a mixture of commercial strategies and practices and governmental policies. A market for ethics is in the making, a process where the (ethical) products, the ethics as well as the consumers themselves are under construction. In this situation, consumers' preferences, their access to information as well as the products available for choice, is heavily influenced by commercial interests and governmental concerns.

Given this entangledness one may even ask, as Christian Coff does, whether in our society, ethical consumption is possible at all (Coff 2005; see also Lang and Heasman 2004:280). Consumers lack the necessary knowledge (due to poor transparency in the food chain), autonomy (due to commercially skewed marketing) and relevant alternatives (due to retailers' and manufactures' power and consumers' habituated practices) to be able to perform intelligent and ethically guided choices. Hence, the notion of political consumption as a way to exercise ethical power in purchasing situations is illusory, according to Coff (pp. 60-61). In his view, the necessary preconditions for most consumers to make ethically guided choices are simply not fulfilled, due to the way food production, manufacturing, distribution and marketing is organized in our societies.

If Coff is at least partly right in his criticism (and it should be obvious from the previous pages that we think he is), we should shift focus from moralizing over consumers choices, to start discussing which moral burdens people *as consumer* can and should be expected to bear. May be some concerns should mainly be addressed through their citizens roles, supporting collective action by means of governmental regulation. For some kinds of problems, this may be the only feasible way to proceed. However, in this neo-liberalistic epoch, to delegate choices to consumers seams to have become more or less the default option for decision-shunning politicians. As such, instead of moralizations over consumers' lacking of virtues, what is needed is a renewed ideological debate on private vs. public responsibilities.

Moreover, in sectors where a "we", through some kind of democratic process, find it reasonable and effective for consumers to take on ethical responsibilities, we should discuss the preconditions for ethically guided consumer choices to be made. How can consumers be given a more defined role as autonomous responsible actors? How can consumers be given relevant and reliable information about the production and distribution methods? And, how can consumers be given real and realistic alternatives, which ethically considered do make a difference? These are fundamental questions at the base of our current political-economic system that have to be discussed if ethical consumption is to be more than mere market differentiation and individual ego-trips.

References

Bowker, G. C. and Star, S. L. (1999) Sorting Things Out: Classification and Its Consequences. MIT Press: Cambridge, MA.

Braithwaith, J. (1997): On Speaking Softly and Carrying Big Sticks: Neglected Dimensions of a Republican Separation of Powers, Univ. of Toronto Law Journal 47:305-61.

Bugge, A. (2005): Middag. En sosiologisk analyse av den norske middagspraksis. Ph.D. dissertation. NTNU, Trondheim, Norway.

Coff, C. (2005): Smag for etikk. København: Tuscalanum forlag.

Dobson, Paul W. (2003): Boyer Power in Food Retailing: The Euriopean Experience. OECD's Conference on Changing Dimensions of the Food Economy: Exploring the Policy Issues, 6-7 February 2003, den Haag, Nederland.

Follesdal, (2004): Introduction, in Micheletti, M., Follestdal, A., and Stolle, D. (eds) (2004). Politics, products and markets. Exploring political consumerism past and present. New Brunswick and London:Transaction Publishers.
Halkier, B. (2004): Handeling Food-related Risks: Political Agency and Governmentality, in M. Lien and B. Nerlich: The Politics of Food, Oxford: Berg.
Holm, L. and H: Kildevang (1996): Consumers' view of Food Quality: A Qualitative Interview Study, Appetite, 27:1-14.
Ind, N. (2003): Beyond Branding, London: Kogan Page
Jordana, J. and D. Levi-Faur (2004): The politics of regulation in the age of governance, in J. Jordana and D. Levi-Faur (eds.) The politics of regulation. Institutions and Regulatory Reforms for the Age of Governace. Cheltenhan UK: Edward Elgar.
Kjarnes, U. and R. Lavik (2006): Paper Eursafe 2006, Oslo.
Kjarnes *et al.* 2006: Forthcomming.
Kotler, P. and N. Lee (2005): Corporate Social Responsibility: Doing The Most Good For Your Company And Your Cause, Hoboken NJ: John Wiley & Sons Inc.
Majone, G. (1994). The rise of the regulatory state in Europe, West European Politics, 17, 77-101.
Marsden, T., Flynn, A. and Harrison, M. (2000). Consuming interests. The social provision of foods. London: UCL Press.
Merlucci (1980): New social movements: A theoretical approach, Social science information.
Micheletti, M. (2003), Political virtue and Shopping. Individuals, consumerism and collective action. New York: Palgrave Macmillan.
Mol, Annemarie (1999): Ontological politics. A word and some questions. I John Law and John Hassard: Actor Network Theory and after. Oxford: Blackwell Publishers/ The Sociological Review.
Jacobsen 2004, in M. Lien and B. Nerlich: The Rhetoric of Food: Foor as Nature, Commodity and Culture, in M. Lien and B. Nerlich: The Politics of Food, Oxford: Berg.
Nordic Council of Ministers (2004): Political consumerism: Its Motivations, Power, and Conditions in the Nordic Countries and Elsewhere.
Ponte, S. and Gibbon, P. (2005): Quality standards, conventions and the governace of global value chains. Ecoonomy and Society Vol. 34, No 1, Feb. 2005:1-31
Roex, J. and M. Miele (2005): Farm Animal Welfare Concerns. Consumers, retailser and Producers. Welfare Quality Reports No. 1, Cardiff: Cardiff University,
Sørensen, M. P. (2004): Den politiske forbruker. København: Hans Reitzels forlag.
Stø, E. and Strandbakken, P. (2005): Eco labels and consumers, in Rubik, F. and Frankl, P. (2005): The Future of Eco-labelling. Making environmental product information systems effective. Sheffield: Greenleaf Publishing Ltd.
Swann, G. M. P. (2002): There's more to the economics of consumption than (almost) unconstrained utility maximisation, in Andrew McMeekin, Mark Tomlinson, Ken Green and Vivien Walsh (eds.) Innovation by demand. New Dynamics of Innovation and Competition, Manchester: Manchester University Press
Terragni, L. and U. Kjærnes (2005): Ethical consumption in Norway. Why is it so low? I TemaNord 2005:517, s.471-485
Torjusen, H., U. Kjærnes, L. Sangstad and K. O'Doherty Jensen (2004). European Consumers' Conceptions of Organic Food : A Review of Available Research. Professional report no. 4-2004. Oslo: SIFO
Von Wright, G. H. (1993): Myten om framsteget (The Myth of Progress), in Swedish.

Safer than ever? Trust and gender in relation to meat provisioning in Norway

Marianne Elisabeth Lien[1] *and Randi Lavik*[2]
[1]*Department of Social Anthropology, University of Oslo. Postbox 1091, 0317 Blindern Norway, m.e.lien@sai.uio.no*
[2]*National Institute for Consumer Research. Postboks 4682 Nydalen, 0405 Oslo, Norway, randi.lavik@sifo.no*

Abstract

Meat is a highly charged element in the nexus that connects food production and food consumption as it evokes issues pertaining to gender, human-animal relations, nutrition and social distinction. In this article we take Norwegian consumers' attitudes towards meat as an entry to the complex relations of dependencies and trust implicated in the systems of provision relating to meat. We focus on the impact of food scandals, notably the BSE (Bovine spongiform encephalopathy) outbreak in the UK on the shifting configurations of trust and attitudes to meat in Norway. Results of surveys conducted in 1997, 2000 and 2004 indicate that the initial distrust sparked by food scandals gradually disappeared and consumers now feel 'safer than ever'. To understand this, we need to take into account the gendered attributions of trust, and the moral legitimacy of the state specific to the Norwegian context. The starting point for our analyses is a situation in which BSE was in the early stages of 'becoming real' in Norway, where the public debate was less intense than in countries that detected BSE, such as the UK and Germany. Our research questions concern the ways in which attitudes changed over time, and the gendered dimension of attitudes and practices in relation to meat.

Keywords: meat, consumer attitudes, food scandals, trust, gender

Introduction

Modern capitalism relies on a distinction between production and consumption which structures the area of food, and is particularly striking in relation to meat (Vialles 1994). In recent years, the aim towards bridging these domains has become more salient in academic debates (cf. Harvey *et al.* 2002, Lien 2004). Systems of provision and institutionalization are now more often called upon to explain aspects of food consumption, while constructions of 'consumers' in food manufacture are often part of the analysis of food production (e.g. Lien 1997). This implies that the spheres of production, consumption and food regulation – although institutionally and empirically separate - are approached as an interactive (and interlocked) systems or relations.
This article explores the shifting attitudes in relation to meat in Norway over from 1997-2004, as expressed through three repeated surveys among representative samples of Norwegian respondents. The surveys are part of a longitudinal research project which also includes a qualitative study, a literature survey and a media analysis. Although the surveys explore attitudes and practices among Norwegian consumers, the production side and regulatory framework provide an important context for the analysis. Furthermore, the gendered division of labor and patterns of consumption in the household constitute yet another context for the analysis, which together with production and regulation comprise the most relevant dimensions of what we may call the 'instituted relations of meat consumption' in Norway.

A significant background for this project was the turbulence around meat in the aftermath of the BSE outbreaks in the UK, and the subsequent ban on British beef in 1996 (Lien, Bjørkum and Bye 1998). Consequently, the surveys reflect shifts in Norwegian consumer attitudes from the time when the possible spread of pathogens from animals to humans had just been exposed, through a period of heightened attention to other food safety issues in Norway and the foot-and-mouth outbreak in the UK, to a period when meat practically disappeared from news headlines. In Norway, where there have been no recorded cases of BSE to date, the public debate has been much less polarized than in other European countries. Nevertheless, the food scandals in other parts of Europe did bring about a heightened awareness both among producers, veterinary and food safety authorities as well as the general public regarding the potential health risks associated with certain types of meat.

Risks 'becoming real'

In her analysis of the outbreak of BSE in the Czech Repbulic in 2001, Haukanes (2004) suggests that we think of risks *becoming real* as a staged process involving (1) a vague awareness of risk, (2) knowledge about the BSE and its potential spread to humans through the variant Creutzfeldt Jacobs disease, and (3) experiencing the consequences of BSE by knowing someone who actually falls ill. While the third stage is clearly restricted to those countries in which cases of BSE were actually detected, the second stage of becoming real could affect a much larger population, including that of Norway. Even if a BSE outbreak is prevented, as in the Norwegian case, the mere exposure of production practices to the general public, and especially the role of meat as a potential 'vessel for pathogens', could bring about irreversible changes in relations of trust and attitudes to meat and cause a temporary or a more permanent 'breakdown' of foundational relations of trust in the food market.

The starting point for our analyses is a situation in which BSE was in the early stages of 'becoming real'. A certain element of distrust and ambivalence in relation to meat is thus to be expected from the outset. Our research questions concern first the ways in which such attitudes changed over time, and especially whether the concerns voiced initially were upheld during the following years. Secondly, we explore the attitudes and practices in relation to meat from the perspective of gender. The provisioning of meat is highly gendered, as men dominate the production sphere, women hold a dominant position as gate keepers to the household, while the consumption of meat (especially red meat) serves as a cultural marker for masculinity (O'Doherty Jensen and Holm 1999). In light of this, we ask to what extent skepticism towards meat is also gendered, and explore whether men and women's attitudes towards of meat and meat eating can tell us something about the gendered mediations of trust and distrust within the household.

Skepticism towards meat

The most salient finding when comparing attitudes over the entire eight year period is a decreased skepticism towards meat. While in 1997, 37% of the respondents reported some sort of concern (Norwegian: *betenkeligheter*) in relation to their meat consumption, in 2004 this was down to only 32%.This proportion is even lower than what was found in a preliminary survey in 1995, *before* BSE became a public issue, when 36% reported some concern in relation to their meat consumption. Similarly, factor analyses conducted for each year indicate that questions related to the factor labeled 'skepticism towards processed meat', decreased from a mean score of 81 in 1997 to 73 in 2004. At the same time, a factor labeled 'trust in meat' which concerns trust in the production system increased from a mean score of 73 in 1997, (down to 68 in 2000) to 77 in 2004. Pointing in the same direction, the factor relating to risk society and a feeling of powerlessness decreased as well, from a mean score of 70 in 1997 to 63 in 2004. Altogether, these changes indicate that rather than weakening relations of trust in the Norwegian systems of provision, successive food scandals in Europe and Norway have in fact been followed by a period of heightened trust and reduced skepticism. This is paralleled by a total increase of meat consumption from 57,3 kg/person in 1997, to 60,8 kg/person in 2004, as well as a significant increase in the frequency of having meat for dinner in the same period.

Gender differences

When it comes to gender, we find throughout the entire period that gender is an important background variable that explains variation in both attitudes to meat and meat consumption. However, this tendency was significantly stronger in 1997 than in 2004, indicating that the polarization of differences between female and male respondents has diminished to some extent. The differences between men and women generally confirm cultural assumptions that associate meat and masculinity. We find that women are far more likely than men to mobilize against meat (e.g. by trying to convince others to eat less meat), and more likely to be skeptical towards processed meat. Furthermore, we find that men are more likely than women to report trust in meat and more men than women see meat as a necessary component of a meal or a diet. While these gender differences are valid for the entire period, the polarization is becoming less sharp. This is the case in relation to meat consumption as well. While in 1997, women, (and especially women without a male partner), reported significantly less frequent consumption of meat than men, this was no longer the case in 2004. Furthermore, if we look at reported concerns *(betenkeligheter)* in relation to one's meat consumption, we find strong and gender differences in 1997 and 2000, but less strong differences in 2004. In 1997, the proportion of men and women reporting concerns in relation to meat consumption was 30% and 42% respectively. In 2004, these differences had diminished as the proportion of concerned men was 29% while the proportion of concerned women was down to 35%. As the figures indicate, the less polarised situation is a result of changes of attitudes among women, who seem to gradually adopt the attitudes that previously characterised men.
How can we explain these changes? At first sight, the results suggest that the anticipation of a crisis that never occurs, such as BSE in Norway, could actually lead to an intensification of trust in the system of meat provisioning. Thus, compared to other countries in Europe, Norway may be seen to represent a special case. However, research from the UK in the aftermath of BSE indicates that even in the case of serious disease outbreak consumer trust in domestic provisioning may be re-established remarkably quickly. Furthermore, there were also food scandals *in* Norway in the same period, such as outbreak of salmonella in 2001, indicating a heightened awareness of meat safety issues in general. The high level of trust three years later could indicate that consumers approved of the way problems were handled. This interpretation is supported by results showing a marked increase in the proportion of respondents who confirm statements relating to institutional trust from 1997 to 2004 (Lavik and Kjørstad 2005).

The low-carbohydrate effect

While these interpretations may to some extent account for the higher levels of trust, they do not account for gender differences, and for the greater acceptance of meat among women. Looking more closely at questions relating to nutrition, we find marked changes in consumers' perceptions of the nutritional value of meat. In 1997, 38% of male respondents and 49% of female respondents would agree that 'meat is not very nutritious' (Norwegian: *kjøtt er ikke spesielt sunt).* In 2004, these figures were down to 30% for men and 33% for women, indicating a dramatic change of attitude among women especially. These changes were not entirely unexpected, as a Norwegian proponent of a low-carbohydrate diet not unlike Dr. Atkins (Fedon

Lindberg) had become a key public figure in the meantime, inspiring a highly publicised debate among Norwegian nutrition experts. His dietary advice emphasised lean meat as an important component of a healthy diet, at the expense of cereal and potatoes. Anticipating these changes, we therefore added a few questions in the 2004 survey related to dietary issues, and found that as much as 32% of female respondents and 21% of all male respondents had recently changed their eating habits as a result of the recent nutrition (low-carbohydrate) debate. Responding to what such changes had entailed, 36% of the female respondents reported eating more unprocessed meat, while 59% reported eating more poultry (defined as meat in the survey). In other words, a massive 'rehabilitation' of meat in relation to its nutritional value had taken place, and with immediate effects upon consumption patterns.

Eating what you trust, or trusting what you eat?

How does this relate to trust and safety issues? As far as the debates were concerned, connections were occasionally made between scandals pertaining to BSE, foot-and-mouth disease and salmonella outbreaks, as they all pertained to food safety issues. However, no such connections were made between the food safety issues on the one hand and the nutritional issues on the other. Nevertheless, we find that exactly as meat is seen as becoming 'more nutritious' and a substantial proportion of women thus make a conscious shift in their everyday diet, the concerns and skepticism towards meat also become less prevalent. Why is this so? We suggest that in order account for the way meat has now become 'safer than ever' we need to look at trust in food more broadly. Our surveys indicate that meat has for a long time been associated with considerable ambivalence, and to some extent it still is (Lien, Bjørkum and Bye 1998). The fact that meat is considered more nutritious than before as a result of the low-carbohydrate trend, does not make food safety and animal welfare issues disappear. However, from the perspective of those who decide to eat meat more frequently, a deep concern for such issues may seem difficult to uphold. This is because the consumption of food requires a fundamental level of trust (Lien 2004). Thus, by way of concluding, we suggest that the causal relation between levels of trust and consumption patterns is not as simple as it may appear. Rather than assuming that changes in levels of trust lead to changes in eating patterns, we suggest that the opposite may also occur. As eating patterns change for whatever other reasons, levels of trust may change as a result. The implications of our findings are, first, that the long-term implications of food scandals in relation to trust are not necessarily irreversible. Secondly, there are indications that the way crises are handled are likely to be significant in relation to future trust. Thirdly, gender must not be overlooked in studies pertaining to meat consumption, attitudes and trust. Fourthly, and most importantly, we suggest that trust and skepticism should not only be seen as independent variables in relation to food consumption. There is also the possibility that they may change as a result of consumption changes that are not, in themselves, related to issues of trust.

References

Harvey, M., Quilley, S. and Beynon, H. (2002). *Exploring the tomato.* Cheltenham, UK: Edward Elgar.

Lavik, R. and Kjørstad I. (2005) *Meat, attitudes and changes 1997 -2004.*

Haukanes, H. (2004). Enjoyment and choice in an age of risk: The case of BSE in the Czech republic. In: *The politics of food.* M. E.

Lien and B. Nerlich (eds.) Oxford: Berg. 101-120.

Lien, M.E. (1997) *Marketing and modernity.* Oxford: Berg.

Lien, M.E., Bjørkum E. and Bye, E.K. (1998) *Meat; Changes in attitudes and consumption.* Oslo: National Institute for Consumer Research.

Lien, M. E. (2004). Introduction. In: *The politics of food.* M. E. Lien and B. Nerlich (eds.) Oxford: Berg. 1-18.

O'Doherty Jensen, K. and Holm, L. (1999). Preferences, quantities and concerns: socio-cultural perspectives on the gendered consumption of foods. *European Journal of Clinical Nutrition*. 53:351-359.
Vialles, N. (1994). *Animal to edible*. Cambridge University Press, Cambridge.

The Citizen goes shopping: What do the peers have to say about it?

Tassos Michalopoulos
Applied Philosophy and Business Economics, Centre for BioSystems Genomics, Wageningen University, Hollandseweg 1, 6700 EW Wageningen, The Netherlands, tassos.michalopoulos@wur.nl

Abstract

The mounting pressure for a minimal (neo-classical) role for neutral (liberal) governments, together with the absence of institutionalized information systems that conform with the evolving set of consumer concerns, create a political deficit in the food market: a space out of reach alike for citizens and for consumers. Therefore, when democratically elected authorities are asked to reduce their regulatory role in the market, the question becomes urgent whether the assumption of 'perfect information' (upon which neo-classical economics bases its promise to maximize the aggregate satisfaction of consumer preferences) is being reasonably met. The present paper argues that in liberal market democracies, when citizens' concerns cannot get sufficiently addressed in the political arena, then appropriate information on non-perceptible food (production) attributes must be provided in the market. Also, a set of suggestions about the use of the term 'appropriate', and about the generation of such information by means of multiple criteria models is being made. Such concerns can relate to environmental risk tolerance, health risk tolerance, metaphysical issues (referring to animal welfare and food naturalness), and socioeconomic issues (referring to human rights and distributive justice).

Keywords: liberal democracy, consumer concerns, food market, food labeling, multiple criteria assessment

Introduction

Let us assume for a moment that there is a food market where no information (other than price) is provided to the consumer. In that occasion, when encountering two *prima facie* substitutable food products, the consumer would be able to distinguish between them only in terms of these features which she could perceive with her senses[91]: e.g. their color, taste, aroma, functionality, and price. Therefore, these features would define the terms that would determine her market choice. Consequently, these would also be the terms in which the market mechanism would evaluate her choice. Then, in a well-operating (competitive) market, one could expect that foods would become attractively colored, increasingly tasty, odorous, convenient, and relatively cheaper. To describe the market preferences of the particular consumer one would probably have to make use of a 'hedonistic' or an 'opportunistic' terminology.

Let us assume now that the particular citizen/consumer is neither predominantly a hedonist, nor predominantly an opportunist. Instead, let us assume that she happens to value environmental, health, metaphysical, and humanitarian principles. Moreover, let us assume that she wishes to live her private life in a way that allows her to tell herself at bedtime that she spent her day in agreement with her privately-valued principles, and that she did her share to proportionally affect the way that related issues are treated in her polity. Would she then have a reason to be unhappy with the absence of relevant information in the market? Not necessarily. Not if food production was uniformly regulated in a way that agreed with her own principles, and if political channels were available for her to shape these regulations. When these conditions would not be

[91] Hereinafter I term such issues as 'perceptible'.

met, the citizen would probably feel oppressed for being deprived of the possibility to live her life in agreement with her principles. She would probably wonder:
Who is taking the decisions about *my concerns* about the environmental, health, metaphysical, and humanitarian implications of food production in the *democracy* that I live in?

Who is taking the decisions?

People that are being told the story that they are living in a 'Liberal Democracy' can reasonably expect two things. One, that they are living in a 'Democracy' and two, that this democracy is 'Liberal'.
The use of the term 'Democracy' implies a situation where for each citizen is acknowledged a right to shape[92] collective decisions about any issue that is relevant to the operation of the polity. Hereinafter, I refer to this democratic commitment with the term 'the right to political autonomy'.
The use of the term 'Liberal' implies that for citizens who maintain reasonable views of the good life is acknowledged a right to live in agreement with their beliefs. Hereinafter, I refer to this liberal commitment with the term 'the right to private autonomy'.
Either for normative reasons (e.g. due to the absence of criteria that allow objective judgments on the validity of reasonable views of the good life) or for practical reasons (e.g. to ensure that disobedient minorities will not threaten the stability of the polity), liberal democracies give priority to the private over the political autonomy of citizens. Consequently, liberal democratic administrations are not comfortable to implement policies that favor or discriminate against existing (reasonable[93]) worldviews (Streiffer and Hedemann, 2004). Therefore, liberal democratic administrations see their primary role to be the regulation of the supply of a set of 'neutral' or 'primary' goods that are supported by an overlapping consensus (goods about which all citizens are concerned regardless of their particular worldview). However, even when regulations enjoy overlapping support, they might be factually constrained by the WTO condition for 'scientifically sound' risk assessment (Kerr, 2003). One may ask what happens with the supply of those goods that do not enjoy either overlapping demand, or a 'scientifically sound risk' support. The empirical observation is that to optimize the supply of such goods administrations rely on the capacity of the market mechanism to self-regulate. When concerned about such goods, for the satisfaction of her private autonomy and for the expression of her political autonomy, *the citizen is being advised by liberal democratic administrations to go shopping.*
The choice to direct citizens to the market when private and political concerns cannot be addressed sufficiently in the political arena is conceivable. If it can remain competitive, then the rational neoclassical system of self-regulated markets seems to work well as a maximization machine: Once it receives its input (the preferences of consumers as they are revealed by their market choices), it efficiently allocates scarce recourses to produce a positive output (the maximal aggregate satisfaction of these preferences).
However, notwithstanding the substantial liberalization of the economy worldwide for the last decades the empirically observed reality allows little optimism that certain citizen concerns are getting satisfied. Worldwide, concerns are raised about the impact of human activity on global climate, the massive loss of biodiversity, the rapid exhaustion of natural resources, the health risks involved in modern food production, the treatment of animals, the moral status of genetic engineering, the growing inequalities in wealth distribution, the rise of global poverty, and the

[92] Usually through her representatives and a 'majority rule'.

[93] Hereinafter, when referring to views of the good life (else: 'worldviews'), and consequently to concerns, Rawlsian reasonableness is assumed. When a worldview, and consequently a concern, can be shown to be 'unreasonable' the argument does not apply.

dramatic decline of life expectancy at the least favored regions of the planet (Africa). One might be led to wonder what does the market maximize.

What does the market maximize?

The market maximizes the satisfaction of consumer/citizen preferences as these are revealed by her market choices. The step from preferences to choices however can only be crossed by means of some assumptions. First, the citizen/consumer must be rational so that she opts for those goods that fit best with her preferences. Second, she has to have adequate income to afford the goods that she prefers (ideally, a 'democratic market' requires equal 'consumption power' among consumers[94]). Third, when it comes to non-perceptible issues, she has to have information.

For a number of citizens' concerns connected to food production, liberal democratic administrations show limited capacity to regulate and information is not institutionally anticipated in the market. Such concerns can relate to environmental risk tolerance, health risk tolerance, metaphysical issues (referring to animal welfare and food naturalness), and socioeconomic issues (referring to human rights and distributive justice). For each of these concerns, the ability of liberal democratic administration to regulate the food market so that it meets the expectations of concerned citizens is defined by the level of support that the respective concern enjoys within the citizenry. Because food naturalness is not regarded to enjoy overlapping support, administrations do not see justice in imposing 'naturalness standards' in the market. The rest of these concerns are acknowledged to enjoy some level of overlapping support, and relevant regulations are not absent. However, either due to scientific uncertainty (for environmental and health concerns), or because the concept of risk might not apply well (for animal welfare, human rights, and distributive justice concerns), regulations that enjoy overlapping support might fail to meet the 'scientifically sound risk assessment' condition, and consequently they are to be sanctioned by the WTO (see *de facto* GM moratorium in EU). For such reasons the autonomy of relevantly concerned citizens is constrained in the political arena. The market mechanism however appears to carry an inbuilt ability to mitigate these shortcomings. By rational consumption, the citizen who can afford it, can maximize the satisfaction of her private autonomy given the available product options. She can also 'vote' to increase the spectrum of preferred product options, and thus express her political autonomy. She can 'steer' the market. As long as she has appropriate information to translate her preferences into product choices, that is. When this information is missing, the question might become:

Why am I directed to the market to address certain concerns when the information required to perform that operation is not institutionally anticipated? Who steers the market on these issues?

Who is steering the market?

There are three popular replies to this question. One, that the market is steered by those actors of the supply chain who concentrate significant market power. In times of crises, this possibility maintains among concerned citizens an attitude of mistrust and cynicism against the agro-food industry. Two, that the market is not steered by anybody in particular. This view maintains that the food market evolves under the influence of a complex variety of factors. The question here becomes whether the current market mechanism is indeed useful to liberal democracies committed to citizens' political and private autonomy, for the optimization of the supply of non-consensual goods. Three, that the market is steered by citizens/consumers. The suggestion

[94] Herein factual income inequality is assumed to constrain the level of democracy that the market could achieve.

here is that emerged voluntary labels provide information which allows citizens to reveal their preferences, if, when, and to the extent they are really concerned.
Insofar as emerged labels are 'binary' and cover a fragment of product options, the third reply falls short of the observed reality. First, binary labels reduce the allocative capacity of the market: They require from producers significant investments to obtain a full certification, they require from citizens/consumers a significant expense to purchase the fully-labeled product, and they do not allow producers and consumers to make tradeoffs between different attribute levels. Second, when labels cover a fraction of existing products, they transfer steering power from consumers/citizens to food producers and to major retailers: Whether concerned citizens will purchase an unlabeled product depends on their image for the particular brand. Through psychology-based promotion producers can affect these images even when brands do not observe relevant concerns. Consequently, advertising can drive consumers to market choices that are, strictly speaking, 'irrational'. Major retailers can affect the opportunities for concerned citizens to cast market votes by deciding on which labeled products will be readily available in their shops.

What information?

If the food market is to meet the commitments of liberal democratic governance then, when the political and private autonomy of concerned citizens/consumers is not observed in the political arena, appropriate information must be anticipated so that autonomy can be observed in the market. To be appropriate, information must allow citizens/consumers to make use of the allocative merits of the market mechanism to optimize the satisfaction of relevant concerns. In other words it must allow for tradeoffs. Therefore, information must not be provided in a binary 'all-or-nothing' format. Moreover, information must be valid. To be valid, information must be justifiable to whom it is addressed. If information is to be useful for observing the autonomy of concerned citizens in the market, then product valuations must proceed in terms that are recognized by the reasonable views of the good life of concerned citizens. Thus, to assess 'how much natural' or 'how much sustainable' a product is, one must refer to a conception of 'naturalness' or 'sustainability' that is agreeable to those that voice relevant concerns. Finally, information must be credible. To be credible, the terms of product assessment must be explicitly articulated so that evaluations are as transparent as possible, and so that they provide the production sector with operational guidance for product development.
'Non-binary' information could be generated with the use of multiple criteria assessment models. Four issues at stake in such a model are: *Who* defines the assessment criteria, *who* provides the relative weights needed for the aggregation of these criteria, *whether* an overall aggregation of different categories of concerns is meaningful or desirable, and *how* does the model adjust to the dynamic nature of societal concerns. For the criteria and relative weights the present analysis suggests that they should obey the relevant perceptions of concerned groups. However, the explicit articulation of criteria demands some level of expertise from the decision makers. Therefore, relevant decisions could either be generated through deliberative citizens/consumers workshops, or be based on the opinion of associated actors from the civil society. The overall aggregation of a food's performance in different categories of concerns is problematic because relevant concerns are incommensurable. Also, this information would only be useful to these societal groups that maintain the full set of identified concerns. Nevertheless, aggregated assessment could be useful to direct the attention of the producers to the most urging societal concerns. It could have 'guiding value'. In that case the relative weights required for the overall aggregation should be provided by the wider citizenry, and not by some particular societal group. Finally, to keep the model in touch with the evolving set of societal concerns, the full set of weights and criteria should be updated in reasonable time intervals. This is a conceivable

position because societal concerns, and the criteria that become relevant, depend on the course of technological innovation and on the occurrence (or likelihood) of catastrophic events.

Concluding remarks

This paper has not argued for or against liberal democracy[95]. Also, this paper has not argued for or against the usefulness of the market mechanism to address constraints in liberal democracies[96]. Finally, this paper has neither argued that the availability of information in the market will drive this world towards a more 'sustainable', 'healthy', 'moral', 'humane', or 'just' future, nor that it ought to be so[97].

Instead, this paper has argued that when people are being told the story that they are living in liberal democratic polities, they can reasonably expect that their administrations acknowledge for them a right to political and private autonomy. Due to factual and normative constraints however, for the particular set of concerns, these rights are not observed when appropriate information is not anticipated in the market. This paper also made a set of suggestions about the use of the term 'appropriate'.

Once, Robert Frost defined a liberal as someone that cannot take his own side in an argument. As true as this might be when referring to policy decisions, a task that surely remains available to politically liberal citizens is the check for consistency between the stories that we live with and the empirically observed reality.

Acknowledgements

I owe to Henk van den Belt, Volkert Beekman, and Jacoba Wassenberg for insightful criticisms, patient editing and support. The full responsibility for content, style, errors, and inaccuracies is mine.

References

Kerr, W.A. (2003). Science-based rules of trade – A mantra for some, an anathema for others. *Estey Centre Journal of International Law and Trade Policy*, 4:86-97.

Streiffer, R. and Hedemann, T. (2005). The political import of intrinsic objections to genetically engineered food. *Journal of Agricultural and Environmental Ethics*, 18:191-210.

[95] On this, provided burdens in judgment, in the author's understanding political liberalism approximates the status of a moral necessity.

[96] It is acknowledged that (once competitive) the current mechanism targets a positive outcome. Whether this outcome is the most preferable depends on the qualities of the feasible alternatives.

[97] When such ideals intolerably fail in the market then a different set of questions must follow.

Mobilising for the animal cause as consumers: An analysis of strategies and practices in Netherlands and Norway

Laura Terragni, Liesbeth Schipper, Hanne Torjusen, Volkert Beekman and Unni Kjærnes
SIFO- the Norwegian National Institute for Consumer Research, P.O. BOX 4682, Nydalen, 0465 Oslo, Norway, Laura.terragni@sifo.no

Abstract

In the last years, there has been an increasing focus on consumption as a means for expressing ethical concerns and for influencing questionable market practices. The issue of the treatment of farm animals has been openly discussed for a long time. The focus of this paper is to analyse varying patterns in how people mobilise for the animal cause through their consumption practices. The comparative data presented in this study indicate that people, as consumers, may rely on different strategies for showing their concerns, and protest, against the ways farm animals are treated. Abstaining of eating products of animal origin (being a vegetarian or a vegan) represents one option. Besides, other strategies may be adopted, like avoiding specific food products or preferring other ones. The data suggest that there are significant differences across countries in the strategies people, as consumers, take on. The national context, in terms of public debate, consumption practices and availability of products, represents an important frame for understanding how consumers mobilise for the animal cause.

Keywords: animal rights, animal welfare, political consumption, food consumption practices

Introduction

The rise of concerns about the living conditions of farm animals and fact that this has become an issue for political activism and public debate (Jesper and Nelkin, 1992), can be regarded as the result of new sensibilities typical of post-industrial societies, characterised by altruism and disentanglement from ones' personal interests (Inglehart, 1977). This kind of political participation deeply interconnects collective and individual identities, where private choices become part of wider forms of mobilisation (Melucci, 1988). Consumption often plays an important role in this context, as it emerges as a specific arena for expressing protest and practicing alternative lifestyles (Micheletti, 2003). This paper addresses the question of how individuals mobilise, as consumers, for the animal cause by presenting data from a comparative study on consumers' attitudes and practices towards animal welfare. The study, conducted in seven Europeans countries is based on six focus groups discussions and a survey among a representative sample of the population of each country. For the purpose of this paper, we shall consider manly two of these countries: the Netherlands and Norway.

Our study indicates that there can be a plurality of ways for showing attention to the welfare of animals or protest against industrial farming. Some of these can more likely be characterised as forms of political consumption, since there is an explicit relation between the aim and the practice (Micheletti, 2003). More often, however, the data from our study indicate the prevalence of ordinary forms of expressing concerns through consumption (Halkier, 2001): consolidated routines, where the memory of why they started tends to vanish.

Our data, moreover tend to suggest that different approaches towards the animal cause, reflect alternative ways of interpreting the human animal relation. While people living in Norway, for instance, tend not to consider killing animals for food as problematic (as a large majority believes that it is acceptable to kill farm animals or to hunt game), the Dutch express, instead, reluctances to fully legitimate the killing of farm animals and openly oppose hunting. On the other side, however, Norwegians tend to show a higher interest in issues related to animal welfare. These finding may indicate the presence- also in our data- of the tension between two poles of debating and coping with the use of animals for human needs: one, stressing the fact that animals have the right to a life of their own and an other, which does not question the killing of animals for food, but wants them to live a decent life.

Practices: Protest and routines

Vegetarians, although limited (but more relevant in the Dutch sample), represent a critical presence, as they indicate the fact that alternative choices are possible. Informants tend to refer to this alternative when wanting to stress their opposition to industrial farming methods. "*I could really become a vegetarian*", "*do we really need to eat meat?*", were among the statements emerging at some stage during the focus groups discussions. Vegetarians are often among the most active among the informants, combining specific consumption choices with other forms of involvement: as being members of animal organisations, being active in internet networks or in local activities. The idea that people, as consumers, have a responsibility and that everyone is able to make a difference often characterises their discourses: "*Not that I sit at home, on the couch and think about animal welfare. But you do have a responsibility. And you can make a choice. You grow in your consciousness but you have to take your responsibility*" (Dutch politically active and vegetarian consumers). During our field work in Norway, we met a group of vegetarians gathering periodically in order to share a dinner. The initiative was taken by one of them, who put an ad in a local shop. The organiser of these reunions said that she consider herself as "*a drop of water in the sea. I try not to convince others to be like me. I just try to show them that I lead a good life. That they can be happy to choose the same type of life as me*".
Vegetarians, however, do neither represent a homogeneous neither nor static group. Together with the ones firmly stating that they "*will never eat their brothers*" (in the words of our Norwegian informants), emerged a group that can be defined as "contextual" vegetarians (Guzman and Kjærnes, 1998). Among these, are the ones who do not eat meat as a form of protest against modern factory farming and the living conditions of animals in intensive production. The main issue, in this case, is not eating meat, but not eating specific kinds of meat. Considerations about the animals' living conditions often play an important role in these decisions: foie gras, meat coming from veal, chicken, lamb or piglets are the most often quoted examples of products one should definitely avoid if caring about animals. Reducing the consumption of meat, furthermore, is described also as a possible option for taking responsibility. This practice is often framed in discourses where price for food is discussed: "when meat is too cheap you should wonder who is paying the costs", tend to be a common opinion. As a remedy, "*we can eat less, pay what is needed for good welfare for animals, and take an extra potato if we are still hungry* (Norwegian rural woman). Buying organic food, as well as buying eggs from free range hens, or choosing "small eggs" instead of "large ones", are often indicated as positive strategies for taking animal welfare into consideration as consumer. Some people, as exemplified by the words of a Dutch informant, seem to have elaborated a rather articulated chart guiding them in the choice of the meat they'd rather prefer to buy or to avoid: "*well I do have that I am being influenced in my choice of meat, I think that chickens and pigs are worse off, I think that sheep and cows just have a better life. (...) Cows also give milk, they are in more ways useful. For that reason I don't take pig meat so much, chicken maybe some more*" (man, 30, young singles).
However, it emerged clearly that avoiding some products, as well as preferring others, has often become a routine. Examples of this kind of process emerged particularly in the Dutch focus groups, where, despite the fact that not many people buy battery eggs (since it is not possible to

buy battery eggs in Dutch supermarkets), clearly for animal welfare reasons, nobody mentions this when asked if they boycott or buycott animal products (Schipper and others, forth). A man told that he only buys outdoor access eggs now that he knows that free-range hens are not so well off as he thought. He also regularly buys organic meat, for animal welfare reasons as well. Yet, when asked whether he boycotts any products, he did not respond. It seems that in many cases the boycotting and buycotting of certain animal products has become part of the daily routine. It is not something that is regarded as so special anymore and not something that people are aware of. The underlying reasons for boycotting and buycotting remain there, but they become somehow implicit.

Consumers agency inside, beside and outside the market

Mobilising as consumers in the market depends on a number of factors (Terragni and Kjærnes, 2003). Availability of alternative products is one of them. If one wants to express protest through choices, choices must be available. In this respect we have observed interesting differences between Norwegian and Dutch consumers in their ways of coping with the choice of alternative products. "Harvesting directly from nature" through hunting and fishing represents for many Norwegians a way for expressing protest towards intensive farming methods (Terragni and Torjusen, 2005). Independency from the conventional market is indicated as a powerful way for expressing consumers' agency: "*we never buy meat in the shops*" said proudly a man living in the Norwegian countryside. Also among young urban people, this kind of alternative provisioning system is regarded as an important option. This may also partially explain (or be explained by) the scarcity, in Norway, of labelled "animal friendly" product, a kind of opportunity, instead, available to a larger degree for the Dutch consumers. Solutions inside the conventional market seem to be mainstreaming in the Netherlands. An example of this is the fact that since 2004 it is not possible to buy battery eggs in the supermarkets, as they have been replaced by free-range ones. This story is often reported as a success of consumer activism (since almost no one was buying battery eggs anymore). Yet it is also a rather disputed story (Schipper *et al.* 2005). The reality of free-range chicken farming, in fact, is in sharp contrast with the image of lucky animals that these labels suggest. To discover this was rather discomforting for many Dutch consumers. According to the focus group discussions, some lost trust in market claims, others moved towards labels they felt could be trusted, as the organic ones. Two participants took part to the "adopt a chicken" initiative, enabling people to adopt an organic chicken paying an amount of money per year receiving in return six eggs per week.
The tendency of being active in the market, as consumers, varies often according to countries, suggesting that political dimensions concerning participation and forms of expressing voice also play a role. According to the data from the survey, the tendency to be active in the market (complaining, boycotting, boycotting) is rather low among Dutch consumers, while Norwegians seem to be more active. However, compared with Norwegians –but also with the other countries participating to the survey- Dutch are more involved in pressure groups or organisations working for improving food. In addition, Dutch, more than Norwegians, state that their view on animal welfare has been influenced by animal protectionist campaigns. These findings may indicate a higher degree of mobilisation on this issue among the Dutch civil society. At that regard it is worth mentioning that in 2003, an "Animal Party" (*Partij voor de Dieren*) was established, and it only barely did not win enough votes to gain a seat in the European parliament. In Norway, on the contrary, issues related to animal welfare are very seldom the object of a public debate. Discourses and policies concerning farm animals are more often dealt with as a field dominated by expertise than by activism.

Concluding remarks

Our analysis indicates that people mobilise for the animal cause as consumers in a variety of ways. We find distinctions between strategies which are more concerned with animal rights vs. animal welfare; conscious or routinised choice; and, finally, market behaviour or civil society activism. The Dutch case seems to be characterised by the presence of strong moral concerns about animals and by the tendency of being more active in the political arena than not in the market. The presence of "routinised practices" may indicate that people actually do alternative choices as consumers without regarding them (anymore?) as political. In Norway, instead, we face the fact that animal cause has not been-at least up to now- a political issue. Concerns towards the welfare of farm animal are present, and are mirrored through heterogeneous practices. These choices, however, do not become politically visible in the market, probably because of the scarcity, in Norway, of products addressing clearly the issue of animal welfare.

Acknowledgments

This paper draws on comparative data collected by the *Welfare Quality* research project, (FOOD-CT-2004-506508), co-financed by the European Commission within the 6th Framework Programme. This paper represents the authors' views and does not necessarily represent a position of the Commission, who will not be liable for the use made of such information. More information about the project is available on www.welfarequality.net.

References

Guzmán, M.A. and Kjærnes, U. (1998). *Menneske og dyr. En kvalitativ studie av holdninger til kjøtt*. Arbeidsrapport nr. 6-1998. SIFO, Oslo.

Halkier, B. (2001). Consuming Ambivalences: Consumer Handling of Environmentally Related Risks in Food. *Journal of Consumer Culture*, vol. 1(2) p.51-65.

Inglehart, R. (1977). *The silent revolution: changing values and political styles among western publics*. Princeton University Press: Princeton.

Jesper and Nelkin (1992). *The animal rights crusade*. The Free Press: New York.

Melucci, A. (1989). *Nomads of the present, social movements and individual needs in contemporary society*. Temple University Press: Philadelphia.

Micheletti, M., (2003). Political virtue and Shopping. Individuals, consumerism and collective action. New York: Palgrave Macmillan.

Schipper, L., Beekman V. and Korthals, M. *Focus group discussions country report*. Forthcoming in the Welfare Quality working paper series.

Terragni, L. and U. Kjærnes (2005). *Ethical consumption in Norway: Why is it so low?*. In *TemaNord 2005:517*, The Nordic Council of Ministers: Copenhagen.

Terragni, L. and Torjusen, H. (2005). *Recognising animals as subjects and enjoying them as food. An investigation of consumers' ideals and dilemmas when eating food of animal origin*. Paper presented at the XXI ESRS Conference: A common European Countryside? Change and Continuity, diversity and cohesion in the enlarged Europe. Keszthely, Hungary, 22-27 August 2005.

Part 13
Animal welfare and food production

Animal welfare differences between Caesarean section and natural delivery in Belgian Blue cattle

Stefan Aerts, Iris Kolkman, Hilde Vervaecke, Jo Vicca and Dirk Lips
Centre for Science, Technology and Ethics, Kasteelpark Arenberg 30, 3001 Leuven, Belgium, Stefan.Aerts@biw.kuleuven.be

Abstract

Over the last decade international resistance to the use of the Belgian Blue cattle breed (BBB) has been growing. Only the last years have these concerns gained (some) acceptance in the production sector. The preliminary results of an ongoing research project show some behavioural differences between cows that deliver through Caesarean Section (CS) versus cows that calve naturally, but it remains to be clarified whether these differences indicate pain and thus reduced welfare. These results seem to contradict the common perception that CS is detrimental to animal welfare and somewhat paradoxically, there is evidence in literature that calves experience more stress during natural calving. A second line of argument against CS is that the routine use of CS in BBB cattle is a sign of excessive instrumentalisation of these animals. We acknowledge the validity of this line of argument in general but argue that this is not specific to the BBB. Moreover, it is not a valid argument in an animal welfare discussion, which should focus on what matters for the animal itself. We conclude that he discussion on CS in the BBB is typically mixing what is important for humans and what matters for the animal. Although both are ethically important, only the latter is relevant to a real animal welfare discussion.

Keywords: animal welfare, animal rights, caesarean section, cattle

Introduction

At the third EurSafe conference in Florence, we have already reported about the growing social pressure on the Belgian Blue cattle breed (BBB) and the validity of ethical objections to the use of Caesarean Section (CS) (Lips *et al.*, 2001). This was inspired by the vivid debate about the ruling of the European Court of Justice (1998) that denied several Scandinavian countries the right to block the import of BBB sperm. The public debate about the issue has abated, but there is still considerable international pressure on the BBB (e.g. Turner, 2002), albeit sometimes based on incorrect or suggestive information (D'Silva, 2002). The most often cited objection against the use of the BBB is the high rate of CS in purebred lines: estimates usually surpass 90% (Coopman *et al.*, 2000; Eurogene AI Services, 2002).
During the last few years, the discussion about the BBB and the use of CS has also been opened in professional circles. There are still considerably different positions (e.g. Nantier, 2005 vs. Hanset, 2004), but the fact that this discussion is possible seems to indicate a growing openness of the BBB selection and production sector to these non-technical considerations, even in Belgium. It is in this changing climate that a first preliminary research project into the selection of the BBB was granted.

The preliminary project

The study we describe is part of a larger research project into the different aspects relating to the high CS incidence in the BBB. We have reported the launch of this project to the international community at an early stage (Aerts and Lips, 2005). Its fourfold goal is: (1) to identify whether there is a minimum pelvic size (relative to calve size) for BBB cows to enable natural calving, (2) to gather information about the growth curve of the female BBB pelvis, (3) to get an indication of the heritability of pelvic sizes in cattle, and (4) to assess the painfulness of a CS for a cow. In the context of this paper, only the latter is discussed. We have used an ethological approach to study possible behavioural indicators of chronic and acute pain in cattle, based on an extensive literature study on pain assessment in cattle. Through focal animal sampling, using The Observer software (Noldus, 2003), we studied the behaviour of 13 naturally calving BBB cows and 17 BBB cows that calved through CS. All animals were housed at the same farm and management did not differ between the two groups. All animals were 'S'-meat type. Observations were carried out on four days: prepartum (at least one month before delivery), the first and third day postpartum and two weeks postpartum. Each observation day, the animals were observed for 45 minutes during the morning, the afternoon and the evening.

In the context of this paper it is interesting to note that there were some significant behavioural differences between the two groups (Vandelook, 2006). On the first day postpartum, cows that calved by CS lied down more often, showed less foot motility, showed more transitions from lying to standing and vice versa and reacted more when left side was touched. Cows that calved naturally reacted more when the vulva was touched. It remains to be further investigated whether these behavioural differences are indicators of pain or discomfort. Cortisol quantification from faecal samples will be carried out in a later phase of this project as a physiological measure of stress.

Ethical questions about the BBB

The first factor determining animal welfare that comes to mind is pain. Vandenheede *et al.* (2001) observed that calves are allowed to suckle from the left side of the cow – the side where the section is performed - as much as calves that are born through a natural calving. This indicates little discomfort in the cow caused by external pressure from the calf. More importantly though, one should ask the right question: the question is not 'Does CS cause pain?', the question is 'Does CS cause *more* pain than natural delivery?'. The latter question must then be answered for acute pain (at the time of calving) and for chronic pain. Our results indicate some significant short-term behavioural differences between the cows that calve naturally and cows that calf by CS, but overall, the differences are subtle.

Animal welfare comprises more than the absence of pain (see e.g. Appleby and Sandøe, 2002) and it should be recognized that suffering or reduced welfare (to be more general) can occur without pain, even in a rather functional approach to animal welfare. We should therefore address the question whether this omnipresent discussion on CS and pain is in fact the most important issue surrounding the BBB. Moreover, a certain level of birth stress may be beneficial, particularly to the calf. Uystepruyst *et al.* (2002) observe positive effects when suspending calves by the hind legs after birth by CS and suggest that CS apparently does not induce *enough* stress in the calf.

In 2001 (Lips *et al.*) we have already discussed some phenotypically adverse traits in the BBB (see also Coopman and Van Zeveren, 2000 and Arthur, 1995). Some of these are genetically determined, but others are caused by a lack of space in the uterus. Coopman *et al.* (2004) for example observed that 20% of the calves in their study weighed more than 59 kg at birth (up to 81 kg), which is quite heavy for dams whose average height at the withers is between 1,32 and 1,38 m (HBBBB, 2006). The deformations that may result from this situation may not have direct painful consequences, but are surely detrimental for the animal's welfare. The lack of (national and international) debate on this topic – possibly due to a lower visibility of these problems – does not render it less important.

If CS is not the key ethical issue regarding the BBB, what then might be the reason that it is the main point of discussion? We will not discuss the economic (protectionist) reasons that might inspire breeders and representatives from some countries to use the CS-argument to block the use of BBB cattle, but there are some genuine ethical concerns that need addressing. The main aspect here is the opposition against excessive instrumentalisation of animals. The public does not want animals to be used purely as machines that produce milk, meat, eggs, wool, etc. The creation of the double muscled BBB that can not procreate successfully without human assistance, appears to be a step beyond what is commonly considered to be an acceptable level of instrumentalisation. Nevertheless, it should be accepted that a certain level is unavoidable in any form of animal production. Although public concern does not equal ethical necessity, we do in this case reach a very fundamental ethical issue. Are we or are we not allowed to use and adapt animals to the point that they, as a breed, are no longer – physically – able to lead a natural life (in which successful procreation is a key factor)? Is dependence from CS fundamentally different from other forms of instrumentalisation in animal production?
The former questions need careful ethical scrutiny, for which space is too limited here, but we can provisionally discuss the latter. The use of CS is evidently not the only form of instrumental use of BBB cattle. Bulls as well as cows are routinely disbudded, in some cases they are held in a tiestall during part of the year and artificial insemination is the rule rather than the exception. Branding and castration on the other hand is rare in Belgian production circumstances. If we compare this with production practices in other species we see that commercial turkeys are in a similar situation (difficult natural procreation) and, more importantly, that other species (chickens, pigs, ...) are faced with far more limiting living conditions. Other cattle production practices can equally be regarded as far-going instrumentalisation (castration in Anglo-Saxon cattle production, culling of steer calves of dairy cows, ...). We can conclude that the BBB is not fundamentally differently or more heavily ' instrumentalised' than other agricultural animal breeds. It is however clear that the routine use of CS in the BBB is a sign of far-going instrumentalisation.

Animal ethics

Through the example of the discussion on CS in the BBB we hope to have demonstrated that animal welfare discussions are often mixing what is important for humans and what matters for the animal. If we carefully analyse the above discussion, we see these two lines of argument: (1) pain and suffering of cow and calve should be avoided (these matter to the animals), and (2) (far-going) instrumentalisation is wrong (as such this matters only to humans – somewhat provocatively stated). This difference resembles the difference in the works of Bentham (1781), Salt (1894), Singer (1975) and Regan (1983). Although their conclusion is similar (vegetarianism or even veganism is the ethically right way of life), they do not use the same line of argument. Bentham's and Singer's utilitarianist approach focuses on (the animal's) pain and suffering, while the rights-based approach (Salt and Regan) deals with the correct human behaviour towards animals, and pain and suffering are in fact of secondary importance.
We consider only the first line of argument to be relevant to animal welfare, which means that we hold a feelings based animal welfare definition. Any procedure, situation or state can only be considered as connected to animal welfare if it is recognized by, felt by (or in any other way matters to) the animal. Thus, any line of argument that uses exclusively human constructions such as rights, integrity and instrumentalisation should be banned of any animal welfare discussion.
This is not to say that we reject the integrity- or rights-based arguments altogether. On the contrary, both lines of argument are ethically important in their own right, and excluding the rights discussion from animal welfare is not equal to dismissing it entirely. We do not hold the

position that anything is allowed as long as the animal does not suffer. We state that presenting these ethical arguments as animal welfare related does not do them right, and effectively cripples both lines of argument.

Conclusion

The routine use of CS in the breeding of BBB is commonly taken to reduce the cow's welfare. We have argued that this can only be true if CS induces more pain (or suffering) in cows than does the alternative: natural calving. It remains to be further investigated whether the observed behavioural differences between cows calving by CS versus cows that calf naturally, are indicative of pain or reduced animal comfort.

This does not preclude other legitimate ethical concerns about the use of CS in the BBB. There are e.g. good arguments to consider this as a sign of too far-going instrumentalisation of these animals. However, we consider these arguments as irrelevant in a real animal welfare discussion.

More generally, the failure to recognize the difference between these two lines of argument (and the underlying axioms) is in our view an important reason for the often heated animal welfare debate. The result is the current flawed discussion in which the focus is continuously shifting away from the real centre: the animals themselves.

Acknowledgements

The BBB project is supported by the Flemish Government (PWO Fonds). We are deeply indebted to the farmer for the opportunity to study his animals and to Jeroen Vandelook for all the observation work.

References

Aerts, S. and Lips, D. (2005). *The Belgian Blue Beef and Caesarean Section.* Poster presented at: *From Darwin to Dawkins: the science and implications of animal sentience.* Compassion in World Farming Trust, London.

Appleby, M.C. and Sandoe, P. (2002). Philosophical debate on the nature of well-being: implications for animal welfare. *Animal Welfare,* 11: 283-294.

Arthur, P.F. (1995). Double muscling in cattle: a review. *Australian Journal of Agricultural Research,* 46: 1493-1515.

Bentham, J. (1781). *An introduction to the principles of morals and legislation.* T. Payne, London.

Coopman, F. and Van Zeveren, A. (2000). Meest voorkomende afwijkingen bij het Belgisch Witblauw vleesvee. *Vlaams Diergeneeskundig Tijdschrift,* 69: 323-333.

Coopman, F., Gengler, N., Groen, A.F., De Smet, S., and Van Zeveren, A. (2004). Comparison of external morphological traits of newborns to inner morphological traits of the dam in the double-muscled Belgian Blue Beef breed. *Journal of Animal Breeding and Genetics,* 121: 128-134.

Court of Justice. (1998). *Judgement of the Court in Case C-162/97, on the interpretation of Article 30 of the EC Treaty and Article 2 of Council Directive 87/328/EEC of 18 June 1987 on the acceptance for breeding purposes of pure-bred breeding animals of the bovine species.* European Court of Justice, Luxembourg.

D'Silva, J. (2002). *Farm animal genetic engineering and cloning.* Petersfield, Compassion in World Farming Trust.

Eurogene AI Services. (2002). *Belgian Blue test results* [online]. Eurogene AI Services. Formerly available at http://www.eurogeneaiservices.com [Date of consultation: 01/03/2002]. (no longer available)

Hanset, R. (2004). *Emergence and selection of the Belgian Blue Breed* [online]. In honour of the Danish B.B. Herd-Book at the occasion of its 25th anniversary. Available at: http://www.hbbbb.be/images/emergenceandselection.pdf [Date of consultation: 31/03/2006].

HBBBB. (2006). *Characteristics of the BBB* [online]. Available at http://www.hbbbb.be/belgianbluebreed.htm [Date of consultation: 04/04/2006].

Lips, D., De Tavernier, J., Decuypere, E. and Van Outryve, J. (2001). Ethical objections to caesareans: implications on the future of the Belgian White Blue. In: *Preprints of EurSafe 2001 "Food Safety, Food Quality and Food Ethics". The Third Congress of the European Society for Agricultural and Food Ethics, 3-5 October 2001, Florence, Italy.* M. Pasquali (Ed.). A&Q, Polo per la Qualificazione del Sistema Agroalimentare, University of Milan, Milan. pp. 322-326.

Nantier, G. (2005). In voortrekkersrol: bekkenbreedte meten een mogelijke aanzet tot natuurlijk kalven bij witblauw. *VeeteeltVlees* 4(2): 16-18.

Noldus. (2003). *The Observer Software Version 5.0.* Noldus Information Technology, Wageningen.

Regan, T. (1983). *The case for animal rights.* University of California Press, Berkeley.

Salt, H.S. (1894). *Animals' rights considered in relation to social progress.* MacMillan & Co, London.

Singer, P.A. (1975). *Animal liberation. A new ethics for our treatment of animals.* New York, New York Review/ Random House.

Turner, J. (2002). *The gene and the stable door: biotechnology and farm animals.* Petersfield, Compassion in World Farming Trust.

Uystepruyst, C. and Lekeux, P. (2002). Sternal recumbency or suspension by the hind legs immediately after delivery improves respiratory and metabolic adaptation to extra uterine life in newborn calves delivered by caesarean section. *Veterinary Research* 33: 709-724.

Vandelook, J. (2006). Een aanzet tot het selecteren van BWB dat natuurlijk kalft: Pijnverschillen tussen natuurlijke partus en keizersnede. MSc Dissertation. Faculteit Bio-ingenieurswetenschappen. Katholieke Universiteit Leuven, Leuven.

Vandenheede, M., Nicks, B., Désiron, A. and Canart, B. (2001). Mother-young relationships in Belgian Blue cattle after Caesarean section: characterisation and effect of party. *Applied animal behaviour science,* 72: 281-292.

Stakeholder's attitudes to and engagement for animal welfare: to participation and cocreation

Stefan Aerts and Dirk Lips
Centre for Science, Technology and Ethics, Kasteelpark Arenberg 30, 3001 Leuven, Belgium, Stefan.Aerts@biw.kuleuven.be

Abstract

A project was conducted in Belgium to investigate different stakeholders' attitudes to animal welfare by letting them engage in a dialogue on animal welfare. In this project consumers, primary producers, other economic actors, scientists, government officials and representatives of relevant NGO's were included. Using the focus group technique, we identified 6 major areas ('insights') that are relevant to the animal welfare debate: image formation, economic reality, ethical considerations, policy, knowledge and technical progress. This has led to the identification of three 'similarities', three 'differences', and six 'dilemmas'. The group of stakeholders selected 5 dilemmas for further discussion: animal welfare versus economy, animal welfare versus ergonomics, ethical necessity versus luxury problem, housing versus nature, and science versus social expectations and politics. This further discussion elicited four values that live among the stakeholders, connected to animal welfare: (1) economic welfare as a cornerstone for a good living quality, (2) care, (3) respect and (4) taking responsibility. From there we have identified seven *social indicators* for animal welfare. This type of project is an uncommon, non-technical approach and it does not have any real "educational" intentions. *Dialogue* as a project technique is uncommon, but it makes this project just as much a *process* as a project. It returns not only scientifically interesting data, but it makes a lasting contribution to future developments in animal welfare.

Keywords: animal welfare, stakeholder, deliberation, dialogue, indicator

Introduction

The EU Action Plan (Anonymous, 2006) and the results of the 2005 Eurobarometer (Anonymous, 2005) clearly show that animal welfare is a high on the social and political agenda. As a consequence, there is quite some research focused on animal welfare or a welfare component is added to research proposals. Increasingly it is recognised that a technical approach is insufficient to address the social concerns about animal welfare. This results in research into these consumer concerns (see e.g. Miele and Evans, 2005). This indeed generates very useful information when working towards a more welfare friendly animal production.
Unfortunately what is often ignored is that there are many more – and equally or more important – stakeholders to be included in such research. Parallel with and independent from the Welfare Quality project, the authors have worked together with several Belgian organisations towards a greater understanding of animal welfare as it is perceived by *all* stakeholders: consumers, farmers and retailers, but also researchers, industry, government officials, opinion makers and pressure group ('NGO') members. This is a first step towards improved communication about agriculture and animal welfare and essential when one is to develop animal welfare improvements or even assessment schemes.
For the sake of brevity we will regularly use 'AW' instead of animal welfare.

Project description and methodology

Within the project "Social indicators for animal welfare" there are several important phases, most importantly the consultation and the dialogue phase. In the former, we first conducted ten focus group interviews (for detailed information: Krueger and Casey, 2000) with socio-demographically and geographically diverse consumer groups. The most important selection criteria were: urban/rural, age, social network and geographical distribution, information on social class and education was collected. Two focus groups were dedicated to special interest groups: vegetarians/ vegans and 'responsible' consumers. We also conducted four identically structured focus groups with the other stakeholder groups mentioned in the introduction. Spontaneous reactions on AW were collected before focusing on farm animal welfare. Ideal AW, problems with, priorities in and importance of AW were addressed, as well as the expected stakeholder roles. All focus groups were recorded and transcribed live (Goris and Baeten, 2005).
The second phase of the project consists of three dialogue days aimed at confronting the participants with each others questions and views. They were asked to engage in a real but structured dialogue (= deliberation, not discussion). Dialogue is defined here as looking for a (deeper) understanding of views, realities and assumptions of *all* actors, clarifying different sorts of uncertainties, making different 'looking glasses' explicit, and implicates equality of ideas and views . This process of dialogue resulted in 6 *insights*, 3 *similarities*, 3 *differences*, 10 *dilemmas*, 4 *values* and 7 *indicators* connected to AW. As it is difficult to generate mutual understanding from accenting differences, we did not proceed from these, but from the dilemmas that were identified. The flow of the project is shown in Figure 1.

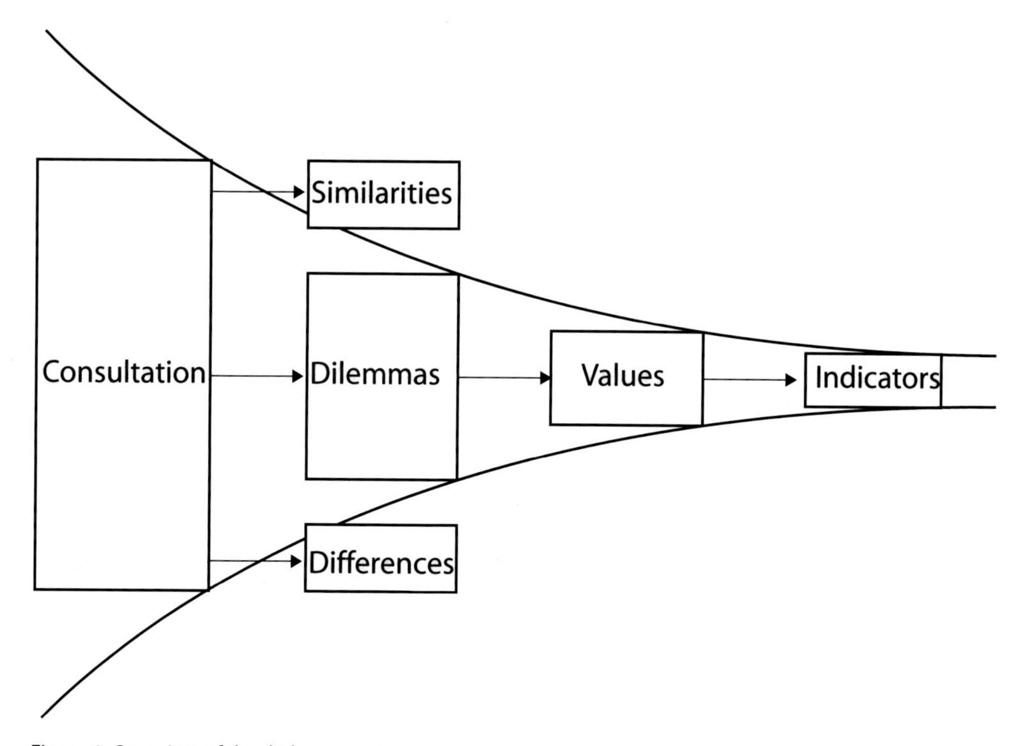

Figure 1. Overview of the dialogue project.

Results

Analysis of the data from the focus groups and the first dialogue day has revealed six important insights about the animal welfare situation and debate. Firstly, the *image* of animal production in the media cools the willingness of economic actors to look for alternatives to current practices, and (therefore?) the existence of animal welfare problems is often denied. Stakeholder's perceptions are usually based on clichés rather than reality. Secondly, AW problems are often due to industrial production systems, for which no single actor is to be held responsible. This *economic reality* can not be changed by individual consumers or by adapting local legislation alone. Thirdly, the consumer's *ethical expectations* are high and not strictly limited to AW. Often there is a strong connection made between animal and human welfare and (welfare friendly) animal products are expected to be available for everyone. Furthermore, matters such as poverty are generally considered more important than AW and all stakeholders put most responsibility with the government, who has to communicate transparently and consistently and has to strive for a greater coherence between different *policy* choices. *Knowledge* about animal production and AW is very limited. This is only partly due to insufficient communication, but more fundamentally with a lack of trust in the knowledge sources. Most suggest that restoring the connection between consumers, producers and animals is better than dissipation of 'objective' knowledge. Lastly, the development of new husbandry systems provides the opportunity to take AW into account, if the economic and ethical choices are correctly communicated. Consumers consider a purely *technical* approach as very threatening (for AW as well as far food safety).
In three areas most stakeholders held similar views. All agree that animals are more than mere production inputs and need to be treated with respect. The dilemmas are commonly recognised as difficult decisions, with no absolute right or wrong. All actors feel that they need more knowledge about the views of others.
Clear differences were observed between actors with respect to their views on how much space animals need for a good life, on what makes housing adequate for animals and most important on the degree of emotional connection with animals.
Table 1 gives an overview of all ten dilemmas that were identified. The attendees selected five of these for further discussion: AW versus economy, AW versus ergonomics, luxury problem versus ethical necessity, housing versus nature, and science versus society and politics. These dilemmas are an intermediate for a further structured dialogue, working towards an explicit formulation of values that people hold with regard to animal production and consumption.
Analysis of the consultation data and the discussions on the dilemmas revealed four primary values connected to AW: (1) economic welfare as a cornerstone for a good living quality, (2) care, (3) respect and (4) taking responsibility. *Economic welfare* represents the importance (for consumer and producer) of a good income in order to insure sufficient living quality. Its primary position does not prevent the existence of other considerations and neither does it mean that everyone strives for a maximum income. *Care* appeals to the moral duty to actively take care of vulnerable entities (humans, animals, ecosystems) within the boundaries of personal possibilities. *Respect* on the other hand is an attitude. This means one should respect animals as living creatures and that animal products are not something to be taken for granted. When problems arise or changes are necessary, all should *take responsibility*. The amount of responsibility depends on everyone's influence and carrying capacity.
The search for indicators to 'measure' the state of AW in contemporary society is based on the framework proposed in works like that of Vemuri (1978) and Østergaard and Hansen (1995). We can not elaborately discuss the different characteristics of an indicator here, but in general it should at least be clearly connected to the topic, have a scientific basis, be quantifiable and give information about solutions or corrections. The following indicators have been proposed (and should be ready for immediate use): (1) % eggs from alternative systems, (2) % government

*Table 1. Overview of dilemmas, values and indicators identified in the project. Dilemmas with * were selected for further discussion by the attendees.*

Dilemmas	Values	Indicators
AW vs. economy*	Economic welfare as a cornerstone for good living quality	% eggs from alternative systems
Regulate vs. market disruption	Care for vulnerability	% government support for animal friendly systems
AW vs. ergonomics*	Respect	Importance of AW in purchase of animal products
Housing vs. nature*	Taking responsibility	Economic lifespan of cows and sows
Luxury vs. ethical necessity*		Animal health (index)
Science vs. society and politics*		Number of animal abuse complaints
Living creature vs. product		Number of sanctioned farms (hormones, residues)
Openness vs. protection		
AW vs. environment		
AW vs. food safety		

support for animal friendly systems, (3) importance of AW in purchase of animal products, (4) economic lifespan of cows and sows, (5) animal health (an index indicator with data on mastitis, rejected pigs at the slaughterhouse and ascites in broilers), (6) number of farm animal abuse complaints, (7) number of sanctioned farms (hormones, residues).
The members of the project agreed on a visualisation system based on the 'radar' or 'spider' diagram (Figure 2), a technique also used in other works (such as Ten Brink, 2000). Applying this diagram year after year will give a clear view of how (attention to) AW is evolving.

Discussion

This type of project yields its results on two levels. There are visible results such as the lists of dilemmas, values and indicators, but long-term effects can also be expected. This project will hopefully influence the nature and practice of the AW debate in Belgium. At a first glance it might seem inappropriate to consider the methodology used in this project as an important realisation. This ignores the reality of most attempts to take further steps in the AW debate by bringing together several actors. Generally these attempts result in little more than the reiteration of everyone's – widely know – views and proposed solutions. This follows from the incorrect interpretation of the AW debate. Although the data from our consultation phase confirms the fairly general lack of knowledge about current agricultural practices, we argue that the debate on AW will neither be greatly facilitated by trying to educate the general public nor by putting experts together to work out a 'solution'. We hypothesize that the major impediment in the animal welfare debate is the lack of trust and familiarity between the important actors (individually and collectively). We propose dialogue to take useful steps ahead in this respect and thus reach long term effects. This paper proves that dialogue can have promising results.
Trust grows by the meeting of all stakeholders, brought together by a 'neutral' party, where all are treated on an equal basis. An essential element is that the people that make up and represent the different stakeholders meet *without* having to reach decisions or solutions. Consequently,

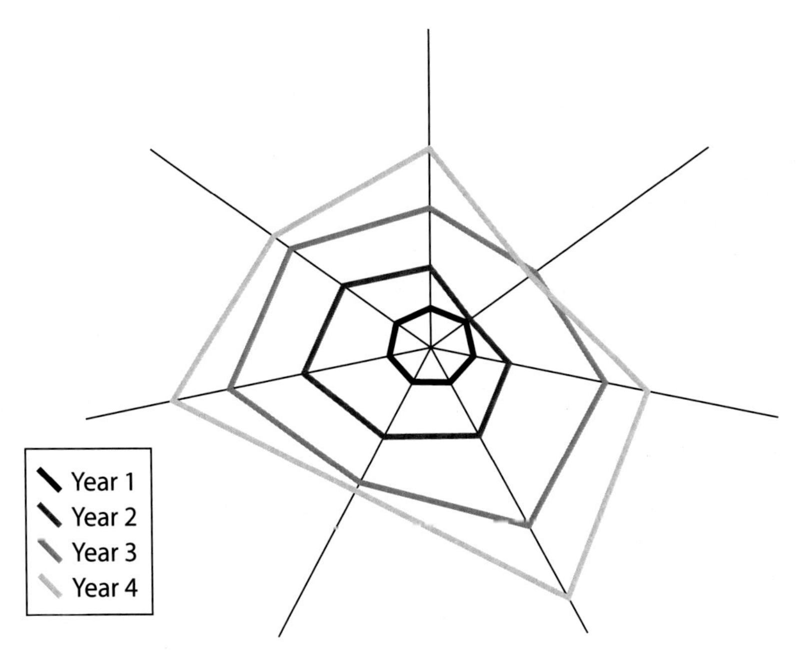

Figure 2. Proposed use of the animal welfare indicator series.

they may see another side of the other party's views. This can only succeed when all parties feel safe to express their own opinions and assumptions. We have tried to create such an atmosphere of safety – presumably successfully – by starting with single stakeholder meetings where they build up the feeling that (information about) part of the puzzle lays with the other actors. Only then will people be ready to withdraw from long-standing disputes and engage in real dialogue. There is no reason why this trust must be absolute (at all), nor does withdrawing from disputes mean that one has to abandon one's identity, views and values.

Inevitable, working through dialogue has disadvantages. Focus groups generate a great amount of qualitative information, but it is important to ensure sufficient diversity in the participating group of people. It is often difficult to find people willing to participate in focus groups, as such projects are unfamiliar and seem to be time consuming and lacking a concrete goal. The proximity of the location seems not to influence the willingness to participate, although this was not explicitely analysed. Contacting people through social networks seems to be the most efficient way, which also avoids attracting only people already interested in AW. It seems that the lack of apparent interest confirms that AW is a high social but not personal priority (for a broader discussion: Lips, 2004).

One should be particularly careful to avoid an elitist approach in which supposed competences prevent the participation of members of certain social groups. Putting to much emphasis on reaching a consensus can also obscure the value of differences and guard the current *status quo* (Heysse, 2006). In general the social debate is facing two important dilemmas (Goorden, 2006): the need for clarity versus a need for room for doubts and the general reflex to emphasise differences while still looking for binding factors. A good participative (or deliberative) process should provide the room for these dilemmas to be out in the open. It should therefore be an inclusive process (all participants, views and arguments are allowed), without being necessarily open-ended. It should be open to the real world (e.g. no abstraction of relative power), but with the possibility to reflect critically on it. Therefore we have chosen to work with the dilemmas rather than with the similarities or differences. We believe to have succeeded in building an inclusive, reflexive and fairly closed-ended project.

Conclusion

Conventional deliberation methods often fail to overcome the dilemmas that accompany complex issues requiring fundamental (social) changes. Dialogue as we have described it is a technique that can assist in dealing with these very difficulties.
Animal welfare is such an issue, and we have elicited what we believe are fundamental values that underlie the normative positions that exist in and across (our) society. It is clear that not all values will be equally important for all, this is indeed the basis for the differences between individuals. We do believe that these four values are present in each of us and that only their relative importance is different. That we have managed to agree on this in a group consisting of consumers, farmers, retailers, researchers, industrial leaders, government officials, opinion makers and pressure group members only strengthens this claim and gives hope for the future.

References

Anonymous (2005). Report of the Special Eurobarometer 229 "Attitudes of consumers towards the welfare of farmed animals". Brussels, European Commission.

Anonymous (2006). Communication from the Commission to the European Parliament and the Council on a Community Action Plan on the Protection and Welfare of Animals 2006-2010. COM(2006) 13 final. Brussels, European Commission.

Goorden, L. (2006). Met nieuwe participatieve praktijken naar een betere afstemming tussen het maatschappelijke en het politiek debat. Communication at the colloquium “Beter beslissen door dialoog”, 16 March 2006, Brussels, King Baudouin Foundation.

Goris, S. and Baeten, J. (2005). Maatschappelijke indicatoren voor dierenwelzijn. Fase 1: Consultatie. Tongerlo, Tri.zone.

Heysse, T. (2006). Principes, mogelijkheden en problemen van het model van de deliberatieve democratie. Communication at the colloquium “Beter beslissen door dialoog”, 16 March 2006, Brussels, King Baudouin Foundation.

Krueger, R.A. and Casey, M.A. (2000). Focus groups: a practical guide for applied research (3rd ed.). Thousand Oaks, Sage.

Lips, D. (2004). Op zoek naar een meer diervriendelijke veehouderij in de 21ste eeuw. Aanzet tot het ontwikkelen van win-winsituaties voor dier en veehouder. Diss. Doct. Faculteit Landbouwkundige en Toegepaste Biologische Wetenschappen, K.U.Leuven.

Miele, M. and Evans, A. (2005). European consumer's views about farm animal welfare. In: Butterworth, A. *Science and society improving animal welfare.* Welfare Quality conference proceedings, 17-18 November 2005, Brussels.

Østergaard, V. and Hansen, J.P. (1995). Indicators - A method to describe sustainability of livestock farming systems. In: Agricultural Sciences for Biodiversity and Sustainability in Development Countries - Proceedings of a workshop. F. Dolberg and P.H. Petersen (Eds.). Tjele, Denmark, National Institute of Animal Science.

Ten Brink, B. (2000). Biodiversity indicators for the OECD. Environmental outlook and strategy. A feasibility study. Bilthoven, Rijksinstituut voor Volksgezondheid en Milieu (RIVM).

Vemuri, V. (1978). Modeling of complex systems - An introduction. New York, Academic Press.

Ethics, bioethics and animal welfare: At what cost

Salvatore Barbera
Dipartimento di Scienze Zootecniche, University of Turin, Via L. Da Vinci 44, 10095 Grugliasco (TO), Italy, salvatore.barbera@unito.it

Abstract

Ethics, bioethics and animal welfare applied to animal production are a source of conflict. Ethics attempts to define that which is right from that which is wrong. Bioethics concerns the ethical questions involving biology. Animal welfare is the care of animals kept in the service of mankind. These three concepts are causing ethical conflicts as a result of a demand for bioethical animal foods. The questions are "Is it ethically justifiable to use food grains as feed or to assure animal welfare, when many humans are still far from having met their own needs? Will the unsustainability of feed grain production, as a result of over exploitation of natural resources, affect the food availability for future generations?". Nevertheless intensive animal production, where animals have little or no contact with their natural surroundings, generates cheap animal foods. In an investigation of 55 pigs bred using semi-extensive (maximum animal welfare) and intensive methods and fed the same diet, the Feed Conversion Index was 5.67 *vs* 4.50 to obtain a final weight of 168 kg. The diet included 68.6% food grain (corn 52.5% and barley 16.1%) and the semi-extensively bred pigs consumed 26% more. Replacing food grains in animal production could help to resolve conflicting ethical obligations, with animal nutrition progress supporting breeders in producing ethical animal foods. Adopting an "Ethical Index" branding animal food, consumers, aware of recommendations to reduce animal food consumption for a healthy human diet, can induce producers to change their policies to ethical animal food production.

Keywords: ethics, bioethics, animal welfare, animal nutrition, pig

Introduction

Ethics (from Greek ethikos) is one of the four major branches of philosophy which attempts to understand the nature of morality and define that which is right from that which is wrong (Wikipedia, 2006). Bioethics concerns the ethical questions that arise in the relationships amid biology, medicine, cybernetics, politics, law, philosophy, and theology. Disagreement exists about the proper scope for the application of ethical evaluation to questions involving biology. Some bioethicists would narrow ethical evaluation only to the morality of medical treatments or technological innovations, and the timing of medical treatment of humans. Other bioethicists would broaden the scope of ethical evaluation to include the morality of all actions that might help or harm organisms capable of feeling fear and pain (Wikipedia, 2006). Animal welfare is the care of animals kept in the service of mankind, so that their wellbeing is provided for, their natural needs are not restricted and their worth and dignity as individuals are recognized. Animal welfare concepts and practices extend to animals reared for food and clothing, for biotechnology products, for research, for work, for recreation and for sport.
Intensive animal production is the practice of keeping animals in large-scale units, where mass production methods are used, where animal have little or no contact with natural surroundings and where they are subject to management designed to shorten their time to slaughter and reduce the cost of production; a feature of agribusiness with an inherent risk of treating animals as disposable resources.

The intensive animal production has given an enormous contribution to the improvement of human wellbeing, at least in the richest countries.
The intensive animal production together with the increasing individual consumption of animal food, are the cause of many and major ethical and real problems. These new and emerging problems are inherent to the environment, human health, and not last, ethics and bioethics.

Environmental problems

Vegetation represents a substantial component of the planet's biomass and provides the major source of human nutrition and a direct and indirect contribution to livestock nutrition. Livestock has had and has a special role in human history. They make effective use of large areas of land that are not capable of producing crops for direct human consumption, such as sparsely scattered vegetation and crop residues. They provide a large component of the fertilizer without which soil productivity would be quickly diminished. Certain species supply draught power for human needs and valuable non-food products such as wool, hides, bones and dung for fuel. Together with their offspring, livestock provide a natural insurance against penury resulting from natural disasters such as drought. In some areas, they are the only viable means of providing a livelihood for the existing population. In harnessing solar energy and nutrients from the major non arable areas of the world's land surface including hills, marginal or wetter areas or crops and industrial residues, livestock are absolutely more efficient. Livestock in certain ecosystems such as the semi-arid zones and high-altitude areas, has a definitive comparative advantage over cropping, as climatic risks (low rainfall, frost or high altitude) preclude safe cropping.
Unfortunately producing animal protein at a global level, demands high inputs of photosynthetic energy and it increases pressure on marginal grain growing land and accelerates soil degradation, which today are not sustainable for human diets based on meat products (McMichael, 1993).

Human health

Countries that have to face food emergencies have as an obvious consequence undernourished populations with sanitary problems and even over nourished people are also having significant negative consequences on health. Overnutrition is associated with chronic diseases like ischemic heart disease, diabetes mellitus, stroke and hypertension and so on. Overnutrition is a real problem in industrial countries and it is spreading further to less developed countries as economic growth proceeds and *per capita* incomes rise. Their consumption habits gradually change from a diet primarily based on staple grains to one consisting more of meat, vegetables, fruit, milk and other dairy products. This trend is really a great dilemma. Data on under and over nourishment related to meat consumption and their space-time trend provide more elements to strengthen the ethic issue (Gardner and Halweil, 2000).
Current recommendations for a healthy diet in more developed countries are directed at a reduction of meat consumption, and will lead to decreases, rather than increases in present levels of consumption.

Ethics and bioethics

The hunter-gatherer's ethical principles include the ideas that one ought to act in harmony with the natural world; and that one must respect the sacrality and power of individual animals. Several moral rules follow including two specific items: man is justified in killing animals to preserve his own life; and man is entitled to kill and eat animals so long as he preserves adequate herd size for future use. The word husbandry means pastoral nomad and farmer engaged in selective breeding and rearing of animals. The husband's ethical principles include

the ideas that one ought to treat animals humanely, caring for their needs and slaughtering them painlessly.
Since the beginning of husbandry agriculture, some 10,000 years ago, domesticated animals and humans have enjoyed a mutually beneficial relationship. In particular humans benefited from the animal food produced by useless agricultural land or crop residues. In the past and still now in emerging countries livestock has been and is complementary to human beings. When possible, lands and vegetables have always been used directly for human needs, leaving livestock crop residues or by-products to obtain animal food and increasing the number of surviving people.
In the richest countries livestock is no more complementary but rather a competitor for the use of natural resources. Livestock is a true competitor as the productivity in terms of human food per hectare drops by a factor of ten. If food value is assessed in terms of protein rather than energy, the results are essentially similar. The protein from a hectare of crops ranges from five to fifteen times that from the same hectare devoted to meat production, depending on whether cereals, legumes or leafy vegetables are grown. A result of this competition is the increasing diversion of land from food to feed grains. World-wide many countries and populations are still far from having met their needs for food grains. Is it morally justified to divert land and other resources, used in the production of food grains into feed grains, to meet the growing demand of over-nourished groups with increasing health problems? Moreover, is it morally justified to divert food grains into growing livestock to assure animal welfare, not meat production? Since ethics is concerned with human behaviour and particularly with its morality, society is now being confronted with the concept of ethics in animal production. Ethics is already involved in social aspects with positive effects on society. Consumers are now sensitive to social morals concerning child labor, mistreatment of workers in developing countries and so on. All these have forced business to change their policies in these and a myriad of other areas.
Another important ethical issue is the sustainability of feed grain production due to the over exploitation of natural resources which will affect the food availability of future generations.

The experiment

In Europe the pig industry relies on a limited number of breeds well adapted to intensive animal production systems. The requests for organic animal products, animal welfare and conservation of old germoplasm are pressing people to pay more attention to old breeds.
In Europe 25 local breeds exist and of those, 5 in Italy. One of them, the Mora Romagnola was used in this experiment. It was fairly common until the mid '50s in Emilia Romagna (NW Italy) and in 1949 there had been more than 22.000 animals. This breed began to disappear as the Large White took over the area; in 1998 only 12 animals survived.
In a number of trials, throughout 4 years, 55 animals of different breeds (19 Mora Romagnola [MR], 7 crossbreeds of Duroc and MR [DMR], 29 crossbreeds of Large White and MR [LWMR]) were reared testing two types of husbandry and two methods of feeding. The daily gain (ADG) and the conversion index of feeds (FCI) were evaluated using diets formulated for the growing-finishing hybrid pigs, as the needs of this old breed were unknown (Table 1). Meals were given mixed with water.

Husbandry

Two types of husbandry (Figure 1) were tested: intensive (low animal welfare) and semi-extensive (high animal welfare). In the intensive practice pigs were reared indoors in 2 slatted floor pens; one of them equipped with a self-feeder. In the semi-extensive practice pigs were reared in two large corrals with a pound in the centre and a hut with straw; one of the corrals was equipped with a self-feeder (Barbera, 2006). All the animals reached the final weight of about

Table 1. Components and composition of diets (% DM).

	Live Weight		
Feeds %	<60 kg	60-90 kg	> 90 kg
Corn	32.5	45.0	55.0
Barley	25.0	18.0	15.5
Soybean meal 44%	20.0	16.0	12.0
Wheat bran	15.0	13.5	11.0
Lineseed meal	2.5	2.5	2.5
Minerals and vitamins	3.55	3.55	3.5
Bentonite	1.0	1.0	-
NaCl	0.3	0.3	0.4
Lysine	0.1	0.1	0.05
Methionine	0.05	0.05	0.05
	87.5	87.6	87.2
DM %			
CP	18.6	17.2	15.5
DE MJ/kg	14.2	14.4	14.7
ME MJ/kg	13.6	13.8	14.1

DM = dry matter

Figure 1. Semi-extensive (high animal welfare, left) and intensive (low animal welfare, right) husbandry.

168 kg and 317 days old. The animals were raised at the Experimental Station of the Facoltà di Agraria - University of Turin (Italy).

Method of feeding

Two feeding systems were compared during the intensive and the semi-extensive husbandry. In the traditional system animals received feeds twice a day *ad libitum*, distributed in two long troughs with a space allowance of 0.33 m/head. In the other system pigs received feed individually by a self-feeder, *ad libitum* but divided in small quantities at meals to simulate natural conditions. This self-feeder is part of a structured self-made feeding station.
Statistical analysis was carried out by SAS/STAT in SAS 9.01 using a three-factorial (husbandry, breed and method of feeding) covariance analysis with the final live weight as covariate (SAS, 2004). Results are presented as LSmeans ± S.E.M.

Results and discussion

Results of the experiment are shown in Table 2. Covariance analysis removes dependent variable variations associated with different final live weights resulting in more precise estimates and powerful tests.
The intensive and semi-extensive husbandries are significantly different as Feed Conversion Index is 4.50 *vs* 5.67 to obtain a final weight of 168 kg and 104 kg live weight gain, during the trials. An intensive pig has consumed 467.6 kg of feed *versus* 589.2 kg of semi-extensive one. According to Table 1, it has consumed on average 68.6% of food grain (corn 52.5% and barley 16.1%), *i.e.* the intensively reared pig 321.0 kg and 404.5 kg the semi-extensive reared pig. The semi-extensive FCI is 1.17 significantly higher than the intensive FCI, which causes a semi-extensive husbandry pig to consume 121.6 kg more *i.e.* 83.5 kg (26%) of corn and barley to ensure better animal welfare. Moreover the ADG is higher in the intensive husbandry than the semi-extensive.
The use of a self-feeder does not make any difference to the FCI. The self-feeder reduces feed competition as each pig is protected during feeding. When moved to traditional feeding they have suffered competition which has caused a significant lower ADG. Moreover the self-feeder is closer to pig's natural way of feeding as they have to look for feed.

Table 2. The effects of different factors on pig productive parameters FCI and ADG (LSMean and SE).

Factors	Levels	FCI	ADG (g)
Husbandry	Semi-extensive	5.67 ± 0.201^{A}	549 ± 20.6^{A}
	Intensive	4.50 ± 0.186^{B}	661 ± 18.4^{B}
Method of feed distribution	Self-feeder	4.89 ± 0.217	637 ± 21.4^{a}
	Self-feeder and traditional	5.38 ± 0.387	535 ± 38.2^{b}
	Traditional	4.99 ± 0.192	644 ± 18.9^{a}
Breed	Duroc x Mora R.	5.13 ± 0.330	642 ± 32.5^{a}
	L. White x Mora R.	4.94 ± 0.229	603 ± 22.6^{ab}
	Mora Romagnola	5.19 ± 0.190	571 ± 18.7^{b}

Means by factors in the same column with different letters are significantly different (a, b: $P<0.05$; A, B: $P<0.01$)
FCI = Feed Conversion Index
ADG = Average Daily Gain

The breed factor seems not to have an effect on FCI but ADG is significantly lower in MR *vs* DMR. To obtain the same live weight gain the MR breed has to eat 11.2% more feed of which 7.7% corn and barley. This is a clear effect of genetic selection but this not absolutely mean the old MR breed is useless as the diet was typical for modern breeds. Using corn and barley as feed is clearly more ethical in the feeding of modern breeds than old ones.
In brief, results show that pig production costs are:

- from an ethical point of view 68.6% of food grain used as feed;
- animal welfare has an additional cost of 26% of food grains is used as feed;
- the conservation of old breeds consumes 7.7% more food grains used as feed.

This data are the origin of ethical and bioethical issues. What to do? Animal production strategy should be to replace food grains by ethical feeds. Already farmers try to replace feed grains with by-products for economical reasons as animal nutrition science already supports them. It could suggest an "Ethical Index" branding animal food, to indicate to the consumers the quantity of ethical feed used to produce animal foods.

Conclusions

Ethics of animal production is more likely to be "ethics stating which type of feed is to be used in animal nutrition, with respect to human beings and animal welfare". Gardner (1996) says that perhaps the greatest way to increase food use efficiency is to reduce the world's consumption of meat. This is not sufficient, would be necessary to replace food grains in animal production by ethical feeds. This could help to reduce or resolve conflicting ethical obligations. The animal nutrition progress can achieve that, as the know-how can support breeders to produce ethical animal foods. Consumers, aware of the recommendations to reduce animal food consumption for a healthy human diet, can induce producers to change their policies to ethical animal food production. Adopting an "Ethical Index" branding animal food, to indicate the quantity of ethical feed used to produce it, can help a conscious consumption.

References

Barbera S. (2006). *Progetto "Beato Porcello"*. Available on http://www.beatoporcello.it 23/04/06.

Gardner, G. (1996). Preserving agricultural resources. In: *State of the World 1996*. Brown L. *et al.* (eds). W.W. Norton, New York. 78-94.

Gardner, G. and Halweil, B. (2000). *Underfed and Overfed: The Global Epidemic of Malnutrition*. Worldwatch Paper 150, Worldwatch Institute, Washington, DC.

McMichael, A.J. (1993). *Planetary overload – Global environmental change and the health of the Human species*. Canto Edition 1995. Cambridge University Press, Cambridge.

SAS (2004). *The SAS System for Windows, Release 9.01*. SAS Institute Inc., Cary, NC, USA

Wikipedia (2006). *Wikipedia The Free Encyclopedia*. Available on http://www.wikipedia.org 13/04/06.

Importance of emotional experiences for societal perception of farm animal welfare: A quantitative study in the Netherlands

Birgit K. Boogaard, Simon J. Oosting and Bettina B. Bock
Animal Productions Systems Group, Wageningen University, P.O. Box 338, 6700 AH Wageningen, The Netherlands, Birgit.boogaard@wur.nl

Abstract

Animal welfare is studied in many disciplines and it may be defined in many ways. Improvements in animal welfare should also resonate with public attitudes and values. Therefore it is important to understand which considerations (and definitions) people have concerning farm animal welfare. The purpose of this paper is to define factors that influence societal perception of farm animal welfare and particularly in relation to people's emotional experiences with animals and farms. Randomly selected Dutch respondents (n = 1074) completed a questionnaire on animal welfare. We analysed 15 propositions (4-point Likert-scale) and through factor analysis we defined four factors of societal perception of animal welfare: Human-Animal Hierarchy (HAH), Use of animals for Human Consumption (HC), Life Quality of farm animals (LQ), and Farmers' Image (FI). Pet owners perceived farm animal welfare as being less positive than people without a pet (p = 0.001). Quite on the contrary, people with farm experiences ('Connection with agriculture through work or relatives' and 'Urbanisation degree' as descriptors) perceived animal welfare as being more positive (p = 0.000 and p = 0.032 respectively). It can be concluded that pet owners have different considerations (and definitions) towards farm animal welfare than people without a pet. Besides, people with connection to agriculture (farm experiences) also have different considerations (and definitions) towards farm animal welfare. These findings are of importance, since nowadays many people have a pet whereas the number of people with farm experiences is decreasing in western European societies.

Keywords: animal welfare, perception, society, pet owners

Introduction

Animal welfare may be defined in many ways (McGlone, 2001) and can be approached from different scientific angles: Animal scientists often look at animal welfare from an instrumental relationship between farmer and farm animal, which is in line with the focus of animal scientists, e.i. improving production (Lijmbach, 1993). However, we followed the philosophy of Hemsworth and Coleman (1998) who stated that "although science has an important role in providing sound defensible information on how animals respond to a specific practice, ultimately it is an ethical decision by the general community that will determine the acceptable standards for farm animals". It is important to understand which considerations (and definitions) people have concerning farm animal welfare. Serpell (2004) emphasizes that improvements in animal welfare should also resonate with public attitudes and values. Therefore the purpose of this paper is to define factors that influence societal perception of farm animal welfare and particularly in relation to people's emotional experiences with animals and farms.

Theoretical framework

People differ in their perception of reality (Jager and Mok, 1999). This is true for reality in general, but also for the perception of issues such as the welfare of farm animals. How people perceive reality depends on their frame of reference, which can be defined in many ways depending on the purpose of use (Dewulf *et al.*, 2005). For example, Aarts and Woerkum (1994) distinguish values and norms, convictions, experiences, and interests as separate elements of a frame of reference. In the present study we differentiate between four elements that together contribute to a frame of reference (Figure 1). The first element concerns *Convictions and Values.* Convictions are generally accepted basic truths, which are not easily questioned and which are closely related to values (Aarts and Woerkum, 1994). Considered a second element, *Interests* are part of a frame of reference, Jager and Mok (1999) emphasise that people derive their frame of reference by individual experiences in life and do not appoint 'interests' as a specific element of frame of reference. Taking into account that few people have an interest in agriculture (and animal welfare) nowadays, we followed Jager and Mok (1999) and left interests out of further consideration. The third and fourth element concern experiences, which we distinguished into two types, i.e.: Experiences through reading or hearing information, so-called *Factual Knowledge* and experiences through real life contact with farms and animals, so-called *Emotional Experiences.* Emotional experiences with animals are influenced by real life contact with farm animals as well as by real life contact with pets.

Materials and method

Data selection

For the present study the dataset of the Rathenau Institute was selected. The Rathenau Institute is an independent Dutch organization with the purpose to inform politics on issues of potential social importance at the interface of science, technology and society. The dataset concerned

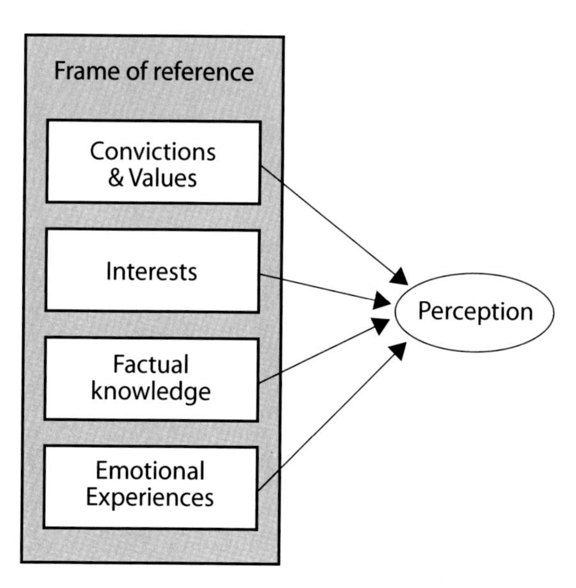

Figure 1. Four elements of frame of reference in relation to perception (adapted from Aarts and Woerkum, 1994).

'Citizens' perception of handling animals in livestock production systems' ('Burgeroordelen over de omgang met dieren in de veehouderij', Verhue and Verzijden, 2003). In 2003, 1074 randomly selected Dutch respondents of the 'TNS NIPO CAPI@HOME'-database filled in a questionnaire on perception of animal welfare. The dataset included variables such as gender, educational level, age, value orientation, possession of a companion animal (pet ownership), connection to agriculture and area of living (urbanisation degree). Pet ownership included all type of animals; cats, dogs, birds, rodents, fishes and 'others'.

Statistical analyses

Statistical analyses (SPSS 12.0.1) consisted of two steps. Firstly, we performed factor analysis of 15 propositions on livestock production and farm animal welfare. Secondly, we used a statistical model to analyse variation in perception per factor in relation to individual characteristics. Five individual characteristics are included as explanatory variables in the full model.

$$Y_{ijklmn} = \mu + A_i + B_j + C_k + D_l + E_m + e_{ijklmn}$$

Y_{ijklmn} = Factor y
μ = Mean
A_i = Value orientation ($i = 1,..., 8$)
B_j = Pet ownership ($j = 0, 1$)
C_k = Urbanisation degree ($k = 1,..., 5$)
D_l = Connection to agriculture ($l = 1, 0$)
E_m = Leaflet information ($m = 1, 0$)
e_{ijklmn} = Error

If factors were normally distributed, we searched for significant main effects of variables on each factor through ANOVA and. If factors had non-normal distributions we performed Kruskal-Wallis nonparametric tests.

Results

We defined four factors of societal perception of animal welfare (Table 1): Human-Animal Hierarchy (HAH), Use of animals for Human Consumption (HC), Life Quality of farm animals (LQ, the actual animal welfare indicator), and Farmers' Image (FI).
We analysed the contribution of five variables to variation in FI, HAH, LQ and HC (Table 2). Emotional experiences with farming and animals (farm animals as well as pets) strongly influenced factor LQ. Notably, pet owners perceived farm animal welfare as being less positive than people without a pet ($p=0.001$). Quite on the contrary, people with farm experiences ('Connection with agriculture through work or relatives' and 'Urbanisation degree' as descriptors) perceived animal welfare as being more positive ($p=0.000$ respectively $p=0.032$).

Conclusion and discussion

In general, Dutch society perceived the life quality of farm animals as slightly positive, because Likert scale scores (1 to 4 with median 2.5) of the five propositions of the actual indicator for animal welfare (factor LQ) had a range from 1.63 to 2.57 (Table 1).We will discuss each factor in relation to emotional experiences on the basis of several theories.
For perception of *Use of animals for Human Consumptions (HC)* we can conclude that in general people agreed that humans are allowed to use animals for consumption. But there are interesting differences between respondents. People in possession of a pet agreed less strongly on HC than

Table 1. Descriptives of 15 propositions divided over four factors.

Propositions per factor	Descriptives			
	Mean[a]	S.D.	*n*	DK[b]
Farmers' Image (FI)				
Most farmers focus too much on management instead of their animals	2.33	0.80	1,005	0.06
Most farmers consider their animals too much as means of production	2.07	0.79	1,042	0.03
Most farmers solely treat animals properly as long as it is financially beneficial	2.27	0.88	1,026	0.04
Most farmers economize on animal welfare	2.49	0.79	966	0.10
Most farmers are actually not interested in animals	3.08	0.81	1,026	0.04
Human-Animal Hierarchy (HAH)				
Human life is of higher value than an animal life	1.84	0.98	1,061	0.01
Humans are more important than animals	1.98	1.00	1,067	0.01
Life Quality of farm animals (LQ)				
Animals on an average farms are better of than animals in nature	2.57	0.79	1,006	0.06
Animals on farms have quite a good life	2.21	0.75	1,038	0.03
Most farmers give their animals a comfortable life	2.13	0.76	1,026	0.04
Most farmers are animal lovers	2.03	0.82	1,026	0.04
Most farmers are in daily contact with their animals	1.63	0.74	1,039	0.03
Use of animals for Human Consumption (HC)				
Humans are allowed to use animals for consumption	1.42	0.60	1,069	0.00
Humans have the right to use animals for consumption	1.72	0.79	1,063	0.01
I don't feel guilty when I eat meat/fish	1.84	0.95	1,068	0.01

[a] Mean score on a 4-point Likert-scale (1 = I agree completely, 4 = I disagree completely)
[b] DK: fraction of total respondents (N= 1074) answered 'Don't Know'

Table 2. Significance of contribution to variation of variables A, B, C, D, and E on factors FI, HAH, LQ, and HC.

Variable			Factor[a]			
Code[b]	Name	*df*	FI[c]	HAH[d]	LQ[c]	HC[d]
A_i	Value orientation	7	ns[e]	ns	**	ns
B_j	Pet ownership	1	***	***	**	**
C_k	Urbanisation degree	4	Ns	*	*	ns
D_l	Connection to agriculture	1	**	ns	***	*
E_m	Leaflet information	1	*	ns	Ns	ns

[a] Factor names: FI: Farmers' Image, HAH: Human-Animal-Hierarchy, LQ: Life Quality of farm animals, HC: Use of animals for Human Consumption.
[b] Full model $Y = \mu + A_i + B_j + C_k + D_l + E_m + e_{ijklmn}$
[c] Normally distributed: Differences explained with ANOVA and Gabriel's Post Hoc test
[d] Non-normally distributed: Differences explained with nonparametric Kruskal-Wallis test
[e] *** P<0.001, **P<0.01, *P<0.05, ns = P>0.05

people without a pet. On the contrary, people with relation to agriculture agreed more strongly on this factor than people without relation to agriculture. For perception of *Human-Animal Hierarchy (HAH)* we can conclude that in general people agreed that human life is of higher value than animal life, however some people agreed less strongly. These were pet owners or living in urbanised areas. The fact that in general people agreed with HAH and HC goes back to the Judean Christian culture in Western European societies, in which animals are considered as organisms that were created to serve human interests (Serpell, 1986).

Serpell's model (2004) of human-animal relations may also be used to explain results of the present study. In this model he describes a controversy between the two primary motivational dimensions 'Affect' (emotional responses to animals) and 'Utility' (instrumental value of animals) as possible cause for many tensions and paradoxes in human-animal relationships. Utility is defined as human interest and on the left side animals are 'detrimental to human interests' (score -1) in contrary to the right side in which animals are 'beneficial to human interests' (score +1). Concerning farm animals, the utility purpose (food production) is obvious, however for possession of a companion animal a different type of purpose is applicable and 'company' is not directly associated with an 'instrumental value'. The other dimension, affect, is defined with 'Love, Sympathy, Identification' on one side (+1) and with 'Fear, Loathing, Disidentification' on the other side (-1). The side of love, sympathy and identification is also described by Teusch (1993), who states that people can perceive an animal as a member of their own species and people can see animals as human-like. This is inline with the fact that animals can sometime have the status of children and friends (Bryant *et al.*, 2005).

Depending on the purpose of the human-animal relationship, people define and evaluate animal welfare differently. In case of pet owners the purpose (high affect, treated human-like) is difficult to unite with the utility purpose of farm animals i.e. production for human consumption. Serpell (2004) describes that "animals with a high utility value often seem to be precluded of becoming the objects of people's positive affections, presumably because such animals are harmed as result of their utility". It seems plausible that due to lack of knowledge and/or experiences with farming, people use their frame of reference concerning the welfare of their pets to judge the welfare of farm animals and farmers. This would not only explain why pet owners agree less strongly about HAH and HC, it could also explain the difference in perception of *Life Quality (LQ)*. It seems plausible that pet owners use a different definition of 'life quality of animals' compared to people without a pet, because pet owners judge from a high affect dimension, which are other evaluation criteria for animal welfare. Moreover, it could even explain why pet owners are less positive about *Farmers' Image (FI)*. It seems plausible that farmers are considered as people with a human-animal relationship based on utilization purposes, instead of affect, because 'production animals' inevitably leads people to think about them from an instrumental value. And as for pet owners the utilization purpose is perceived to be in conflict with the dimension of affect, pet owners perceive farmers as people who cannot care about their animals in an affective way.

In addition to the above mentioned results, this study showed that people with connection to agriculture perceived farmers' image (FI) more positively, were more strongly convinced that humans are allowed to use animals for consumption (HC) and were more positive about the life quality of farm animals (LQ). This suits abovementioned model, in which people with relation to agriculture rather judge from a utility frame of reference, based on their experiences with farms, than from the affect dimension. Consequently, definition of 'life quality' can be different, because due to the utility purpose, people tend to put different demands on animal welfare (compared to companion purpose), which could explain why people with connection to agriculture are more positive about animal welfare than people without connection to agriculture. These findings show the importance of emotional experiences with farming through real life contact with farms. People can gain real life contact with farms through farm visits.

It is interesting to cite the author of the European study 'Attitudes of consumers towards the welfare of farmed animals', that "visits to farms seem to increase the awareness and concern for animal welfare" (Eurobarometer, 2005). The Norwegian project 'The Farm as a Pedagogical Resource' is another interesting example. It demonstrates that school children visiting farms as part of their education achieve a greater understanding of the processes in nature and farming (Jolly *et al.*, 2004). Moreover, Serpell (2004) confirms the importance of childhood interactions with animals to develop stronger utility orientations.

These findings are of importance, since nowadays many people have a pet whereas the number of people with farm experiences is decreasing in western European societies.

References

Aarts, M.N.C. and Woerkum, C.M.J.v. (1994). *Wat heet natuur? de communicatie tussen overheid en boeren over natuur en natuurbeleid.* Landbouwuniversiteit, Wageningen. 125 p.

Bryant, J., Baggott la Velle, L. and Searle, J. (2005). *Introduction to bioethics.* Wiley, Chichester. 240 p.

Dewulf, A., Gray, B., Putnam, L., Aarts, N., Lewicki, R., Bouwen, R. and Woerkum, C. (2005). Disentangling approaches to framing: mapping the terrain, Conference of the International Association for Conflict Management (IACM), Seville, Spain.

Eurobarometer (2005). *Attitudes of consumers towards the welfare of farmed animals*, Special Eurobarometer 229. European Commission.138 p.

Hemsworth, P.H and, Coleman, G.J. (1998). *Human-livestock interactions: the stockperson and the productivity and welfare of intensively farmed animals.* CAB International, Wallingford. 152 p.

Jager, H.d. and Mok, A.L. (1999). *Gezichtspunten en begrippen. Grondbeginselen der sociologie.* EPN Educatieve Partners Nederland, Houten. 351 p.

Jolly, L., Krogh, E., Nergaard, T., Parow, K. and Verstad, B. (2004). The Farm as a Pedagogical Resource, *6th European Symposium on Farming and Rural Systems Research and Extension*, Vila Real, Portugal.

Lijmbach, S. (1993). The Human-Animal Relationship as an Object of Social Research. In: Hicks, E.K. (Ed.), *Science and the human - animal relationship: proceedings of the international conference on 'Science and the human-animal relationship'*, March, 1992. Siswo, Amsterdam. 105-113 p.

Serpell, J. (1986). *In the company of animals: a study of human-animal relationships.* Blackwell, Oxford. 215 p.

Serpell, J.A. (2004). Factors influencing human attitudes to animals and their welfare. *Animal Welfare* 13. S145-151.

Teutsch, G.M. (1993). Traditional Sociology and the Human-Animal Relationship. In: Hicks, E.K. (Ed.), *Science and the human - animal relationship: proceedings of the international conference on 'Science and the human-animal relationship'*, March, 1992. Siswo, Amsterdam. 67-79 p.

Verhue, D. and Verzijden, D. (2003). Burgeroordelen over de veehouderij. Uitkomsten publieksonderzoek. Veldkamp, Amsterdam. 73 p.

Animal welfare in animal husbandry: How to put moral responsibility for livestock into practice

Herwig Grimm
Institut TTN (Institute Technology, Theology and Natural Sciences), Marsstraße19, 80335 Munich, Germany, herwig.grimm@elkb.de

Abstract

On the basis of a well-founded moral obligation, every farmer has to care about his animals. When it comes to the question of the individual farmer's responsibility for animal welfare on his/her farm, the problem arises how ethicists can support ethical decision making in a realistic way. Therefore, the aim of this paper is to structure and frame the livestock keeper's responsibility in order to prepare the field for ethical decision making. Questions beyond the keeper's responsibility are not addressed in this paper. As it is claimed that every farm animal experiences pain a pathocentric approach will be founded and clarified. Respecting the fundamental "five freedoms" are the means to put this moral responsibility for the animal's well being into practice. Since every livestock farmer works within legal and economic constraints and is bound to moral obligations, these aspects have to be woven into the method for ethical reflection and decision making. Accordingly, a pathway is presented which follows several steps: After estimating the intensity of infringements of the five freedoms (step 1) the expected benefits are identified (step 2) and then balanced against the infringements (step 3). Finally, alternatives are taken into consideration (step 4) and viability restrictions come into focus (step 5), which limit the possible alternatives. In summary, interferences with the "five freedoms" can only claim moral justification, if they are legal, do not overstress the animal's adaptability, they result in relevant benefits, and they can not be achieved by other less harmful practices or types of livestock housing.

Keywords: animal welfare, farmer's responsibility, pathocentric approach, practical approach, decision making

Introduction

Although there is a generally held consensus that livestock husbandry is morally acceptable, this basic assumption does not assist in solving concrete problems. Particularly, animal welfare and economic constraints are perceived as an area of tension. Frequently, economic ends are related to highly questionable practices in terms of poor animal well-being and occurrence and/or infliction of pain and sufferings. Accordingly, ethicists are increasingly challenged to give concrete moral advice for ethical decision making and to give answers to the question of whether certain practices are justifiable or not. Considerable thought has been given to concepts such as "factory/industrial farming" and "animal friendly farming", as well as more value-laden subjects such as "animal welfare", "dignity of creatures" or their "integrity". Unfortunately, the given answers often appear to be too vague or too abstract so they do not help or support actual decision making of the farmers concerned. Since it is extremely unlikely that industrial agriculture can be fully reversed, it is therefore of crucial importance to weave animals' welfare aspects into the design of animal husbandry; system modifications need to progress in order to meet animal's needs. The most promising way to promote this progress is a mixed strategy. This paper pursues the strategy to define the farmers' realm of moral responsibility and to clarify how they can reflect on their practice and make their standpoint transparent.

Firstly, the paper suggests that moral responsibility to care for animal welfare, has to be based on morally relevant reasons. Secondly, a plausible understanding of "animal welfare" will be presented. Thirdly, the paper explores the crucial aspects of the farmer's area of responsibility. Finally, it brings the developed aspects in a pragmatic order and presents a pathway for ethical decision making.

Moral responsibility for "animal welfare"

When it comes to ethics in animal husbandry, the vital idea is that moral subjects have moral responsibility to assure animal welfare. In the first place these moral subjects are the farmers. It seems commonplace that there is such a moral obligation and how it is to be understood. But since we know that such attitudes can change, we should bring the reasons to mind in order to examine this attitude on rational grounds. Consequently, the first question we have to tackle is: "How can we justify the moral responsibility to assure animal welfare in animal husbandry?" The intuitively plausible claim to respect animal welfare as morally relevant can be supported by the following syllogism, which follows the essential aspects of Badura's (2001) reasoning:

P1: The ability to experience pain and sufferings serves as a well established criterion of moral relevance within the realm of interactions among humans.

P2: The ability to experience pain and sufferings can be plausibly attributed to our farm animals.

P3: Principal of equal treatment: Equal treatment for equals – unequal treatment for unequals.

From these premises we can infer:

c: If we take the ability to experience pain and sufferings as a criterion of moral relevance in the case of interactions among humans, which we do (P1), *and* we can plausibly attribute this ability to farm animals (P2), *then* we should respect this ability in our dealing with farm animals (P3).

The conclusion, therefore, formulates an obligation to assure animal welfare for moral reasons in a pathocentric sense. To date this has not been enough to guide moral decision making in practice, and consequently, leads us to the second question, what "animal welfare" stands for.

Animal welfare and the five freedoms

When we talk about animal welfare, we focus on the animal's needs and what influences its well-being. The animal's fundamental five freedoms provides a comprehensive framework of facets, which influence animal well-being (Farm Animal Welfare Council, 2006). It facilitates deliberations on animal welfare without proposing abstract concepts like "dignity" or "integrity" which are difficult to put into practice.[98] Furthermore, they can clarify how the farmer's specific responsibility can be understood:

1. Freedom from Hunger and Thirst - by ready access to fresh water and a diet to maintain full health and vigour.
2. Freedom from Discomfort - by providing an appropriate environment including shelter and a comfortable resting area.
3. Freedom from Pain, Injury or Disease - by prevention and/or rapid diagnosis and treatment.
4. Freedom from Fear and Distress - by ensuring conditions and treatment which avoid mental suffering.

[98] These normative concepts are of great importance when e.g. breeding goals are questioned, but in the case of animal husbandry animal welfare appears to be the key concept (Grimm, 2005a).

5. Freedom to Express Normal Behaviour - by providing sufficient space, proper facilities and company of the animal's own kind.

These freedoms define ideal states rather than standards for acceptable welfare. However, to use such a list in ethical reflection has the advantage that interferences with these freedoms cause observable effects on animal based parameters (Unshelm, 2002):

- Anomalous behaviour (behaves in repetitive stereotypical ways such as bar-biting etc.).
- Performance of individual animals (irregular gain in weight, anomalous milk quantity, signs of malnutrition etc.).
- Physiological parameters (anomalous pulse, temperature, fitful respiration, blood count etc.).
- Medical evidence (abrasion, infections, illnesses, intestinal parasites etc.).
- Mortality rate and causes.

The longer and the more intense (according to the animal based parameters) the five freedoms are infringed, the bigger the problem from an ethical point of view (Busch and Kunzmann, 2006). It lies in the farmer's responsibility to aim at the highest possible degree of animal welfare corresponding to the animal's five freedoms. However, limits and constraints within the farmer's area of responsibility have to be taken into account.

Vital aspects of livestock husbandry

The farmer's area of responsibility significantly differs from other areas in animal ethics (like companion animals or laboratory animals). It is important to make those differences explicit to see which aspects need to be taken into consideration (Grimm 2005a). The following list provides us with relevant aspects which are always bound to livestock husbandry:

a. Using animals and their products as means to economic ends (this includes controlled and intended reproduction of animals, selective breeding, enhancement of productivity and utility).
b. Influences on the animals' well-being (interferences with the animal's five freedoms at least by confining them to a particular place).
c. Legal requirements (animal protection act, directives of the European Parliament and the Council).
d. Moral obligation to the animal's well being (to keep interferences with the five freedoms as low as possible).

The question arises, as to how the farmer can balance all these aspects in an appropriate way. Especially the tension between "economic ends" (a) and the "moral obligation to assure the animal's well being" (d) appear problematic. Unfortunately, farming animals is always accompanied by interferences with the five freedoms (b), also managing animals extensively is no guarantee of welfare, but this is not to say that any practice can be justified. "Legal requirements"(c) have to be met in any case under practical circumstances. *Beyond* legal requirements the question arises whether the practice is morally justified.

The elaborated aspects will now be brought together in a pathway of ethical decision making which integrates legal, moral, and viability considerations and presents them in a pragmatic order. This pathway helps to guide moral deliberations in the farmer's area of responsibility. Questions beyond this scope are not addressed.

Pathway of ethical reflection in livestock husbandry

Together the following five steps present a checklist which frames moral deliberations in animal husbandry, and questions whether the farmer pays his dues in his area of responsibility *beyond legal requirements* (Busch and Kunzmann, 2006; Grimm 2005b).

Step 1 – Estimation of intensity

The first challenge of every ethical consideration in this field should be to estimate whether and how a (legal) *practice* or *type of livestock housing* interferes with one or more of the five freedoms. This will give a picture of the problematic aspects. Basically, any infringement of the five freedoms is in need of justification. However, less intense infringements will not be as problematic as very intense ones. Therefore, the first step is to estimate the intensity of the interference. This provides us with a reasonable starting point for further considerations.

The intensity of any interference will increase in accordance with its duration and depth. Whereas we can estimate the duration with regard to the animals total lifespan quite easily (temporary, medium term, long term, lifelong), the depth's estimation is more problematic. This question can only be settled comparatively. For example, the impact of an injection will be estimated far less problematic than a sow's antenatal fixation in a full confinement farrowing crate until the piglets are weaned. For example, to administer an injection can count as "short time" and "low depth" infringement in terms of pain, distress and fear, while the farrowing crate – which massively restricts the sow's movements for several weeks and inhibits social contact with the piglets – can count as "medium term" and "high depth" in terms of restriction of behavioural needs and discomfort. Another example would be the castration of piglets which lasts only a few seconds but is of high intensity in terms of pain, injury, distress, and fear. There can be cases where the impact over stresses the animal's adaptability. Whether the adaptability is over stressed or not is open to debate and will always be bound to scientific data and differing moral views and beliefs. No ethical approach can get around the question of "adaptability"; the approach presented in this paper has the advantage to bring the debate down to observable influences on animal based parameters.

Step 2 – Estimating benefits

Only if the interferences with the five freedoms are acceptable, are we justified to move on to the next question of, what are the expected benefits of a *practice* or *type of livestock housing?* Such benefits can arise in various extents for those involved. It may arise for the *animal itself* (medical treatment, claw trimming etc.), for other *animals in the herd, flock etc.* (dehorn calves or fawns to prevent injury), *for the farmer* (dehorn calves to prevent injury, fattening bulls on slatted floors to reduce the amount of work), for *other people* or the *environment*.

When the expected benefit of the practice or housing system in question is estimated, it is important to prove that it has some relevant benefit. If there was no such benefit or reason at all, of course, no such infringement could gain moral justification.

Step 3 – Balancing infringements and benefits

When expected benefits have been identified, we face the question as to whether the expected benefits can justify the interference with the five freedoms. Whenever we deal with the issue of incommensurability of values we lack the measure to weigh or compare burdens/sufferings of the animal and expected benefits. We would be comparing apples with oranges.

At this point we have already seen that the practice or housing system has passed several stages: It has proved legal, does not cause unacceptable infringements of the five freedoms and results in relevant benefits. The problem of incommensurability can be transformed and thereby overcome by a search for alternatives. In order to keep interferences with the five freedoms as low as possible a farmer has to prove that there is no better realisable alternative in his area of responsibility.

Step 4 – Search for alternatives

Is it possible to accomplish a comparable benefit with fewer burdens to the animal? Basically, possible alternatives can be found on two levels: Alternatives to the *practice* in the sense of actions of lower impact, and alternatives on the level of *types of livestock housing*. The characteristic of alternatives on the first level – such as using anaesthetics or improved management or reducing the stocking rate – is that their implementation will not need structural alterations. If there are no such alternatives, a farmer should rethink whether there are alternatives in the latter sense which will demand such alterations. If there are no alternatives on either level, improvement is not in the scope of the farmer's responsibility. If, however, alternatives are found, they should be implemented. But as mentioned above, the farmer works under certain constraints which are outside his influence. Such structures are societal, geographical, and of course economical. These factors may make it possible or inhibit the implementation of identified alternatives. At this point, the farmer's viability restrictions come into focus.

Step 5 – Realisation of alternatives

Is it possible for the farmer to implement the identified alternatives under the given structural circumstances (which he can not change)? If a farmer can prove that viability restrictions force him to refrain from realising alternatives at this point the farmer should not be accused of wrongful action. On the contrary, he has done everything possible within his area of influence and responsibility to assure animal welfare.

Conclusion

In conclusion, a farmer can claim moral justification if he is able to show that the *practice* or *type of livestock housing* is legal, does not overstress the animal's adaptability, does result in relevant benefits and can not be achieved by other less harmful alternatives within his area of responsibility. This is not to say that we can always identify and draw a clear line, since every decision is open to debate and will always be bound to scientific data and differing moral views and beliefs. No ethical approach can get around that problem. This paper integrates moral aspects with the farmer's area of responsibility in a practical way.

This ethical pathway makes transparent the evidence and illustrates where alterations are required in animal husbandry. For example not to risk birth injury by choosing the right breeds for insemination or administer anaesthetics when dehorning calves. On a wider scale a farmer may consider to alter or rebuild the livestock housing system. Although the economy will always play a role in agricultural decision making processes, the presented pathway limits its influence. This approach is one of many strategies to advance animal welfare in livestock husbandry. Nonetheless, it highlights an essential aspect of the issue, the farmer's own responsibility.

References

Badura, J. (2001). Leidensfähigkeit als Kriterium? Überlegungen zur pathozentrischen Tierschutzethik. In: *Den Tieren gerecht werden – Zur Ethik und Kultur der Mensch-Tier-Beziehung*. Manuel Schneider (ed.). Universität Gesamthochschule Kassel, Kassel. 195 – 210.

Busch, R.J. and Kunzmann, P. (2006). *Leben mit und von Tieren – Ethisches Bewertungsmodell zur Tierhaltung in der Landwirtschaft (2nd ed.)*. Herbert Utz Verlag. München.

Farm Animal Welfare Council (2006). 5 Freedoms. Available on http://www.fawc.org.uk/freedoms.htm [date of consultation: 18.04.06]

Grimm, H. (2005a). Ethical Issues in Agriculture. In: *Interdisciplinary and Sustainability Issues* (ed. Olaf Christen). In *Encyclopedia of Life Support Systems* (EOLSS), Developed under the Auspices of the UNESCO. Eolss Publishers. Oxford,UK. [http://www.eolss.net].

Grimm, H. (2005b). Praxisorientierte Ethik in der Nutztierhaltung, In: Chancen und Grenzen einer tiergerechten Nutztierhaltung. Freiland Verband (ed.). Wien. 7 – 12.

Unshelm, J. (2002). Indikatoren für die Tiergerechtheit der Nutztierhaltung. In: *Umwelt- und tiergerechte Haltung von Nutz-, Heim- und Begleittieren*. W. Methling and J. Unshelm (eds.). Parey Buchverlag im Blackwell Wissenschafts-Verlag. Berlin. 242 – 248.

Competing conceptions of animal welfare and their ethical implications for the treatment of non-human animals

Richard P. Haynes
Department of Philosophy, University of Florida, P. O. Box 118545, Gainesville, Florida 32611-8545, USA, rhaynes@phil.ufl.edu

Abstract

Animal welfare scientists have appropriated the concept of animal welfare, claiming to give a scientific account that is more objective than the "sentimental" account given by *animal liberationists*. This strategy plays a major role in supporting merely limited reform in the use of animals and runs the risk of assuming that there are conditions under which animals may be raised and slaughtered for food that are ethically acceptable. Reformists do not need to make this assumption, but they tend to conceptualize animal welfare is such a way that death does not count as harmful to the interests of animals, nor prolonged life a benefit. In addition to this *prudential value assumption*, some members of this community have developed strategies for defending suitably reformed farming practices as ethical even granting that death and some other forms of constraints are harms. One such strategy is the fiction of a *domestic contract*. However, an accurate conceptualization of animal welfare, using an extension of L. W. Sumner's account to domestic animals, has different implications for which uses of animals should be regarded as ethically acceptable. Animal breeders, as custodians of the animals they breed, have the ethical responsibility to help their animal wards achieve as much autonomy as possible in choosing the form of life made available to them and to provide that life. Attempts to avoid these implications by alluding to a contract model of the relationship between custodians and their wards fail to relieve custodians of their ethical responsibilities of care.

Keywords: animal welfare, farm animals, ethics

Introduction

Since its inception, the welfarists have tried to give a scientifically objective but minimalist account of animal welfare. The farm animal science part of this movement beginning in the early 1970s in reaction to the 1966 Brambell Report, continued in this tradition, stressing suffering as the main factor in evaluating the welfare of animals in intensive confinement systems (Mench 1998). To correct this trend, Mench proposes a broader operational definition of animal welfare incorporating the general goals identified by Fraser (1993): a high level of biological functioning, freedom from suffering, and positive experiences. Those who have progressed beyond these minimalist criteria, in their conceptualization of animal welfare, such as Mench, Duncan, Fraser, Appleby, and Sandoe, to name a few representatives of the "animal welfare science community," though clearly arguing for greater reform in the way animals are kept in intensive systems, have still "under-conceptualized" welfare, with the result that limited reform seems to be all that is ethically required. One source of conceptual confusion in this attempt to "appropriate" the concept of animal welfare is an unclear or mistaken account of human welfare, confusing welfare with "subjective" elements of a good life that are instrumentally valuable to it or "objective" elements that are standardly beneficial to humans, or, in the case of non-humans, to members of particular species (Sumner, 1996). A second source of confusion is a skepticism about choosing a correct account of welfare (Appleby and Sandoe, 2002). A third source of confusion is the way in which ethical questions are intertwined with scientifically based judgments about the quality of life that animal caretakers owe their wards. I will give some example of each of these confusions, and offer clarifications and corrections.

What is welfare

As Sumner (1996) argues, we say that someone is faring well when they are justifiably satisfied with the way their life is going. "Justifiably" means that their choice of criteria used to assess their life is based on freedom rather than on oppressive socialization. Objective accounts of welfare confuse lists of things that are typically beneficial with the criteria actually chosen by the subject. Hedonistic and preference satisfaction accounts of welfare, though subjective do not include the element of autonomy. We may take pleasure in situations or prospects that we misconstrue, or have preferences among overly narrow lists of possibilities. Applying Sumner's account to animals whose lives are under our control, promoting their welfare is like promoting the welfare of the children under our control. The autonomy requirement is replaced by a judicious paternalism that requires us to educate our wards to realize their best capabilities (Nussbaum, 2004), enabling them to make informed choices about the type of life they want to live, having developed the skills needed to live that sort of life. Animal wards, if they are to fare well, must be given the same sort of capabilities training, and then placed in an environment that gives them the opportunities to exercise those capabilities that they find most rewarding. Just what sort of "flourishing enabling" capabilities our children or our animals wards have and how they can be developed is an "empirical question," and there well may be "scientific" studies that could shed light on the sorts of things that are typically beneficial to members of that particular species, and "scientific" or "professional" techniques for discovering how individual variations are to be treated. But these empirical-scientific questions are necessarily infused with ethical questions, such as what resources are caregivers ethically required to make available to their wards. Attempts to identify and address these ethical questions in some of the animal welfare science literature reveal other sources of confusion in framing the ethical questions. One source of confusion is reflected in the debate about how animal welfare should be conceptualized, when, in practice, the debate is carried out in terms of what sorts of things or conditions are standardly beneficial. Supporting this confusion is an expressed skepticism about whether we can justify choosing among contending conceptions.

Is there a correct conception of welfare?

Appleby and Sandoe (2002) express skepticism about the ability of philosophy to provide a definitive answer to the question of what animal welfare is, partly because of a similar inability to conclusively say what human welfare is. To make this point, they identify three "theories of human welfare." These theories correspond to the theories that Sumner rejects. Objective theories confuse a list of what is standardly beneficial to members of a species (or to humans) with the role these things might play in being satisfied with one's life with a theory of what welfare is. The subjectivist "theories" also fail as theories for the reasons Sumner give. Duncan and Fraser (1997) show a similar confusion. This confusion may be endemic to the discourse of the animal welfare science community and its divisions into several competing schools. It is not clear whether these schools are to be understood as general theories of welfare, comparable to the theories that Sumner identifies, and should, therefore, also apply to humans, or whether they are simply different views about the sorts of things that a science of animal welfare can discover about what types of things are standardly of prudential value to different animals of different types. The same confusion appears in Duncan's analysis of the role of "feelings" in welfare. It is unclear whether Duncan is arguing for a subjectivist conception of welfare that is

hedonic, or a subjectivist conception that is appraisive (one's welfare is determined by how one *feels* about one's life), or simply maintaining that good and bad feelings (pains and pleasures) are standardly beneficial or harmful to most animals' lives.
It is also unclear what role value judgments allegedly play in an animal scientist's views about the welfare of the animals they study. Aside from judgments about which methodologies scientists should use to warrant their findings, I can see three roles: One role are the judgments that an investigator needs to make about what is of prudential value to the subject of study. But these seem to be judgments about what is of prudential value to the subject rather than judgments about what the subject should prudentially value (unless, of course, the scientist is adopting the perspective of someone evaluating the authenticity of how the subject in fact seems to appraise their current life). The second role, which is the one that Duncan seems to have in mind, is an ethically related one. What level of welfare are animal users ethically required to provide. but scientists should not be letting these ethical judgments interfere with their efforts to find out what is standardly beneficial and harmful to animals of the type they are studying. A third role of value judgments does seem appropriate. In order to find out what is standardly instrumentally beneficial and harmful to a subject of a particular type, a scientist must put themselves in the role of an *authenticator* of the subject's actual goals and ask to what extent these goals have been chosen from a reasonably available range of goals. By "reasonably available," I mean something like this. One could imagine that a subject (including a human subject), if given a series of scenarios of the types of lives that they might choose to live, would choose a life that required the availability of instrumental goods that are never or rarely available given the condition of the world as it is or is ever likely to become. So to find out what is truly in the interest of the subject in question, we would have to narrow the scenarios to those that can reasonably be expected in the best of worlds likely to become available through ethically motivated efforts to improve it. Applying this to the scientist's problem, the scientist, as an authenticator, must try to find out not only what choices the scientist's subject would make under existing conditions, but what choices they might make under a range of improved conditions and the degree to which these choices are authentically made. Since the range over which these expanded choice opportunities could be considerable, scientists trying to perform the authenticator role must, for practical reasons, narrow the range by choosing those that seem most reasonable, and such choices would require value judgments of several sorts, some of which are ethical.

What are the ethical issues?

The primary ethical question that livestock breeders – animal custodians – face is how much welfare do they owe their wards, or, how well off should we strive to make our animal wards. Sandoe *et al.* (2003) express this question in this way: What is the baseline standard for morally acceptable animal welfare? They then rephrase this question by asking, "What is a fair deal for farm animals?" By rephrasing the question in this way, the authors are suggesting that we employ a contract model in making these decisions. The contract model has been appealed to by a number of writers on this topic (e.g., Rollin, 1995; Appleby, 1999; Larrére, and Larrére 2000, and Lund *et al.*2003) and the theme is developed by Budiansky (1999).
One thing objectionable about the contractual model is that it is used to justify only limited reform rather than abolition. This seems to be the role that Rollin puts it to when he talks about the husbandry contract between pre-intensive farming farmers and their livestock. In earlier systems, there were mutual benefits exchanged between the farmer and the livestock. Although there is some evidence to support the claim that at some point in time, dogs, or their early ancestors, chose to associate with humans, it is more controversial to claim that this "choice" was made by all domesticated animals or even that a young dog raised from birth makes this choice. Even if we do accept the neoteny thesis developed by Coppinger (1983) and Budiansky, the thesis

fails to show that individual animals, who are bred into captivity, choose their domestication, nor do dogs or other pets born into domestication make this choice. So the contract model would only serve its justificatory purposes (it can be used as a tool for urging reform of existing practices and as one for justifying some reformed version of existing practices) if we can argue that domestication gives a "fair deal" to livestock and so they ought to choose it. As caretakers and domesticators (we socialize and otherwise train animals to aspire to a certain sort of life, thus limiting their aspirations to other sorts of lives), then, we are entitled to *construct consent* only on the grounds that we are helping them achieve autonomy. This model of constructed consent could be said to apply to the parent child relationship. Good parents are entitled to be "paternal-maternal" because, in knowing what is ultimately good for their child, they can make decisions in its behalf. But this model does not seem to work when the ward is to be used in ways that are ultimately to its disadvantage (think of a parent raising its child to remain dependent on it when it becomes an adult in order to ensure that the adult-child continue to provide the companionship that care-dependency provides, as we often do with our pets).

The main obstacle to a constructed consent model to justify restricting the liberty of livestock is that it is hard to see how including an early death in the contract can be seen as consentful. Appleby (1999) tries to argue that it can be on the grounds that the life of the individual animal would not have been engendered by its caretaker, so it is better for the ward to have lived a short but happy life than not to have lived at all. I do not see how this comparison can be made since we cannot make sense of the choice: would you rather live a short happy life (a year or two) or not live at all. If the only condition under which livestock species can be preserved is that they be raised to play the traditional role of livestock, then I see no grounds for agreeing with Appleby's argument that it is better for them to live a short life than none at all.

Conclusion

The caretaker model seems a more appropriate way of modeling the relationship between the producer and livestock, but this model, when used to determine the duties a parent has to its child, seems to imply that the caregiver make the sorts of choices for its ward that are consistent with the ward's future autonomy (or its constructed autonomy). As Rollin puts it (1992), the fact than an animal owes its existence to its breeder no more makes it the breeder's property than a human child is the property of its parents. We do not think that a parent is justified in killing their child just because the child owes its existence to its parents' breeding choices.

If we use this version of the custodial model, there is still an important role for animal welfare scientists, namely to find out what sorts of things are standardly beneficial (and harmful) to the various species of animals that humans breed, so that caretakers can provide authentically good lives for their wards. To perform this information-providing role, animals welfare scientists must go beyond preference testing for a narrow range of environmental conditions, and expand these conditions in much the same way parents are expected to find out what is in the best interest of their children when they become adults, and thus prepare them for adult living.

References

Appleby, M.C. (1999). *What Should We Do About Animal Welfare?* Blackwell Science: Oxford.

Appleby, M.C. and Sandøe, P. (2002)."Philosophical debate on the nature of well-being: implications for animal welfare." *Animal Welfare* 11 (2002), 283-294.

Duncan. I.J.H. and Fraser, D. (1997). "Understanding animal welfare." Ch. 2 (pp. 19-31) in M. C. *Animal Welfare.* M.C. Appleby and Hughes, B.O. (eds.).CABI: Wallingford, UK.

Coppinger, R.P. and Smith, C.K. (1983). "The domestication of evolution." *Environmental Conservation* 10: 283-92.

Larrére, C and Larrére, R. (2000). "Animal rearing as a contract?" *Journal of Agricultural and Environmental Ethics* 12(1): 51-58.

Lund, V., Anthony, R. E. and Rocklinsberg, H. (2004). "The ethical contract as a tool in organic animal husbandry," *Journal of Agricultural and Environmental Ethics* 17(1); 23-49.

Mench, J.A. (1998), "Thirty Years after Brambell: Whither Animal Welfare Science?" *Journal of Applied Animal Welfare Science* 1(2): 91-102.

Nussbaum, M.C. (2004) "Beyond 'Compassion and Humanity': Justice for Nonhuman Animals." Pp. 299-320 in *Animal Rights. Current Debates and New Directions*. C. Sunstein and Nussbaum, M.C. (eds.) Clarendon Press: Oxford.

Rollin, B.E. (1992). *Animal Rights and Human Morality*. Revised Edition. Prometheus: Buffalo, NY.

Rollin, B.E. (1995). *Farm Animal Welfare. Social, Bioethical, and Research Issues*. Iowa State University Press: Ames, IA.

Sandøe, P., Christiansen, S.B. andAppleby, M.C. (2003). "Farm animal welfare: The interaction of ethical questions and animal welfare science." *Animal Welfare* 12: 469-78.

Animals and the harm of death

Frederike Kaldewaij
Ethics Institute, Utrecht University, Heidelberglaan 2, 3584 CS Utrecht, The Netherlands,
F.E.Kaldewaij@ethics.uu.nl

Abstract

Is painless death against the interests of animals? The answer to this question is an important factor in determining the moral acceptability of the production and consumption of animal products such as meat. Some philosophers have argued that the harm of death lies in the thwarting of a subject's desires regarding her future existence, and that animals are not harmed by death because they cannot have such desires. This view, however, seems to imply an implausible prudential value theory, the view that something can only be in your interest if you actually desire it. Thomas Nagel argues that death is harmful because it deprives an individual of the goods that continued life would have brought her. Nagel's view is compatible with a more plausible subjective value theory and gives an explanation of why death is such a serious harm for all conscious beings with reasonable prospects in life. Is death much less harmful for animals than for human beings? Are the goods involved in animal lives qualitatively inferior to those in human lives, or do animal lives contain fewer goods than human lives? I argue that qualitative arguments are built on shaky ground, and quantitative arguments have only limited force. There are good reasons to assume that the animals that are killed for the production of food are seriously harmed by death. We may therefore have to reconsider the view that it is morally permissible to kill such animals provided they are treated humanely during their life.

Keywords: killing animals, harm of death, prudential value theory

Introduction

In Western societies, it is generally acknowledged that animals have an interest in not suffering. This is often considered to have ethical implications, one of which is that animal welfare should be a concern in animal agriculture. Whether painless death harms animals is a more controversial issue, and this is an important factor in determining the moral acceptability of the production and consumption of animal products like meat and dairy. (The dairy industry indirectly leads to killing of animals, for instance the male calves of dairy cows). In this paper, I will defend Thomas Nagel's view of the harm of death, which can be applied to conscious animals, against the view that animals are not harmed by death as they cannot have desires regarding their future existence. Next, I will consider different arguments that have been used to support the claim that animals are harmed much less by death than human beings. I will conclude that conscious animals such as those killed in the food industry are seriously harmed by death.

Can animals be harmed by death?

The question of whether death is harmful for human beings is a well-known philosophical problem. Epicurus (341-270 B.C.) points out that when we are alive, we are not yet harmed by death, and when we are dead, there is nobody to be harmed. If there is no (possibly unpleasant) afterlife, we need not fear death (Rosenbaum, 1993). However, death may not be a misfortune because it is (paradoxically) a bad state to be in, but rather because of what it takes away from us.

Thomas Nagel's account of the harm of death has been called "the most popular anti-Epicurean view" (Feldman, 1993). Nagel (1993) argues that death is harmful because it deprives us of the goods that our future life would have contained.

There is an alternative view of the harm of death, according to which death is harmful because it thwarts an individual's desire to stay alive and other desires she may have for her future. Animals, according to this argument, are not harmed by death as they cannot have such desires, which involve sophisticated beliefs and concepts, such as the concept of one's own mortality. Two advocates of this view are Peter Singer and Ruth Cigman. Singer (1993) argues that only rational and self-conscious beings can have preferences regarding their future existence. Cigman (1981) argues that only beings who have the capacity of having "categorical desires" can be harmed by death. As Cigman explains it, while conditional desires (such as a desire to eat when one is hungry) presuppose being alive, categorical desires (e.g. to write a book or raise children), answer the question of whether one wants to remain alive.

Prudential value theory is concerned with the (nonmoral) good or interests of individuals. In subjective value theories, an individual's good is determined by reference to her own desires or values. Objective value theories, in contrast, claim that some things are valuable in themselves. Singer and Cigman's views imply a subjective value theory. An objective theory may be seen to be problematic, because it is unclear why certain things, such as accomplishment or knowledge, are in an individual's interest if she does not value them at all. Subjectivism especially seems to be a plausible theory within the context of animal ethics: it is usually said that a necessary condition for having interests is having feelings and desires (e.g. Feinberg, 1980; DeGrazia, 1987). While we do not say it is against a plant's interest if its leaves are pulled off, we do say that an animal is harmed when its leg is caught in a leg-trap: unlike the plant, it will suffer as a result of its injuries.

While an objective value theory may not be desirable, a theory that determines the subject's good solely in terms of the satisfaction of her actual desires is not either. Steve Sapontzis (1987) and Tom Regan (2004) have both, in the context of animals and the harm of death, pointed out that having an interest is not the same thing as taking an interest (in Sapontzis' terminology). In other words: your desires may not always reflect what is good for you. For example, children may not like eating vegetables, and addicts may crave cocaine. But how can we say that something can be in an individual's interest even if she does not desire it, without falling into the pitfalls of an objective value theory? This problem is often solved by appealing to the desires an individual would have, if she were fully informed and rational. A problem with this view with respect to the harm of death for animals is that animals are not *capable* of being rational and informed enough to understand that life is in their interest. Neither, for that matter, are very small children able to understand the health benefits of vegetables. But is this a reason to assume that these things are not in their interest?

Cigman argues that you must have the *capacity* to have categorical desires to be able to be harmed by death (she uses the term "misfortune" for "harm"). Even a suicidal person can be harmed by death: "X's misfortune must either be something which X did not want; or it must be something that X *should not* have wanted, because it so obviously conflicted with his interests" (1981, italics hers). According to Cigman, it can be unfortunate for human beings if they fail to have categorical desires, but animals cannot be pitied for not desiring something they cannot desire, i.e. life. However, Cigman does not, and cannot, explain *why* it "obviously" conflicts with human beings' interests to die, if they do not actually desire to live.

A good explanation for why something can be in our interest even if we do not desire it is that it is *instrumentally valuable* for something we do value. Vegetables are good for small children in the same way riboflavin is good for adults who have never heard of this vitamin, and eventually this can be explained in terms of their future well-being and desire-satisfaction. It is arbitrary to make a distinction between those who have the capacity to understand something is instrumentally valuable for them and those who do not. A fundamental aspect of Nagel's theory is that life has instrumental, not intrinsic value. As he notes, organic survival has no value for a permanently comatose patient (1993). Human beings and conscious animals are harmed by death because they are deprived of goods *that they would have valued, enjoyed, found desirable.* Nagel's theory of the harm of death is compatible with a subjective value theory, and offers an explanation of why people with reasonable prospects in life have an interest in life. We usually want to prevent people from committing suicide, because they may overcome their psychological problems and live many years filled with goods they will value.

An additional benefit of Nagel's theory is that it can explain the magnitude of the harm of death: death takes away the possibility of ever experiencing, doing or accomplishing anything you value again. According to Singer and Cigman, death is harmful because it thwarts the desires we have at the time of death. But is someone who just enjoys life as it comes harmed much less by death than someone who has his life completely planned out already? Cigman's example of the desire to eat when hungry as a conditional desire may seem compelling: you do not need to eat when you are dead. However, it is misleading: what about eating for pleasure as well as to avoid starvation, or valuing a meal shared with family and loved ones? Unlike writing a book, or raising children, we do not value such things as part of a "life plan". But not everybody may have such plans, while they may still enjoy the goods in their lives on a day-to-day basis. Animals do not only have desires like eating when they are hungry, desires to alleviate frustration. They actually like eating, grooming, rolling in the mud, playing, etc. Such desires give their life instrumental value, just like whatever we find valuable makes our lives worth living.

There is one way in which being rational and self-conscious is relevant to the harm involved in *some* deaths. Human beings who come to understand they will die soon, face a kind of psychological suffering that animals are probably not capable of. However, I am only concerned here with the *basic* harm of death, not with suffering related to death and dying. Even when you are killed suddenly, in your sleep, you can suffer this basic harm of death. For this harm, you need not have an abstract concept of your own mortality.

Is human life much more valuable than animal life?

Is the harm of animal death trivial compared to the harm that human beings suffer when they die? There have been two strategies to support this claim: according to the first, the goods associated with animal life are qualitatively less valuable than those involved in human life, and according to the second, animal lives contain fewer goods than human lives. I will discuss each of these arguments in turn.

Quality

The most well-known defense of the qualitative argument is Mill's claim that "[i]t is better to be Socrates dissatisfied than a fool satisfied, better to be a human being dissatisfied than a pig satisfied" (1957). According to Mill, someone acquainted with both "higher pleasures" (e.g. intellectual pleasures) and "lower pleasures", (e.g. sensual pleasures), will always give a preference to the "higher pleasures". Singer (1993) gives a similar argument, claiming that if one were in the position to be a horse, a human being, and in a third state, in which one could compare both, one would always choose the life of the most rational and self-conscious being. This kind of existence is "intersubjectively" more valuable.

This argument fails. To begin with, one can question whether people acquainted with both kinds of pleasure would always choose "higher pleasures". But the real problem is that the "being of higher faculties" is not competently acquainted with the life of someone with "lower faculties". It would be frustrating for Socrates to live a pig's life, but does that mean that a satisfied pig enjoys the goods involved in its life less than Socrates values the goods of his own life? The view that Socrates' life is better is made from our point of view, not from an intersubjective point of view. Saying, alternatively, that human goods are objectively better than a pig's is problematic, for what makes it objectively true that one individual's values and enjoyments are superior to another's?

Quantity

Regan seems to give a quantitative argument when he says that humans have "more sources of satisfaction" available to them than animals, such as aesthetic or scientific interests. But someone who has more sources of satisfaction may not lead a more satisfactory life. Someone may lead a simple life in the country, intensely enjoying a few goods, while another person in the city may mildly enjoy many goods. Both lives contain the same total amount of satisfaction. Is the second kind of life obviously better? Besides, as Sapontzis (1987) points out, many animals have sources of goods we do not have, such as a dog's sense of smell or a bat's echolocation. We cannot claim that the goods associated with our capacities are better than theirs, this would be a qualitative claim.

But perhaps human lives always contain more goods in total than animal lives? As an empirical fact, it is true that human beings tend to live longer than other animals, with the exception of for instance certain varieties of turtles. Besides this, DeGrazia (1996) says we can perhaps assume that the mental lives of animals like trouts and alligators are pretty dim. If a cognitively very simple animal's behavior can mainly be explained in terms of stimulus and response, this may be a reason for thinking it is not very conscious. But the animals we are mainly concerned with in animal agriculture are mammals and birds. Even fish turn out to have social intelligence (Leland, Brown and Krause, 2003), and healthy pigs are in fact curious, playful animals. These animals do not live their lives just going through the motions, it is most plausible to assume that they vividly experience their lives.

Conclusion and final remarks

Both human beings and conscious animals are harmed by death, because this deprives them of the goods that continued life would have brought them. Nagel's theory is compatible with a subjective value theory, because life is instrumentally valuable for those who enjoy and value the goods involved in their lives. There do not seem to be good reasons to assume that the harm that death causes animals that are kept for food production is much less serious than the human harm of death.

I have not yet argued what our moral obligations to animals are, or to what extent animal interests should be taken into account. For now, however, I can point out that increasingly, animal welfare is accepted as a legitimate moral concern. Implicit in this view is the idea that suffering is against the interests of animals. If we take animal interests seriously, we should, in the moral evaluation of the practice of raising and killing animals for food production, not only consider the welfare of animals during their lives, but also take into account the deprivation caused by premature death.

References

Cigman, R. (1981). Death, Misfortune and Species Inequality. *Philosophy and Public Affairs*, 10: 47-64.

DeGrazia, D. (1996). *Taking animals seriously. Mental life and moral status*. Cambridge, Cambridge University Press.

Feinberg, J. (1980). *Rights, Justice, and the Bonds of Liberty: Essays in Social Philosophy*. Princeton, Princeton University Press.

Feldman, F. (1993). Some Puzzles About the Evil of Death. In: *The Metaphysics of Death*. Fischer, J.M. (ed). Stanford, Stanford University Press. 305-326.

Leland, K.N., Brown, C. and Krause, J. (2003). Learning in fishes: from three second memory to culture. *Fish and fisheries*, 4: 199-202.

Mill, J.S. (1957). *Utilitarianism*. Priest, O. (ed). Indianapolis, The Bobbs-Merrill Company. (Original work published 1861).

Nagel, Th. (1993). Death. In: *The Metaphysics of Death*. Fischer, J.M. (ed). Stanford, Stanford University Press. 59-69.

Regan, T. (2004). *The Case for Animal Rights* (2nd edition). Berkeley and Los Angeles, University of California Press.

Rosenbaum, S.E. (1993). How to Be Dead and Not Care: A Defense of Epicurus. In: *The Metaphysics of Death*. Fischer, J.M. (ed). Stanford, Stanford University Press. 117-134.

Sapontzis, S.F. (1987). *Morals, Reason, and Animals*. Philadelphia, Temple University Press.

Singer, P. (1993). *Practical Ethics* (2nd edition). Cambridge, Cambridge University Press.

Tail docking in horses: Tradition, economy, welfare and the future of the Belgian draft horse

Dirk Lips and Stefan Aerts
Centre for Science, Technology and Ethics, Kasteelpark Arenberg 30, 3001 Leuven, Belgium

Since 2001 the Belgian law prohibits tail docking (caudotomy) of horses. In 2004 a proposition to change legislation back fired a fierce (and ongoing) debate on the subject. Arguments against tail docking are typically based on welfare considerations (chronic pain, vulnerability to insects, ...). Arguments in favour also cover welfare issues (wounds during mounting, hygiene), but are dominated by references to cultural traditions ("tail docking has been done for centuries", "it is typical for the breed"), practical constraints ("the horse will be uncontrollable if the tail gets caught in the line") and economic benefits ("undocked horses are worthless", "export will collapse"). This paper will start with a overview of the background of this debate and of the international context. We will analyse the different (categories of) arguments for their validity and their ethical importance. We will also address the (lack of) focussed scientific research into this matter, which influences the ethical evaluation of the situation. Based on this evaluation we will argue that tail docking of horses is an ethically unsound practice when considered from a zoocentric or moderate antropocentric approach. From this we will plead for research efforts into this subject that will clarify the remaining blind spots, and for a European regulatory move, even if the economic consequences are moderate (at most) compared to other agricultural branches. We will conclude with some remarks on the future of the Belgian draft horse.

Keywords: animal welfare, animal production, sustainability, food production

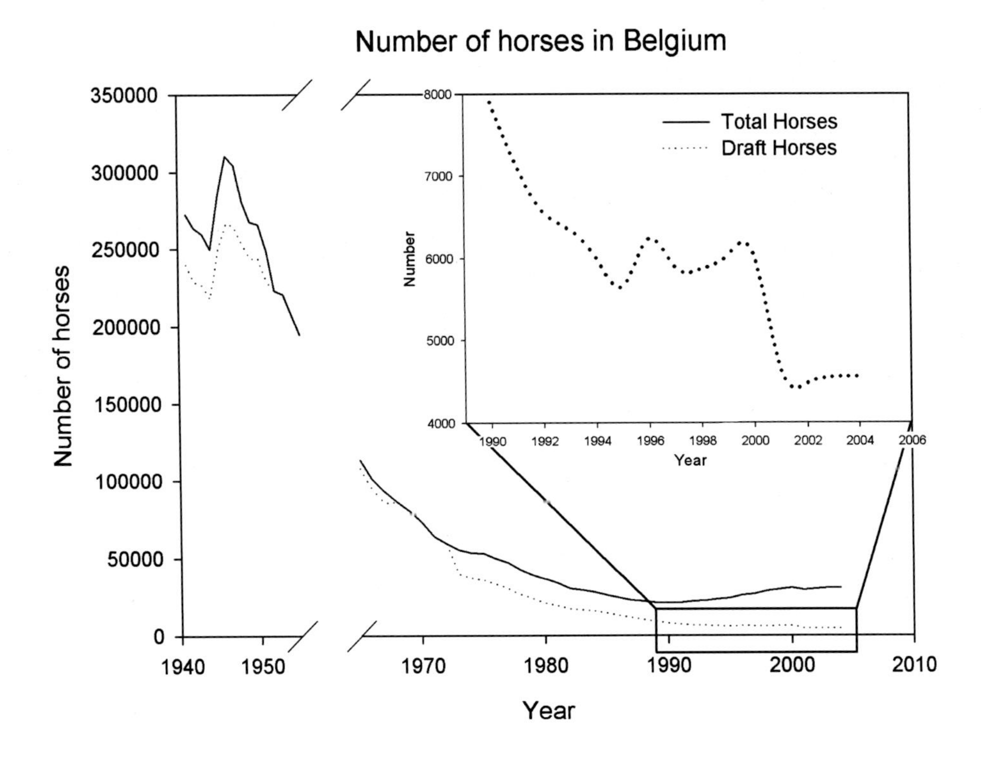
Number of horses in Belgium
Number of horses
350000
300000
250000
200000
150000
100000
50000
0
1940
1950
1970
1980
1990
2000
2010
Year
Total Horses
Draft Horses
Number
8000
7000
6000
5000
4000
1990
1992
1994
1996
1998
2000
2002
2004
2006
Year

All animals are equal but some animals are more equal than others: A cross-cultural comparison of human-animal relationships in Netherlands and Norway

Liesbeth Schipper, Hanne Torjusen, Volkert Beekman, Laura Terragni, Michiel Korthals
Applied Philosophy Group, Wageningen University, Hollandseweg 1, 6706 KN Wageningen, The Netherlands, liesbeth.schipper@wur.nl

Abstract

Food consumption, preparation and shopping practices may reflect and embody taken-for-granted normative assumptions about human-animal relationships. Consumers' self-reported descriptions of these practices may furthermore point to instances of ambivalence with respect to using and killing animals for meat products. This paper presents a cross-cultural comparison of Dutch and Norwegian strategies for dealing with the tension between people's affectionate feelings towards animals and their practices of eating these same animals. This comparison is based on the results from Dutch and Norwegian qualitative focus group discussions and a quantitative survey. Although Norwegian and Dutch respondents largely used the same (implicit) tactics and arguments to deal with ambivalence, the results also show differences. A key finding is that in both countries people shared the idea that the more an animal is perceived to have had a good life; the more morally permissible it is to eat this animal. Respondents in the two countries, however, judged differently about what types of animals are suitable to kill for food. For Norwegians the reference point was nature and game was therefore considered to be the most ideal sort of meat. A majority of the Dutch respondents was strongly opposed to hunting and animal welfare was therefore connected to the context of farming. Both similarities and differences shed light on human-animal relationships in the context of food consumption.

Keywords: ambivalence, animal welfare, food consumption, human-animal relationships

Introduction

Food consumption, preparation and shopping practices may reflect and embody taken-for-granted normative assumptions about human-animal relationships. Consumers' self-reported descriptions of these practices may furthermore point to instances of ambivalence with respect to using and killing animals for meat products. The consumption of meat and other animal products is connected to death and suffering of animals. Since people also have the ability to empathise and sympathise with animals, this may imply that they love animals but also eat them. This does not go well together and could lead to feelings of ambivalence. People have different ways to deal with such discomforting feelings. This paper presents a cross-cultural comparison of Dutch and Norwegian strategies for dealing with the tension between people's affectionate feelings towards animals and their practices of eating these same animals.

Methodology

This paper is largely descriptive. The empirical results that inform this paper stem from qualitative and quantitative studies as conducted in the context of the European project 'Welfare Quality'. This project organised focus group discussions in seven European countries to study consumer concerns about animal welfare. As a follow-up of these focus group discussions, a quantitative survey with 1500 respondents in each of the seven study countries has been conducted. This paper is based on the Norwegian and Dutch results from these qualitative and quantitative studies. The focus group discussions focused on consumers' concerns about animal welfare and on their views about information-provision strategies with respect to animal welfare. Focus group discussions are a frequently used method of qualitative social scientific research. They operate through a loosely structured discussion about a topic that is of specific relevance to the participants. The participants in a focus group have some common characteristics that are relevant for the topic under discussion, i.e. food consumption practices in relation to animal welfare (Krueger, 1988). In both countries the same recruitment guide and interview guide have been used for these focus group discussions. The results from the focus group discussions have been used as input for a follow-up quantitative survey among a representative sample of the population.

Results and comparative analysis

Feelings of ambivalence with respect to the consumption of meat and other animal products were present in the Dutch and the Norwegian focus group discussions, albeit in different ways and to different degrees. Ambivalence can be understood as the tension between the following two feelings: 1) I like animals and believe that they deserve better treatment than is currently provided; and 2) I like to eat cheap and tasty animal products. Basically, five different situations of ambivalence with respect to the consumption of meat and other animal products could be identified in the focus group discussions. First, in some cases ambivalence was not obviously present. Second, in some cases ambivalence was resolved by changing consumptive practices. Third, ambivalence was avoided. Fourth, ambivalence was denied. Fifth, ambivalence was resolved through selectivity in people's attachment to animals. This section describes these different strategies for dealing with ambivalence as observed in the different focus group discussions. The subsequent comparative analysis is based upon results from both focus group discussions and survey.

A good life for animals

A key finding in both countries is that the more an animal was perceived as having had a good life; the more acceptable it was considered to eat this animal. There was no, or not so much, reason for feeling ambivalent in cases of animals having had a perfectly good life. In Norway this was very clearly illustrated by the generally accepted idea that meat from moose can be eaten with a clear conscience. The living conditions of this wild animal, which has been born, living and killed in nature, were regarded and often emphasised as an ideal situation: "They are really what you would call free-range. I eat it with a clear conscience, as I know that the moose have to be 'harvested' every year or there will be too many of them [...] This gives me a good feeling" (Norwegian female senior). The same kind of statement recurred across all Norwegian focus group discussions.

In comparison, Dutch focus group discussions not so much focused on wild animals as an ideal, or on moose or some other form of game as the ideal sort of meat. On the contrary, Dutch participants to the focus group discussions widely shared a strong opposition towards hunting. This can be explained by reference to The Netherlands as a densely populated country. As a consequence, the whole notion of naturalness is quite different from a country like Norway with plenty of nature. Any nature that is still left in The Netherlands is something that should be left alone according to many Dutch participants to the focus group discussions. It is also important in this respect that hunting is considered to be an elite game in The Netherlands. For Dutch participants to the focus group discussions, in contrast to Norwegians, animal welfare was in the first place about farming methods. Since edible animals were considered to be farmed animals,

and not wild animals, animal welfare was connected to the context of farming and not to the context of nature management for the Dutch participants.
The sharp contrast between Norwegian and Dutch respondents regarding game could also clearly be observed in the survey results. Whereas only 6% of the Norwegian respondents did not find venison appealing at all, a stunning 56% of the Dutch respondents agreed to this statement. Similarly, only 6% of the Dutch respondents found venison very appealing, whereas 52% of the Norwegian respondents found venison very appealing. Since organic farming was regarded as the ultimate animal-friendly way of keeping animals by a majority of the participants to the Dutch focus group discussions, organic meat comes closest to being the equivalent of the Norwegian moose as a meat that you can eat with a clear conscience. The difference is, however, that moose seems to be seen as a much more guilt-free type of meat for the Norwegians than organic meat for the Dutch consumers.

Conversion

Since ambivalence is a consequence of feeling uncomfortable about the death and (ab)use of animals for their meat and other products, it is not surprising that in both countries vegetarian participants in the focus group discussions did not experience ambivalence to the same degree as people that did eat meat. Vegetarians, or participants to the focus group discussions that only bought organic animal products, showed far less ambivalent feelings than other participants. This could be witnessed among politically active and vegetarian consumers but also among the other focus group participants. Some vegetarians seemed to adopt a position of moral superiority. They were taking responsibility, why were others not doing the same thing? These people seemed to be quite aware that their vegetarianism had the consequence that they lacked much of the disturbing feelings of ambivalence and even turned these feelings into the positive feeling of having taken responsibility. This positivism was reflected by the terminology that quite some of these participants used to indicate that they become happy from buying or eating certain products: "Suppose I eat steak, organic from the supermarket Albert Heijn, then I am aware of the rightness of the choice. I also sometimes tell the other meats that I am happy that I can eat them [...] I would like to accomplish something greater with buying organic meat but it is about the small things. I also get happy when it is about small farmers in a poor country receiving a good price for their products. I do not buy Max Havelaar but I do get happy from that" (Dutch male empty nester). One of the means to overcome experienced ambivalence could thus be to change behaviour: become a vegetarian or start eating only 'good' animal products.

Avoidance

Ambivalence may also be resolved by avoiding the dilemma. This was a common strategy for participants in the Norwegian and Dutch focus group discussions. Quite some different ways existed to avoid the dilemma. Often participants claimed to avoid information about animal welfare in general, and about the living conditions of the animal before it was meat in particular. They indicated that they did not want too much information about the animal and its living conditions. One Norwegian woman said: "I think that the label must not say too much either. It is not really necessary that it is said: 'Lille Kolle' was a happy pig before he was brought to the slaughterhouse. Because then it is not food anymore. Then it is the animal" (Norwegian female empty nester).
Quite some participants did not want to recognise the animal in the meat products that they are eating or cooking, which is another instance of avoiding the ambivalence: "I do not think about the animal in itself, to be honest. I think that this is pork, for example, and have a rather vague idea of the animal. I do not think of the animal in a way that gives it form, personality. I do not visualise it as a specific pig, not at all. When I am standing there frying the meat, I do not consider whether it has been a free-range pig or not. I do not check the label on the package

to find that out. I just do not think 'I am frying a little pig here' when I am in the kitchen. It all seems rather removed somehow, almost as if the pig only existed theoretically" (Norwegian urban mother). The picture emerged from the focus group discussions that Dutch consumers find it harder to deal with the idea that the meat has been an animal than Norwegians, since more Dutch participants referred to the idea that they did not like to be reminded that the meat they are preparing or eating comes from an animal. This picture has been confirmed by the survey data. Less Dutch (33%) than Norwegian respondents (67%) strongly disagreed with the statement that they do not like to think, when eating, of meat coming from animals. A higher percentage of the Dutch (24%) than of the Norwegian respondents (13%) agreed with this statement. The same picture emerged for the statement 'I find it unappealing to handle raw meat'. Only 53% of the Dutch but 78% of the Norwegian respondents strongly disagreed with this statement (14% and 8% respectively strongly agreed).
The Norwegian respondents, on the other hand, more explicitly emphasised the idea that keeping and slaughtering animals is natural: "It is the most natural thing in the world that you have animals and that they are slaughtered. That is our livelihood [...] That is how it is" (Norwegian female empty nester). This justification for using and keeping animals is also sometimes referred to in the Dutch focus group discussions, most often by the older participants. It is interesting that Norwegian and Dutch participants to the focus group discussions regularly connected this idea of naturalness to the idea that humans have to take good care of the animals that they are justified to keep and kill. A strong moral codex thus seemed to be included in the notion that humans may use animals to meet our needs. For some Norwegians this was related to a claim that a qualified 'need' should exist before an animal was viewed as acceptable to kill and use. Some Norwegian participants voiced a concern that this qualification lost forces over time.

Denial

Denial of the dilemma is another means, and arguably the most radical, that could be observed as a way to resolve ambivalence. In the Dutch focus group discussions the animal welfare conditions in husbandry systems were sometimes trivialised, when it was suggested that animals are quite well off in Dutch livestock production. Furthermore, the animal welfare principles that were proposed by the animal scientists from 'Welfare Quality' were ridiculed. Some respondents assumed that animals are relatively well off, with the consequence that they did not feel guilty about their behaviour: "When I read the ten principles of animal welfare proposed by 'Welfare Quality' animal scientists, I think that those animals will have better lives than humans", "My guinea pigs live like this", "Yes, and elderly people do not" (Dutch rural women).

Selectivity

A final means to deal with ambivalence is selective attachment or selective offence in relation to animals in general and animal welfare in particular. People choose not to eat young animals, or not to eat certain animals, e.g. horses. This is a consequence of the distinction that is made between animals that are suitable for production purposes and animals that are not. In order to like animals but at the same time be able to eat them, people classify animals differently. Pets are loved and not to be eaten, non-pets are suitable for the latter purpose. Of course, people need to be clear about which animal is a pet and which one is not. They need to take good care that animals do not transgress the categories to which they belong, otherwise that may give rise to feelings of ambivalence.

Conclusion

Qualitative and quantitative results show that, although Dutch and Norwegian consumers solve or mask ambivalence in generally the same ways, remarkable differences prevail between the employed means to deal with ambivalence. In the Norwegian focus group discussions meat from wild animals was regarded as the ideal sort of meat from an animal welfare perspective. In contrast, for the vast majority of Dutch consumers animal welfare was connected to the context of farming. Another important difference is that Dutch consumers were more concerned with concealing the animal origin of meat. They were more likely to state that they did not want to be reminded of the meat as having been an animal, and also found it less appealing to handle raw meat. Dutch consumers, furthermore, were more likely to ridicule the animal welfare conditions in husbandry systems in order to deny the existence of the dilemma. On the other hand, the Norwegians were more likely to justify that animals are used and killed for their meat and other products by referring to the 'naturalness' of this human-animal relationship. In short, not all animals are equal but what animals are more equal than others is a matter of culture-specific considerations.

Acknowledgement

This paper presented results from the 'Welfare Quality' research project. This project has been co-financed by the European Commission, within the 6th Framework Programme, contract no. FOOD-CT-2004-506508. The paper represents the authors' views and does not necessarily represent a position of the Commission, who will not be liable for the use made of this information.

References

Krueger, R.A. (1988). *Focus groups; A practical guide for applied research.* Sage Publications, California.
Serpell, J. (1986). *In the company of animals; A study of the human-animal relationship.* Blackwell, Oxford.

Vethics: Vets for ethics in food

Erik Schmid
Landhaus, 6901 Bregenz, Austria, erik.schmid@vorarlberg.at

Abstract

Veterinarians traditionally feel committed to the wellbeing of animals. Animal welfare is a natural key issue of their profession. Corresponding to the specialisation into pet animal surgeons and farm animal practitioners the public awareness of animal welfare problems and general attitude diverged. While companion animals as social partners are getting personalised, the fate of farm animals is drifting towards being forgotten or displaced. This development offers benefits for the pet animal profession. On the other hand the economic gap for farmers is narrowed and automatically puts pressure on the veterinarians, as expensive treatments are not financed any more. The situation is not only dramatic and almost unrealistic from an economic point of view, it additionally gets unbalanced referring to ethical aspects. The public takes the final decision on animal welfare, either as pet owner or as consumer by purchasing food with different ethical burden. But to take this responsibility they need thorough information and objective advice. This has to be done by Veterinarians, who still have high social credits as reliable experts in animal husbandry. Animal welfare is a very strong and decisive argument in quality of food. The new EU-legislation is putting animal welfare onto the same level of importance as hygiene and processing. That is a matter of fairness and transparency. Veterinarians could play a decisive role in this process of truth and trust. They should do so, not only because of their intrinsic compassion to animals but also for long lasting prosperity of their own profession.

Keywords: animal welfare, food & feed, quality production labelling, veterinary profession

Introduction

The veterinary profession traditionally feels to be committed to the wellbeing of animals. Veterinarians often argue to be the authentic advocates of the animals. The annual meeting of the Protestant Academy Bad Boll in 1992 stated in the conclusions of the conference, that veterinarians are the experts interpreting and translating the feelings and needs of animals to the public and legislation. In 1998, the Institute for Animal Husbandry and Animal Welfare was founded at the University for Veterinary Medicine in Vienna offering better skills and education for students as well as the possibility for post graduate specialisation.

Religious, ethical aspects

Kaplan (2000) is raising the central question: How is it possible to treat animals with no biological difference which could be morally relevant i.e. dogs and pigs on the one hand so well and so badly on the other? Psychoanalytical investigations in the psychological development showed that children are led to eat meat by early education whereas they simultaneously are getting a positive emotional attitude towards animals. This is a basic contradiction in human-animal relationship. The conflict arises when children realise that you have to kill the beloved animal if you want to eat meat. Parents have a strong influence telling their children between 5 and 7 years of age, that it is completely normal to kill animals for nutrition. Mechanisms of defence like displacement, splitting and rationalisation help not being aware of this contradictory behaviour. Children identify with the behaviour of their parents. This ordinary inadequate situation is

gaining mirrored in legislation. The new Animal Welfare law of Austria bans the killing of dogs and cats for getting meat or other products (§6 (2) BGBl I 2004/118). This regulation refers more to a moral obligation to the high social status of dogs as companion animals, than to a real ethical difference. Otherwise, knowing well the ethical level of dogs and pigs to be the at least the same, the obligation to treat them in the same way should be considered also by legislation. As a consequence this ethical chain reaction would end in mandatory vegetarianism.
Another fundamental legal principle is the need for a sensible reason for killing any animal. Eating meat is still well accepted in western societies as a condition for healthy nutrition or even for spontaneous performance of life finally basing on the argumentation of self-defence. But similar to poison, it is a matter of dosage. Can the average meat consumption of nearly 93 kg per person and year be reasonable? Considering the fact that more than half of our diseases of modern civilisation are directly correlated to the level of meat consumption, you have to agree that this is simply unhealthy, as well as uneconomical. Therefore killing pigs for such a high level of consumption can never be ethically justified.
Altner (1999) brings another aspect into discussion: "creative dignity". For him as biologist and theologian every creature on our planet has got his own creational dignity as a gift and humans are obliged to respect this dignity. The anthropocentric approach is limited to aspects of humans. If you consider the intrinsic value of creature you have to use the bio centric approach, the environmental ethics, looking after sustainable use of nature. Accepting the permanent situation of competition in nature, described as: "I am life within life that wants to live" he pledges the responsibility of mankind as most intelligent beings for respecting the dignity of all creatures as intentional beings. Implementation on the living conditions of pigs this would mean an immediate stop of intensive farming.
Knörzer (1998) states, that everyone who kills (an animal) has the obligation for absolute justification against the creator and preserver of all life. He interprets the "theological sight" to be a reflection about common faith and world view. There is no killing in paradise and every killing is against the will of god. In daily life he sees no difference between Christians and non-Christians, both eat meat from the same pig, raised in the same intensive farming, slaughtered in the same big abattoir after a cruel long distance transport. The concept of reverence for life contains a principal and escheatable respect for the individuality of every being, which is subject of a will. Therefore the killing of an animal leads to an unavoidable conflict. An absolute ban of killing without context of circumstances and ethical consideration of the principal of avoiding suffering is not free of contradictions. There are ethical reasons for killing an animal like euthanasia to avoid further pain and damage. Slaughter for food is different but also has to be justified by moral consideration.
Bartussek (2001) supports this religious point of view and summarises that there is no mandatory vegetarianism in the New Testament. On the other hand we are obliged to treat animals in a respectful way. This could lead to general social benefits within society which would not be possible if you stop any using of animals.

Economical aspects

Dogs are becoming a major economic factor: the estimated overall turnover of the dog associated sector in Austria is about € 884,650,000 per year. The costs for feeding a dog of 25 kg of weight are € 650 per year; costs for healthcare are € 225. Together with the costs for medication this results in a summarised turnover for veterinarians in Austria of € 210 Mio. 60% of the veterinarians make their income as pet animal surgeons (Kotrschal *et al.*, 2004). This tendency is even more evident in students, as over 80% of them are females and nearly 2/3 of all want to become pet animal practitioners. So not only is the market growing, but also the participants, which means that the pieces of the cake are getting smaller. Unfortunately the whole cake in large animal practice is getting smaller. The economic gap for farmers is narrow and automatically puts pressure on the veterinary profession also, as expensive treatments are not financed any more. The period of usage of e.g. a milk-cow is decreasing permanently reaching an actual level of 4.2 years; that means 2.5 periods of lactation on average. There is simply no time left for long acting therapy. Additional problems result in strict Food & Feed-regulations, as animals with potential pharmaceutical residuals of drug using therapy are not allowed to enter the food-chain any more. Animals with severe illness and doubtful result of treatment cannot be slaughtered any more; they have to be killed not only for reasons of food safety but also for animal welfare.

The pressure of free trade world market on European agriculture is argued as need for consequent further intensification of production. But lowering production costs by growing in scale could become a vicious circle. Permanent pressure and highest level of performance is threatening immunological capacity of reaction and increases the risk of illness and damage for both, animals and farmers. Common Agricultural Policy (CAP) has to decide if Europe still wants to compete with world market or should better concentrate strictly on the home market.

Achterbosch (2006) from the Agricultural Economic Research Institute (LEI) gave an economic perspective on upgraded animal welfare standards in the EU at the EU Animal Welfare Conference in Brussels. He proved the common argument "increasing requirements regarding product quality and animal welfare raise the costs of production" to be false. The cost factors of competitiveness are space/animal density, feed, technology/equipment, labour, healthcare/veterinary medicine and efficiency. In a scenario of costs of animal welfare on farm he was able to point out that good farming welfare does not lead to additional costs, neither for producers nor for consumers, whereas an anonymous label increases both by 20% and organic rises production costs by 88% and consumer costs by 53%.

The market perspectives show clearly that 75% of the market is fresh products (meat, table eggs). As animal welfare standards are not an isolated requirement on one hand and offer options for marketing of upgraded products at premium prices at the same time. More stringent rules on farm animal welfare will not automatically evoke low quality imports from third countries as their competitive advantage in production costs is not primarily based on regulatory standards but on other factors. Of course the gap widens with higher standards, but as EU consumer markets are expanding especially in processed products, this is also a real opportunity, not only a threat.

Concerning consumer expectations from regulators he documented that they are concerned about process and methods in food production. Regulators should guarantee minimum standards and guide trough the wide variety of labels.

The role of consumers and the self-regulative power of the market must not be underestimated. Only 12 to 15% of the overall costs of livelihood are spent on nutrition, so everyone could afford high quality/price food. If you have a look in detail e.g. at the prices of meat, you will see that pork is cheaper than bread and cheaper than pet-food (not feed). 93 kg of meat of average € 4 means € 372 per person and year (remember the feeding costs of an average dog with € 650 per year!) This situation is not only dramatic and almost unrealistic from an economic point of view; it additionally becomes unbalanced when referring to ethical aspects, i.e. if you look at justified reasons for the killing of an animal.

It is the ethical responsibility of the consumer who pays for the products (milk, meat, eggs) and his care (in pets) and makes the final decision at the point of sale. It is the ethical responsibility of the veterinarians to help all animals, dogs and pigs (underdogs). There is neither a split ethical value between animals of the same evolutionary development nor a strictly divided responsibility between consumers and veterinarians. As they have a thorough economic partnership (direct in pets, indirect in farm animals), they should face this ethical challenge together.

The Special Eurobarometer 229 (2005) concerning attitudes of consumers towards the welfare of farmed animals found out that the European consumers` perspective reveal very distinctive realities with regard to the animal welfare and protection of farm animals within the Union. They observed that the welfare and the protection of farm animals is considered to be superior in the EU, but judged differently for each species (poorer in laying hens). Only a slight majority of citizens do not take account of animal welfare when buying food as the identification of animal welfare friendly production systems seems very difficult. Three quarters of citizens believe they have the capacity to influence the welfare of farmed animals by their purchasing behaviour. People seem willing to accept up to 25% price increase for products sourced from animal welfare friendly systems. Visits to farms seem to increase the awareness and concern for animal welfare.

Consumers take the final decision by purchasing food with different ethical burden. But to take this responsibility they need thorough information and objective advice. This should be done by veterinarians, who still have high social credits as reliable experts in animal husbandry. The definition, control and inspection of "Welfare Quality" is a major task for veterinarians, especially if you prefer to use animal based evaluation systems to farm based ones. Despite legal regulations consumers want a fair and transparent system of labelling of products. If they should be willing to pay the higher price for better "Welfare Quality" (EU-project), they have a right of guarantee that the animals the products come from were kept under conditions respecting their needs and not only the minimum legal standards. That is a matter of fairness and transparency. Veterinarians could play a decisive role in this process of truth and trust. They should do so, not only because of their intrinsic compassion to animals but also for long lasting prosperity of their own profession.

The EC Animal Welfare Action Plan 2006-2010 puts forward several proposals to use the possibilities highlighted in the Eurobarometer study. One of the most important actions probably is the EU-Animal Welfare Label. Recognising animal welfare to be a very strong and decisive argument in quality of food, also the new EU-legislation (food & feed) is putting animal welfare onto the same level of importance as hygiene and processing. The strategy of strict implementation of minimum standards and support for labelled high quality products (stick and carrot) in combination with using all the opportunities of the free market could get a story of success of CAP. Frankly it is the only way to keep the financial framework.

Conclusions

- The ethical differentiation between animals of similar evolutionary level is not justified.
- Pet animals are endangered of being over supported (nutrition medical care), often social stressed.
- Intensive farming and the actual level of meat consumption is not ethical and unhealthy.
- Veterinarians play a very important role in information of the public (owner, consumer).
- As classical service providers they are only advisors; together with the consumer and owner they have to take their part of the responsibility.
- The consumer makes the final decision at the point of sale, but he needs clear and honest information to be able to make a good one (sustainable and ethical).
- The EC Animal Welfare Plan and the animal welfare label will facilitate the upgrading of animal. welfare standards in the EU market (perhaps also in WTO).

References

Achterbosch, T. (2006). Animal Welfare in the Market. An Economic Perspective on Upgraded Animal Welfare Standards in the EU. Agricultural Economics Research Institute (LEI) International Conference on Animal Welfare 30.03.2006 Brussels.

Altner, G. (1997). Mitgeschöpflichkeit und Würde der Tiere. Tierschutz vor Gericht. Protokolldienst Evangelische Akademie Bad Boll1/98, 30-140.

Bartussek, H. (2001). Ist Fleischkonsum ethisch vertretbar? 8. Freilandtagung Wien, 34-37.

Community Action Plan on the Protection and Welfare of Animals 2006-2010, SEC(2006) 65 Commission Working Document Brussels 23.01.2006.

Kaplan, A. (2000). Zum Verhältnis von Mensch und Tier. Dissertation Naturwissenschaftliche Fakultät Universität Salzburg.

Knörzer, G. (1999). Verantwortung des Menschen für das Töten von Tieren aus theologischer Sicht. Ehrfurcht vor dem Leben. 12-19.

Kotrschal, K., Bromundt, V., Föger, B. (2004). Hunde als Wirtschaftsfaktor, Zahlen und Fakten. In: Faktor Hund, 16-34.

Special Eurobarometer 229/Wave 63.2 – TNS Opinion & Social. Attitudes of consumers towards the welfare of farmed animals. Publication June 2005.

Robustness of laying hens: An ethical approach

Laura Star[1], *Esther Ellen*[2], *Koen Uitdehaag*[2] *and Frans Brom*[2]
[1]*Adaptation Physiology Group, Wageningen University*
[2]*Animal Breeding and Genetics Group, Wageningen University, P.O. Box 338, 6700 AH Wageningen, The Netherlands, Esther.Ellen@wur.nl*

Abstract

Until recently, animal breeding was mainly directed at improving efficiency. Although it is broadly recognized that health and welfare of production animals is important, it is difficult to develop economically viable production systems without affecting animal health and welfare. In poultry, for instance, problems like feather pecking, cannibalism and mortality due to diseases are of major concern in all types of poultry production systems. For solving these problems several strategies are needed. We present a research project that aims to improve animal health and welfare by breeding more 'robust' animals. At the start of the project we defined robustness as 'the potential to keep functioning (physiological, behavioural, and immunological performance) and take short periods to recover under varying environmental circumstances'. One of the aims of the project is to distinguish layer lines that will function well in a specific environment. This will be done by searching for behavioural, immunological and/or physiological parameters and to propose ways to implement selection for robustness in breeding programmes. Improving the health and welfare of production animals by increasing animal robustness raises several ethical questions. An analysis of these questions is an integrated part of the project. We provide an overview of ethical issues raised by our project: Is it acceptable to adapt animals to the production environment by selective breeding, rather than changing their environments? Should animals be adapted to all environments, even the worst? Does selection for robustness affect the integrity of the animal? We argue that it is possible to develop a breeding programme to improve robustness which will result in better health, decreased expression of feather pecking and cannibalism in suitable environments, without affecting the integrity of the laying hen.

Keywords: robustness, animal welfare, ethics, integrity, laying hen

Introduction

Until recently, animal breeding was mainly directed at improving efficiency. Although it is broadly recognized that health and welfare of production animals is important, it is difficult to develop an economically interesting production system without endangering animal health and welfare. In poultry, for instance, problems like feather pecking, cannibalism and mortality due to diseases are of major concern in all types of poultry production systems.
Since a couple of years breeding strategies are shifting from production traits towards traits concerning health and behaviour to improve welfare. Farmers and poultry breeding organisations prefer animals that do not show abnormal behaviour and animals that are able to cope with changes in their environment. They are interested in animals that perform well independent of internal or external disturbances, so called 'robust' animals.
The aim of this paper is to define robustness and to describe the problems, doubts and questions arising in the ongoing project 'The genetics of robustness in laying hens'. Robustness is related to health, welfare and integrity of animals. Each of these terms will be evaluated in general, followed by an application towards the project. Improving robustness by selection of production animals raises several ethical questions: Is it acceptable to adapt animals to the production

environment by selective breeding, rather than changing their environments? Should animals be adapted to all environments, even the worst? These and other ethical questions concerning robustness that raise public concern will be evaluated. Finally, we argue that it is possible to develop a breeding program to improve robustness which will result in better health, decreased expression of feather pecking and cannibalism in suitable environments, without affecting the integrity of the laying hen.

Robustness

Robustness is becoming a key trait in animal breeding. Knap (2005) defined robustness in pigs as 'pigs that combine high production potential with resilience to external stressors, allowing for unproblematic expression of high production potential in a wide variety of environmental conditions'. Kitano (2004) mentioned that robustness is often misunderstood, to mean staying unchanged regardless of stimuli or mutations. He defined robustness as 'maintenance of specific functionalities of the system against perturbations, and it often requires the system to change its mode of operation in a flexible way'. At the start of our project, robustness was defined as 'an animal that has the potential to keep functioning (physiological, behavioural and immunological performance) and take short periods to recover under varying environmental conditions'.
Main characteristics informative for robustness seem to be production, adaptation, and environmental conditions. Besides, the nature and needs of animals must be considered. According to Rollin (1989) this is about the *telos* of an animal. He defined *telos* as 'the unique, evolutionarily determined, genetically encoded, environmentally shaped set of needs and interests which characterize the animal in question'. Each animal has a *telos* that is unique to its species, it can be seen as the 'chickenness of the chicken' or the 'pigness of the pig', which are essential to their well-being. This does not mean that the *telos* of an animal cannot be changed (Rollin, 1989).
Selection for robust laying hens will lead to hens that can adapt to different environmental conditions, without decreasing production or showing abnormal behaviour. A 'new' laying hen might be developed, which causes a change in the *telos* of the hen. Will it be morally wrong to select robust hens while changing their *telos*?
Vorstenbosch (1993) mentioned two categories of ethical concerns related to the animals themselves. These are health and welfare, and the integrity of the animal and are interrelated with robustness. A better understanding and a clear definition of health, welfare and integrity have to be evaluated for a better understanding of robustness.

Health

Animal health could be considered as 'the physical condition required to achieve well-being at an acceptable level' (Rutgers, 1993). In the past, animal breeding was directed at improving production, and hardly at resistance to diseases which is important for physical condition and health. Health can be enhanced by selection for resistance to diseases. Enhancement of immune mechanisms by genetic selection may be effective in resistance to a wide range of pathogens and may an effective strategy to protect laying hens under a variety of environmental conditions (Lamont, 1998).
Breeders have the responsibility to select for the full range of existing production systems and farmers should aim for a level of production that animals can sustain given the conditions on a farm (Ten Napel *et al.*, unpublished). Conditions in many animal production systems are controlled, and management is aiming at limiting threats and disturbances. Nevertheless, animals housed in controlled conditions have difficulties to cope with fluctuations of temperature and disease pressure (Ten Napel *et al.*, unpublished). Within the project 'The genetics of robustness

in laying hens' we aim for a system where animals are able to cope with disturbances through adaptation. Adaptation mechanisms include behavioural, physiological and immunological performance at animal level and a stable micro-flora at herd level. The ability to cope successfully requires sufficient genetic potential, building up experience and getting the opportunity and resources to adapt (Ten Napel *et al.*, unpublished).

Welfare

Animal welfare is a public concern and subject of scientific research. The public concern about animal welfare is that 'the animal is suffering because it can not live the life where it is biologically made for'. Two core elements in definitions of welfare are quoted in this statement; experiences of animals and their biological functioning. Stafleu *et al.* (unpublished) integrated these elements into two concepts for welfare; the concept of experience and of functioning. Although the concept of experience could be used in scientific studies it is rather difficult, because of the impossibility to ask an animal directly in which situation it feels comfortable. The concept of functioning can be used as a tool to make animal experiences more applicable. 'Good functioning' can be linked to scientific theories or models. For instance, 'coping' can be defined using scientific theories about behavioural, physiological and immunological mechanisms (Stafleu *et al.*, unpublished).

Welfare definitions are often based on the concept of functioning, like the definitions given by Broom (1993) 'welfare of an animal is reflected by the success of its attempt to cope with its environment' and by Siegel (1995) 'welfare depends on physiological ability to respond properly in order to maintain or re-establish homeostatic state or balance'. Combining these two definitions give a comprehensive definition 'welfare of an animal is reflected by the success of its attempt to cope with its environment, and depends on its physiological ability to respond properly in order to maintain or re-establish its homeostatic state or balance'. This definition corresponds with the following definition 'the potential to keep functioning (physiological, behavioural, and immunological performance) and take shorts periods to recover under varying environmental circumstances'. The latter definition is formulated for the project 'The genetics of robustness in laying hens' and is taken as a practical definition for robustness. Taking the definition for welfare and robustness, it seems that these terms are closely related. The distinction between animal welfare and robustness is that welfare is based on animal experiences and it is often measured by an animals' response to a current stressor, whereas robustness is based on the possibility to respond adequately to a stressor and is aiming at less disturbed functioning by challenge with a stressor. To measure robustness (and welfare) of an animal is difficult, because good indicators are lacking. But when robustness can be increased it will help to increase welfare.

Integrity

Integrity has been described as the 'wholeness and intactness' of the animal and its species-specific balance, as well as the capacity to sustain itself in an environment suitable to the species. Integrity of animals leads to considerations and arguments beyond animal health and welfare (Grommers *et al.*, 1995). Where health and welfare can be considered for empirical and objectified testing and research, integrity points us back to our own moral position, purposes and perspectives with regard to animals (Vorstenbosch, 1993). This makes that integrity has no obligatory definition (Brom, 1997). For example, breeding blind chickens will reduce common problems like feather pecking and cannibalism (Ali and Cheng, 1985). These chickens seems to be well adapted to their situation and, assuming that blind chickens do not suffer in any other way, they may live a better life than chickens which are able to see. Many people, however, believe that this is a morally wrong approach to improve welfare (Sandøe *et al.*, 1999).

The concept of integrity, seems to be most appropriate in contexts where there is an entity with (relatively) well-defined boundaries (like a body), that can be affected or violated (Vorstenbosch, 1993). But integrity may be considered violated by non-invasive biotechnologies which will not necessarily pose any risk to welfare of animals (Seamark, 1993). Maybe in time, techniques will be used to create new species of animals with different needs and preferences. How are we to think about welfare and integrity of these animals? (Vorstenbosch, 1993). Are we also creating new species of animals if we adapt animals to their environment by selection instead of changing the environment? In that case, changing the animals so that they are better suited for intensive farming instead of adapting the farming system may be considered violations of animal integrity (Christiansen and Sandøe, 2000).
Animal breeding has an effect on genetic diversity and often does not fit the ideas people have about animal integrity (Kanis *et al.*, 2004). When breeding animals we should aim at increasing the sum of that which contributes to the equality of life of animals, but the increase should not be achieved at the cost of the individual animal or by depriving the animals of natural abilities, such as being able to see. Breeding should not aim at the loss of abnormal coping behaviours but at animals which are able to cope with unfavourable or unfamiliar conditions by normal adaptive behaviour. Selection for robust laying hens should result in better welfare, better immune system and animals that have more social skills. Such adaptations are morally recommendable.

Project 'The genetics of robustness in laying hens'

For the project, robustness was defined as 'the potential to keep functioning (physiological, behavioural, and immunological performance) and take short periods to recover under varying environmental conditions'. Integrating the definitions of health, welfare and integrity, we can define robustness in a broader perspective 'a robust animal has the potential to maintain functioning (physical, mental, and social well-being), producing and homeostatic state, and take short periods to recover from internal or external (environmental) disturbances, without affecting the animals integrity'.
It is difficult to measure robustness in a direct way. Sørensen *et al.* (2001) developed an assessment system for indirect measurement of an animals' welfare, which is based on four sources of information: system, system applications, animal behaviour and animal diseases. These sources of information provide useful tools for measurement of robustness, especially when they are related to traits of interest of the breeding company: do not peck with intact beaks, and keep functioning at high temperatures, at high disease challenge and with changing feed quality. Taken together, these sources of potential parameters for robustness are useful for the project. From a scientific point-of-view robustness as a concept is well founded if we have knowledge about the individual sources of information and the interaction between the sources. It is, however, difficult to define an optimal level of robustness, especially, since opinions and motivations of breeders and farmers play a major role. In the end, breeders and farmers are in the position to make the choices. This means that the principle aspects of robustness may be different for each individual breeder or farmer, but also reference values can change. Besides, in the future other issues may arise that have to be taken into account when utilising robustness in animals, because ideas of people about robustness may change.

Conclusion

The overall aim of increasing robustness by breeding is not simply to adapt birds to a particular environment but, rather to increase their ability to interact successfully with the physical and social environment and thereby to make them more able to adapt to whatever husbandry systems are considered appropriate. The ability to adapt successfully requires sufficient genetic potential, building up experience and getting the opportunity and resources to adapt. Selection for robust laying hens will result in better welfare, improved disease resistance and animals with more social skills. Therefore breeding for robust animals is morally desirable.

Acknowledgement

This research is part of a joint project of Hendrix Genetics, Nutreco and Wageningen University on 'The genetics of robustness in laying hens' which is financially supported by SenterNovem.

References

Ali, A., Cheng, K.M. (1985). Early egg production in genetically blind (rc/rc) chickens in comparison with sighted (Rc+/rc) controls. *Poultry Science*, 64: 789-794.

Brom, F.W.A. (1997). *Onherstelbaar verbeterd: biotechnologie bij dieren als moreel probleem*. Van Gorcum & Comp. B.V., Assen, The Netherlands. 292 p.

Broom, D.M. (1993). Assessing the welfare of modified or treated animals. *Livestock Production Science*, 36: 39-54.

Christiansen, S.B., Sandøe, P. (2000). Bioethics: limits to the interference with life. *Animal Reproduction Science*, 60-61: 15-29.

Grommers, F.J.,Rutgers, L.J.E., Wijsmuller, J.M. (1995). Welzijn - intrinsieke waarde - integriteit: ontwikkeling in de herwaardering van het gedomesticeerde dier. *Tijdschrift voor Diergeneeskunde*, 120: 490-494.

Kanis, E.,van den Belt, H.,Groen, A.F.,Schakel, J., de Greef, K.H. (2004). Breeding for improved welfare in pigs: a conceptual framework and its use in practice. *Animal Science*, 78: 315-329.

Kitano, H. (2004). Biological robustness. *Nature Reviews, Genetics*, 5: 826-837.

Knap, P.W. (2005). Breeding robust pigs. *Australian Journal of Experimental Agriculture*, 45: 763-773.

Lamont, S.J. (1998). Impact of genetics on disease resistance. *Poultry Science*, 77: 1111-1118.

Rollin, B.E. (1989). *The unheeded cry: animal consciousness, animal pain and science*. Oxford University Press, Oxford. 308 p.

Sandøe, P.,Nielsen, B.L.,Christensen, L.G., Sørensen, P. (1999). Staying good while playing god - the ethics of breeding farm animals. *Animal Welfare*, 8: 313-328.

Seamark, R.F. (1993). Recent advances in animal biotechnology: welfare and ethical implications. *Livestock Production Science*, 36: 5-15.

Siegel, H.S. (1995). Stress, strain and resistance. *British Poultry Science*, 36: 3-22.

Sørensen, J.T.,Sandøe, P., Halberg, N. (2001). Animal welfare as one among several values to be considered at farm level: the idea of an ethical account for livestock farming. *Acta Agriculturæ Scandinavica*, 30: 11-16.

Stafleu, F.R.,Grommers, F.J., Vorstenbosch, J.M.G. (unpublished). Animal welfare: a hierarchy of concepts.

Ten Napel, J.,Bianchi, F.B., Bestman, M. (unpublished). Utilising intrinsic robustness in agricultural production systems.

Vorstenbosch, J.M.G. (1993). The concept of integrity. Its significance for the ethical discussion on biotechnology and animals. *Livestock Production Science*, 36: 109-112.

Attitudes of organic pig farmers towards animal welfare[99]

Marjolein van Huik and Bettina Bock
Rural Sociology Group, Wageningen University, P.O. Box 8130, 6700 EW Wageningen, The Netherlands, Marjolein.vanhuik@wur.nl

Abstract

This paper presents partial results of the Welfare Quality research project, which aims to identify motivations and barriers for farmers to the development of animal friendly production. Using semi-structured, in-depth interviews with 13 Dutch organic pig farmers, we studied their attitudes towards animal welfare and animal welfare regulations. As described in literature, different types of organic farmers can be identified, based on their motivation to become an organic farmer. Within the sample five 'pragmatically motivated farmers' and eight 'ideological motivated farmers' could be identified. 'Pragmatic' and 'ideological' farmers differ in their definition and appreciation of animal welfare, and in their view on animal welfare regulations. Whereas the 'ideological' farmers mainly regard animal welfare to be important due to intrinsic motivations and define animal welfare by referring to the ability for the animal to perform natural behaviour, 'pragmatic' farmers tend to involve economic conditions and zoo-technical performance in their description and valuation of animal welfare. 'Pragmatic' farmers define animal welfare mainly as the absence of negative factors. In this 'pragmatic' organic pig farmers resemble conventional pig farmers. This means that being an organic pig farmer is an insufficient indicator for farmer's ethics regarding animal welfare. We need to look closer into farmers' main motivation to become an organic farmer. Why farmers converse to organic farming, matters for their attitude towards animal welfare, and thus organic farmers cannot be perceived as a homogeneous group, sharing the same beliefs and ethical standards.

Keywords: animal welfare, organic, pig farming, attitudes

Introduction

The Welfare Quality program aims to improve animal welfare and to link animal husbandry practices to informed animal product consumption (www.welfarequality.net). A part of this program is to identify what constrains and stimulates farmers to adopt animal friendly production methods (Van Huik and Bock, 2006; Bock and Van Huik, 2006). Farmers who practice organic pig farming, have to adopt special measures for animal welfare, as is regulated by international standards (EC Regulation 1804/99). It is also part of the more general organic philosophy. One might thus expect that organic farmers are more animal friendly in their attitudes and behaviour per se. From earlier studies (Schoon and Te Grotenhuis, 1999; Darnhofer *et al.*, 2005) it is known that there are different farmer types in organic farming, based on their original motivation to adopt organic farming. There are the 'ideological' farmer type, who became an organic farmer out of ideological motives, and the 'pragmatic' farmer type, who is mainly motivated by the continuation of his farm and economic arguments (Schoon and Te Grotenhuis, 1999; Darnhofer *et al.*, 2005). The Dutch policy of stimulating transition to organic farming with financial incentives,

[99] The present study is part of the Welfare Quality research project which has been co-financed by the European Commission, within the 6th Framework Programme, contract no. FOOD-CT-2004-506508. The text represents the authors' views and does not necessarily represent a position of the Commission who will not be liable for the use made of such information.

may have attracted farmers of the 'pragmatic' type to organic farming. This study examines how the 'ideological' and 'pragmatic' farmers differ in norms and values concerning animal welfare, and their willingness to adopt more stringent animal welfare measures. More generally, the aim is to understand if stimulating organic pig farming by financial incentives supports the conventionalisation (Guthman, 2004) of organic agriculture and eventually dilutes the animal welfare ethics of organic farmers. As a result organic farmers might not play the pioneering role in animal welfare that we expect them to play. If indeed this is the case, it may also be a more general warning that stimulating animal-friendly production methods by financial incentives alone, might unintendedly have undermined the sector with the highest animal welfare ethics.

Material and methods

To explore farmers' beliefs, attitudes and motivations, we choose a qualitative approach, using semi-structured in-depth interviews with a sample of 62 pig farmers. The farm type and the type of animal welfare scheme these farmers participated in, served as selection criteria. In this paper we zoom in on a subgroup of 13 organic pig farmers. The pig farmers were interviewed about their own farm practices, animal welfare in general and animal welfare legislation. The author and two hired interviewers conducted the interviews, visiting farmers on their farm for a face-to-face interview that lasted on average 1,5 hours. The answers of the farmers were written down and analysed. The analyses focused on the uncovering of issues and concerns, on the understanding of motivations and reasoned behaviour, and on detecting specific differences or similarities between farmers with an 'ideological' or 'pragmatic' motivation for conversion to organic farming.

Typology of organic farmers

To determine whether organic farmers were of the 'ideological' or 'pragmatic' farmer type, farmers were asked for their original motivation to converse to organic farming, and if and when they would consider giving up organic pig farming. Farmers were considered to be an 'ideological' farmer when their conversion towards organic farming was mainly driven by ideological motives, and when farmers wouldn't easily give up organic farming. Some farmers even indicated that they would rather stop being a pig farmer than to go back to conventional pig farming.

> *'[I changed] for ideological reasons. In the beginning of the 90's, we started to have doubts about how our pigs were housed. In the farming practice, the animal was no longer taken into account. We visited a free-range farm and then we really thought we were doing things wrong. And we already used minimum amounts of antibiotics. So we transformed to organic pig farming. We really needed to do things differently.'*

Organic farmers were considered to be a 'pragmatic' farmer when their decision to become an organic farmer was mainly driven by economic arguments, including the continuation of their farm, and when staying an organic farmer was made dependent on the profitability of pig production.

> *'Eventually, I'm in it to gain a higher income. If the pigs are better off, than that's fine, but it happens to serve my income. That's what it is all about. If there is no profit, I'll switch back to conventional in no time.'*

The groups of 'ideological' and 'pragmatic' farmers also differ also in other characteristics. Table 1. shows the characteristics of the total sample of organic farmers and of the subgroups of 'ideological' farmer and 'pragmatic' farmer.

Results

Farmers' perception on animal welfare

Table 1. Sample characteristics.

	Total organic farmers	Ideological farmers	Pragmatic farmers
Number of farmers	13	8	5
Number of female farmers	2	2	-
Average age (years)	45.1	47.5	41.2
Average herd size by organic fattening pigs	582	425	817.5
Average herd size by organic sows	111	63.4	203
Number of specialised pig farms	4	2	3
Number of farmers that combine organic and conventional farming	3	-	3
Number of part-time farmers	2	2	-

The 'ideological' farmers mainly defined animal welfare as the possibility for the animal to express natural or normal behaviour, while 'pragmatic' farmers emphasized the absence of negative factors, such as fear, stress, pain and injuries when they described animal welfare. The quotations below may serve as an illustration.

'[Animal welfare is] that the animal can do as it likes, that it can express natural behaviour. It must be able to choose where it stays, indoors or outdoors, in the sun or shadow. That they are able to go into the fields, also at night. ...They are perfectly capable to regulate themselves... to follow their own rhythm...'

'When an animal is without stress, then it can grow. No fear, no pain and no physical injuries, then the animal welfare is good. But it is hard to determine animal welfare as a human. Things like surface, light, time, it is very hard to say something about it. And if you look at organic pigs, they are very well off for six months of the year. That's when everyone comes and watches them. The other six months it is cold, and than there is no-one.'

Not only did the subgroups differ in their description of animal welfare, they also differed in the importance they attached to animal welfare and how they assessed the animals' welfare. All 'ideological' farmers regarded animal welfare as important or very important, while the 'pragmatic' farmers considered animal welfare to be important, but felt limited to invest in animal welfare by the profitability of their farm. In assessing animal welfare 'ideological' farmers judged animal welfare by behaviour and physical appearance of their animals, while 'pragmatic' farmers used behaviour and zoo-technical performance of their animals as evaluation criteria for animal welfare. The reason that animal welfare was considered important was mainly because of the sense of responsibility the farmer felt for his animals, but 'pragmatic' farmers added that animal welfare was important for its impact on the performance of the animals.

'We don't have animals to abuse them, but for production. Production must be high and that depends on animal welfare.'

Farmers' perception on animal welfare regulations

In general organic farmers accepted the organic regulations for animal welfare, and for most 'ideological' farmers and half of the 'pragmatic' farmers it was no problem to comply with these regulations. The other half of the 'pragmatic' farmers had a problem with the restrictions on the use of medication or with the prohibition on teeth grinding. Some 'ideological' farmers thought that the regulations could be made more practical, and that some regulations should become stricter, for instance they asked for minimum requirements for the use of straw, since they felt that some competing organic farmers did not provide enough straw for their animals.

The farmers were confronted with eight specific animal welfare measures. The way in which the two groups of farmers appreciated them, is shown in Table 2.

Comparing 'Ideological', 'Pragmatic' organic farmers and conventional farmers

Van Huik and Bock (2006) demonstrate that conventional pig farmers in the Netherlands primarily define animal welfare by the fulfilment of biological needs, to be judged on the basis of high zoo-technical performance. The 'pragmatic' organic farmers take one more step, focussing on the absence of fear, stress and injuries next to the fulfilment of basic biological needs. But they resemble conventional farmers when appreciating animal welfare for its impact on the performance of animals and when assessing animal welfare by zoo-technical performance. 'Ideological' farmers define animal welfare by the freedom an animal has to express natural behaviour and don't look at zoo-technical performance in their appreciation or evaluation of animal welfare.

'Pragmatic' organic farmers think that group housing for non-suckling sows improves animal welfare, but negatively affects zoo-technical performance. Conventional farmers reject group housing for the same reason. Conventional farmers regard physical mutilations like tail docking

Table 2. Appreciation of 'Ideological' and 'Pragmatic' farmers for specific animal welfare measures.

	Ideological farmers	Pragmatic farmers
Group housing for non-suckling sows	Majority in favour, since the measure improves animal welfare	Divided. Group housing improves animal welfare but also leads to more labour, more aggression among sows and decreases zoo-technical performance
Ban on tethering sows	All in favour	All in favour
Minimum weaning age of 21 days	Is too early for the piglets	Is OK
Straw bedding	All in favour	Majority is in favour
Outdoor housing	Majority is in favour	Not possible on all farms
Ban on castration	All in favour, 50% sets condition that is must not influence the sale of pig meat	All in favour, 50% sets condition that is must not influence the sale of pig meat
Ban on tail docking	Majority is in favour of ban on tail docking	50% Does not object to tail docking
Ban on tooth grinding	Majority is in favour of ban on tooth grinding, minority wants possibility for curative tooth grinding	50% Wants the possibility for curative tooth grinding

and tooth grinding as a minor inconvenience for the animals, which serves to prevent more serious problems with animal welfare. 'Ideological' farmers are in favour of banning these mutilations and don't deem them necessary. A part of the 'pragmatic' farmers stand in between these two positions: they do not regard physical mutilations like tail docking and tooth grinding as a minor inconvenience, but do not object to them and prefer to have the possibility to use these means for curative reasons.

Conclusion

'Pragmatic' organic farmers were drawn to organic farming by the favourable financial prospects of organic farming, either caused by governmental support or by higher prices for organic products or both. They are ready to give up organic practices when more financial gain can be accomplished with other ways of pig farming. Although 'pragmatic' and 'ideological' organic farmers both comply with the same regulations for animal welfare, and do not differ in their behaviour in this regard, this study shows that they do differ in attitude towards animal welfare. The 'ideological' farmer type focuses on the animal itself in defining, assessing and valuing animal welfare. The attitudes of the 'pragmatic' farmer type towards animal welfare are diluted by a focus on performance levels, and thus indirectly by the importance of production efficiency. This can eventually conflict with the interests of animal welfare. When the financial profits of organic farming will decrease, it is likely that these farmers will switch back to conventional farming and will reduce the level of animal welfare of their animals. This means that a strategy that induces farmers to improve the level of animal welfare on their farm by using financial incentives, might only be successful as long as a higher financial gain for the farmer is maintained – either by governmental support or higher prices in the market – as it does not change the farmer's animal welfare ethics. Because of the focus of the 'pragmatic' farmer on profitability, the debate, started by Guthman (2004), about blurring ethics in organic agriculture and conventionalisation, is not imaginary. The results of this study show that conventionalisation takes place also in organic pig farming. When 'pragmatic' farmers will exert a 'light' version of those organic standards that are not exhaustively described and regulated, for instance the use of straw, this can ultimately lead to unfair competition between 'pragmatic' and 'ideological' farmers. This can force 'ideological' farmers to reduce their ideological practices in order to survive in organic farming. Eventually, that can put consumers' trust in organic products at risk.

References

Bock, B.B. and Van Huik, M.M. (2006, forthcoming). Pig farmers and animal welfare. A study of beliefs, attitudes and behaviour of pig producers across Europe, In: Farm Animal Welfare Concerns, consumers, retailers and producers (working title). J. Roex and M, Miele (eds). Cardiff University: Cardiff Paper series no 2, Cardiff.

Darnhofer, I., Schneeberger, W. and Freyer, B., (2005). Converting or not converting to organic farming in Austria: Farmer types and their rationale. *Agriculture and Human Values*, 22: 39-52.

Guthman, J. (2004). The trouble with 'Organic Lite' in California: A Rejoinder to the 'Conventionalisation' Debate. *Sociologia Ruralis*, 44: 301- 316.

Huik, M.M. van and Bock, B.B. (2006, forthcoming). Visie van Nederlandse Varkenshouders op Dierenwelzijn. Welfare Quality Project. Wageningen University, Wageningen.

Schoon, B. and Te Grotenhuis, R., (2000). Values of Farmers, Sustainability and Agricultural Policy. *Journal of Agricultural and Environmental Ethics,* 12: 17-27.

Defining the concept of animal welfare: integrating the opinion of citizens and other stakeholders

Els Van Poucke, Filiep Vanhonacker, Griet Nijs, Johan Braeckman, Wim Verbeke and Frank Tuyttens
Institute for Agricultural and Fisheries Research, Animal Science Unit, Animal Husbandry and Welfare, Burg. Van Gansberghelaan 115, 9820 Merelbeke, Belgium, els.vanpoucke@ilvo.vlaanderen.be

Abstract

Animal welfare is a complex and multidimensional concept and there is little consensus about how it ought to be defined. This project aims to develop a definition of farm animal welfare based both on science and on consensus among citizens and stakeholders such that animal welfare becomes a more workable concept in politics and society. The study consisted of both qualitative and quantitative research. Firstly, a list of 73 aspects, considered to be important for animal welfare, was produced based on focus group discussions with citizens and on a literature review. Secondly, these aspects were condensed into a set of five mutually exclusive welfare-dimensions and the relative weight of these dimensions was quantified based on a survey among Flemish citizens. Thirdly, the list of welfare dimensions and their weights was discussed with stakeholders during focus group discussions and in-depth interviews. This study gives insight into the attitudes of the different stakeholders in society towards the welfare of farm animals. Moreover, the list of dimensions and their relative weights is believed to be invaluable for the aggregation of various welfare indicators into a comprehensive assessment of the general state of welfare.

Keywords: animal welfare, farm animals, definition, citizens, stakeholders

Introduction

Animal welfare is a complex and multidimensional concept (Mason and Mendl, 1993; Fraser, 1995). Many definitions have already been conceptualised. In general, distinction can be made between objective and subjective definitions of welfare (Bracke, 2001). The former relates welfare directly to measurable parameters and emphases the importance of biological functioning (e.g. Broom, 1986); the latter defines welfare in terms of subjective emotional states of animals (e.g. Duncan, 1996). Dispute between the two views still exists, however there is overall agreement about sentience being a necessary condition for welfare (Bracke, 2001) and feelings being an important part of the concept of welfare (Anonymous, 2001).
The probably best known description of what animal welfare is all about, originates from the Farm Animal Welfare Council (1992) and is based on five dimensions, 'the five freedoms': freedom from hunger and thirst; freedom from discomfort; freedom from pain, injury or disease; freedom to express normal behaviour and freedom from fear and distress. As these freedoms overlap, Capdeville and Veissier (2001) subdivided them into 16 mutually exclusive basic needs. This corresponds to Bracke (2001) stating that animal welfare can be decomposed into a set of needs. In addition, Keeling and Veissier (2005) defined a set of 12 mutually exclusive dimensions (or 'welfare criteria') which were selected by scientists and discussed with stakeholders. This set of dimensions meets a number of theoretical and practical requirements, based on Bouyssou (1990): (i) the set must be exhaustive, i.e. containing every important viewpoint; (ii) the set must be minimal, i.e. containing only necessary dimensions; (iii) the dimensions should be

independent of each other; (iv) the set should be agreed by the stakeholders; and (v) the number of dimensions should be kept to a minimum.
Under the authority of the Flemish government this project aims to develop a definition of animal welfare based both on science and on consensus among Flemish citizens and stakeholders such that animal welfare becomes a more workable concept in politics and society. The first objective of this study was to list a complete set of aspects of animal welfare based on qualitative research and a literature review. The second objective was to condense these aspects into a comprehensive list of mutually exclusive dimensions and to quantify the relative weights Flemish citizens allocate to each of these welfare-dimensions.

Methods

The study consisted of both qualitative and quantitative research. Qualitative research took place during March 2006 through four focus group discussions with 29 participants in total. Each focus group consisted of six to nine participants of the same sex but differing in age and place of residence (rural/urban). Meat consumption was homogeneous within each group. Three groups consisted of females (vegetarians, heavy meat users and low meat users) and one group of males (vegetarians). This exploratory research gave insight into the different aspects citizens consider as important for the welfare of farm animals. The list of aspects, mentioned during the focus group discussions, was revised and completed based on literature review. For this purpose papers, focusing on animal welfare assessment on-farm (Striezel *et al.*, 1994; Bartussek, 1999; Anonymous, 2001; Bracke, 2001; Capdeville and Veissier, 2001; Sørensen *et al.*, 2001; Whay *et al.*, 2003; Keeling and Veissier, 2005)) and on welfare problems during transport and slaughter (Hall and Bradshaw, 1998; Broom, 2000; Grandin, 2000; Knierim and Gocke, 2003; Delezie *et al.*, 2005; Keeling and Veissier, 2005; Simmonds, 2005), were selected. Each aspect in the list was phrased such that it was comprehensible to the average citizen. Moreover, the list was discussed with the project steering committee consisting of scientists and members of the Ministry of the Flemish Community. This resulted in a list of 73 aspects.
This list was used as input for the quantitative research by means of a questionnaire-type survey with 1081 Flemish citizens (i.e. citizens from Dutch-speaking Belgium) during April 2006. Prints and also a digital version of this questionnaire were distributed among friends and acquaintances, who on their turn distributed it further. In addition the questionnaire was also distributed among train passengers. First, the respondents were asked to score each of the 73 aspects, according to how important they considered them for animal welfare, on a five point Likert-scale ranging from 1 (not important at all) to 5 (very important). Second, we inquired for respondents' perception of the current state of welfare of farm animals in Flanders in general and for broiler chickens, laying hens, pigs, beef cattle and dairy cattle separately. Answering possibilities ranged from very bad to very good on a seven point Likert-scale. Third, information was gathered about the respondents' consumption of pork, beef, poultry, eggs, meat substitutes, fish and dairy products. Fourth, affiliation with animal husbandry was estimated by means of six yes/no questions. Principal component analysis was used to aggregate the aspects into a set of mutually exclusive welfare-dimensions. For this, the requirements of Bouyssou (1990) were taken into account. The relative weights of these dimensions were quantified by confirmatory factor analysis (LISREL).
Qualitative research was carried out once again in May 2006. Three focus group discussions were conducted for producers, retailers and animal rights organisations separately. In addition in-depth interviews were performed with ethicists. Besides talking about what is important for farm animal welfare, participants were also confronted with the survey results. This qualitative research gave insight into the perception of animal welfare by other stakeholders (producers,

retailers, animal right organisations, ethicists) and the differences in the opinions of these stakeholders and citizens.

Results

Focus group discussions (citizens)

Participants' knowledge of current animal husbandry practices was limited. Animal welfare was rarely mentioned spontaneously as an important food selection criteria, still most participants agreed that too little was known about it "... *that is the problem, information is not sufficiently available* ...". A minority expressed no interest in animal welfare. 56 aspects, mentioned by at least one of the participants as important and relevant for the welfare of farm animals, were listed. Most of the participants however mentioned that they had insufficient knowledge of what really matters to the animal from the animal's point of view. This list was revised into a list of 73 aspects following a review of relevant literature on farm animal welfare. This final list was used for the survey.

Survey

Data were collected during the first three weeks of April 2006, resulting in a gross response of 1081 respondents and a valid response of 834, of which respectively 27% and 73% via printed and digital versions. At the time of writing, the analysis of the survey results was not finished, so only some preliminary results are mentioned. Final results were presented and discussed at the conference.

The following preliminary results were based on 423 digital questionnaires. The age of respondents ranged from 17 to 75, with a mean of 34.5 years. The number of female and male respondents was nearly equal (respectively 45.5% and 54.5%). 39.5% of respondents had an urban place of residence, 60.5% lived on the countryside.

Based on the first question, the list of 73 aspects could be condensed into five main dimensions, i.e. *(i) natural behaviour and housing, (ii) transport and slaughter, (iii) feeding and housing climate, (iv) animal suffering* and *(v) human-animal relationship.*

Respondents' perception of the current state of farm animal welfare in Flanders was moderate to rather good. The perception differed significantly between men and women ($P < 0.001$) and was better for male then for female respondents. Perception also differed depending on the place of residence ($P < 0.001$), with city dwellers perceiving the welfare of farm animals as worse compared with people from rural areas. No correlation was found between the age of respondents and their perception of farm animal welfare ($r = 0.091$; $P = 0.06$). Perception of animal welfare also differed according to species and was lowest for poultry (rather moderate to moderate; laying hens lower then broiler chickens), and highest for cattle (rather good to good; dairy cattle higher then beef cattle). Pig welfare was scored intermediately (moderate to rather good). Men and city dwellers gave higher scores then women ($P < 0.001$) and rural people ($P \leq 0.001$) respectively.

Focus group discussions (other stakeholders) and in-depth interviews

At the time of writing, the focus group discussions and in-depth interviews were not yet completed. Results were presented and discussed at the conference.

Discussion

Based on the focus group discussions 56 aspects found important for animal welfare, could be listed. After literature review the list was revised into a list of 73 aspects. So, although participants stated that they only had limited knowledge of what matters to the animal, a large part of the final aspects was mentioned by them.

The preliminary results indicate that farm animal welfare can be divided into five main dimensions: *(i) natural behaviour and housing, (ii) transport and slaughter, (iii) feeding and housing climate, (iv) animal suffering* and *(v) human-animal relationship.* These first results indicate that citizens strongly associate the expression of natural behaviour with the housing conditions of the animals. This makes sense as the expression of natural behaviour is more likely when animals are housed in a way that resembles their natural living conditions. Animal suffering and human-animal relationship were grouped as separate dimensions, which is in line with the recently developed scientific view of animal feelings being important when considering animal welfare (Anonymous, 2001).

In general, the citizens' perception of farm animal welfare in Flanders was moderate to rather good. However, welfare was perceived to differ between the five species, with the welfare of laying hens being worst, dairy cows best and pigs in between. In the Eurobarometer (2005) similar results were found: 26% of the Belgian respondents gave a positive opinion about the welfare of laying hens, 79% of them perceived dairy cow welfare as positive and 46% had a positive perception of pig welfare. City dwellers were more critical about farm animal welfare in comparison to people living in a rural area. This was also the case for dairy cows and pigs, but not for laying hens, in the Eurobarometer (2005). Women were more negative then men which correspond to Köhler (2001) who reported that women rate the acceptability of farm animal treatment lower then men. No relationship between respondents' age and their perception of farm animal welfare was found.

After completing the study, insight will be gained into the attitudes of the different stakeholders in society towards the welfare of farm animals. Moreover, assessing welfare concerns evaluating different dimensions (or 'criteria') of welfare and the first step in creating a welfare assessment model is to define this set of dimensions on which the model can be built (Bouyssou, 1990). Consequently, the final list of dimensions and their relative weights is believed to be invaluable for the aggregation of various welfare indicators into a comprehensive assessment of the general state of welfare.

Acknowledgements

This research (ALT/AMS/2005/1) was funded by the Ministry of the Flemish Community.

References

Anonymous. (2001). Scientists' assessment of the impact of housing and management on animal welfare. *Journal of Applied Animal Welfare Science*, 4 (1): 3-52.

Bartussek, H. (1999). A review of the animal needs index (ANI) for the assessment of animals' well-being in the housing systems for Autrian proprietary products and legislation. *Livestock Production Science*, 61: 179-192.

Bouyssou, D. (1990). Building criteria: a prerequisite for MCDA. In: *Readings in multiple criteria decision-aid.* Bana e Costa CA (ed). Springer Verlag, Heidelberg. 58-80.

Bracke, M. B.M. (2001). *Modelling of animal welfare. The development of a decision support system to assess the welfare status of pregnant sows.* PhD thesis, IMAG, Wageningen. 150p.

Broom, D.M. (1986). Indicators of poor welfare. *British Veterinary Journal*, 142: 524-526.

Broom, D.M. (2000). Welfare assessment and welfare problem areas during handling and transport. In: *Livestock handling and transport. 2nd Edition.* Grandin, T. (ed). CAB International, Wallingford. 43-61.

Capdeville, J. and Veissier,, I. (2001). A method of assessing welfare in loose housed dairy cows at farm level, focusing on animal observations. *Acta Agriculturae Scandinavica Section A Animal Science,* Suppl. 30: 62-68.

Delezie, E., Lips, D., Lips, R. and Decuypere, E. (2005). Mechanical catching of broiler chickens is a viable alternative for manual catching from an animal welfare point of view. *Animal Science Papers and Reports,* 23 Suppl. 1: 257-264.

Duncan, I.J.H. (1996). Animal welfare defined in terms of feelings. *Acta Agriculturae Scandinavica Section A Animal Science,* Suppl. 27: 29-35.

Eurobarometer. (2005). Attitudes of consumers towards the welfare of farmed animals. *Special Eurobarometer 229, June 2005.*

Farm Animal Welfare Council. (1992). FAWC updates the five freedoms. *Veterinary Record,* 17: 357.

Fraser,D. (1995). Science, values and animal welfare: exploring the 'inextricable connection'. *Animal Welfare,* 4: 103-117.

Grandin, T. (2000). Handling and welfare of livestock in slaughter plants. In: *Livestock handling and transport. 2nd Edition.* Grandin, T. (ed). CAB International, Wallingford. 409-439.

Hall, S.J.G. and Bradshaw, R.H. (1998). Welfare aspects of the transport by road of sheep and pigs. *Journal of Applied Animal Welfare Science,* 1 (3): 235-254. (abstr.)

Keeling, L. and Veissier, I. (2005). Developing a monitoring system to assess welfare quality in cattle, pigs and chickens. *Proceedings of the Welfare Quality conference 'Science and society improving animal welfare', 17-18 November, Brussels.*

Knierim, U. and Gocke, A. (2003). Effect of catching broilers by hand or machine on rates of injuries and dead-on-arrivals. *Animal Welfare,* 12: 63-73.

Köhler, F. (2001). Consumer concerns about animal welfare and the impact on food choice. Report on national survey – Germany. *Report of project CT98-3678, March 2001, EU FAIR-CT98-3678.*

Mason, G. and Mendl, M. (1993). Why is there no simple way of measuring animal welfare? *Animal Welfare,* 2: 301-319.

Simmonds, N. (2005). Perspectives on controlled atmosphere stunning. *Animal Science Papers and Reports,* 23 Suppl. 1: 255-256.

Sørensen, J.T., Sandøe, P. and Halberg, N. (2001). Animal welfare as one among several values to be considered at farm level: the idea of an ethical account for livestock farming. *Acta Agriculturae Scandinavica Section A Animal Science,* Suppl. 30: 11-16.

Striezel, A., Andersson, R. and Hörning, B. (1994). Tiergerechtheitsindex für legehennen. In: *Tiergerechtheitsindex 200 – Ein Leitfaden zur Beurteilung von Haltungssystemen.* Sundrum, A., Andersson, R., Potsler, G. (eds). Koellen, Bonn. 73-112.

Whay, H.R., Main, D.C.J., Green, L.E. and Webster, A.J.F. (2003). Animal-based measures for the assessment of welfare state of dairy cattle, pigs and laying hens: consensus of expert opinion. *Animal Welfare,* 12: 205-217.

Part 14
Other contributions

Comparative analysis of ethical assessment of agricultural biotechnology: Committees functioning and philosophical presuppositions

Catherine Baudoin
Institut National de la Recherche Agronomique / ACTA, Unité TSV, 65, boulevard de Brandebourg, 94205 Ivry-sur-Seine cedex, France, Catherine.Baudoin@ivry.inra.fr

Keywords: ethics committee, biotechnology, agriculture, Europe

Introduction

The development of biotechnology applied to plants and animals, for the last twenty years, has led stake-holders to reflect on the related ethical issues. States or research organisations have established ethics committees to debate on the subject. These committees are advisory deliberation groups of experts that write reports on particular issues. Depending on cases, their remit and the way they work differ in some degree. Ethics committees aim at helping decision-makers, accompanying researchers' reflection or at stimulating public debate. They may ask experts for further research or legal reports, or alter the first version of a report according to queries or positions that were expressed during public consultation. The composition of the ethics committees results in choices concerning the represented communities: science, private sector, religion representatives, citizens or consumers advocacy groups.

Objective

The purpose of the study is to question the differences in the way ethics committees work and in their conclusions, and to analyse the philosophical presuppositions of ethical assessment. The study concentrates on ethics committees that deal with agricultural biotechnology, in Europe with an extension to Canada. We will examine some of the following ethics committees:

- the Finnish National Advisory Board for Biotechnology;
- the Finnish Board for Gene Technology;
- the Swedish Gene Technology Advisory Board;
- the Danish BioTIK Expert Group;
- the British Nuffield Council on Bioethics;
- the British Food Ethics Council;
- the Netherlands Subcommittee "Ethics and social aspects" of the Committee on Genetic Modification;
- the French Ethics and Safety Committee on Agricultural Research Applications;
- the Swiss Ethics Committee on Non-human Gene Technology;
- the Spanish Advisory Committee on Ethics of Scientific and Technical Research;
- the Quebec Science and Technology Ethics Committee;
- the Canadian Biotechnology Advisory Committee.

The work will result in a description of the way ethics committees function, a description of the development of the assessments, and a reflection on ethical meanings that are engaged in the procedures and decisions of the committees.

Method

Publications, annual and research reports, discussion papers and public consultation reports will be analysed. When possible, we will observe ethics committees meetings, and make non directive interviews with members. Legal and official documents related to the work of the ethics committees will be analysed. Non members' views will also be taken into account.

Hypothesis

Can we consider that certain committees reason according to consequentialist ethics, by weighing the consequences of an action? Can we consider that other committees reason according to deontological ethics, by evaluating the rightness of an action according to its conformity to a rule? Is a rational risk assessment enough for an ethical assessment? Are there any cultural differences that can not be directly translated in philosophical differences? Do certain ethics streams, other than the main stream of moral philosophy - like animal and environmental ethics - play a role in these assessments? How are biotechnologies dealt with: do recommendations concern plants, animals or biotechnology in general? What are the consequences of considering living organisms as distinct domains? How do committees come to a consensus? How are disagreements taken into account? What are the procedures used: reflexive equilibrium, coherentism...?

References

Carr, S. and Levidow, L. (2000). Exploring the links between science, risk, uncertainty and ethics in regulatory controversies about genetically modified crops. *Journal of agricultural and environmental ethics*, 12: 29-39.

Langlois, A. (2004). Comités d'éthique. In: *Dictionnaire d'éthique et de philosophie morale*. Canto-Sperber, M. (eds). Presses Universitaires de France, Paris. 323-327.

Langlois-Lafitte, A. (1992). *La régulation de l'expérimentation sur l'homme : des Comités d'éthique de l'Assistance publique de Paris aux Comités de protection des personnes, Thèse de doctorat, Philosophie, Paris 10.* 693 p.

Larrère, C. (1997). *Les philosophies de l'environnement*. Presses Universitaires de France, Paris. 124 p.

Larrère, R. (2004). Organismes génétiquement modifiés. In : *Dictionnaire d'éthique et de philosophie morale*. Canto-Sperber, M. (eds). Presses Universitaires de France, Paris. 1378-1381.

Parizeau, M.H. (2001). Comité d'éthique. In: *Nouvelle encyclopédie de bioéthique, Médecine – Environnement - Biotechnologie*. Hottois, G. and Missa, J.N. (eds). De Boeck Université, Bruxelles. 191-196.

Verweij, M., Brom, F.W.A. and Huibers, A. (2000). Do's and dont's for ethics committees: practical lessons learned in the Netherlands. *HEC Forum*, 12(4): 344-357.

Wynne, B. (2001). Expert discourses of risk and ethics on GMOs: the weaving of public alienation. *Politeia*, XVII(62): 51-76.

Pigs breeding outdoors and weighing stress: an automatic station for feed distribution and weight taking

Paolo Cornale and Salvatore Barbera
Dipartimento di Scienze Zootecniche, Università degli Studi di Torino, via L. Da Vinci, 44 - 10095 Grugliasco (TO), Italy, paolo.cornale@unito.it

Nowadays it is not possible exclude from *animal productions* the concepts of *animal health* and *welfare* and *food safety* and *quality*.
The increasing sensibility on animal welfare and ethics at farm have a major impact in the product quality definition leading to the development of extensive production systems.
Production of food from animal includes real and perceived attributes. Physical, chemical, microbiological, and organoleptic measures are real quality attributes, directly and objectively measurable in a given product. In an opposite way, environmental impact, animal welfare, food traceability, and safety aspects of the production systems are "secondary" (but no less important) quality attributes related to the consumer perception of products. Providing a measure of such attributes is a trivial task.
Actually, despite outdoor production do not appear to modify chemical, physical, and organoleptic qualities of products, it is perceived by the consumer as a more sustainable, traditional, animal-friendly and family-based farming system.
Reduced space and barren environment are considered to be social stresses causing abnormal and aggressive behaviors. The most relevant advantage of outdoor breeding is giving animals a greater space and environmental diversity, allowing expression of a wide range of behavior patterns.
From an economical point of view, outdoor pigs breeding is characterized by lower establishments and management costs w.r.t. intensive systems.
Nevertheless, in this non-conventional farms, which try to satisfy animal's physiological and behavioral needs some activities remain (e.g. transportation, medication, weighing, etc.) where the human-animal interaction may be a source of stress (Lensink, 2002). Aversive handlings are negative experiences for animals. They lead to fear reactions toward humans, with possible consequences on the animal's welfare, on the production, and finally on the quality of meat (Hemsworth and Coleman, 1998).
Weight taking allows controlling weight gain and furthermore is an essential parameter for feeding softwares used by automatic feeding systems. Weighing handling can be a common factor of stress. Traditionally, the farmer directs his animals in a forced course ending with a balance where he can weight each animal one at a time. Auguspurger and Ellis (2002) suggested that feed intake was reduced and feeding patterns substantially modified during weighing days, even in those subjects being frequently handled. Wolter *at al.* (2002) did not find this relationship. Nevertheless their results suggest that frequent weighing of animals can results in an increased rate of mortality and morbidity.
The aim of our experiment is the evaluation of a system allowing to sample live weight data of each animal without interrupting its normal daily routine with a tangible well-being advantage.
At the Experimental Station of the Facoltà di Agraria of the University of Torino (Italy) were raised nine subjects of the Italian Mora Romagnola breed, involved in a conservation program, and ten Mora Romagnola x Large White crossbreeds. All 19 individuals were dived in two mixed balanced groups and reared outdoors in two pens with a pond in the center and a hut with straw.

Inside the pen of first group we realized an automatic station for feed distribution and animals weighing under the control of an analog camera. Furthermore, each pig was equipped with a transponder placed in the auricle like an eartag. The second group received feed *ad libitum* twice a day instead, and moreover these subjects were weighed fortnightly.
By entering in the feed station a subject triggers the weighing system and after the electronic identification it receives a portion of feed and water (Figure 1). During each feeding session the pig weight is recorded using a balance (bars at load cells) placed in front of the trough. Using this approach it is possible to obtain an updated and more precise set of data. Moreover this set can be inserted in the feeding software allowing a more correct use of feed. The weighing operations in traditional group required that the operators have got to capture free animals. Every time it involved a tiring job for stockpersons and risks for animlas' health (injuries, legs' fractures, etc.) and consequences on welfare. Behavioural observations suggested that pigs of first group were quieter and less fearful in comparison with second one. At the entrance of humans in pen with automatic station it was possible to note that animals were calm and curious while approach and entry in second pen led fear reaction toward human outsider.
From the welfare point of view handlings due to weighing operations are completely removed. Weight taking, from post weaning to slaughter day can be daily realized without being a stress factor.

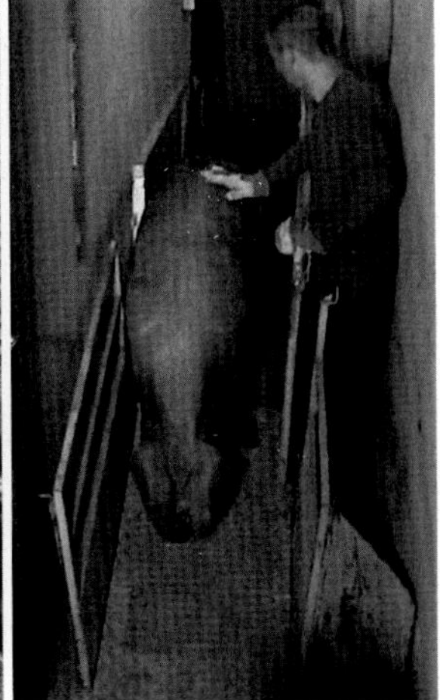

References

Augspurger, N.R. and Ellis, M. (2002). Weighing affects short-term feeding pattern of growing- finishing pigs. *Can. J. Anim. Sci.*, 82: 445-448.

Hemsworth, P.H. and Coleman, G.J. (1998). Human-Livestock Interactions: The Stockperson and the Productivity and Welfare of Intensively-farmed Animals. *CAB International*, Oxon UK.

Lensink, B.J. (2002). *The human-animal relationship in animal production.* In: First Virtual Globe Conference on Organic Beef Cattle Production. September, 02 to October, 15 – 2002 – Via Internet.

Wolter, B.F., Ellis, M., DeDecker J.M., Curtis S.E., Hollis, G.R., Shanks, R.D., Parr, E.N. and Webel, D.M. (2002). Effect of double stocking and weighing frequency on pig performance in wean-to-finish systems. *J. Anim. Sci.*, 80: 1442-1450.

Dairy cows systems and natural resources utilization for a sustainable and ethical production in NW Italy

Marcella Gentile, Salvatore Barbera and Luca Battaglini
Dipartimento Scienze Zootecniche, Università degli Studi di Torino, Via L. Da Vinci 44, Grugliasco, Italy, luca.battaglini@unito.it

During last century the role of the animals has changed: if in the past they would represent an additional food, exploiting resources not accessible for human kind, nowadays they are in competition with the latter for the utilization of some important vegetal origin food, mainly grains. At present, talking about bioethics in breeding systems, relationships among human beings, livestock and production means to consider welfare, hygienic conditions, shelters and an appropriate activity space for animals. Nevertheless, one aspect is not often considered being represented by the utilization of vegetal resources from an ethical point of view. Researches that analyse dairy cows breeding systems (including environment, breeds and feeding resources) are needed to verify the sustainability of livestock production systems in relation to different feeding resources availability (e.g. grains, conserved fodders, pasture); moreover, ethical aspects concerning feeding resources used in different environments should be considered. Side by side, it is well known that the variation of peculiar qualitative milk parameters is strictly related to the animal feeding and definitively to the environment. These parameters could be chosen to promote livestock management variations, in particular concerning feeding systems and limiting as possible the grains use. Alpine breeding systems are an example of sustainable integration between land management and productive processes; the inherent exclusive forage exploitation has characterized and modified landscape and environment. Moreover, alpine pasture has increased its importance for the multifunctional features attributed in the recent years to mountain productive activities. In these last years, particular attention has been dedicated to milk fatty acids composition and conjugated linoleic acid (CLA) content, and to the relevant correlations with diets based on grass: in fact, it is well known that milk products originating from high alpine grazing of cows, seem to provide an extraordinarily high proportion of CLA and polyunsaturated fatty acids, allowing these products to be considered as functional foods with an expected high value for human health (Hauswirth *et al.*, 2004). This study refers on milk quality of two grazing autochthonous dairy cows breeds, during the pasture season 2003 in different realities of NW Italian Alps: Aosta cattle in Aosta valley and Pustertaler-Barà cattle in Torino valleys. On individual morning milk samples have been determined: individual daily milk yield, fat, proteins content, somatic cells (SC) count, fatty acid profile (FA) including CLA. The pasture season showed evident effects on milk yield and quality due to the climate trend and to the particular dryness of 2003 in NW Alps: in Aosta breed cows, the negative effects of this period in terms of milk quality are evident: in fact, in these alpine pastures frequently characterised by a good botanical composition, the effect of climate determined a generalized modest grass quality. Previous results of Battaglini *et al.* (2003), always in Aosta region and Aosta cattle show an evident increase of PUFA and CLA percentage in high pastures, in case of more favourable climate conditions. It is well known that a pasture in advanced vegetation maturity conditions, characterized by higher proportions in saturated fatty acids and reduced in linolenic acid, determine a minor amount of CLA (Loor *et al.*, 2002). Anyway, Aosta cattle produced, on average, high total CLA milk contents, double and more than the average value of cosmopolite breeds (e.g. Brown, Holstein as reported by Bugaud *et al.*, 2001). Pustertaler-Barà cattle, a recently recovered double purpose breed original of the Eastern Alps, represents an interesting opportunity for several reasons: rusticity, dairy productivity and veal meat. Observing its results in Torino valleys, at similar parity and lactation stage of

Aosta cattle in Aosta valley, yields are on average higher, with lower fat and protein percentages but with a better somatic cells count. By the lipids profile point of view, Pustertaler-Barà cows produce a good quality milk: even in a harsh climate year, at pasture, the proportion of PUFA in this breed is slightly lower than in Aosta cattle milk but with an higher CLA concentration. These results confirm the evident enrichment of CLA in milk produced by cows grass-only fed at pasture (Dhiman *et al.*, 1999) (Table 1). Global results of these researches indicate that, beyond the influence of the well known factors (cattle breed, diet regime, physiological features, season and climate trend), an evident variation of fat contents seems to be related to the grass diet. The favourable effects of the alpine pasture on milk fatty acids profile and CLA in alpine cattle, even during particularly critical grazing seasons, confirm the convenience of breeding autochthonous breeds, less exposed to climate variations. It is anyway important choosing a correct pasture management considering any feeding interventions but limiting as possible the supplementation mainly represented by grains. Ethically correct strategies for employing natural forage resources should produce effects in different application fields such as: environment, products traceability, sustainability, products quality, natural resources recovering and human health. An interest for the natural resources utilisation by an ethical approach is to be hoped: in this sense research on animal origins food production should be orientated, as long as possible, to reduce man-animal feeding competition.

Table 1. Milk yield, composition, fatty acid profile and fatty acids ratio of Aosta and Pustertaler-Barà dairy cows.

		Aosta	Pustertaler-Barà
Milk yield	kg h^{-1} d^{-1}	7.88	10.51
Fat content	%	4.10	3.76
Protein content	%	3.64	3.39
SCC	no. 1000 ml^{-1}	432	277
Total saturate FA (SFA)	% total FA	54.32	57.50
Total monounsaturate FA (MUFA)	% total FA	39.81	37.13
Total polyunsaturate FA (PUFA)	% total FA	5.89	5.46
SFA/(MUFA+PUFA)		1.19	1.35
Total CLA	% total FA	2.92	3.12

References

Battaglini, L., Lussiana, C., Mimosi, M., Ighina, A. and Linty, C. (2003). Milk yield and quality of Aosta Chestnut dairy cows grazing on alpine pastures. *Atti 38° Simposio Internazionale di Zootecnia*, 123-130.

Bugaud, C., Buchin, S., Coulon, J.B., Hauwuy, A. and Dupont, D. (2001). Influence of the nature of alpine pastures on plasmin activity, fatty acid and volatile compound composition of milk. *Lait*, 81: 401-414.

Dhiman, T.R., Anand, G.R., Satter, L.D. and Pariza, M.W. (1999). Conjugated linoleic acid content of milk from cows fed different diets. *Journal of Dairy Science*, 82: 2146–2156.

Hauswirth, C.B., Scheeder, M.R.L. and Beer, J.H. (2004). High omega-3 fatty acid content in alpine cheese - The basis for an Alpine Paradox. *Circulation*, 109, 103–107.

Loor, J.J., Herbein, J.H. and Polan., C.E. (2002). Trans 18:1 and 18:2 isomers in blood plasma and milk fat of grazing cows fed a grain supplement containing solvent-extracted or mechanically extracted soybean meal. *Journal of Dairy Science*, 85: 1197-1207.

Pesticide residues in drinking water versus other beverages: a case of an unjustified discrepancy?

Khara Grieger and Stefan Trapp
Vaernedamsvej 5a, 1819 Frederiksberg, Denmark, kgrieger@hotmail.com

Abstract

Dietary sources, like drinking water and foods, are major routes for public exposure to pesticide residues. Although pesticides in drinking water are regulated relatively well, foods like beverages receive much less attention in both regulation and monitoring. In fact, the maximum residue limits (MRLs) for pesticides in beverages may be several orders of magnitude greater than permissible in drinking water, particularly in the EU. Furthermore, drinking water standards incorporate the precautionary principle to further protect public health, whereas standards for beverages may not be sufficiently protective, especially for children. Therefore, there is an unethical and unjustifiable discrepancy in the regulation of pesticide residues in foods such as beverages, especially when compared to legal standards set for drinking water.

Keywords: beverages, drinking water, maximum residue limit, pesticide, precautionary principle

Introduction

Drinking water and food safety authorities set maximum residue limits (MRLs) for pesticides in drinking water and foods in order to protect public and consumer health from potential adverse health effects associated with exposure to these chemicals. However, when comparing the legal standards of pesticide residues between drinking water and beverages (EC, 1998, 2004b; FAO/WHO, 2001, 2005; Hamilton *et al.*, 2003), there are large inconsistencies not only in the mere existence of MRLs, but also in the actual permissible residue concentrations. According to consumption data, beverages comprise a large portion of our diet, and therefore the legal standards set for pesticides in these dietary sources play a key role in determining our overall chemical exposure.

Results and discussion

Although 30-50% of both drinking water and beverage samples contain detectable pesticide residues, concentrations in beverages are much greater than those in drinking water (EC, 2002, 2004a). This corresponds to the legal standards set in these dietary sources, and as seen in the following table (Table 1.) for some frequently-detected pesticides, the permissible concentration of a pesticide in drinking water is significantly lower than in beverages such as juice, wine, and milk. Council Directive 98/83/EC set a sweeping 0.1 μg/L limit for individual pesticides and 0.5 μg/L for total pesticide concentrations in European drinking water (EC, 1998). This Directive was also based upon the addition of the precautionary principle to drinking water standards which went beyond the available toxicological data in protecting human health (Swanson and Vighi, 1998). Unfortunately, food safety authorities have not reacted in a similar manner and have not taken precautionary steps in the regulations of pesticides in food.

Furthermore, most MRLs for beverages do not exist in themselves, although milk appears to be more thoroughly regulated. Rather, the MRLs for beverages, particularly juice and wine, are extrapolated from those set for raw agricultural commodities.

Table 1.

Pesticide	MRL				ADI	CRR values		
	Water	Juice*	Wine*	Milk		Water-drinking Adult	Juice- drinking Child	Wine-drinking Adult
Carbaryl	0.0001	1 - 5	3	NA	0.008	0.0002	6.3 -31.3	5.4
Carbendazim	0.0001	0.1 - 5	2	0.1	0.03	5.5 E-5	0.2 -8.3	0.9
Dimethoate	0.0001	0.02 - 2	0.02	NA	0.002	0.0008	0.5 - 50	0.1
Endosulfan	0.0001	0.05 - 1	0.5	0.004	0.006	0.0003	0.4 - 8.3	1.2
Iprodione	0.0001	0.02 - 10	10	0.05	0.06	2.8 E-5	0.02 8.3	2.4
Permethrin	0.0001	0.05	0.05	0.05	0.05	3.3 E-5	0.5	0.1

MRL= Maximum residue limit (mg/kg or mg/L), ADI= Acceptable Daily Intake (mg/kg body weight/day), NA= Not available. * MRLs do not exist in most cases, and therefore the MRL for the raw agricultural crop is used instead.

To make matters worse, the legal standards set for pesticides in beverages may not fully protect individuals, especially children and above-average consumers of wine, as revealed by calculated Consumers Risk Ratio (CRR) values (taking into account consumption, body weight, MRL values, and ADI). Values > 1 indicate that the legal daily uptake may be far above the toxicologically-sound uptake. As seen above, some CRR values can reach up to 31.3 and 50 for juice-consuming children, which is considerably alarming since children are more vulnerable to chemical exposure than adults.

Conclusions

- There are great inconsistencies in the regulation of pesticide residues in foods like beverages, especially in comparison to drinking water regulation.
- Legal standards for pesticides in beverages are several orders of magnitude greater than those permissible in drinking water.
- European authorities have included the precautionary principle in drinking water standards, while food safety authorities have neglected this inclusion.
- Given the current high consumption patterns of beverages, this discrepancy is unethical and unjustifiable, and legal standards may not fully protect individuals, particularly children.
- The question arises "Who is responsible in the case of harm"?

References

European Commission (EC). (1998). *Council Directive 98/83/EC of 3 November 1998, on the quality of water intended for human consumption.* Available on http://europa.eu.int/eurlex/lex/LexUriServ/LexUriServ.do?uri=CELEX:31998L0083:EN:HTML

EC. (2002). *Synthesis report on the quality of drinking water in the Member States of the European Union in the period of 1996-1998*. Available on www.eu.int/comm/environment/water/water-drink/pdf/report96_98.pdf

EC. (2004a). *Monitoring of Pesticide Residues in Products of Plant Origin in the European Union, Norway, Iceland, and Liechtenstein, 2002 Report- Summary*. Available on http://europa.eu.int/comm/food/fvo/specialreports/pesticide_residues/summary_2002_en.pdf.

EC. (2004b). *Informal coordination of MRLs established in Directives 76/895/EEC, 86/362/EEC, 86/363/EEC, and 90/642/EEC: MRLs sorted by crop group*. Available on http://europa.eu.int/comm/food/plant/protection/resources/08-04-2.pdf

FAO/WHO. (2001). *Code of hygienic practice for bottled/packaged drinking waters (other than natural mineral waters)*. Available on www.codexalimentarius.net/download/standards/369/CXS_227e.pdf

FAO/WHO. (2005). *CODEX Alimentarius: Pesticide Residues in Food*. Available on http://faostat.fao.org/faostat/collections?subset=FoodQuality

Hamilton, D., Ambrus, A., Dieterle, R., Felsot, A., Harris, C., Holland, P., Katayama, A., Kurihara, N., Linders, J., Unsworth, J., and Wong, S. (2003). Regulatory limits for pesticide residues in water (IUPAC Technical Report). *Pure and Applied Chemistry*, 75(8): 1123-1155.

Swanson, T. and Vighi, M. (1998). *Regulating chemical accumulation in the environment: the integration of toxicology and economics in environmental policy-making*. Cambridge University Press, Cambridge, UK. 42 p.

Presentation of the computer programme 'Animal Ethics Dilemma'

Tina Hansen, Trine Dich, Anne Algers, Alison Hanlon, Hillar Loor and Peter Sandøe
The Royal Veterinary and Agricultural University, Institute of Food and Resource Economics, Rolighedsvej 26, DK-1958 Frederiksberg C, Denmark, tih@kvl.dk

Abstract

This poster explains how to play the computer programme 'Animal Ethics Dilemma'. In connection to the poster there will be access to a computer, so that it will be possible to test the programme. 'Animal Ethics Dilemma' is a computer supported learning tool developed primarily for veterinary students. The aim of the programme is to the students aware of ethical issues that arise in animal use. The programme is constructed as a computerized role game with five case studies that the students can play or explore. The poster may usefully be read together with to the paper "Animal Ethics Dilemma – a computer supported learning tool".

Keywords: animal ethics, computer game, learning

Introduction

The computer supported teaching programme, 'Animal Ethics Dilemma' (Hanlon, *et al.*, 2006) is a completely new tool launched in June 2006. It is a learning tool to improve the student's ability to understand and relate to ethical issues that arise in animal use. The programme has been developed in a joint project between the institutions to which the authors are attached: The Royal Veterinary and Agricultural University, Denmark, Swedish University of Agricultural Sciences, University College Dublin, Ireland, Imcode Partners AB, Sweden. The research group "Flexibility in Learning/Creative Environments", Malmö University, Sweden was involved in designing and evaluating an early version of the programme. The programme is freely available on the internet at www.aedilemma.net

Presentation of the programme

When the student first enters to the programme she is presented with a pre-test, to determine her 'personal profile'. The pre-test comprises 12 questions with a choice of answers. The answers are taken to reflect ethical theories. One question in the pre-test is: *"Is it OK to keep animals for slaughter?"* The student can choose between five responses. One of them is: *"Yes, if they are kept and slaughtered in a welfare friendly way"*. Another opportunity is: *"No, because animals should not be kept and killed only to satisfy humans."* The first response is taken to reflect a utilitarian view and the second one an animal rights view. Once the student has selected her/his choice of answers in the pre-test, a "Personal Profile" is generated in the programme. The "Personal Profile" is visually shown as a diagram telling the student that his/hers profile is for example 50% utilitarian, 25% rights view and 25% respect for nature. The profile is used to navigate the game by using an opposing view to that of the profile, as the opening question for the case.

Now the student can begin to play the programme. In the main-menu the student can choose among five cases: "Blind Hens", Dog Euthanasia", ANDi the GM monkey", "Slaughter", and "Rehabilitation of Seals". In each case the student will experience a role play. After choosing a case the student will enter into the first level of the chosen case. Each case is based on four levels. Within the first three levels, there is a statement followed by four or five responses. For

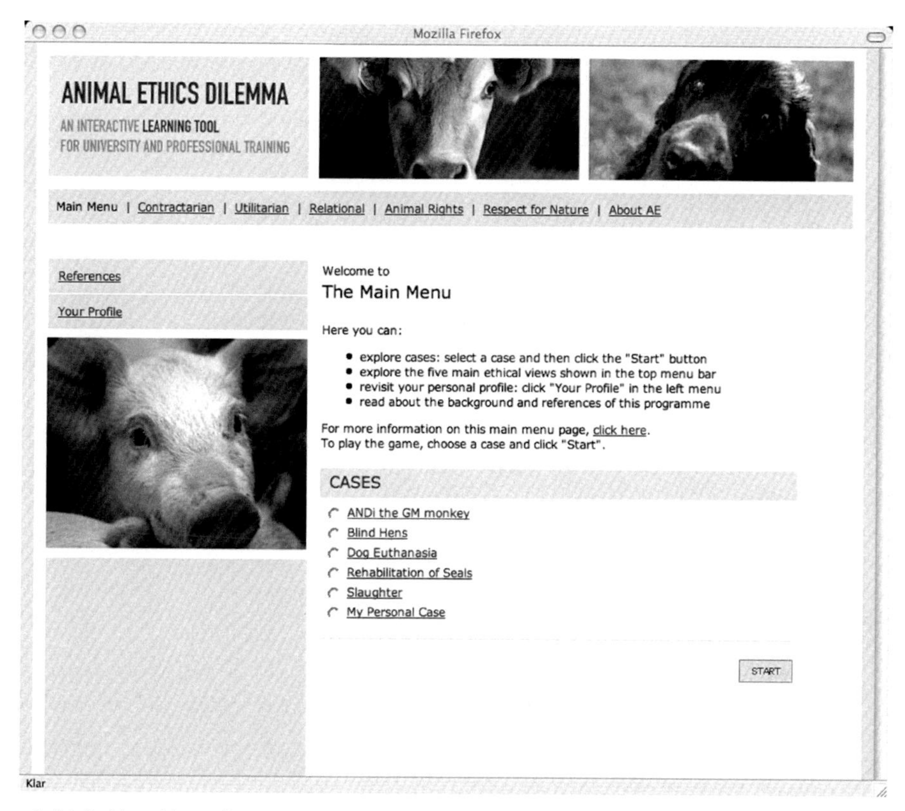

Figure 1. 'Main Menu' from the computer programme "Animal Ethics Dilemma".

example a student wants to play the "Rehabilitation of Seals". At the first level she/he can meet this statement:

While reading the statement the student can choose to get more information about the highlighted words ('wildlife rehabilitation' and 'suffering') in the text. By clicking at the highlighted word a 'pop-up box' will provides more information. After reading the statement the student will have to choose among five responses:

1. *"Every animal deserves the right to life. While there is a chance, we should try to treat this animal"*
2. *"We should euthanase. Resources should only be directed to those that will survive."*
3. *"Looking into the eyes of seal cubs, we feel a strong sense of connection between them and us. We should respect this relation and treat them accordingly. Euthanasia should only be used as the last resort."*
4. *"Unfortunately it would be best to euthanase. It is wrong to allow an animal to suffer, and in any case it is not cost-effective to spend time and money on an animal that will not survive."*
5. *"The cub should be euthanased. This animal is weak. It would not survive under normal conditions. We should not have rescued it in the first place"*

Each answer is taken to reflect an ethical theory, e.g. answer number one reflects the animal rights view.

Having selected an answer the student will in level two and three be presented for new statements. The fourth level serves to round up the case by bringing the story to an end based upon the previous responses.
The story in each case will develop differently dependent of the choice of responses. After playing a case, the "Personal Profile" will be updated, and it is possible to compare the new "Personal Profile" with the profile from the pre-test.
The student can select between different ways to play the game. For example she/he can navigate the programme using the "Assist me" option, which explains the basis of the assignment of ethical theories to the various answers by providing pop-up boxes. For example will the answer:"*Every animal deserves the right to life. While there is a chance, we should try to treat this animal*" be followed by: *(Animal Rights: Theory/Why?)*. The student is told that the answer reflects an animal rights view and by clicking on *Theory* there will be a pop-up box providing the student information on this ethical viewpoint. In the *why* pop-up box the student can read why the answer is taken to reflect the animal rights view.
The cases include visual information such as photos, diagrams and video clips in order to support the text information and utilise the possibilities of the media. Background text and glossary of terminology, "References", are available for the student to explore via links in the cases and in from the main menu. The programme should be considered as a tool to support the learning process, not as a stand alone tool for teaching animal ethics.

References

Hanlon, A.J., Dich, T., Hansen, T., Loor, H., Sandøe, P., Algers, A. (2006). *Animal Ethics Dilemma – an Interactive Learning Tool for University and Professional Training* [on line]. Available on www.aedilemma.net [24/04/2006].

A project looking for novel perspectives in agricultural science communication

Matias Pasquali
Society in Science, Branco Weiss Fellow, AGROINNOVA, University of Torino, Current address: University of Minnesota, Dept. Plant Pathology COAFES - 495 Borlaug Hall, 1991 Upper Buford Circle, St Paul, MN 55108-6030, USA, matias.pasquali@unito.it, pasqu016@umn.edu

Keywords: participatory research, ethics, scientists involvement

Agricultural research is constantly shaping and influencing food production. From functional foods to high production varieties, scientific research is defining what will be the future of our alimentary habits, often with very few interactions with society at large. Moreover agriculture is a systemic science, thus it modifies the context of its own research (Alroe and Krinstensen, 2002). The result is that the objectives of research are defined by values never officially declared, often on the contrary, denied by scientists.
The past and more recent events related to food innovations (see the example of GMO) have shown that acceptability can be achieved only with truly participatory public confrontation that is excluded by the Mertonian rules that have been the theoretical basis for academic science development and communication (Ziman, 1987).
Because the traditional scientific ethos has abolished the ethical considerations from the scientific practice, participation of scientists in the public arena has always been focused on purely "factual" communication, that has been often used to impose undeclared values.
Participatory research in agriculture has been explored in different forms in the past 30 years (Buhler *et al.*, 2002) and have questioned methodologies of scientific research as well as research objectives.
A participatory research initiative that represent a continuation of the approach proposed by Pasquali and Korthals (2004), connected to a molecular biology work, is outlined: it requires not only involvement of farmers and other stakeholders in the scientific process, but also the construction of a different attitude by scientists. The project is trying to change scientific communication strategies. Public analysis in every discussion on food/agricultural innovation rely on ethical arguments. So the idea is that integration of scientific communication (in the agricultural sector) and ethical reflection will radically modify science perception, challenging and redefining also scientists responsibility towards society (Pasquali and Korthals, 2004). Scientists can -have to- play a fundamental role in communicating their work and aims, thus ethics is an essential tool (Ziman, 1998; Hayry, 2003). Ethics deals with values that can be debated at all levels of society and ethical arguments may help also in scientific discussions among colleagues.
The objectives of the initiative are:
-to identify scientists willing to challenge the structure of scientific communication and willing to build a common background for ethical communication within science. Some personal initiatives by single scientists (Schjoining *et al.*, 2004; Scholthof, 2000) testify the actual awareness and interest of a part of scientific community on ethical aspects.

- to introduce ethical consideration in a scientific article, also by direct ethical confrontation in close interaction with stakeholders involved in the research;
- to identify editors that show interest in a novel communication approach;
- to form a community of scientists interested in novel communication strategies, wishing to increase their public activity.

Society in Science - Branco Weiss Fellowship, (http://www.society-in-science.ethz.ch) is supporting this project. It aims at finding novel ways for integrating sciences into societal context sustaining life scientists that are willing to pursue this objective.
The project presented, carried out by a biotechnologist with interests in social sciences, will require interaction of life scientists and social scientists in order to build participatory methods that redefine the traditional scientific language.
Through the interaction with stakeholders, publishers and scientists the project aims at the redefinition of scientific rules of communication.
The outcomes of the formulated communication methods will be evaluated consulting scientists organizations and other stakeholders.
The project will be divided in different procedural steps trying to answer to some questions: is it possible to modify traditional arguments and style in scientific article? Can agricultural scientists effectively interact with social scientists and society (farmers, consumers) as part of their daily work? Are scientific editors willing to change their attitude towards science communication? Are there specific research subjects where the ethical discourse is firmly refused?
Concluding, if ethics has to play a crucial role for integrating innovation in everyday life, also through out ethical engagement of the public as suggested by Miah (2005), thus science will certainly have to challenge the structure of communication, engaging itself with ethics. Including ethical discourses in scientific communication may be an important step in this direction. This project is trying to promote this approach in the agricultural scientific community. If a positive result will be reached it will depend also on the level of support that the community of social scientists may grant.

Acknowledgements

This work was made possible by the Branco Weiss Fellowship support.

References

Alroe, H. F. and Krinstensen, E. S. (2002). Towards systemic research methodology in agriculture: rethinking the role of values in science. *Agriculture and Human values,* 19: 3-23.

Buhler, W., Morse, S., Arthur, E., Bolton, S. and Mann, J. (2002). *Science, Agriculture and Research. A compromised participation.* Earthscan publications, London, UK. 163 p.

Häyry, M., (2003). Do bioscientists need professional ethics? *In:* Häyry, M. and Takala, T. eds. *Scratching the surface of bioethics*, Rodopi, Amsterdam, 91- 97.

Miah, A. (2005). Genetics, cyberspace and bioethics: why not a public engagement with ethics? *Public Understanding of Science,* 14: 409-421.

Pasquali, M., Korthals, M. (2004). New communication scheme for biotechnology innovations. 5th *Eursafe meeting*, Leuven 1-3 September 2004.

Scholthof K.B.G. (2000). From Science and Society to Scientists in Society. *Phytopathology News,* 34: 11-13.

Schjonning, P., Elmholt, S. and Christensen, B. T. (2004). *Soil Quality Management. Concepts and terms. In:* (Eds) P. Schjonning, S. Elmholt and B. T. Christensen *Managing Soil Quality Challenges in Modern Agriculture* 344 p.

Ziman, J. (1987). *An Introduction to Science Studies: The Philosophical and Social Aspects of Science and Technology.* Cambridge Press, UK. 222 p.

The ethical Delphi. A new tool for ethical assessments[100]

Erik Thorstensen and Matthias Kaiser
The National Committees for Research Ethics, Prinsensgate 18, Box 522 Sentrum, 0105 Oslo, Norway, post@etikkom.no

Abstract

This poster presentation exhibits partial results from an EU project ("Ethical BioTA Tools") that had a focus on developing methodological tools for ethical assessments of biotechnology in the food sector. The well-known Delphi-method, stemming from technology assessment, was adapted to perform ethical assessments in a practical setting. The approach was tested in relation to the question of genetically modified (growth-enhanced) salmon. The method is designed to work with selected groups of experts from a wider range of specialities. Though several practical and conceptual issues remain, the method proved useful to get an overview over pertinent ethical issues and how they are conceived and weighed from different expert perspectives. A parallel study was performed by our consortium partners (Centre for Applied Bioethics, University of Nottingham) in the UK.

Keywords: Delphi, practical ethics, technology-assessment, biotechnology, aquaculture

Introduction

Assume you are a body that is regularly asked to provide ethical advice on new technologies or other scientific issues. Here is a situation you might find yourself in:

- Expert input is required for policies under review or development
- Issues are uncertain, controversial and complex
- Many diverse research communities and stakeholders have concerns
- Outcomes from the process should have an impact on several issues, including future policy-making
- There is a need for cross-sectorial scientific debate.

Bodies that deal with technology assessment have for a long time already employed so-called Delphi-method for these kinds of situations. Our research question was therefore whether this method could be adapted to answering ethical issues that come up in a similar way.

The case

The case to test the strengths and weaknesses of an Ethical Delphi was the question of introducing genetically modified salmon into aquaculture for production purposes. The background for the selection of this case was that the Norwegian Board for the Ethics of Patents ("Den etiske nemnda for patentsaker") was asked to evaluate a pending application for patent of a growth enhanced salmon. Two aspects were prominent for this evaluation: a) the animal welfare aspects of the genetic modification, and b) the expected environmental impacts of introducing the gm-salmon into production.

[100] Research for this project was part of the Ethical BioTA Tools project, consisting of a multi-national consortium, funded by the EC, and conducted in close cooperation with the Centre for Applied Bioethics, School of Biosciences, University of Nottingham.

The method

The Ethical Delphi aims to map expert opinion on the ethical dimensions of the use of a novel technology. For this purpose, a group of experts is selected. The selection of participants is a key stage of the process. It is therefore important that participants: (i) feel directly involved in the problem of concern; (ii) have pertinent information to share; (iii) are motivated to include the Delphi task in their schedule of competing tasks; (iv) feel that the aggregation of judgements of a respondent group will include information which they too value and to which they would not otherwise have access.

The experts are then in a first round confronted with a list of questions formulated by the organizers. All experts answer electronically, and none has knowledge of the identity of the other experts on the panel.

The answers of the experts are then summarized and played back to the experts in the second round. The experts are asked to review these inputs, modify their answers in the light of them or comment upon the answers as they see fit. In principle this iterative process could continue for many rounds, theoretically until a consensus on some statements is reached. However, since consensus was not the foremost task of the exercise, rather than the exploration of ethically relevant considerations, their knowledge status and weightings in regard to reaching a decision, we only conducted three such rounds.

Conducting an Ethical Delphi can be represented by the following steps:

- identification of the participants;
- recruitment and selection of expert panel;
- preparation of introductory material;
- preparation of the questionnaires and weighing and analysis methods;
- promotion;
- the running of the Ethical Delphi;
- final report printing and dissemination;
- evaluation.

Results

The tool highlights issues, as well as divergence and convergence of views and values. However, the method will not provide decision-makers with judgements or overall opinions. In line with this is also the hesitation that some of the participants of our study showed towards concluding when it came to "Statement 4: Should the marketing and production of a fast-growing and sterilised GM salmon be permitted in Europe? Please justify your response by indicating the key issues that inform your views". This hesitation should be interpreted as an indication that the Ethical Delphi might be more suitable for mapping positions than to provide definite answers. According to our group of experts priorities should be made as represented in the following table. Note that he list just encompasses topics/themes with an importance mean of less than 2.5. The scale used a range from 1 (very important) to 5 (unimportant):

Statement 1

One of the future options for European fish producers is the production and marketing of GM salmon. Please list key issues that should be considered as part of any assessment of this option.

1. safety aspects related to GM food (1,14);
2. consumer acceptance (1,29);
3. Accept in society and in the opinion forming organisations (1,29).

4. GM salmon must be sold and purchased at will and separately from conventional salmon (1,29);
5. uncertainty, ignorance and complexity (1,40);
6. secondary GM-products related to breeding – like GM fish fodder and GM vaccines (1,43);
7. cost / benefit (1,86);
8. long term health issues related to the consummation of GM food (2,00);
9. the effect upon the salmon's health (2,29).

Statement 2

Please indicate significant advantages or disadvantages of the production and use of GM salmon (e.g. benefits/risks to animal health and welfare; food safety/consumer acceptance; environment effects; economic interests, etc)

1. uncertainty related to environmental consequences (1,43);
2. consumer acceptenace (1,43);
3. uncertainty related to environmental consequences (1,43);
4. uncertainty relating to food safety (1,86);
5. production of sterile fish that excludes cross-breeding (1,86);
6. fish welfare / fish health (2,00);
7. fair distribution or risks and benefits (2,00).

Statement 3

Please indicate any additional ethical issues raised by the patenting and use of GM salmon

1. market power (1,20);
2. the consummation in general in the industrial part of the world (1,67);
3. the distribution of increased food production (1,71);
4. covering the global need for food through production of GM food (2,29);.

Statement 4

Should the marketing and production of a fast-growing and sterilised GM salmon be permitted in Europe? Please justify your response by indicating the key issues that inform your views

1. with the technology of today – No! (1,57);
2. not before the risk assessment and the mitigation strategies have been accepted by the people (2,00).

References

Kate Millar, Sandy Tomkins, Erik Thorstensen, Ben Mepham, Matthias Kaiser (2006), *Ethical Delphi. Manual.* LEI, The Hague. (available also online under Ethical BioTA Tools Project).

Ethical BioTA Tools Project (2003-2005), *Ethical Tools.* Available on http://www.ethicaltools.info/ [date of consultation: 01/05/2006].

Defining animal welfare from a citizen and consumer perspective: Exploratory findings from Belgium

Filiep Vanhonacker, Els Van Poucke, Griet Nijs, Johan Braeckman, Frank Tuyttens and Wim Verbeke
Ghent University, Faculty of Bioscience Engineering, Department of Agricultural Economics, Coupure Links 653, 9000 Gent, Belgium, Filiep.Vanhonacker@Ugent.be

Keywords: animal welfare, Belgium, consumer, information, principal component analysis

Abstract

Animal welfare is more than ever prominent in livestock production and consumption debates (Verbeke and Viaene, 2000; Harper, 2001; European Commission, 2002; Ingenbleek *et al.*, 2004; OIE, 2004; Welfare Quality, 2005; Eurobarometer, 2005; Federal Ministry of Health and Women, 2006). A specific issue pertains to how people define or perceive the subjective and multi-dimensional concept of farm animal welfare in their role as citizen or consumer. The aim of this study was to explore Belgians' perceptions of farm animal welfare and to detect the main dimensions of animal welfare as perceived by individuals. The analysis was based on a cross-sectional data set including 628 respondents. Data was collected through a questionnaire-type survey in four waves, each with a one-year interval from April 2000 until April 2003. First, to explore consumers' perception of farm animal welfare, a list of 23 aspects derived from focus group discussions was presented to the respondents. In a first question, the degree of association as expressed by consumers between each parameter and farm animal welfare was measured on a five point-Likert scale, ranging from "absolutely no association" to "very strong association". A second question measured the perceived current impact of each aspect on farm animal welfare in Belgium, ranging from "very negative impact" to "very positive impact". Furthermore, questions sounding out attitude towards information concerning farm animal welfare were included, using five point-Likert scales.

Based on the first question, the 23 aspects were reduced to six main dimensions using principal component analysis. Figure 1 shows the mean scores on both questions for each of the 23 aspects, which are already grouped per dimension. As citizens' perceptions, the factor *housing* was most strongly associated with farm animal welfare, followed by *climate, transport and slaughter, feeding* and *breeding*. Actual practices with respect to housing and climate were positively evaluated with respect to animal welfare, whereas feeding and transport practices were negatively evaluated. Particularly the gap between the two profiles is of importance, since a large gap is most likely to originate from a parameter which is highly associated with animal welfare but poorly judged. The aspects yielding the highest perceived current problems in Flemish agriculture are *transport, (un)loading, stocking density, cage size* and *the use of growth enhancers*. As consumers' perceptions, the attribute animal welfare when purchasing meat was deemed less important as compared to health, trustworthiness, quality and safety, indicating that animal welfare is rather perceived as a prerequisite to more important characteristics such as health, safety and quality. The origin of the meat on the other hand appeared less important than animal welfare, which is in line with the low interest and significance of traceability in other studies (Verbeke and Ward, 2006).

Respondents largely agreed that there is insufficient information available with respect to animal welfare in order to help them make informed choices. Consumers also expressed considerable doubts about the reliability of available information. About 40% of the respondents did not know whether to believe the reporting or not and another 35% indicated not to believe animal

welfare info at all. Furthermore, 80% of the sample was in favour of more severe controls on animal welfare. This number, combined with 71% of the sample being in favour of an animal welfare label (which is in line with the EU animal welfare labelling plans) and with 60% of the sample claiming willingness to pay a premium price for reliable animal welfare guarantees, opens perspectives for an animal welfare label in the near future.
In addition, these exploratory data will be complemented with additional quantitative data gathered in April 2006, originating from a project funded by the Flemish government. This project intends to develop a definition of animal welfare based both on science and on consensus among citizens and stakeholders such that animal welfare becomes a more workable concept in policy making and societal debate. A comprehensive list of 73 aspects was the result of focus group discussions and a profound literature review. Next, a questionnaire-type survey was performed in order to group these aspect into several welfare-dimensions, using principal component analysis, and to allocate weight to these welfare-dimensions.

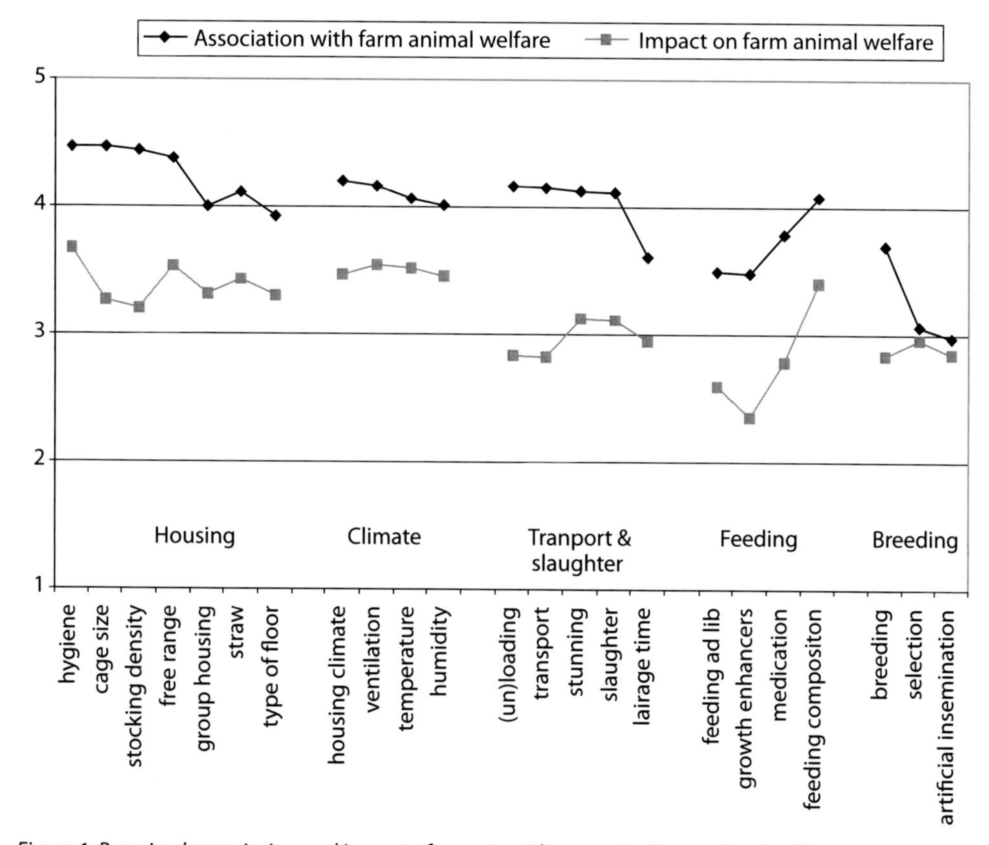

Figure 1. Perceived association and impact of aspects with respect to farm animal welfare.

References

European Commission (2002). *Report of the scientific committee on animal health and animal welfare*. Available on http://www.europa.eu.int/comm/food/fs/sc/scah/out71_en.pdf.

Federal Ministry of Health and Women (2006). International Conference on Animal Welfare, Brussels, 30 March 2006.

Harper, G. (2001). *Consumer concerns about animal welfare and the impact on food choice. Proceedings of the dissemination seminar, EUFAIR CT98-3678.* Available on http://www.agric-econ.uni.kiel.de/Abteilungen/agrarmarketing/EU/harper.pdf.

Ingenbleek, P.T.M., Binnekamp, M., Van Trijp, J.C.M. and de Vlieger J.J. (2004). *Dierenwelzijn in de markt. Een drieluik van consumenten, retailers en belangenorganisaties in Europa.* Available on http://www.lei.dlo.nl/publicaties/PDF/2004/5_xxx/5_04_11.pdf.

OIE (2004). *Global Conference on Animal Welfar, Paris, 23-25 February 2004.*

Verbeke, W. and Viaene J. (2000). Ethical challenges for livestock production: meeting consumer concerns about meat safety and animal welfare. *Journal of Agricultural and Environmental Ethics*, 12: 141-151.

Verbeke, W. and Ward, R. (2006). Consumer interest in information cues denoting quality, traceability and origin: An application of ordered probit models to beef labels. *Food Quality and Preferences*, 17: 453-467.

Welfare Quality. (2005). Science and society improving animal welfare. *Proceedings of the Welfare Quality conference, 17-18 November, Brussels.*

Ethical questions, technological solutions?

Ineke Widdershoven-Heerding
Commission on Genetic Modification, Subcommittee Ethics and Social Aspects, P.O. Box 578, 3720 AN Bilthoven, The Netherlands, Ineke.Widdershoven@rivm.nl

The public debate on gm-crops is lost. A new debate should commence, on the acceptability of cisgene plants. Cisgenesis is the new technological answer to public skepticism and opposition to genetic modification in (food) plants. This position is taken up in the Netherlands by leading researchers in green biotechnology. Several cisgenic crops - strawberry, apple, and potato - are being developed.

What is so special about cisgenic plants, compared to other gm-plants? Though they are manufactured with methods from gene technology, the inserted genes originate from the same or an interbreeding species, not from other organisms. This is the reason cisgenesis is also called intragenesis.

Why develop this particular type of genetic modification? Why expect a larger public enthusiasm for cisgenesis? What is the appeal?

To know why cisgenesis is considered an answer, we must first analyze the question, in this case the ethical question(s) concerning gm-crops.

Opposition against gm-crops originates from different ethical styles of reasoning about our actions towards nature: consequentialism and deontology.

From a consequentialistic point of view - where we consider the rightness of our actions by their possible consequences - objections against gm-crops concentrate on the possible and unknown risks to biodiversity, human or animal health, exploitation of farmers, wasting the resources of the earth, etc.

From a deontological point of view - where our actions are right when they are in accordance with our moral principles - objections against gm-crops focus on notions as the violation of the integrity of nature or of plants, intrinsic value, the transgression of borders between species (set by God), (un)naturalness, etc.

In the public debate about gm-crops both types of argument are represented and sometimes even combined.

Cisgenesis does provide some answers to hesitations both from consequentialism and from deontology. On a consequentialistic level, proponents of cisgenesis claim their products do not fundamentally vary from 'normal' conventionally bred plants, and thus bring no new risks or harms. Cisgenic plants have numerous advantages, which should be taken up in a consequentialistic calculation.

On the level of deontology, some arguments, for example from the organic agriculture movement, are not applicable to cisgenesis. Others are, however. Cisgenesis respects the limits between species, since in cisgenic plants the inserted genes stem from the same or interbreedible species. And, as social research shows, consumers regard GM with plant own material as less unnatural than GM with genes from other organisms.

So we may conclude that cisgenic plants are possibly more acceptable from an ethical perspective, than other gm-plants.

Will cisgenesis dissolve the public distrust of GM in (food) plants? What determines the success or failure of this type of biotechnological innovations? Public acceptance is not mobilized by technical discussions of the characteristics of cisgenesis. It rather requires a debate on a broader scale with stakeholders and lay participants. Of course the safety of the new application of gene technology must be trusted. However, other and broader concerns, ethical or otherwise, on food, nature, naturalness, risks, etc. ought to be taken seriously as well. Both the hesitations and the promises – profit, innovation, efficiency, and sustainability – should come to the fore in the debate. Technically speaking, cisgenesis may be able to do away with some traditional counterarguments to GM. This does not imply however, that there is no room or no need for debate.

My conclusion is that it is indeed wise to open the debate on the acceptance of alternative applications of GM in agriculture. But there are no guarantees as to the outcome.

Author index

W

Y

Keyword index

Printed in the United States
by Baker & Taylor Publisher Services